SEPARATION SCIENCE AND TECHNOLOGY

VOLUME 11

A reference series edited by

SATINDER AHUJA

EVALUATING WATER QUALITY TO PREVENT FUTURE DISASTERS

VOLUME 11

Edited by

SATINDER AHUJA

Ahuja Consulting for Water Quality, Calabash, NC, United States

Academic Press is an imprint of Elsevier
50 Hampshire Street, 5th Floor, Cambridge, MA 02139, United States
525 B Street, Suite 1650, San Diego, CA 92101, United States
The Boulevard, Langford Lane, Kidlington, Oxford OX5 1GB, United Kingdom
125 London Wall, London, EC2Y 5AS, United Kingdom

First edition 2019

Copyright © 2019 Elsevier Inc. All rights reserved.

No part of this publication may be reproduced or transmitted in any form or by any means, electronic or mechanical, including photocopying, recording, or any information storage and retrieval system, without permission in writing from the publisher. Details on how to seek permission, further information about the Publisher's permissions policies and our arrangements with organizations such as the Copyright Clearance Center and the Copyright Licensing Agency, can be found at our website: www.elsevier.com/permissions.

This book and the individual contributions contained in it are protected under copyright by the Publisher (other than as may be noted herein).

Notices
Knowledge and best practice in this field are constantly changing. As new research and experience broaden our understanding, changes in research methods, professional practices, or medical treatment may become necessary.

Practitioners and researchers must always rely on their own experience and knowledge in evaluating and using any information, methods, compounds, or experiments described herein. In using such information or methods they should be mindful of their own safety and the safety of others, including parties for whom they have a professional responsibility.

To the fullest extent of the law, neither the Publisher nor the authors, contributors, or editors, assume any liability for any injury and/or damage to persons or property as a matter of products liability, negligence or otherwise, or from any use or operation of any methods, products, instructions, or ideas contained in the material herein.

ISBN: 978-0-12-815730-5
ISSN: 1877-1718

For information on all Academic Press publications
visit our website at https://www.elsevier.com/books-and-journals

Publisher: Susan Dennis
Acquisition Editor: Kathryn Eryilmaz
Editorial Project Manager: Ruby Smith
Production Project Manager: Vignesh Tamil
Cover Designer: Mark Rogers

Typeset by SPi Global, India

This book is dedicated to people around the world who are concerned about water quality and are doing their best to prevent future water disasters.

Satinder (Sut) Ahuja, PhD
President, Ahuja Academy of Water Quality

Contents

Contributors xi
Preface xiii

1. Overview: Evaluating Water Quality to Prevent Future Disasters
SATINDER AHUJA

1. Introduction 1
2. Complexities of Aquatic Endocrine Disruption in a Changing Global Climate 5
3. Impact of Persistent Droughts on the Quality of Middle East Water Resources 6
4. Present and Potential Water-Quality Challenges in India 6
5. Arsenic Contamination in South Asian Regions— The Difficulties and Challenges 7
6. Cyanobacteria and Their Toxins 7
7. Educational Partnerships Combined With Research on Emerging Pollutants for Long-Term Water-Quality Monitoring 7
8. Impacts of *Deepwater Horizon* Oil and Dispersants on Various Life Stages of Oysters 8
9. Analytical Methods for the Comprehensive Characterization of Produced Water 8
10. Innovations in Monitoring With Water-Quality Sensors: Applications for Floods, Hurricanes, and Harmful Algal Blooms 8
11. Developing a Sensitive Biosensor for Monitoring Arsenic in Drinking Water Supplies 9
12. Investigating the Missing-Link Effects of Noncompliance and Aging Private Infrastructure on Water-Quality Monitoring 9
13. GenX Contamination of the Cape Fear River, North Carolina: Analytical Environmental Chemistry Uncovers Multiple System Failures 10
14. Analysis of GenX and Other Per- and Polyfluoroalkyl Substances in Environmental Water Samples 10
15. Sustainable Magnetically Retrievable Nanoadsorbents for Selective Removal of Heavy Metal Ions From Different Charged Wastewaters 10
16. Lessons Learned From Water Disasters 11
17. Conclusions 11
References 11

2. The Heat Is On: Complexities of Aquatic Endocrine Disruption in a Changing Global Climate
B. DECOURTEN, A. ROMNEY, AND S. BRANDER

1. Introduction 13
2. Hypoxia 15
3. Salinity 19
4. Acidification 23
5. Temperature 25
6. Multiple Stressors 30
7. Transgenerational and Epigenetic Effects 30
8. Population Level Effects 32
9. Community Level Effects 33
10. Conclusions and Future Research Directions 34
Acknowledgments 38
Author Contributions 38
References 39
Further Reading 47

3. Impact of Persistent Droughts on the Quality of the Middle East Water Resources
Y. SHEVAH

1. Introduction 51
2. Drought Impact on Aquatic Ecosystems 54

viii CONTENTS

3. Impact on Water Resources Availability and Quality 59
4. Development of Nonconventional Water Resources 67
5. Regional Cooperation on Transboundary Water 72
6. Improving the Knowledge Base 74
7. Conclusions and the Way Forward 76
Acknowledgments 77
References 77
Further Reading 84

4. Present and Potential Water-Quality Challenges in India
SANJAY BAJPAI, NEELIMA ALAM, AND PIYALEE BISWAS

1. Introduction 85
2. Quantity and Quality of Water 85
3. Enlistment of Existing Water Quality Challenges 87
4. Major Water-Quality Problems Prevalent in India 87
5. Evaluating Water Quality to Prevent Disasters 109
6. Conclusions 110
Acknowledgments 111
References 111
Further Reading 112

5. Arsenic Contamination in South Asian Regions: The Difficulties, Challenges and Vision for the Future
SUKANCHAN PALIT, KSHIPRA MISRA, AND JIGNI MISHRA

1. Introduction 113
2. Health Problems due to Arsenic Contamination 113
3. Country-Wise Status of Arsenic Contamination in South Asia 115
4. Difficulties and Challenges Associated With Arsenic Contamination in South Asia 116
5. Remedial Measures for Removal of Arsenic in Groundwater 117
6. Conclusions and Vision for the Future 121
References 121

6. Cyanobacteria and Their Toxins
J.S. METCALF AND N.R. SOUZA

1. Introduction 125
2. What Are Cyanobacteria? 125
3. What Are Cyanobacterial Toxins? 127
4. What Conditions Make Cyanobacteria Bloom and Under What Circumstances? 131
5. How Do Cyanobacteria Affect Waterbodies? 132
6. How Can We Detect Cyanobacterial Toxins? 134
7. How Can We Remove Cyanobacterial Toxins? 139
8. How Do We Reduce the Likelihood of Exposure to Cyanobacteria and Their Toxins? 140
References 141
Further Reading 148

7. Educational Partnerships Combined With Research on Emerging Pollutants for Long-Term Water-Quality Monitoring
JULIE R. PELLER, LAURIE EBERHARDT, ERIN ARGYILAN, AND CARRIE SANIDAS

1. Introduction 149
2. Experimental 152
3. Results and Discussion 156
4. Conclusions 166
Acknowledgments 167
References 167

8. Evaluation of the Toxicity of the *Deepwater Horizon* Oil and Associated Dispersant on Early Life Stages of the Eastern Oyster, *Crassostrea virginica*
JULIEN VIGNIER, ASWANI VOLETY, PHILIPPE SOUDANT, FU-LIN CHU, AI NING LOH, MYRINA BOULAIS, RENÉ ROBERT, JEFFREY MORRIS, CLAIRE LAY, AND MICHELLE KRASNEC

1. The *Deepwater Horizon* Oil Spill Incident 170
2. Chemical Aspects 171
3. The Eastern Oyster *Crassostrea virginica* 174
4. Material and Methods 176
5. Key Findings 182
6. Conclusions 192
Acknowledgments 192
References 192
Further Reading 198

CONTENTS

9. Analytical Methods for the Comprehensive Characterization of Produced Water

TIFFANY LIDEN, INÊS C. SANTOS, ZACARIAH L. HILDENBRAND, AND KEVIN A. SCHUG

1. Introduction   199
2. Analysis of Produced Water   204
3. Conclusions   212
References   213

10. Innovations in Monitoring With Water-Quality Sensors With Case Studies on Floods, Hurricanes, and Harmful Algal Blooms

DONNA N. MYERS

1. Introduction   219
2. Background and History of Continuous Water-Quality Monitoring to 1972   220
3. Innovations in Water-Quality Sensors   225
4. Standardized Procedures for Continuous Water-Quality Measurements   245
5. Sensor Platforms and Deployments   250
6. Data Acquisition, Storage, and Telemetry   259
7. Data-Processing Software, Water-Information Systems, and Web Applications   262
8. Case Studies on Monitoring Floods and Hurricanes With Water-Quality Sensors   266
9. Advantages and Limitations of Water-Quality Sensors   272
10. Summary and Conclusions   274
References   276

11. Biosensors for Monitoring Water Pollutants: A Case Study With Arsenic in Groundwater

JASON BERBERICH, TAO LI, AND ENDALKACHEW SAHLE-DEMESSIE

1. Introduction   285
2. Monitoring of Arsenic in Groundwater   293
3. Biosensors for Monitoring Arsenic in Water   308
4. Contaminants of Emerging Concern and Future Priorities   318
5. Conclusions   321

Disclaimer and Acknowledgment   321
References   322

12. Investigating the Missing Link: Effects of Noncompliance and Aging Private Infrastructure on Water-Quality Monitoring

ADAM COOPER, ALEXA FORTUNA, AND SATINDER AHUJA

1. Introduction   329
2. Case Study: Community-Based Lead Testing   331
3. Methods   333
4. Results and Discussion   335
5. Conclusions   336
Acknowledgments   338
References   338

13. GenX Contamination of the Cape Fear River, North Carolina: Analytical Environmental Chemistry Uncovers Multiple System Failures

LAWRENCE B. CAHOON

1. Introduction   341
2. PFASs as Emerging Contaminants   342
3. PFAS Analyses in the Cape Fear River   343
4. Discovery of GenX in the Cape Fear River   345
5. PFAS Discharges From Fayetteville Works   348
6. How Did the GenX Problem Happen?   348
7. System Failures   352
Acknowledgments   353
References   353
Further Reading   354

14. Analysis of GenX and Other Per- and Polyfluoroalkyl Substances in Environmental Water Samples

QIN TIAN AND MEI SUN

1. Introduction   355
2. Sample Collection, Storage and Conservation   356
3. Sample Preparation   357

CONTENTS

4. Miscellaneous Ways to Improve Precision and Accuracy 358
5. Targeted Analysis 358
6. Nontargeted Analysis 359
7. Total PFAS Analysis 359
8. Case Study: GenX and Emerging PFEA in North Carolina, USA 361
9. Conclusions 367
Acknowledgment 367
References 367

15. Sustainable Magnetically Retrievable Nanoadsorbents for Selective Removal of Heavy Metal Ions From Different Charged Wastewaters

SRIPARNA DUTTA AND R.K. SHARMA

1. Introduction 371
2. Heavy Metal Poisoning Case Studies 378
3. Strategies Devised to Combat Metal Contamination Issues 385
4. Adsorption: Ultimate Choice for Heavy Metal Removal From Wastewater 387

5. Synergistic Integration of Adsorption Technology With Nanotechnology 388
6. Magnetic Nanoadsorbents 390
7. Need for Large-Scale Industrial Applicability: Introducing Reactors for Online and Cyclic Recovery of Heavy Metals 409
8. Conclusions and Future Outlook 411
References 412
Further Reading 416

16. Lessons Learned From Water Disasters of the World

SATINDER AHUJA

1. Arsenic Contamination of Groundwater 417
2. Lead Contamination of Drinking Water in Flint, Michigan 421
3. GenX Contamination of Drinking Water in Wilmington, North Carolina, and Other Counties in the State 423
4. Conclusions 426
References 426

Index 429

Contributors

Satinder Ahuja Ahuja Consulting for Water Quality, Calabash, NC, United States

Neelima Alam Technology Mission Division, Department of Science and Technology, Technology Bhawan, New Delhi, India

Erin Argyilan Department of Geosciences, Indiana University Northwest, Gary, IN, United States

Sanjay Bajpai Technology Mission Division, Department of Science and Technology, Technology Bhawan, New Delhi, India

Jason Berberich Department of Chemical, Paper and Biomedical Engineering, Miami University, Oxford, OH, United States

Piyalee Biswas Technology Mission Division, Department of Science and Technology, Technology Bhawan, New Delhi, India

Myrina Boulais University of North Carolina Wilmington, College of Arts and Sciences & Center for Marine Science, Wilmington, NC, United States; Laboratoire des Sciences de l'Environnement Marin (UMR 6539-LEMAR), IUEM-UBO, Technopole Brest Iroise, Plouzané, France

S. Brander Department of Environmental and Molecular Toxicology, Oregon State University, Corvallis, OR, United States

Lawrence B. Cahoon Department of Biology and Marine Biology, University of North Carolina Wilmington, Wilmington, NC, United States

Fu-lin Chu Virginia Institute of Marine Science (VIMS), College of William and Mary, Department of Aquatic Health Sciences, Gloucester Point, VA, United States

Adam Cooper Stetson University, DeLand, FL, United States

B. DeCourten Department of Environmental and Molecular Toxicology, Oregon State University, Corvallis, OR; Department of Biology and Marine Biology, University of North Carolina Wilmington, Wilmington, DE, United States

Sriparna Dutta Department of Chemistry, Green Chemistry Network Centre, University of Delhi, New Delhi, India

Laurie Eberhardt Department of Biology, Valparaiso University, Valparaiso, IN, United States

Alexa Fortuna Stetson University, DeLand, FL, United States

Zacariah L. Hildenbrand Affiliate of Collaborative Laboratories for Environmental Analysis and Remediation, The University of Texas at Arlington, Arlington; Inform Environmental, LLC, Dallas, TX, United States

Michelle Krasnec Abt Associates, Boulder, CO, United States

Claire Lay Abt Associates, Boulder, CO, United States

Tao Li National Risk Management Research Laboratory, Office of Research and Development, U.S. Environmental Protection Agency, Cincinnati, OH, United States

Tiffany Liden Department of Chemistry and Biochemistry, The University of Texas at Arlington, Arlington, TX, United States

Ai Ning Loh Department of Marine and Ecological Sciences, College of Arts and Sciences, Florida Gulf Coast University, Fort Myers, FL; University of North Carolina Wilmington, College of Arts and Sciences & Center for Marine Science, Wilmington, NC, United States

J.S. Metcalf Brain Chemistry Labs, Institute for Ethnomedicine, Jackson, WY, United States

Jigni Mishra Department of Biochemical Sciences, Defence Institute of Physiology and Allied Sciences, Delhi, India

Kshipra Misra Department of Biochemical Sciences, Defence Institute of Physiology and Allied Sciences, Delhi, India

Jeffrey Morris Abt Associates, Boulder, CO, United States

Donna N. Myers U.S. Geological Survey, Scientist Emeritus, Denver, CO, United States

Sukanchan Palit Department of Chemical Engineering, University of Petroleum and Energy Studies, Dehradun; 43, Judges Bagan, Post-Office-Haridevpur, Kolkata, India

Julie R. Peller Department of Chemistry, Valparaiso University, Valparaiso, IN, United States

René Robert Ifremer, Unité Littoral, Centre Bretagne - ZI de la Pointe du Diable - CS 10070, Plouzané, France

A. Romney Department of Anatomy, Cell Biology, and Physiology, University of California Davis, Davis, CA, United States

Endalkachew Sahle-Demessie National Risk Management Research Laboratory, Office of Research and Development, U.S. Environmental Protection Agency, Cincinnati, OH, United States

Carrie Sanidas Willowcreek Middle School, Portage, IN, United States

Inês C. Santos Department of Chemistry and Biochemistry; Affiliate of Collaborative Laboratories for Environmental Analysis and Remediation, The University of Texas at Arlington, Arlington, TX, United States

Kevin A. Schug Department of Chemistry and Biochemistry; Affiliate of Collaborative Laboratories for Environmental Analysis and Remediation, The University of Texas at Arlington, Arlington, TX, United States

R.K. Sharma Department of Chemistry, Green Chemistry Network Centre, University of Delhi, New Delhi, India

Y. Shevah TOHL, Civil and Environmental Engineering Inc., Atlanta, GA, United States

Philippe Soudant Laboratoire des Sciences de l'Environnement Marin (UMR 6539-LEMAR), IUEM-UBO, Technopole Brest Iroise, Plouzané, France

N.R. Souza Brain Chemistry Labs, Institute for Ethnomedicine, Jackson, WY, United States

Mei Sun Department of Civil and Environmental Engineering, University of North Carolina at Charlotte, Charlotte, NC, United States

Qin Tian National Research Center for Geoanalysis, Beijing, China

Julien Vignier Department of Marine and Ecological Sciences, College of Arts and Sciences, Florida Gulf Coast University, Fort Myers, FL, United States

Aswani Volety Department of Marine and Ecological Sciences, College of Arts and Sciences, Florida Gulf Coast University, Fort Myers, FL; University of North Carolina Wilmington, College of Arts and Sciences & Center for Marine Science, Wilmington, NC, United States

Preface

Worldwide water disasters are unfortunately all too common. They do not necessarily occur just in underdeveloped and developing countries of the world. As a matter of fact, a greater number of water catastrophes have been recorded in developed countries than in underdeveloped countries. Enjoyment of modern comforts exposes us to more of these. The disasters in developed countries relate to oil spills, chemical contaminations, and, at times, microbial contamination. The underdeveloped world, on the other hand, has suffered more from arsenic and fluoride contaminations from mother Earth and microbial contaminations that are due to poor sanitation and hygiene. A number of water disasters are waiting to happen because of climate disruptions (droughts or excessive rain that leads to flooding), cyanobacterial toxins, microplastics, pharmaceuticals including endocrine disruptors, personal care products, pesticides, herbicides, and a large variety of new chemicals that are being utilized by industries.

Most of these calamities can be avoided if we take adequate steps. This will require detailed quality evaluations of water that is used for drinking, cooking, and other human uses that can affect our health. Detailed investigations require employment of separation methods such as chromatography and combine them with mass spectrometry to detect contaminants at ultratrace levels. Separation methods can also help us purify contaminated water.

A broad overview of what we need to do to prevent water disasters is provided in Chapter 1. Chapters 2–5 describe the impact of various events that could lead to water-quality catastrophes. Chapter 2 discusses the complexities of aquatic endocrine disruption in a changing global climate. The impact of drought on water quality in the Middle East is covered in Chapter 3. Chapter 4 discusses at some length the present and potential water-quality disasters in India. Arsenic and heavy metal groundwater contamination in South Asia is covered in Chapter 5.

Chapters 6–8 provide valuable information on potential disasters. The impact on water quality from cyanobacteria and their toxins is covered in Chapter 6, and Chapter 7 describes educational partnerships combined with research on emerging pollutants such as microplastics for long-term water-quality monitoring. The effect of the explosion of the *Deepwater Horizon* and the resulting contamination by oil and dispersants on various life stages of oysters *Crassostrea virginica* are discussed in Chapter 8.

A number of chapters are devoted to water-quality monitoring. Large volume of water is used for oil and gas extraction activity (fracking). Chapter 9 describes analytical methods used for the comprehensive characterization of produced water. Chapter 10 describes innovations in monitoring with water-quality sensors, and Chapter 11 discusses development of a sensitive biosensor for monitoring arsenic in drinking water supplies. Investigations of the missing link effects of noncompliance and an aging private infrastructure on water-quality monitoring for the presence of lead are described in Chapter 12.

Chapters 13–15 discuss how analytical/environmental chemistry helped delineate some recent water disasters. Chapter 13 describes how analytical/environmental chemistry helped uncover multiple system failures that led to GenX contamination of the Cape Fear River in North Carolina. Chapter 14 provides more details on the analysis of GenX and other per- and polyfluoroalkyl substances in environmental water samples. A discussion of sustainable magnetically retrievable nanoadsorbents for removal of lead and palladium ions in wastewater is presented in Chapter 15.

Finally, Chapter 16 discusses lessons that can be learned from water-quality disasters caused by arsenic, lead, and GenX. Interestingly, a basic theme evolves, which suggests that water-quality evaluation can help in the prevention of water disasters.

I want to thank the contributors for their valuable contributions that will help researchers, academicians, and regulatory scientists and engineers involved in evaluating water quality to prevent future disasters.

Satinder Ahuja
December 19, 2018

CHAPTER 1

Overview: Evaluating Water Quality to Prevent Future Disasters

Satinder Ahuja*

Ahuja Consulting for Water Quality, Calabash, NC, United States
*Corresponding author: E-mail: sutahuja@atmc.net

1 INTRODUCTION

Water disasters are occurring worldwide repeatedly. Disaster is defined as a sudden contamination event bringing great damage or loss. It generally affects a large number of people. A quick Internet review provided the following list of 13 water-quality disasters around the world.

1. Valdez, Alaska: Exxon oil spill
2. Lanzhou, China: benzene
3. Woburn, Massachusetts: TCE, PCE
4. Elk River, West Virginia: MCHM
5. Gulf of Mexico: *Deepwater Horizon* oil spill
6. Ghana, West Africa: cyanide
7. Yamuna River, India: sewage, garbage, chemicals
8. Mutare, Zimbabwe: Cr, Ni, bacteria
9. Flint, Michigan: Pb
10. Hinkley, California: Cr
11. Camp Lejeune, North Carolina: PCE, TCE
12. Walkerton, Ontario: *E. coli*
13. Minneapolis, Minnesota: TCE

TCE = trichloroethylene
PCE = perchloroethylene
MCHM = 4-methylcyclohexanemethanol.

It should be noted that water-quality disasters occur in many areas other than the underdeveloped world. The above listing includes more than seven disasters that occurred in the United States. And even though only the Yamuna River contamination is mentioned above, a large number of rivers are contaminated worldwide. Oil, chemicals, and bacteria are among the common threats. The above list is by no means complete or in chronological order. For example, it does not list the most horrendous water disaster in recent history: arsenic contamination of groundwater in Bangladesh that was discovered in the 1980s. Also, it does not mention the GenX (ammonium salt of hexafluoropropylene oxide dimer acid) contamination reported in 2017 that affected three counties in North Carolina. We will return to these issues a little later.

Drinking water comes mainly from rivers, lakes, wells, and natural springs. These sources

Separation Science and Technology, Volume 11
https://doi.org/10.1016/B978-0-12-815730-5.00001-6

Copyright © 2019 Elsevier Inc. All rights reserved.

are exposed to a variety of hazardous situations that can lead to water contamination (Ahuja, 2008, 2009, 2013a,b, 2017a; Ahuja and Hristovski, 2013). The failure of safety measures relating to production, utilization, and disposal of thousands of inorganic and organic compounds can cause pollution of our water supplies. A number of contaminants can arise from the materials we use frequently to improve the quality of life:

- Coal combustion
- Detergents
- Disinfectants
- Endocrine disruptors
- Fertilizers
- Gasoline combustion products and additives
- Herbicides
- Insecticides/pesticides
- Perfluoro compounds
- Personal care products
- Pharmaceuticals
- Phthalates
- Radionuclides

A US Geological Survey (USGS) conducted in 2002 in the United States found pharmaceuticals (hormones and other drugs) in 80% of the streams sampled in 30 states. A very large amount of antimicrobials and antibiotics are administered to healthy animals on US farms each year and this can end up in our water supplies being contaminated when the animal waste is not handled properly. There are a number of other threats to drinking water: volatile and semivolatile compounds, improperly disposed chemicals, heinous terrorist actions, wastes injected underground, and naturally occurring substances. Similarly, drinking water that is not properly treated or disinfected or travels through an improperly maintained distribution system may also pose a health risk. A variety of pollutants can come from wastewater because it is generally recycled to surface water or groundwater after some processing. Wastewater can originate from many places: households, industries, commercial developments, road runoff, etc. As diverse as the sources

of wastewater are, so too are their potential constituents. In addition, we have to contend with nonpoint sources of pollution.

Two recent water disasters—contamination of water in Flint, Michigan (Ahuja, 2017b) and Wilmington, North Carolina (Ahuja, 2018)—are grim reminders of what can go terribly wrong. To save money, the city of Flint began drawing its water from the local river in April 2014 instead of buying Lake Huron water from Detroit. Residents started complaining about burning skin, hand tremors, hair loss, and even seizures. In 2015, high levels of lead were found in the water supply of Flint. Unsafe levels have also turned up in the past in tap water in Washington, DC, in 2001; in Columbia, SC, in 2005; in Durham and Greenville, NC, in 2006. In 2015, this same type of problem was encountered in Jackson, MS, and Sebring, OH.

The GenX water disaster, recently reported from Wilmington, North Carolina, shows how emerging contaminants can impact our water quality. DuPont (parent company of Chemours) introduced GenX in 2009 to replace perfluorooctanoic acid (PFOA), a compound used to manufacture Teflon and also coatings for stain-resistant carpeting and waterproof clothing. GenX has been detected in the drinking water in New Hanover, Bladen, and Brunswick counties, and in surface waters in Ohio and West Virginia. Levels of GenX in the drinking water of the Cape Fear Public Utility Authority—in Wilmington—averaged 631 ppt (parts per trillion) according to a study published by Sun et al. (2016). The United States Environmental Protection Agency (US EPA) has set the drinking water standard for PFOA at 70 ppt; however, they have not yet set a standard for GenX. The water utility companies in this area cannot effectively filter out GenX.

Another looming disaster arises from harmful cyanobacterial algal blooms. In aquatic environments worldwide, they are likely to be a major problem, causing a significant adverse impact on public health and ecosystems. Cyanobacteria produce a wide variety of secondary metabolites and, most importantly, potent toxins called cyanobacterial toxins or cyanotoxins; these toxins are known to affect a wide range of living organisms, including humans. Cylindrospermopsin (CYN) is one of the most widely distributed cyanotoxins in bodies of water. CYN is produced by a large group of cyanobacteria that are highly adaptive and invasive and have been detected in tropical, subtropical, and even temperate areas. Compared with microcystins, CYN can be actively released into the environment, leading to higher extracellular toxin concentrations than the intracellular levels. Unregulated Contaminant Monitoring Rule 4 was set by the EPA in 2016; it includes cyanotoxins.

There is still another impending water disaster, e.g., plastics, especially microplastics, are proving to be a major problem. Microplastics can stunt fish growth and alter their behavior. The extensive use of plastics and their careless disposal have led to pollution of various water bodies. Large parts of the Pacific Ocean are referred to as "plastic oceans," where enormous gyres about the size of Texas are covered with plastic debris. The Pacific is the largest ocean realm on our planet, approximately the size of Africa—over 10 million square miles—and it is the home of two very large gyres. The Atlantic Ocean contains two more gyres and other plastic oceans exist in other bodies of water. It was gratifying to see volunteer activities to rescue reefs and mangroves in La Paz, Mexico by Pablo Ahuja and his team of divers and others. The efforts have been going on in the area for almost 5 years and improvements are significant.

1.1 Water Pollution Regulations

The Federal Water Pollution Control Act of 1948 was the first major US law that was passed to address water pollution. In response to public concern about degraded water quality and a widespread view that pollution of our rivers and lakes was unacceptable, the water act became law in 1972 in the United States. Control of point -source contamination, traced to specific "end of pipe" points of discharge, or outfalls, such as factories and combined sewers, was the primary focus of the Clean Water Act (CWA). As amended in 1972, the law became commonly known as the CWA. The 1972 amendments:

- Established the basic structure for regulating pollutant discharges into the waters of the United States.
- Provided the EPA the authority to implement pollution control programs such as setting wastewater standards for industries.
- Maintained existing requirements to set water-quality standards for all contaminants in surface waters.
- Made it unlawful for any person to discharge any pollutant from a point source into navigable waters unless a permit is obtained under its provisions.
- Funded the construction of sewage treatment plants under the construction grants program.
- Recognized the need for planning to address the critical problems posed by nonpoint-source pollution.

Safe Drinking Water Act (SDWA) was passed in 1974 in the United States, giving the EPA the responsibility for monitoring and enforcing the public drinking water safety. The EPA sets standards for drinking water quality and with its partners, implements various technical and financial programs to ensure drinking water safety. In 1996, SDWA was amended to assure that the EPA will continually seek, identify, and monitor potentially harmful or relevant contaminants that may be currently unregulated. Other nations adopted similar measures and have seen improvement in point-source contamination as well. In 2010, a United Nations resolution declared the human right to "safe and clean drinking water and sanitation."

A simple definition of potable water is any water that is clean and safe for drinking. National primary drinking water regulations control water quality in the United States. Besides chemical contaminants, a large number of aquatic microorganisms can infect or parasitize humans, and these pathogens are responsible for considerable morbidity and mortality worldwide. The strategies and methods for studying these organisms may be found in chapter 8 in Ahuja (2009), including molecular techniques and microbial source-tracking approaches. In addition, the risks posed by microbial biofilms and sediment pathogen reservoirs are discussed as emerging problems.

1.2 The Role of Separation Science and Technology in Handling Water Disasters

It is important to recognize that a large number of inorganic/organic compounds that cover the entire range of the alphabet, from A to Z (arsenic to zinc), can cause contamination of our water supplies (Ahuja, 2006). Separation science can help us resolve these compounds that can then be quantified by utilizing suitable detectors. Examples of these separations and quantifications are methods based on gas chromatography and mass spectrometry (GC–MS) and high-pressure liquid chromatography and mass spectrometry (HPLC–MS or simply LC–MS).

Not all chemicals are harmful; for example, zinc in small amounts is desirable; however, arsenic at concentrations as low as 10 parts per billion (ppb) is quite harmful. Over 700 different chemicals have been found in drinking water in the United States—when it comes out of the tap! The EPA classifies 129 of these chemicals as being particularly dangerous, and it sets standards for approximately 90 contaminants in drinking water. Those standards, along with each contaminant's likely source and its health effects, are available at www.epa.gov/safewater/mcl.html. It should be noted even though tap water can be unsafe at times, there is no assurance that bottled water is any safer.

It is clear that wastewater contains pollutants that have to be removed and/or reduced to safe levels before it is directed to a surface water source (river, ocean, bay, lake, etc.) or to groundwater. A better understanding of wastewater constituents and their abundance at different stages is a first step in recognizing appropriate opportunities for pretreatment. The composition of wastewater affects not only the treatment processes applied but also their source recovery opportunities.

In short, separation science and technology can play a significant role in detecting a potential water disaster by offering a variety of analytical methods that can evaluate water quality to identify the cause of a particular disaster. The following discussion covers the role of ultratrace analysis based on chromatography in monitoring various contaminants. Once we know the nature of contaminants, it is possible to design separation methods that can remove them.

1.3 Evaluating Water Quality for Ultratrace Contaminants

Ideally, we should look for all pollutants at trace or ultratrace levels (at or below ppb levels) in water. This means that our methodologies should separate all contaminants based on their molecular weight, using techniques such as size-exclusion chromatography, and characterize them by infrared, nuclear magnetic resonance, and mass spectrometry or various hyphenated techniques that combine chromatography with spectroscopy. Alternatively, we can separate the pollutants on the basis of whether they are organic or inorganic and do additional tests to determine if they are polymers. It is important to remember that water has to meet certain requirements to be called potable (see Section 1). Real-time monitoring can enable a quick response to water-quality concerns that arise from natural or malicious contamination and allow the greatest protection of public health (see chapter 14 in Ahuja, 2013a).

In the 1978 Metrochem meeting, I presented a paper "In Search of Femtogram." A femtogram is 10^{-15} g, or 1 part per quadrillion—a phantom quantity at that time. It was pointed out that it was essential to analyze very low quantities of various contaminants to fully understand the impact of an assortment of chemicals on the human body. For example, dioxin (2,3,7,8-tetrachloro-dibenzodioxin) can cause abortion in monkeys at the 200 ppt (parts-per-trillion) level, and PCBs at 0.43 ppb level can weaken the backbones of trout. It has been known for some time now that water that we call potable may contain many trace and ultratrace contaminants (McNeil et al., 1977). For more detailed discussion, you may also want to see Ahuja (1986, 2006, 2008, 2009, 2013a,b, 2014, 2017a) and Ahuja and Hristovski (2013).

A large variety of methods are available for monitoring point-source pollutants because we can reasonably assess which pollutants may be present. By contrast, this task becomes much more difficult with nonpoint-source pollution. The overwhelming majority of water-quality problems are now caused by diffuse nonpoint sources of pollution from agricultural land, urban development, tree harvesting, and the atmosphere. The nonpoint-source contaminants are more difficult to effectively monitor, evaluate, and control than those from point sources (such as discharges of sewage and industrial waste).

It is important to assure that the sample used for ultratrace analyses (at or below the ppb level) should be representative of the "bulk material." The major considerations are (Kratochvil and Taylor, 1981):

1. Determination of a "whole population" from which the sample is to be drawn.
2. Procurement of a valid gross sample.
3. Reduction of the gross sample to a suitable sample for analysis.

The analytical uncertainty should be reduced to one-third or less of sampling uncertainty (Youden, 1967). Poor analytical results can be obtained because of reagent contamination, operator errors in procedure or data handling, biased

methods, and so on. These errors can be controlled by the proper use of blanks, standards, and reference samples. It is also important to determine the extraction efficiency of the method.

Preconcentration of the analyte is frequently necessary because the detector used for quantification does not have the necessary detectability, selectivity, or freedom from matrix interferences (Karasek et al., 1981). Significant losses can occur during this step because of very small volume losses to glass walls of recovery flasks or disposable glass pipettes and other glassware. However, with suitable precautions, preconcentration of metals at concentrations down to 10^{-12} g/g for copper, lead, and zinc, and 10^{-13} g/g for cadmium have been successfully demonstrated in a typical polar snow matrix (Wolff et al., 1981). It is crystal clear that ultratrace analyses can help us detect potential contaminants; however, when they do get into water, it becomes necessary to purify the water.

This book discusses water-quality evaluations to prevent future water disasters. Discussed below are various approaches that can help us achieve this goal.

2 COMPLEXITIES OF AQUATIC ENDOCRINE DISRUPTION IN A CHANGING GLOBAL CLIMATE

Planet Earth is in the midst of a shift in global climate norms because of the anthropogenic addition of heat-trapping greenhouse gases to the atmosphere. As such, it is important to consider the significant alterations to environmental conditions occurring in aquatic ecosystems when evaluating the responses of aquatic organisms to endocrine-disrupting compounds (EDCs). Chapter 2 summarizes the current body of literature on concurrent exposure to abiotic factors associated with a global climate change (GCC), e.g., temperature or acidification and selected EDCs in fish and invertebrates. Evidence suggests that the assessment of risk to aquatic organisms should be expanded to include evaluations across temperature, oxygen, pH, and salinity gradients since responses differ depending on the environmental conditions at which exposures are conducted. Ultimately, fundamental homeostatic processes are challenged as environmental conditions become more stressful, reducing the ability to cope with chemical exposure, while chemical exposure simultaneously interferes with the ability to maintain homeostasis. Many effects depend on life stages, with an early life stage being more sensitive to exposure than the adult stage in many cases. Numerous organisms undergo physiological trade-offs when exposed at the same time to GCC-related stressors and EDCs, sometimes possessing the resources to deal with only the most immediate threat. Furthermore, responses depend on environmental fate, which can be influenced by factors associated with GCC.

3 IMPACT OF PERSISTENT DROUGHTS ON THE QUALITY OF MIDDLE EAST WATER RESOURCES

In the Middle East, the most water-stressed part of the world, water resources and quality are severely affected by climate change and global warming, leading to frequent droughts and depleted water reserves (Chapter 3). Desertification and low rainfall cause serious political instability, exacerbating the likelihood of failed states in this politically volatile region. The extreme climate, coupled with mismanagement of water resources and inadequate infrastructure, deprives the growing population of safe drinking water, food, shelter, and stability. Ongoing wars in Iraq, Syria, and Yemen cause migration and dislocation of populations with widespread consequences. Against this background, the region needs to promote sustainable management of water resources; incorporate advanced and innovative technologies to support water demand management; adopt good governance, social empowerment, and the

protection of aquatic ecosystems. Advanced water conservation, consumption administration, data management, and public participation have yet to be practiced on a large scale region-wide. For such a desirable vision to materialize, cross-border and upstream-downstream linkages among riparian states must be addressed, improving capacities and sharing experiences. Water management systems fostered by the UN, such as IWRM, 2030 Agenda and NBS, should be implemented to overcome climate change and adapt to its consequences. The causes, impacts, and adaptation to climate change, along with mitigation measures, providing an integrated picture of water management and technological challenges are discussed in this chapter.

4 PRESENT AND POTENTIAL WATER-QUALITY CHALLENGES IN INDIA

Water scarcity is one of the most severe problems in India in the 21st century. While population increase and economic developments are major factors, environmental changes also contribute to water-quality deterioration (Chapter 4). The impact of geogenic contamination is exacerbated by human factors. Water quality in India has been affected mainly through contamination by bacteria, fluoride, salt, and arsenic. Of late, contamination by nitrates, silica, uranium, lead, etc., has also received attention because water-quality evaluation has shown their increased presence in water. The use of pesticides, pharmaceuticals, personal care products, etc., coupled with inadequate disposal mechanisms, has led to their increased presence in water. The consumption of drinking water containing these contaminants presents diverse challenges to human health. The main challenge is to detect these contaminants present in low and trace levels accurately and to devise appropriate mitigation strategies. This chapter discusses the spread and prevalence of these contaminants in various parts of

5 ARSENIC CONTAMINATION IN SOUTH ASIAN REGIONS—THE DIFFICULTIES AND CHALLENGES

the country and attempts to deduce their potential adverse impacts based on severity levels. Novel water-quality evaluation, monitoring capabilities, and research-based technology solutions for mitigation of these contaminants have also been discussed.

5 ARSENIC CONTAMINATION IN SOUTH ASIAN REGIONS—THE DIFFICULTIES AND CHALLENGES

Arsenic contamination of groundwater poses a threat to the availability of clean drinking water in various countries around the globe (Chapter 5). Regions in South Asia, especially in Bangladesh, India, Cambodia, and Vietnam, are reportedly the worst hit, with the rural and remote areas being almost completely deprived of arsenic-free water. Dependence on this arsenic-laden water leads to serious health problems, namely, arsenicosis, skin cancer, and respiratory problems. The immediate need is to design economic remedial technologies which the underdeveloped as well as the developing countries can effectively adopt for removal of arsenic from water. With this perspective, it can be stated that research is being conducted all over the world to overcome this detrimental problem of arsenic poisoning. This chapter provides a picture of arsenic contamination in various countries of South Asia. Additionally, technologies that are either indigenous to the affected countries or developed abroad are described.

6 CYANOBACTERIA AND THEIR TOXINS

Cyanobacteria, common inhabitants of marine and freshwaters are photosynthetic bacteria that under appropriate conditions have the potential to produce mass populations called blooms (Chapter 6). They can be aesthetically unpleasant and are capable of producing a wide range of low-molecular-weight, highly toxic compounds that have previously resulted in adverse human and animal health events, including deaths. With pressures from human populations and GCC, as examples, cyanobacterial blooms are increasing in frequency and geographic extent. In order to prevent human exposure to cyanobacterial toxins, challenges related to catchment management, water treatment, toxin analysis, and legislation require addressing. This chapter helps us better understand the various challenges involved in tackling the problems posed by cyanobacterial toxins.

7 EDUCATIONAL PARTNERSHIPS COMBINED WITH RESEARCH ON EMERGING POLLUTANTS FOR LONG-TERM WATER-QUALITY MONITORING

The Lake Michigan coastal community of Portage, Indiana, is an especially incentivized region for fresh surface water protection and effective water-quality monitoring because of the geographic location, demographics, and economics of the area (Chapter 7). The reported watershed project represents a creative and needed collaborative program between universities, local schools, government agencies, and nonprofits to provide science-based monitoring and heightened awareness and understanding of local and emerging threats to surface water. As a result of the collaborations, the acquisition of critical water-quality data was demonstrated. In addition, procedures to study microplastics, an emerging water pollution problem, were initiated. For localities where no long-term consistent programs exist, the model described in this chapter can be a solution for scientifically rigorous water-quality monitoring. Additional benefits of these types of programs include heightened awareness and stewardship and, ultimately, an ability to promote informed decisions and actions for communities committed to protecting freshwater. A significant database of water-quality values is now available for two sites in the Salt Creek, Indiana, watershed, and methods for microfiber detection and quantification were developed to add to future data collections. The watershed monitoring for these sites in Salt Creek is now sustainable and is expected to lead to long-term data acquisitions.

8 IMPACTS OF *DEEPWATER HORIZON* OIL AND DISPERSANTS ON VARIOUS LIFE STAGES OF OYSTERS

The explosion of the *Deepwater Horizon* (DWH) drilling unit in April 2010 led to the largest marine oil spill in the US history up to that time (Chapter 8). For 87 days, unprecedented amounts of crude oil and dispersants were released into the Gulf of Mexico (GoM), coincident with the spawning and recruitment season of the oyster, *Crassostrea virginica*. The effects of acute exposures to surface-collected DWH oil, dispersed oil, and dispersant alone (Corexit 9500A®) on various life stages (gamete, embryo, larvae, or spat) of oysters were evaluated in the laboratory. Oil and dispersant adversely affected all life stages tested, from sublethal responses ranging from depressed fertilization, abnormal embryo development, inhibited larval growth and settlement, and feeding disruption of spat, to lethality. Detrimental effects of oil and associated dispersant on the reproduction and early development of *C. virginica* could impact the recruitment and consequently decimate the oyster natural populations in the affected areas.

9 ANALYTICAL METHODS FOR THE COMPREHENSIVE CHARACTERIZATION OF PRODUCED WATER

During the unconventional oil and gas development stimulation process, referred to as hydraulic fracturing, millions of gallons of freshwater can be used, as well as the production of millions of gallons of wastewater with a highly variable, chemically rich composition. This waste stream can contain a mixture of organic and inorganic additives, transformation products, as well as inorganic and organic substances from the formation (Chapter 9). Unfortunately, there is limited knowledge with regard to the composition of chemical additives used during hydraulic fracturing. Again, any methods currently approved by the US regulatory agencies for the evaluation of produced water do not exist. Collectively, these two factors have led to significant knowledge gaps with respect to exposure, hazard data, and analytical methods. As a result, there is an inadequate understanding of the environmental and human exposure implications during contamination events. Furthermore, pursing a deeper understanding of the chemicals that could be present in produced water can provide valuable insight when monitoring the efficacies of various treatment technologies for the reuse of produced water. This chapter will discuss the analytical methods that have been used to evaluate five key constituent categories found in produced water: bulk measurements, organics, biological, inorganics, and naturally occurring radioactive material. In addition, the chapter will touch on the outstanding concerns and remaining knowledge gaps.

10 INNOVATIONS IN MONITORING WITH WATER-QUALITY SENSORS: APPLICATIONS FOR FLOODS, HURRICANES, AND HARMFUL ALGAL BLOOMS

High-temporal frequency monitoring with in situ water-quality sensors was made possible by innovations in four major areas of science and technology: the instrument revolution in chemistry that led to the development of water-quality sensors; digital electronic data devices to record, store, and satellite systems to transmit data to monitoring organizations; computerized databases and software systems with which to store, manage, and retrieve data; and the Internet to deliver and disseminate data in a timely manner (Chapter 10). The first water-quality monitors measuring two parameters were installed in 1955 on the Delaware River estuary. Sensors and digital electronic data storage devices quickly evolved so that by 1966 up to 10-continuous variables could be measured and their analog signals could be digitally recorded and then machine processed for computer entry. In 1971, the first national computer system for input, management, storage, and retrieval of continuous data came into use. In 1975, geostationary satellites began to continuously receive sensor readings from data-collection platforms and transmit those readings to ground stations from which data were relayed to computerized databases. By 1994, water-quality sensor data became available over the Internet within 4h or less of data collection. In 2018, continuous water-quality data are used to support all important water uses that rely on quality as well as fundamental research in earth system science.

11 DEVELOPING A SENSITIVE BIOSENSOR FOR MONITORING ARSENIC IN DRINKING WATER SUPPLIES

Population growth and unsustainable agricultural and industrial activities have led to decreases in water quality and availability, causing ecological risks and challenging drinking water treatment plants to provide clean water (Chapter 11). Many techniques are available for monitoring known water contaminants, but they are typically expensive and cumbersome, requiring trained scientists or technicians to use them correctly. Biosensors provide the simple-to-use, disposable, or continuous tests for monitoring many of the common and emerging contaminants that water-quality personnel are facing today. This chapter discusses the changing profile of water pollution and sources of regulated and emerging contaminants. A detailed discussion is provided on the monitoring of arsenic in groundwater and the current field-monitoring approaches. An introduction to biosensors is provided, along with a discussion on arsenic biosensors that are developed for field applications. The future of biosensors for emerging contaminants is also discussed in this chapter.

12 INVESTIGATING THE MISSING-LINK EFFECTS OF NONCOMPLIANCE AND AGING PRIVATE INFRASTRUCTURE ON WATER-QUALITY MONITORING

Because of the "lead ban" amendments to the Safe Drinking Water Act of 1986, new infrastructure has resulted in a drastic reduction in aqueous lead levels (Chapter 12). However, many systems installed prior to this legislation, as exemplified by the Flint, Michigan, crisis can deliver water with high levels of lead. There are disparities in drinking water quality among older communities with aging water infrastructure. This led to a study of water quality in DeLand, Florida, where the concentration of lead is just under the minimum reporting level for federal and state drinking water-quality standards. Their poor sample-collection participation may be indicative of governmental distrust, which is found disproportionately in poorer neighborhoods with older infrastructures, and thus a higher risk of contamination. The current lack of policies and funding in place for the replacement of lead-based infrastructure gives little opportunity for affected populations to replace their infrastructure and improve the quality of their drinking water. The first step in solving this issue is to gather data on impacted communities. A community-based approach has been piloted in the Spring Hill neighborhood in DeLand. Lessons learned from this approach can be helpful in determining further community interventions and shaping policy.

13 GenX CONTAMINATION OF THE CAPE FEAR RIVER, NORTH CAROLINA: ANALYTICAL ENVIRONMENTAL CHEMISTRY UNCOVERS MULTIPLE SYSTEM FAILURES

The discovery of GenX along with numerous other perfluorinated and poly alkylated substances (PFASs) in the Cape Fear River, North Carolina, and the fact that these compounds are not removed by standard drinking water treatment processes has prompted scientific, regulatory, and legal responses (Chapter 13). GenX replaced PFOA as an intermediate in production of fluorinated materials. PFOA production in the United States was phased out beginning in 2006 because of toxicological and epidemiological concerns. The history of PFOA production and regulation and the alarms raised by the discovery of GenX in public drinking water highlight the weaknesses in the legal and regulatory systems in managing the challenges of GenX, an "emerging contaminant." This chapter demonstrates the critical importance of major advances in analytical environmental chemistry in detection and public disclosure of harmful compounds, despite failures in the regulatory process.

14 ANALYSIS OF GenX AND OTHER PER- AND POLYFLUOROALKYL SUBSTANCES IN ENVIRONMENTAL WATER SAMPLES

The presence of per- and polyfluoroalkyl substances (PFASs) in surface water, groundwater, and finished drinking water is receiving increased attention because of their toxicity and carcinogenicity (Chapter 14). While low toxicity thresholds of these compounds and the evolving regulations require sensitive analytical techniques capable of detecting PFAS concentrations in sub-ng/L levels, the ubiquitous presence of PFASs in lab supplies and instrument parts makes it crucial to prepare samples with minimal contamination. Meanwhile, the fluorochemical manufacturers shift their products from legacy PFASs to alternative species; as a result, more emerging PFAS species have been identified in recent years. Routine PFAS quantification has relied on liquid chromatography-mass spectrometry (LC–MS). Although LC–MS has been a useful tool for analyzing legacy PFASs, it is limited by the availability of analytical standards or even structural information of the emerging PFAS species. As a result, nontarget analyses using high-resolution mass spectrometry, as well as other techniques for total PFAS analysis, have been developed. With these tools, significant efforts have been made to identify and quantify infrequently reported PFAS species. This has revealed the occurrence of many novel species. For example, per- and polyfluoroalkyl ether acids (PFEAs) are a family of emerging PFASs that are discharged into the environment as replacement products made by the fluorochemical manufacturers and as industrial by-products. An important species in the PFEA family, GenX, has been detected in water bodies in multiple countries. The occurrence of GenX and other structurally related PFEA species in North Carolina is discussed as a case study to demonstrate how analytical advances promoted the awareness and removal of these compounds from the environment.

15 SUSTAINABLE MAGNETICALLY RETRIEVABLE NANOADSORBENTS FOR SELECTIVE REMOVAL OF HEAVY METAL IONS FROM DIFFERENT CHARGED WASTEWATERS

A reliable access to potable and safe drinking water represents one of the biggest global challenges of the 21st century. Among the several globally pervasive and emerging reasons behind water-quality deterioration, accumulation of heavy metals in the aquatic environment has been considered to be one of the chief causes (Chapter 15). In a quest to explore sustainable solutions for combating this problem, attempts have been directed toward the effective integration of nanotechnology with superior adsorption technology. With this perspective, it is important to mention that researchers have developed enhanced nanoadsorbents that are magnetically retrievable and have shown great prospects of effectively removing, while simultaneously recovering heavy metals from different charged wastewaters. This chapter provides an extensive overview of the different types of magnetic nanoadsorbents for heavy metal remediation designed by the scientific fraternity to date. In order to enlighten readers about the large-scale industrial applicability of these materials, there is special emphasis on the design of novel reactors for the cyclic recovery of these metals. In addition, a critical investigation on the applicability of developed nanomaterials is attempted, considering the economic viability, environmental safety, and sustainability of the proposed processes.

16 LESSONS LEARNED FROM WATER DISASTERS

Numerous water disasters have occurred worldwide over the years. Chapter 16 describes how a variety of contaminants can enter our water supply and lead to significant crises that can affect human health. Unfortunately, the pollution of freshwater (drinking water) is a problem faced by about half of the world's population. Each year there are about 250 million cases of water-related diseases, with roughly 5 million to 10 million deaths. Even an advanced country like the United States is facing a water crisis. Three disasters relating to contamination from arsenic, lead, and GenX are discussed in some detail. Discussions

cover why a given disaster occurred, its impact, and what lessons can be learned from it. A common thread emerges which suggests water-quality evaluations can help prevent some future water disasters.

17 CONCLUSIONS

A wide range of water disasters have occurred over the years. There are a number of impending disasters that can happen. Three disasters relating to contamination from arsenic, lead, and GenX are discussed in some detail. Discussions cover why a given disaster occurred, its impact, and what lessons can be learned from it. A close review of these disasters suggests water-quality evaluations could help prevent some of the future water disasters.

References

Ahuja, S., 1986. Ultratrace Analysis of Pharmaceuticals and Other Compounds of Interest. Wiley, New York.

Ahuja, S., 2006. Assuring water purity by monitoring water contaminants from arsenic to zinc. In: American Chemical Society Meeting, Atlanta, March 26–30. .

Ahuja, S., 2008. Arsenic Contamination of Groundwater; Mechanism, Analysis, and Remediation. Wiley, New York.

Ahuja, S., 2009. Handbook of Water Purity and Quality. Elsevier, Amsterdam.

Ahuja, S., 2013a. Monitoring Water Quality: Pollution Assessment, Analysis, and Remediation. Elsevier, Waltham, MA.

Ahuja, S., 2013b. Comprehensive Water Quality and Purification. vol. 1–4. Elsevier, Kidlington, Oxford.

Ahuja, S., 2014. Water Reclamation and Sustainability. Elsevier, Amsterdam.

Ahuja, S., 2017a. Chemistry and Water: The Science Behind Sustaining the World's Most Crucial Resource. Elsevier, Amsterdam.

Ahuja, S., 2017b. Monitoring water quality and infrastructure to prevent future flints. In: American Chemical Society Meeting. Fall, Washington, DC.

Ahuja, S., 2018. C&EN. Feb. 12, p. 2.

Ahuja, S., Hristovski, K., 2013. Novel Solutions to Water Pollution. American Chemical Society, Washington, DC.

Karasek, F.W., Clement, R.E., Sweetman, J.A., 1981. Preconcentration for trace. Anal. Chem. 53, 1050.

Kratochvil, B., Taylor, J.K., 1981. Sampling for chemical analysis. Anal. Chem. 53, 924 A.

McNeil, E.E., Otson, R., Miles, W.F., Rahabalee, F.J.M., 1977. Determination of chlorinated pesticides in potable water. J. Chromatogr. 132, 277.

Sun, M., Arevalo, E., Strynar, M., Lindstrom, A., Richardson, M., Kearns, B., Pickett, A., Smith, C., Knappe, D., 2016. Legacy and emerging perfluoroalkyl substances are important drinking water contaminants in the Cape Fear River watershed of North Carolina. Environ. Sci. Technol. Lett. 3, 415–419.

Wolff, E.W., Landy, M., Peel, D.A., 1981. Preconcentration of cadmium, copper, lead, and zinc in water at the 10–12 g/g level by adsorption onto tungsten wire followed by flameless atomic absorption spectrometry. Anal. Chem. 53, 1566.

Youden, W.J., 1967. The role of statistics in regulatory work. J. Assoc. Off. Anal. Chem. 50, 1011.

CHAPTER

2

The Heat Is On: Complexities of Aquatic Endocrine Disruption in a Changing Global Climate

B. DeCourten[a,b], A. Romney[c], S. Brander[a,*]

[a]Department of Environmental and Molecular Toxicology, Oregon State University, Corvallis, OR, United States [b]Department of Biology and Marine Biology, University of North Carolina Wilmington, Wilmington, DE, United States [c]Department of Anatomy, Cell Biology, and Physiology, University of California Davis, Davis, CA, United States
*Corresponding author: E-mail: susanne.brander@oregonstate.edu

1 INTRODUCTION

Exposures to environmentally relevant concentrations of endocrine disrupting compounds (EDCs) in aquatic ecosystems are long established to be a threat to ecological health (Brander, 2013; Colborn and Thayer, 2000; Schug et al., 2016; Windsor et al., 2018). Numerous studies published over the past several decades demonstrate that EDCs, which include pharmaceuticals, pesticides, many industrial compounds, and some metals, can amplify or antagonize the effects of endogenously produced hormones such as estradiol, testosterone, and/or thyroid hormone. These perturbations, often occurring at environmentally relevant concentrations, result in a wide array of downstream effects at the molecular, organism, and even community levels. These include altered gene expression or protein production, developmental defects, changes in growth or maturation rate, and altered sex ratio (Brander et al., 2016a; Guillette, 2006; Jobling et al., 1998; Pait and Nelson, 2002). Population level implications of exposure to EDCs demonstrated both empirically and theoretically, range from reduced fecundity to total extinction (Harris et al., 2011; Kidd et al., 2007; White et al., 2017).

Given that the planet is in the midst of a shift in long-established global climate norms due to the anthropogenic addition of heat-trapping greenhouse gases to the atmosphere, it is important to consider the context in which organisms that are susceptible to EDC-effects are exposed, now and into the future (IPCC, 2013). Mounting evidence suggests that co-exposure to EDCs and altered temperatures, salinity, aqueous pH, hypoxia, and lesser-considered factors related to global climate change (GCC) such as food availability or susceptibility to disease, can result in increased sensitivity to directly exposed organisms and even their progeny (DeCourten and Brander, 2017; Noyes and Lema, 2015). This can occur either via a GCC-altered abiotic factor that increases an organism's sensitivity to a chemical, or by a chemical increasing organismal sensitivity to altered climactic conditions (Hooper et al., 2013). Another concern related to GCC is that although the use of many endocrine disruptive persistent organic pollutants (POPs; e.g., PCBs, DDT, PBDEs) has declined over the past several decades, predicted changes in temperature and associated atmospheric circulation patterns may result in the reintroduction of legacy POPs that have been trapped in ice or frozen soil, as those substrates thaw (Ma et al., 2016). Current and future chemical use must also be considered, since many industrial compounds, pesticides, and pharmaceuticals used today are known EDCs (Brander, 2013) that may be mainstreamed in consumer products prior to thorough evaluation (DeWitt, 2015) and because predicted changes in arable land and disease occurrence will likely influence land and chemical use patterns (Schiedek et al., 2007), potentially increasing usage and thus inputs to waterways.

Likewise, with sea-level rise and the resultant changes predicted for connectivity between ground and surface waters, as well as predicted increases in extreme weather events (e.g., hurricanes) leading to increased severity of rain events as well as longer periods of drought (Hansen et al., 2012), the bioavailability and distribution of pollutants will undoubtedly be influenced (Faust et al., 2016). Predictions made via hydrological modeling suggest that concentrations of estrogenic compounds commonly detected in waterways will increase as much as twofold by the year 2050, as a result of a combination of population growth and climate-induced changes in water flow (Green et al., 2013). Compounding these issues are the increased incidences of coastal hypoxia and changes in

salinity regimes (Diaz and Rosenberg, 2008; Noyes and Lema, 2015). These abiotic factors must also be taken into consideration when assessing the impact of EDCs, particularly since hypoxia is demonstrated to alter sex ratios independently of chemical exposure (Thomas and Rahman, 2011), and because changes in salinity can influence the water solubility and hence the environmental fate of and potential bioavailability of POPs in aquatic systems (Saranjampour et al., 2017).

In any case, GCC and its influence on chemical distribution and toxicity call into question currently used approaches for risk assessment (Hooper et al., 2013), considering that the stable exposure conditions currently used in laboratory settings to determine acceptable levels of legacy and currently used chemicals do not mimic the reality of current and near-future pollutant exposure scenarios in the field (Fig. 1). That being said, the potential for wild populations to adapt to GCC-related environmental stressors must also be taken into consideration, given that plasticity and thus adaptive responses to stress can differ between populations and even among life stages of the same species, and that resistance to stressors is sometimes heritable (Komoroske et al., 2014; Major et al., 2017; Meyer et al., 2002; Schulte, 2014). It has also been suggested that subtle changes associated with GCC may even antagonize one another, resulting in some predicted negative impacts being offset (Lange and Marshall, 2017).

This review will consider exposures and responses of aquatic organisms to EDCs and mixtures in the context of four major GCC-related abiotic environmental stressors: temperature, hypoxia, altered salinity regimes, and acidification. We will review recent findings on physiological and developmental impacts, as well as compare impacts across different taxa, life stages, and aquatic habitat types (e.g., freshwater vs marine), since it is established that responses at smaller scales, such as the organism or individual level, will be related

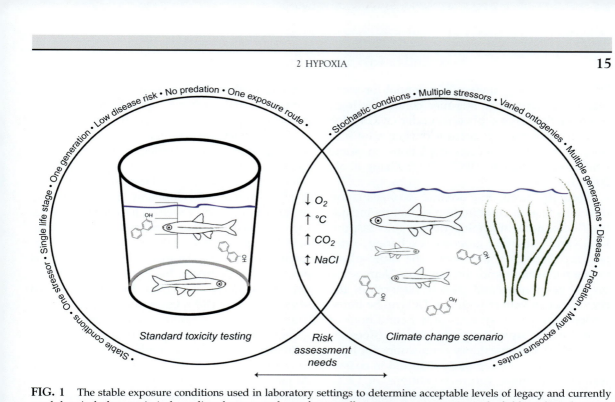

FIG. 1 The stable exposure conditions used in laboratory settings to determine acceptable levels of legacy and currently used chemicals do not mimic the reality of current and near-future pollutant exposure scenarios in the field involving hypoxia, increased temperature, acidification, and altered salinity regimes, which can influence responses as well as the bioavailability of contaminants. Laboratory assessments of endocrine-related endpoints must better account for the influence of global change if risk is to be accurately measured.

to trends observed in populations or across entire ecological communities (Brander et al., 2015). Ultimately, the aims of this review are to summarize the current state of knowledge on exposure to EDCs in a shifting climate, to highlight current gaps in knowledge and to offer direction going forward for scientists and managers in this age of uncertainty.

2 HYPOXIA

Dissolved oxygen (DO) concentrations in the open ocean have been steadily decreasing for at least the past 50 years (Breitburg et al., 2018; Diaz and Rosenberg, 2008; Schmidtko et al., 2017; Stramma et al., 2008). This deoxygenation coincides with the expansion of oxygen minimum zones (OMZ) both of which occur frequently near estuaries and other coastal aquatic systems (Breitburg et al., 2018; Schmidtko et al., 2017). These two processes are often occurring near each other; however, they originate from different sources, thus making their management much more difficult. Increases in oceanic hypoxia (≤ 2 mL of O_2/L) can be attributed to an overall global increase in atmospheric greenhouse gas emissions. However, OMZs more directly result from increased average water temperatures and increased nutrient and pollutant runoff from local watersheds (Altieri and Gedan, 2015; Breitburg et al., 2018; Du et al., 2018; Schmidtko et al., 2017). Under future scenarios of GCC, altered frequencies of drought or heavy rainfall, along with coinciding increases in oxygen stratification and warming, will likely promote an overall trend of increases in oceanic hypoxia (Barth et al., 2018; IPCC,

2013). Conversely, there is a potential for some regions to experience stormier conditions (e.g., more hurricanes or typhoons) under GCC which may increase mixing, disrupt oxygen stratification, and reduce oxygen depletions in some benthic environments (IPCC, 2013; Zhang et al., 2010a,b).

Naturally occurring oceanic deoxygenation develops from a combination of factors including fluctuations in photosynthesis and respiration, and altered gaseous exchange rates between surface waters with the atmosphere or deeper waters and the benthos. Depending on the geography and climate patterns of the region, ocean waters with low DO can be transported into coastal systems under conditions of wind-forced upwelling dynamics (Roegner et al., 2011). Ocean-estuary connections for hypoxic conditions are regularly detected in coastal waters and under future conditions of GCC, increasing frequency, and severity of low DO events are predicted (IPCC, 2013; Roegner et al., 2011). Patterns of water circulation may increasingly contribute to stressful conditions in local ecosystems including both animal and plant life in coastal estuaries.

Most aquatic species are obligate aerobes with only a moderate few that differ in their tolerance to low DO. Additionally, species that demonstrate unique sensitivities do so at varying timescales and ontogenies (Bickler and Buck, 2007; Pörtner et al., 2005; Ramirez et al., 2007). The impacts range from whole ecosystem to the molecular level of biological systems. Specific examples in recent studies have demonstrated how hypoxia can impact biodiversity (Hughes et al., 2015) reproduction (Shang et al., 2006), growth (Altenritter et al., 2018), and gene expression (Lai et al., 2016; Thomas et al., 2007).

2.1 Early Life Stages Effects

Embryonic development in most aquatic organisms is characterized by the encapsulation of the embryo within a protective yet semipermeable chorion, which is protective of environmental stressors prior to the development of the physiological capacity for homeostasis. The impact that occurs when these processes are impaired is complicated to predict, seemingly species specific, and may manifest at later life history stages (Hamdoun and Epel, 2007). Animal development is an already oxygen-demanding process characterized by an increased sensitivity to even brief periods of hypoxia that can result in death or abnormal developmental outcomes (Dunwoodie, 2009; Giaccia et al., 2004). Embryos, having limited mobility, are immediately subjected to environmental stress from low DO (Hamdoun and Epel, 2007). This can be exacerbated when parents deposit gametes in secluded regions of aquatic habitat with little to no water recirculation, in an attempt to reduce predation (Diaz and Breitburg, 2009). At later life stages, when embryos are no longer encased within a chorion, larvae can avoid hypoxic regions by swimming away. Under longer exposures to hypoxia, vertebrates such as fish and amphibians may experience reduced metabolic performance and premature hatching that can lead to physiological impairments that persist throughout their lifespan (Anderson and Podrabsky, 2014; Barrionuevo et al., 2010; Hassell et al., 2008; Mueller et al., 2011; Seymour et al., 2000).

Both hypoxia tolerance and the cellular signaling pathway that occurs in response to hypoxia stress are known to be affected as a result of environmental exposure to EDCs. Specifically, EDCs that act as aryl hydrocarbon receptor ligands can cause impairment of the cellular stress response (HIF—hypoxia inducible factor) gene pathways (Vorrink and Domann, 2014). HIF-1 is a master regulator transcription factor that controls the expression of many downstream genes that increase O_2 uptake and decrease O_2 demand (Hochachka et al., 1996). Gene induction by hypoxia involves a dimer of HIF-1α and the nuclear cofactor HIF-1β, also called the aryl hydrocarbon receptor nuclear translocator

(ARNT). ARNT (HIF-1β) is also involved in induction of genes in response to aryl hydrocarbon exposure (xenobiotics). ARNT and HIF-1β share identical protein sequences and originate from a single gene (Wang et al., 1995), and they recognize a common core DNA sequence that is found in both hypoxia and xenobiotic response elements (HREs and XREs; Swanson et al., 1995; Wenger, 2000). In embryos of zebrafish, *Danio rerio*, cross talk occurs between receptors for hypoxia stress and dioxin toxicity such as that from 2,3,7,8-tetrachlorodibenzo-*p*-dioxin (TCDD) (Prasch et al., 2004). Following exposure to both hypoxia and TCDD, multiple studies have shown defects in cardiovascular development in a variety fish species which consistently include a reduction in peripheral blood flow, hemorrhage, and edema of the pericardial and yolk sacs (Henry et al., 1997; Prasch et al., 2004; Tanguay et al., 2003; Walker and Peterson, 1994).

Responses to multiple stressors was also investigated using fluoranthene, a polycyclic aromatic hydrocarbon (PAH), that is found in locations contaminated with petroleum products or urban runoff (Doong et al., 2000). At ecologically relevant concentrations, (100–500 μg/L), fluoranthene can cause severe developmental toxicity in *D. rerio*, with abnormalities such as pericardial effusion and lordosis in young larvae co-exposed to environmental hypoxia of 7.3% DO (Matson et al., 2008; Fig. 2). Regardless of the cellular mechanisms responsible for this interaction, embryo-toxicity resulting from the co-exposure of one of the most abundant PAHs, fluoranthene, at environmentally relevant concentrations along with moderate hypoxia presents a significant challenge for environmental risk assessments of aquatic environments, considering these gaps in knowledge.

2.2 Adult Effects

Many studies have focused on the sensitivity of embryos to low environmental DO; however, adult stages also demonstrate sensitivities to

O_2 deprivation. In many fish species, exposure to hypoxia during development can impact the adult phenotype and alter morphology, physiology, and behavior. Acclimation to toxic or environmental stressors independently during early life can result in increased metabolic performance that requires an increased expenditure of energy; as a trade-off, this can reduce the energy resources that are available for growth and reproduction later (Gupta, 2011; Heath, 1995; Kooijman, 1998). The duration and severity of exposure is particularly important to distinguish between outcomes on adult physiology. Across vertebrates, hypoxia induces metabolic and ion channel suppression that can ultimately lead to a broad range of effects from behavior to cellular physiology (Bickler and Buck, 2007). Depending on the adaptive capacity of the organism, the ability to recover or tolerate hypoxia is difficult to predict (Bickler and Buck, 2007). Alterations in sex steroid production and sex determination have been repeatedly identified across a diversity of fish species to result from chronic hypoxia exposure (Landry et al., 2007; Shang et al., 2006; Thomas and Rahman, 2011). For example, hypoxic conditions have been shown to reduce the expression of the G protein-couple estrogen receptor (GPER) in the Atlantic croaker, resulting in increased ovarian follicular atresia and a reduction of the number of healthy oocytes (Ondricek and Thomas, 2018). The GPER pathway which is important for maintaining meiotic arrest and allowing for oocyte maturation can be also be influenced by exposure to EDCs; however, the potential for synergistic effects of these stressors is yet to be determined (Thomas, 2017). In work by Ivy et al. (2017), acute embryonic exposure of zebra fish at 36 hpf, for a duration of time that does not affect embryo survival, induced a homeostatic challenge that lead to long-term endocrine and behavioral consequences. These included depressed expression of gonadal aromatase and an increase in dominant and aggressive phenotypes in adults (Ivy et al., 2017).

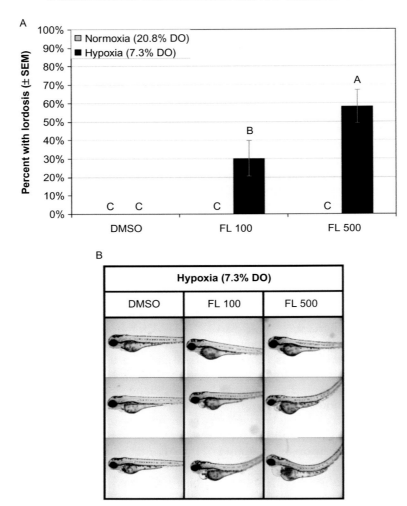

FIG. 2 (A) Occurrence of lordosis in developing zebrafish larvae exposed to hypoxia (7.3% DO) and 0, 100, or 500 μg L^{-1} fluoranthene (FL). Groups not sharing a common letter are significantly different ($P \leq 0.05$), based on Bonferroni-corrected-post hoc multiple comparison following ANOVA. (B) Images of 96 hpf zebrafish larvae showing the range of trunk abnormalities observed for coexposed embryos. While embryos exposed only to hypoxia and carrier control (DMSO) showed no evidence of trunk abnormalities, a significant proportion of embryos coexposed to hypoxia and FL developed lordosis. *Printed with the permission from Matson, C.W., Timme-Laragy, A.R., Di Giulio, R.T., 2008. Fluoranthene, but not benzo [a] pyrene, interacts with hypoxia resulting in pericardial effusion and lordosis in developing zebrafish.* Chemosphere, 74 *(1), 149–154.*

The sensitivities to hypoxia that exist during early life stages are not absent in later life stages. Adults can be just as sensitive to low DO, only that they often have the ability to relocate if possible, to less stressful environments. Adaptations exist across all aquatic organisms that provide them the toolkit to withstand or avoid threatening hypoxic conditions. However, such physiological or behavioral changes are dramatically impacted by co-exposures with EDCs. In work that followed in adult Atlantic killifish, *Fundulus heteroclitus*, a species known to be both dioxin and hypoxia tolerant, researchers examined the glycolytic enzyme production in result of hypoxia and dioxin (PCB) exposure in the liver of adults (Kraemer and Schulte, 2004). Glycolytic enzymes are known to be upregulated during hypoxia to provide a degree of protection against oxygen-limited metabolic stress. PCB exposure caused significant decreases in glycolytic enzyme activity suggesting prior PCB exposure could reduce their tolerance of environmental hypoxia, while also creating trade-offs in terms of resource intensive processes such as growth and reproduction (Kraemer and Schulte, 2004).

2.3 Conclusions

Organisms at the interface of coastal environments are subject to co-occurring stress from hypoxia and pollutants such as EDCs. When challenged by hypoxia alone, fundamental homeostatic processes are threatened including metabolism and ion channel regulation as well as the increased potential for cellular injury from an increased production of reactive oxygen species. Hypoxia can independently cause endocrine disruption: Consistent evidence across multiple laboratories has shown that chronic hypoxia exposure during embryonic development can disrupt the balance of sex steroids (Thomas and Rahman, 2011; Thomas et al., 2007). These alterations can lead to persistent downregulation of aromatase expression that further leads to altered ratios of sex determination within a population.

When coupled with EDCs, a diversity of outcomes and long-term impacts can occur. Furthermore, the complex interactions between GCC, hypoxia, and contaminants such as PAHs, dioxins, and PCBs are predicted to increase through biomass burning and may alter and remobilization of legacy persistent organic pollutants (Holmstrup et al., 2010; Hooper et al., 2013; Noyes et al., 2009). Effects of co-exposure are interactive and dynamic: hypoxia alters the toxicity EDCs and as well, EDCs have the capacity to alter any tolerance to hypoxia. However, there are such interactions that can occur that alter enzymatic activity to provide a degree of protection against future low DO exposure (Kraemer and Schulte, 2004; Semenza, 2000; Wenger, 2000). Nonetheless, it is becoming an increasingly urgent global problem that has already caused a severe decline in species and major shifts in aquatic ecosystems.

3 SALINITY

Warming temperatures and the accompanying changes in precipitation are modifying patterns of salinity throughout oceanic and freshwater environments (Durack et al., 2012; Held and Soden, 2006; Knouft and Ficklin, 2017; Wentz et al., 2007). According to future GCC predictions modeled from previous decades of hydrological patterning, these conditions will intensify aquatic environmental salinity in a way that can be simply described as "salty ocean regions getting saltier and fresh regions are getting fresher"; a mechanism first introduced by Chou et al. (2009). These global dynamics in water cycling have large impacts on the salinity patterns of local ecosystems in which organisms have adapted to stable patterns of water conditions (Cloern et al., 2011; Durack et al., 2012; Hilton et al., 2008; Moyle et al., 2018). Changes in salinity impose risks to aquatic communities in complicated and unpredictable ways that depend on whether organisms can accommodate (1) to overcome species invasions and competitions that accompany when populations undergo range shifts in an attempt to seek refuge form salinity stress (discussed later in this review) or (2) the physiological challenge of salt water regulation beyond what they have evolved to perform.

3.1 Early Life Stages Effects

The impact of salinity on the biology of aquatic organisms is often dependent on life stage. For example, in both invertebrates and vertebrates that reproduce via external fertilization, the released gametes undergo cellular activation (that returns cells from a state of dormancy back into the cell cycle) prior to fertilization that is sensitive to environmental conditions of ionic composition and salinity (reviewed in Finn, 2007; Horner and Wolfner, 2008). Upon egg activation, the outer chorion expands mainly due to an influx of external water that swells to form the perivitelline space around the plasma membrane in a manner dependent on external osmotic conditions (Coward et al., 2002; Finn, 2007).

After fertilization yet prior to the development of major osmoregulatory organs (gills, kidney, and gastrointestinal tract), pre-hatch invertebrate, and vertebrate embryos are also vulnerable to salinity stress as they are unable to migrate away when their immediate environmental conditions become stressful. In the later developmental stages of aquatic organisms, changes in salinity can alter developmental timing, hatching rates, and produce malformations in newly hatched larvae (Pérez-Robles et al., 2016; Zhang et al., 2010b). Following hatch in many species, newly emerging vertebrate and invertebrate larvae become passive recipients of the immediate water conditions (Armstrong and Nislow, 2006; Charmantier, 1998; Varsamos et al., 2005). In general, the prospects of survival for a species impacted by salinity changes in response to GCC is complicated and can be dependent on life stage and osmoregulatory patterns, for example, organisms that actively regulate their internal osmotic composition (regulators) may be less susceptible than those that passively equilibrate with the environment (conformers); and euryhaline organisms (broad salinity tolerance) may be more tolerant than those that are stenohaline (narrow tolerance) (Finn, 2007; Pérez-Robles et al., 2016).

Estuarine communities are important sites for research on impacts of salinity stress in combination with exposure to EDCs. Invertebrates, and specifically crustaceans, have long been used as resident sentinels for assessing estuary health and their development and metamorphosis have direct relationships with their salinity and endocrinology. Many estuaries that provide ecologically and economically are also located near agricultural fields associated with abundant pesticide and herbicide application and as a result receive substantial chemical runoff (Ramach et al., 2009). Many current and legacy pesticides including synthetic pyrethroids and organophosphate insecticides, as well as DDT and its degradates have been found within the eggs of the several crab species including, *Callinectes sapidus*, *Hemigrapsus oregonensis*, and *Pachygrapsus crassipes* (Goff et al., 2017; Smalling et al., 2010). Recent work by Goff et al. (2017) found that at increased salinities, both fipronil and its photodegraded form, fipronil desulfinyl, altered juvenile growth in the blue crab, *C. sapidus*, and also interfered with the expression of genes that underlie molting, growth, and reproduction. They suggested that because of having a reduced solubility at higher salinities (Saranjampour et al., 2017), fipronil may have been more bioavailable, allowing for greater accumulation in tissue of *C. sapidus* (Goff et al., 2017). In fish, Moreira et al. (2018) discovered that exposure to the herbicide diuron at higher salinities caused decreased expression of thyroid and growth hormone receptors concurrent with decreased condition factor and body mass. As ecotoxicological work continues to explore the underlying mechanisms that contribute to the declining health of these communities, it is clear that changing salinity regimes and EDCs will have impacts during transitions through early life history stages including growth and metamorphosis.

3.2 Adult Effects

Adaption to salinity is well characterized conceptually in later life history stages when the physiological capacity to tolerate stress has fully formed. However, physiological acclimation to alternative salinity regimes alone cannot remediate the potential for toxic interactions between salinity and EDCs. This is because salinity has a unique effect on the solubility properties of chemical stressors that may be present in aquatic systems that can lead to alterations in the chemical uptake and biological disposition for organisms of those environments (Hooper et al., 2013). In addition to the differential chemical behavior of organic compounds between fresh and saltwater, it is established that aquatic organisms inhabiting saltier waters must intake more water in comparison to freshwater species, in order to maintain equal osmolality (salt content) with their surroundings (Evans, 1980). This may mean that the relative aqueous exposure of marine and estuarine organisms to aqueous pollutants is higher as a result of to the larger volume of water they ingest and reduced urination relative to freshwater organisms.

Not only can salinity alter the toxicity of EDCs but recent work identified that alternative salinities can change the water solubility characteristics of different compounds. Saranjampour et al. (2017) estimated median lethal concentration LC50 values using Estimation Programs Interface (EPI) Suite for a range of aquatic pollutants (pesticides, PAHs) in both fresh and sea water, and found that after 96h, LC50 values were lower (more toxic) for all compounds in seawater. Factors including the biota, organic matter, suspended sediment, and particulate matter along with water conditions such as pH, temperature, and turbulence can all have deleterious effects on the bioavailability of toxic compounds for marine or estuarine organisms. The impact of salinity is magnified for more hydrophobic compounds, also demonstrated by findings from Yang et al. (2016), which demonstrated that increased salinity pushed a variety of EDCs from the dissolved to solid phase, increasing their propensity to accumulate in sediments and tissue. This work altogether highlights those impacts of co-occurring stress from salinity and EDCs needs to be more reliably predicted by incorporating the physical and chemical characteristics of an ecosystem.

In fact, such a relationship between exposure salinity and endocrine-related responses was recently identified in a meta-analysis performed on a large volume of primary studies investigating-EDCs in fish (Bosker et al., 2017). Their analysis included 12 species of either freshwater or saltwater fish under alternative salinities and co-exposure of two model estrogens, 17α-ethinylestradiol and 17β-estradiol (E2), and three androgens (17β-trenbolone, 5α-dihydrotestosterone, and 17α-methyltestosterone). They found from the 59 studies they included in the analysis that the most sensitive endpoints in fish exposed to both estrogenic and androgenic EDCs are E2 levels and altered fecundity in females under alternative salinities (Bosker et al., 2017; Fig. 3). Not only did their work assess the influence of salinity on reproductive effects in fish exposed to common environmental EDCs, but it also revealed minor differences in the health outcomes between species that would otherwise be difficult to observe with investigations of individual taxa. In another study from Bosker's group with mummichogs (*F. heteroclitus*), which compared reproductive endpoints in response to dihydrotestosterone (DHT) across two salinities, DHT-exposed adults had lowered egg production compared to controls at 16 ppt, but not at 2 ppt (Glinka et al., 2015). These manuscripts highlight the need for future investigations that develop a clearer understanding of salinity interactions with EDCs in order to work toward more comprehensive environmental risk assessments.

Estuarine organisms other than vertebrates have also been identified as susceptible to the

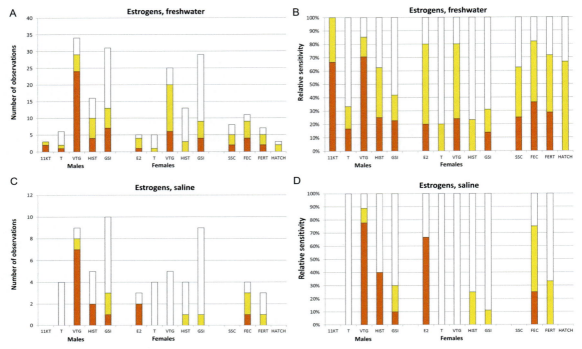

FIG. 3 Endpoint sensitivity for estrogens under fresh (A and C) and saline (B and D) exposure conditions. The number of experiments reported with lowest observed effect concentration (LO) is shown in *orange*. The number of studies in which observed effects were reported above the LOEC within the same study is shown in *yellow*. The number of studies in which no observed effects was reported at concentrations higher than the maximum tested concentration is shown in *white*. Printed with the permission from Bosker, T., Santoro. G., Melvin, S.D., 2017. Salinity and sensitivity to endocrine disrupting chemicals: a comparison of reproductive endpoints in small-bodied fish exposed under different salinities. Chemosphere 183, 186–196.

effects of EDCs and salinity stress. Bifenthrin, like many pesticides designed with specificity to insects, are similarly toxic to crustaceans because of their conserved arthropod physiology. Work by Hasenbein et al. (2018) examined the toxicity of this potent and widely used insecticide, to the epibenthic amphipod *Hyalella azteca*. They found that a slight increase in salinity readily decreased survival and reduced the swimming performance of *H. azteca* in the face of exposure to nanomolar concentrations of bifenthrin (Hasenbein et al., 2018). As well, they found that co-exposure of higher salinities and bifenthrin resulted in a reduced expression of heat shock protein genes associated with cellular stress response, suggesting a lack of protection against cellular damage and oxidative stress (Hasenbein et al., 2018). Therefore, even mild environmental stress such as that predicted for the near future, and that is still well within physiological limits of a species, may synergistically increase toxicological sensitivity.

3.3 Conclusions

Responses to changes in salinity with combined exposure to environmental pollutants are demonstrated to be complex (Finn, 2007). Furthermore, the importance of a broader ecotoxicity assessment inclusive of estuaries and coastal areas is heightened because bioconcentration factors of organic contaminants (many of them EDCs), can increase with escalating salinity. This may be due to increased partitioning (salting-out) of organic compounds into biota and sediments (Eganhouse and Calder, 1976; Hashimoto et al., 1984; Jeon et al., 2011; Saranjampour et al., 2017; Yang et al., 2016). As such, estuarine and marine organisms may be differentially affected by aquatic contaminants depending on the salinity of exposure (Goff et al., 2017; Hasenbein et al., 2018; Jeon et al., 2010a,b). In particular, the bioavailability and associated toxicity of a wide array of aquatic pollutants may be higher for estuarine and marine organisms in comparison to those exposed in freshwater. The interactions of salinity on the chemical nature and reactive activity require an intensive investigation in order to assess the risks that are predicted under future GCC scenarios. Additionally, current habitats of concern for conservation management are those that are directly subject to future alterations in salinity, and a continued investigation of the biological impacts of co-exposure with EDCs will improve our ability to predict and potentially reduce stressful conditions.

4 ACIDIFICATION

The pH of global ocean waters has been steadily declining over the last 200 years in large part as a result of the absorption of increased concentrations of carbon dioxide (CO_2) from the atmosphere (Raven et al., 2005). It is predicted that surface ocean waters will continue to decrease in pH by approximately 0.3–0.5 units by the year 2100 (Caldeira and Wickett, 2005; Dore et al., 2009). Ocean acidification (OA) occurs when absorbed atmospheric CO_2 is chemically converted to carbonic acid (H_2CO_3) which then dissociates into bicarbonate (HCO_3^-) and hydrogen (H^+) ions. A reduction in the resulting abundance of free carbonate ions (CO_3^{2-}) can be deleterious for many marine invertebrates that use calcification to build exoskeletal structures (Hoegh-Guldberg et al., 2007). In addition to constraining calcification, OA also disrupts processes that rely on acid–base chemistry like those specific to cellular metabolism and ion regulation (Claiborne et al., 2002; Henry et al., 2012).

In addition to OA resulting from increased atmospheric CO_2, the pH of coastal waters is further impacted by runoff from local lands with known toxic and endocrine disruptive effects (Brander, 2013; Doney et al., 2007; WHO, 2012). This can be incredibly harmful to organisms that live in near-shore habitats such as estuaries and coastal oceanic environments that experience increased exposure to acidification. While interactions, specifically between EDCs and OA are underrepresented in the literature, there are a few studies that have examined them in combination (Maulvault et al., 2015).

4.1 Early Life Stages Effects

Embryonic stages of both invertebrates and vertebrates demonstrate dynamic responses to OA whereby some species show a robust tolerance (Byrne et al., 2009; Munday et al., 2009; Tseng et al., 2013) and others show a sensitivity to changes in pH (Baumann et al., 2011; Chambers et al., 2014; Forsgren et al., 2013). Even a subtle decrease in pH (0.5 units) was shown to significantly reduce fertilization in the Pismo clam, *Tivela stultorum*, from Baja California (Alvarado-Alvarez et al., 1996); a region predicted to be severely impacted by OA in the near future (Halpern et al., 2009). Reductions in pH resulted in decreased egg sizes and delayed hatching in the perch, *Perca fluviatilis* (Vinogradov, 1985). Studies that determined a limited tolerance to acidification measured attributes such as hatching success and survival of the embryos (Vinogradov, 1985). In larval or post-hatch stages, studies demonstrate that tolerance to acidification is largely dependent on the ontogenetic state of regulatory organs (gut, gill, skin, and kidney) that are critical for maintaining acid–base equilibrium and ion exchange via cellular proton pumps (Pörtner et al., 2004; Vinogradov, 1985).

Such toxic effects of OA alone can impact a wide array of processes including organismal physiology and cellular metabolism (Pörtner et al., 2004), of which all have the potential to interact synergistically with added insult from exposure to chemical pollutants such as EDCs. Recently, interactions between aquatic chemical pollutants and OA were investigated in embryos of the marine amphipod *Gammarus locusta* (Loganimoce, 2016). Under hypercapnic conditions (a reduction of 0.5 units), exposure to the synthetic steroid, levonorgestrel, resulted in significantly lower growth rates (mm per day; Loganimoce, 2016). Hypercapnia alone is known to suppress metabolic processes causing delayed development, however, Loganimoce (2016) suggested that considering levonorgestrel is a synthetic progesterone derivative, it is likely that even at low concentrations (10 ng/L) in their aqueous habitat, it could have potentially interfered and disrupted the endocrine axis governing molting activity for *G. locusta* (Loganimoce, 2016).

In general, early life stages are found to be more sensitive to ocean acidification than later stages because the exposure precedes when the important osmoregulatory organs are operational (Forsgren et al., 2013; Frommel et al., 2014; Melzner et al., 2009; Pörtner et al., 2004). In addition to demonstrating reduced survival and growth at these stages, it is likely that negative impacts on these regulatory systems in early development leads to long-lasting effects on animal performance and fitness at later life stages.

4.2 Adult Effects

For most adult marine species of fish, acid–base imbalances are tolerated via homeostatic adjustments of bicarbonate accumulation and ion exchange across the gills (Claiborne et al., 2002). Teleosts, in general, can adapt to prolonged elevation of CO_2 saturations through acid–base regulation and by increasing ventilation frequencies, thereby avoiding internal acidosis (Ishimatsu et al., 2005, 2008). However, not all fish species demonstrate the same capacity of tolerance or compensation for pH imbalance, nor do all life stages. Regardless of a species' tolerance to lower pH or the capacity to compensate for any internal acid–base disturbances, acidic environmental conditions generally result in a depressed metabolic performance in both invertebrate and vertebrate organisms (Fabry et al., 2008; Pörtner et al., 2004). Features of acid–base and gas properties of arterial blood were examined in the squid, *Illex illecebrosus*, and it was found that pH was a critical driver in the relationship between O_2 binding, oxygen partial pressure (PO_2), and blood pH (Pörtner et al., 2004). Effects of OA can be particularly harmful to marine calcifiers that rely heavily on pH for baseline physiological performance and survival (Pörtner et al., 2004). For example, the blue mussel *Mytilus edulis*, regulates internal pH by suppressing metabolism and dissolving their carbonate shell when exposed to increased CO_2 levels (Bibby et al., 2008; Gazeau et al., 2007). Additionally, the work by Bibby et al. (2008) found that with increased acidity, the cellular immune response mechanism, hemocytic phagocytosis, was significantly reduced in *M. edulis*. With the potential for future OA to be harmful for invertebrates by itself, the interaction with other anthropogenic factors, including EDCs could likely alter their toxicity and change the sensitivity of organisms toward these stressors (Fabry et al., 2008; Freitas et al., 2015).

Interactions between toxicants and ecologically relevant levels of OA have been understudied in aquatic ecosystems. With GCC-induced OA becoming a major and present concern, more mechanistic studies are needed to address the possible interactions between EDCs and environmental stressors. Analysis on biologically harmful metals by Millero et al. (2009) found that decreases in concentration of

OH^- and CO_2^- ions can directly affect chemical and physical properties including solubility, adsorption, toxicity, and rates of redox reactions metals in seawater. Such outcomes will have a great potential to alter the availability and toxicity of metals to marine organisms (Millero et al., 2009). Processes associated with contaminant exposure such as bioaccumulation and elimination of endocrine disruptive flame retardants have been shown to be increased acidic conditions in two estuarine bivalve species, the Japanese carpet shell clam *Mytilus galloprovincialis* and *Ruditapes philippinarum* (Maulvault et al., 2018b). However, it is hypothesized that the majority of the physiological effects are due to exposure to increased aquatic CO_2 saturation (environmental hypercapnia) rather than lower ambient pH (Ishimatsu et al., 2004).

Preus-Olsen et al. (2014) proposed that the ecological context may amplify climate-EDC impacts markedly among inbred species and populations that occupy narrow niches with limited genetic diversity (e.g., threatened or endangered species) or those that display environmental sex determination. They examined combined exposure to perfluorooctane sulfonic acid (or sulfonate; PFOS) and elevated levels of CO_2 saturation in juvenile Atlantic cod (*Gadus morhua*). PFOS is a persistent organic pollutant (POP) that is detected globally and associated with numerous adverse health effects, including endocrine disruption (Kannan, 2011; Lau et al., 2007). Preus-Olsen et al. (2014) measured concentrations of sex steroid hormones and estrogenic responses and found that combined PFOS and hypercapnia exposure (0.3% increase in CO_2) produced increased effects on steroid levels as compared to hypercapnia alone (Preus-Olsen et al., 2014; Fig. 4). What is yet to be fully understood is the mechanisms that underlies any additive relationship between OA and chemical pollutants. Further multi-stressor studies should be conducted for conservation management and a better understanding of ecophysiology in marine ecosystems.

4.3 Conclusions

Not only predicted for the future but under current conditions, CO_2 represents an abiotic stressor that remains relatively constant in most of the global ocean. Only until recently have investigations been published on the physiological interactions and associated mechanisms between chemical pollutants and GCC-induced OA. Likely for all members of aquatic ecosystems, CO_2 poses a threat to metabolic performance. Long-term effects include reduced growth and reproduction which may result negatively for the health of even tolerant species (Fabry et al., 2008; Pörtner et al., 2004). Trade-offs have yet to be identified from the literature but will likely present impacts on physiological adaption and fitness in aquatic communities that can moreover affect population dynamics or even result in extinction (Pörtner et al., 2004). Although the changes in water chemistry that results from the oceanic uptake of anthropogenic CO_2 are well characterized over most of the ocean, the biological impacts of ocean acidification and the potential interaction of EDC on aquatic communities are only just beginning to be understood.

5 TEMPERATURE

Of the environmental changes attributed to anthropogenic climate change, increased temperature is the most well-represented factor within the scientific literature. In the next century our global climate is expected to warm as a direct result of greenhouse gas emissions (IPCC, 2013). To put this into perspective, over the past few decades the surface of our planet has warmed 0.11°C on average (1971–2010). This warming trend is expected to continue at an accelerated rate over the next century, impacting our climates on a global scale. The most dramatic increases in temperature are expected at mid-high northern latitudes, impacting many organisms adapted to

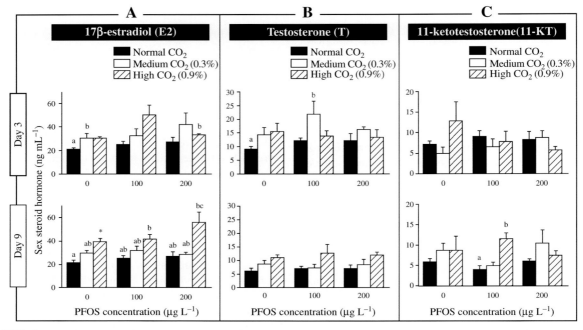

FIG. 4 Muscle tissue concentration of 17β-estradiol (E2: A), testosterone (T: B), and 11-ketotestosterone (11-KT; C) in juvenile Atlantic cod (*Gadus morhua*) after exposure to the various combinations of PFOS (0, 100, and 200 μg L^{-1}) and altered water CO_2 saturation (0%, 0.3%, and 0.9% increase in CO_2). Steroid hormones were analyzed in fish sampled at day 3 and 9 into the CO_2 exposure period. Steroids were extracted from fish muscle and concentrations correspond to 533 mg tissue/mL extraction volume. Data are given as mean values ± standard error of the mean (SEM). Different letters indicate significant differences between exposure groups ($P<0.05$). Asterisk (*) denotes borderline significance ($0.10<P<0.05$), $n=5$ in all groups. *Reprinted with permission from Preus-Olsen, G., Olufsen, M.O., Pedersen, S.A., Letcher, R.J., Arukwe, A., 2014. Effects of elevated dissolved carbon dioxide and perfluorooctane sulfonic acid, given singly and in combination, on steroidogenic and biotransformation pathways of Atlantic cod. Aquat. Toxicol. 155, 222–235.*

a temperate climate. As a result, organisms in polluted waterways will have the added stress of elevated temperature mitigating their response to chemical stressors. Warming in the winter months alone has been linked to reduced reproductive success in fish (Farmer et al., 2015). Additionally, lack of nighttime cooling in combination with contaminants can result in higher mortality of exposed organisms (Hallman and Brooks, 2015). In addition to thermal stress, temperature may also influence how organisms respond to exposure to endocrine disruptors in aquatic habitats.

Temperature can dictate the environmental fate and availability of EDCs entering watersheds, with longer half-lives observed at lower temperatures (Cox et al., 2017). Although rising temperatures may accelerate degradation of these chemicals in the environment, they can also influence the way organisms interact with these chemicals. In poikilothermic species, rates of metabolic function are highly dependent on ambient temperature (Clarke and Fraser, 2004), thus increasing the uptake and metabolism of EDCs in surrounding media. Additionally, the metabolites of some EDCs can be more potent than their parent compounds (DeGroot and Brander, 2014). As a result, EDCs may be more potent at elevated temperatures if compounds are converted to the metabolite more rapidly and efficiently.

5.1 Early Life Stage Effects

Studies have shown that EDCs, particularly those with estrogenic activity, can lead to developmental deformities such as craniofacial and spinal deformations (David et al., 2012; DeCourten and Brander, 2017; Jin et al., 2009; Warner and Jenkins, 2007; Zha et al., 2008). In some cases, parental exposure to environmental contaminants can have effects on the development of subsequent generations with offspring exhibiting spinal, craniofacial, yolk sac, swim bladder, and pericardial deformities (Corrales et al., 2014). Likewise, elevated temperature alone can to cause deformities in fish (Wang and Tsai, 2000). Increased temperature and bifenthrin exposure have been linked to notochord deformities in fish independently (Jin et al., 2009; Linares-Casenave et al., 2013). Little is known about how temperature may affect development of skeletal abnormalities following EDC exposure at early life stages. There is also evidence that temperature can increase the occurrence of skeletal deformities after exposure to estrogenic compounds, in the offspring of exposed parents (DeCourten and Brander, 2017).

Exposure to estrogenic EDCs during early life stages has been shown to affect the development of gonads following exposure to environmentally relevant concentrations resulting in skewed sex ratios (Bergeron et al., 1994; Ohtani et al., 2000). In some cases, as in the early life stages of caimans, sex of individuals can be reversed after differentiation following exposure to environmentally relevant concentrations of BPA (Stoker et al., 2003). Increases in temperature have been shown to amplify the effects of estrogenic EDCs such as EE2 on gonadal development leading to greater skews in sex ratios (Brown et al., 2015; Luzio et al., 2016). Levonorgestrel, which can act as an androgen in fishes, can decrease the rate of ovary maturation with a stronger effect observed in fish exposed under a 3°C temperature increase (Cardoso et al., 2017). The effects of EDCs on gonadal development are of even greater concern for organisms with physiology that can be influenced by temperature, such as those with temperature-dependent sex determination (TSD). TSD has evolved in multiple taxa including fish, amphibians, and reptiles and is governed by a set of molecular mechanisms that control the production of the enzyme aromatase, ultimately altering the production of sex steroids and gonadal development (Matsumoto and Crews, 2012). Plasticity in sexual development allows organisms to rapidly respond to changes in environmental factors to optimize the fitness of the population regardless of genotype of individuals (Conover and Kynard, 1981; Strüssmann et al., 2010). A study by Duffy and colleagues found that individuals in populations with plasticity in sexual development are more susceptible to altered gonadal development following early life stage exposure to E2, than those with in populations with primarily genotypic sex determination (Duffy et al., 2009). Studies have shown that the expected pattern of gonadal development can be altered or reversed following EDC exposure, with a greater effect observed at higher temperatures (DeCourten and Brander, 2017; Luzio et al., 2016).

Growth rates of ectothermic organisms are highly dependent on ambient temperature, with elevated temperatures typically increasing growth rates of early life stage organisms (Russell et al., 1996). However, accelerated growth rates in response to elevated temperatures can result in developmental abnormalities (Wang and Tsai, 2000). EDCs have been shown to increase the growth rates of organisms exposed at early life stages in some cases (Bell, 2004; B.M. DeCourten et al., unpublished data), and impede growth in others (Hayes et al., 2006) depending on the mode of action. Studies have shown that exposure to diuron, which interferes with the HPT-axis, can impair the growth of larval fish, contrary to typical growth patterns at higher temperatures (Moreira et al., 2018). The effects of EDCs and temperature on growth rates of larval organisms could impact survival, as larval quality has been linked to size (Berkeley et al., 2004). Results from the studies above demonstrate that EDC exposure can alter physiological responses of ectothermic organisms to changes in temperature. A study measuring growth rates of crustaceans at two different temperatures during levonorgestrel exposure found that the EDCs had a stronger effect on growth at 18°C than it did at 22°C, with organisms exposed at lower temperatures growing slightly faster (Cardoso et al., 2018).

5.2 Adult Effects

Reproduction of aquatic organisms can be highly influenced by environmental cues such as seasonal fluctuations in temperature. It is possible that reproductive physiology and phenology will be disrupted in response to rising temperatures associated with GCC. Seasonal temperatures are an important environmental cue for many organisms to become reproductively active. High temperatures can decrease reproduction in fish by activating germ cell degeneration, inhibiting gametogenesis, curtailing spawning, and skewing sex ratios of the population (Strüssmann et al., 2010). It has been suggested that organisms will adjust their reproductive phenology as a mechanism to cope with a changing climate. However, longer lived species with sensitive reproductive strategies, such as turtles, are unlikely to be able make this adjustment, resulting in reproductive failure (Telemeco et al., 2013). Studies have shown that relevant temperature increases can lead to a reduction in eggs production (Bonner et al., 1998; DeCourten and Brander, 2017). This suggests that ectothermic organisms have a limited range of temperatures in which egg production is optimized, outside of which reproduction may be curtailed. In addition, temperature increases of 5°C can reduce sperm size and motility, which could decrease fertilization rates (Breckels and Neff, 2013). These findings suggest that elevated temperatures predicted under GCC scenarios may hinder reproduction in organisms that use environmental cues, such as temperature and photoperiod, to initiate spawning in the warmer months. Exposure to estrogenic EDCs has also been shown to reduce egg production and fertilization rates, leading to diminished reproductive success (Brander et al., 2016b; Nash et al., 2004; Thorpe et al., 2009). Because both temperature increases and EDC exposure can have profound effects on reproductive success, understanding the combined effects of these stressors is important for determining overall impact. For example, the number of gametes produced by organisms can be influenced by exposure to elevated temperature and EDC exposure, with the combination of the two producing greater effects than either stressor alone, suggesting an additive response (Cardoso et al., 2017). Sublethal concentrations of perchlorate, which primarily disrupts thyroid function, were found to decrease egg production with consistently increasing in severity over 3°C temperature increments (Lee et al., 2014). These findings suggest that fecundity of organisms in already contaminated waterways will be further inhibited by rising temperatures over the next century.

Reproductive success of organisms is also depended on the survival of offspring. Increasing temperature may induce mortality of early life-stage aquatic organisms decreasing recruitment after exposure to elevated temperature (Houde, 1989; Pepin, 1991; Pörtner et al., 2001). Exposure to contaminants can increase embryo mortality by inducing premature hatching and decrease the survival of larvae following parental exposure (Corrales et al., 2014). Increases in temperature can drastically reduce the number of viable offspring produced in fish (Brown et al., 2015), with EDC exposure also reducing offspring production, and with effects being most pronounced following early life stage exposure of parental fish (DeCourten and Brander, 2017, Fig. 5). Individuals that do survive to adulthood may experience further mortality from chronic exposure to elevated

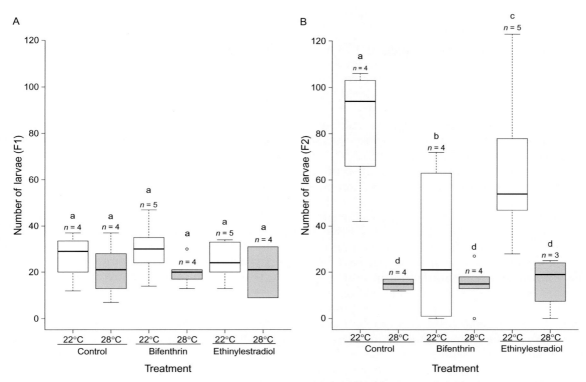

FIG. 5 Box plot of the number of larvae assessed at 21 dph produced in the F1 and F2 generations. *Boxes* denote the interquartile range and whiskers represent the upper and lower quartiles, with median denoted *solid bars*. Treatments that share the same lowercase letters are not significantly different ($P < 0.05$). Lowercase "*n*" represents the number of replicate tanks in each treatment. *Reprinted with permission from DeCourten, B.M., Brander, S.M., (2017). Combined effects of increased temperature and endocrine disrupting pollutants on sex determination, survival, and development across generations. Sci. Rep. 7 (1), 9310.*

temperatures. This is especially true for tropical species that typically have a narrow thermal range, outside of which small increases in temperature can cause increased mortality (Rodgers et al., 2018). The independent and combined effects of increased temperature and EDC exposure on survival lower the overall fitness of the population potentially leading to long-term population declines.

5.3 Conclusions

Understanding the effects of elevated temperature on EDC concentration is vital to future conservation efforts as the planet continues to warm. Temperature can influence organismal order processes, such as vitellogenin production, following exposure to a mixture of estrogenic EDCs (Brian et al., 2008). As discussed above, temperature can influence the effects of EDCs across organization levels, from molecular to organismal impacts. More research is necessary to determine the potential for long-term population level effects that can be caused by a combination of EDC exposure and temperature particularly in long-lived ectothermic animals that will experience decadal temperature changes. In addition to temperature, other climate change factors (e.g., salinity, pH, DO changes) must be incorporated into studies to better understand the effects of climate change on the toxicity of pollutants.

6 MULTIPLE STRESSORS

As discussed above, the activity of EDCs can be heavily influenced by abiotic changes associated with GCC. Understanding how combinations of multiple stressors associated with climate change can affect the toxicity of these chemicals is crucial in assessing risks and implementing appropriate management strategies for long-term conservation (Furlan et al., 2017). For example, the bioavailability and absorptions rates of endocrine disruptors can be influenced simultaneously by abiotic stressors such as temperature, pH, and salinity (Borrirukwisitsak et al., 2012), leading to synergistic, additive, or antagonist effects (Todgham and Stillman, 2013). Contaminants such as copper have been shown to cause hatching failure in limpet egg masses when exposed in combination with higher temperatures and UV light exposure (Kessel and Phillips, 2018). A study by Blewett et al. (2013) found that exposure to a range of temperatures and salinities had an effect on 17β-ethinylestradiol (EE2) uptake showing additive effects of temperature and salinity. Additionally, they found that uptake and metabolic rates increased with temperature and were highest at intermediate salinities (16 ppt), with a reduction in uptake observed after transfer to freshwater or seawater (Blewett et al., 2013). Changes in salinity and temperature can also exacerbate the effects of EDCs on the production of thyroid hormones ultimately impacting growth of exposed organisms (Moreira et al., 2018). These findings suggest that salt water intrusion, associated with sea-level rise, and temperature increases could increase the uptake of contaminants in freshwater organisms in an additive manner. As organisms encounter multiple abiotic stressors, it is possible for the effects of those stressors to impact the toxicity of endocrine disruptors in a synergistic manner. Similarly, food restriction combined with an 8°C temperature increase was found to impact the effects of estrone (E1) on sexual development, hematocrit, and liver size (Shappell et al., 2018). There is evidence that behaviors can also be influenced by multiple stressors, with one study finding that venlafaxine exposure in combination with acidification and warming decreased shoaling behaviors (Maulvault et al., 2018a). A study incorporating levonorgestrel, pH, and a 4°C temperature increase with crustaceans found that warming had the strongest effect, with the introduction of a third stressor (acidification) mitigating impacts on survival (Cardoso et al., 2018, Fig. 6). This suggests that it is possible for the effects of climate change and EDC exposure to act antagonistically, reducing the overall effect on organisms. Another study testing the effects of salinity and temperature on small invertebrates exposed to a pyrethroid pesticide, found that survival decreased with salinity at lower temperatures, however, this effect was not observed in the treatments with a 2°C temperature increase (Hasenbein et al., 2018). These findings illustrate the complexity with which multiple stressors may affect organisms exposed to EDCs in situ as the effects of GCC intensify. More research focused on the combined effects of abiotic factors influenced by GCC is needed to improve our understanding of environmental impacts and interactions of anthropogenic stressors.

7 TRANSGENERATIONAL AND EPIGENETIC EFFECTS

It is now established that nongenetic inheritance, sometimes induced by rapid environmental changes such as those associated with GCC or anthropogenic pollution, can facilitate relatively rapid phenotypic change (Bonduriansky et al., 2012; Shama and Wegner, 2014; Thor and Dupont, 2015). This rapid change may allow organisms to adapt to longer-term changes in abiotic conditions, such as increasing temperature or acidification (Munday, 2014), or to evolve tolerance to other related stressors (De Schamphelaere et al., 2010). However,

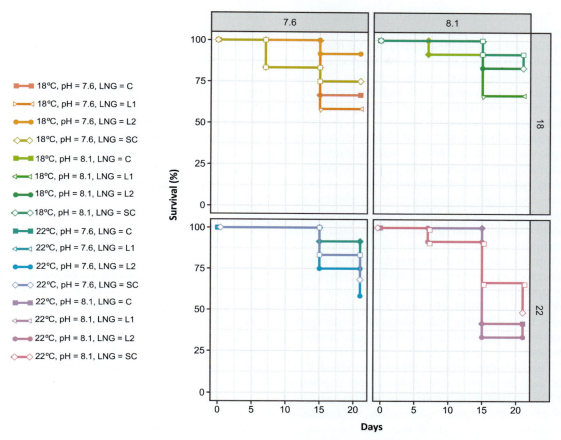

FIG. 6 Survival (%) of *G. locusta* exposed to different combinations of temperature (18°C and 22°C) and pH (7.6 and 8.1) under distinct LNG concentrations ($n=12$ per treatment). C, control; SC, solvent control; L1, LNG (10 ng/L); and L2, LNG (1000 ng/L). *Reprinted with permission from Cardoso, P. G., Loganimoce, E. M., Neuparth, T., Rocha, M. J., Rocha, E., & Arenas, F. (2018). Interactive effects of increased temperature, pCO2 and the synthetic progestin levonorgestrel on the fitness and breeding of the amphipod* Gammarus locusta. *Environ. Pollut. 236, 937–947.*

shorter term or stochastic changes may result in having offspring with maladaptive phenotypes or reduced genetic diversity (Day and Bonduriansky, 2011; Ward and Robinson, 2009), and to date there is limited evidence of anticipatory parental effects that increase the survival of offspring (Uller et al., 2013). Although studies on the transgenerational effects of co-exposure to GCC-related stressors and EDCs are limited, several investigations have been completed. Recent studies indicate that parental exposure to environmentally detected levels (low ng/L) of EDCs such as bifenthrin and ethinylestradiol at higher temperatures, results in organism-level impacts on embryonic development and skewed sex ratio in indirectly exposed offspring in vertebrates with TSD. Furthermore, in the same study, co-exposure to increased temperature and EDCs caused a reduction in fecundity (fewer viable offspring), as well as increased developmental deformities in the indirectly exposed generation and alterations in gene expression (DeCourten and Brander, 2017; DeCourten et al., 2019).

Notably, organisms with environmentally mediated sex determination (TSD, such as certain reptile and fish species, appear to be markedly sensitive to altered temperatures associated with GCC (Consuegra and Rodríguez López, 2016), as well as co-exposure to EDCs and climate change-related factors (Brown et al., 2015). In invertebrates such as *Siphonaria autralis*, co-exposure to increased temperature, UVB light coupled, and aqueous copper resulted in reduced embryonic viability (Kessel and Phillips, 2018). Alternatively, parental conditioning, defined as exposure during gametogenesis in invertebrates and most fish species, could enhance transgenerational plasticity and allow for transgenerational acclimation (Munday, 2014), allowing organisms to rapidly adapt to environmental change, as was observed recently with sea urchins in response to upwelling conditions associated with GCC (Wong et al., 2018). However, it remains to be seen how effective this adaptive tendency in organisms exposed to multiple simultaneous stressors.

The transgenerational effects described thus far may be attributable to the direct exposure of the primordial germ cells within parents, which produce the F1 generation, to combined stressors. However, there may also be changes in DNA structure and regulation contributing to observed responses. Epigenetic tags, such as methyl or acetyl groups, modify the structure of DNA. Epigenetic modifications are vital during embryonic development with regards to cellular differentiation, but epigenetic modifiers also respond to environmental stress (Brander et al., 2017; Head et al., 2012). Such tags are now known to play a pivotal role in controlling the expression of genes. A number of recent papers highlight the importance of epigenetics in terms of evaluating responses to aquatic pollutants and demonstrate that epigenetic mechanisms are important both within the lifetime of an organisms as well as in subsequent generations, since epigenetic modifications to gametes can be inherited through multiple generations (Bhandari, 2016; Corrales et al., 2014; Head, 2014; Voisin et al., 2016). Findings from these initial multigenerational studies highlight the importance of evaluating effects within and across generations when it comes to multiple stressors related to GCC, particularly since the effects of low-dose exposures in parents may not be observed until the F1 or F2 generation and may be carried through to subsequent generations. It has become evident that considerations of effects across generations must begin to be incorporated into risk assessment (Shaw et al., 2017).

8 POPULATION LEVEL EFFECTS

Climate change scenarios projected into the next century have the potential to drastically reshape global species distribution (Chen et al., 2011). Much of the areas organisms currently inhabit are likely to become inadequate for sustaining populations under projected climate parameters (Hart et al., 2018). Organisms that are endemic or unable to migrate to more suitable habitats have been shown to experience population declines in response to warming, especially those whose current habitats reach temperatures near their thermal limits (Komoroske et al., 2014; Pounds et al., 1999). Organisms may be most susceptible to the effects of climate change (salinity and temperature changes) depending on the life stage of the organism, indicating that the timing of these events is important to determining the overall impact on the population (Byrne, 2012; Komoroske et al., 2014). High mortality of early life stage offspring has been linked to population declines in species impacted by changing temperatures (Kiesecker et al., 2001). EDC exposure may also lead to population declines through similar mechanisms of causing early life stage mortality (DeCourten and Brander, 2017). For example, population declines in Atlantic salmon have been linked to exposure to 4-nonylphenol (4-NP) causing disruption of hormones regulating the parr-smoltification transformation, resulting in high juvenile mortality (Brown and Fairchild, 2003). However, because EDCs can also lead to reduced gamete production or egg production, skewed sex ratios, and altered reproductive behaviors the effects on population can be more severe over short time scales (Brander et al., 2016b; Brown et al., 2015; Salierno and Kane, 2009). Exposure to these stressors can result in measurable effects on offspring, demonstrating the potential for long-term effects from short-term exposures (Meyer et al., 2002). Organisms chronically exposed to EDCs over multiple generations have been shown to select for genetic mutations that increase resistance to these stressors (Major et al., 2017). These findings suggest that the complex nature of these stressors requires further testing to establish long-term population level consequences.

Maintaining a reproductively healthy population is vital for sustaining a population. TSD has evolved as a mechanism to optimize fitness of a population by adjusting the phenology of different sexes (Conover, 1984). As global temperatures increase, organisms with this life strategy may experience dramatic population declines as sex ratios are skewed to the point of inhibiting reproduction (Telemeco et al., 2013). One study that modeled the projected sex ratios of three sea turtle species under climate change scenarios and found that only 0.4% of the populations are expected to be born male by the year 2090 (Laloë et al., 2016). Another study using forecast models found that, in addition to high rates of clutch mortality, turtle clutches may reach 100% female sex ratios under projected climate change scenarios (Telemeco et al., 2013). The potential for drastic sex ratio skews is especially important for animals, like silversides (*Menidia* species), that produce male skewed sex ratios at warmer temperatures as population growth is determined by female fecundity (White et al., 2017). A study by Duffy et al. (2009) found that a population of silversides with TSD are more

susceptible skewed sex ratios following EE2 exposure than a population with GSD, thus organisms with plasticity in sex determination are more likely to be influenced by EDC exposure (Duffy et al., 2009). A skew in sex ratio can be measured across generations following exposure to clotrimazole, which was shown to decrease population viability (Brown et al., 2015). Effects measured in laboratory studies have been incorporated in population models confirming that small skews in sex ratios persisting through multiple generations can result in long-term populations declines (White et al., 2017). This highlights the importance for testing the combined effects of climate change and EDCs across multiple generations to better understand the potential population level consequences of these stressors.

9 COMMUNITY LEVEL EFFECTS

Contaminants can move throughout aquatic food webs, often biomagnifying to have the greatest effect on higher trophic levels (Borgå et al., 2004). This is especially well documented for the biomagnification of methyl mercury, which has reached levels of concern in some seafood intended for human consumption (Baishaw et al., 2007). There is evidence that EDCs may bioaccumulate in a similar manner, affecting all organisms in the community from benthos to vertebrates (Takahashi et al., 2003). Similar to methyl mercury, EDCs can biomagnify across trophic levels, resulting in secondary consumers containing the higher levels of EDCs (Ruhí et al., 2016). Aquatic organisms face an elevated risk as EDCs have been shown to bioaccumulate more rapidly through exposure in surrounding media as opposed to dietary exposure (Al-Ansari et al., 2013). Thus, organisms exposed via contaminated water and dietary exposure are at an increased risk. A number of factors can determine the bioaccumulation of EDCs including metabolic rate, life history, and trophic level leading to variable bioaccumulation factors (Liu et al., 2010; Ross et al., 2000; Takahashi et al., 2003). EDCs with lipophilic characteristics, including synthetic hormones and many pesticides, show increased accumulation in organisms with higher lipid contents. The behavior of these chemicals results in an increased risk to organisms that are higher in marine food webs and typically have high lipid content, such as Pacific killer whales, that have been found to have high levels of endocrine disruptors detected in tissue samples (Ross et al., 2000).

As temperature increases and ice thaws, legacy POPs that have been stored in ice can be reintroduced to these ecosystems, creating a new source of chemicals which are no longer used because of environmental concerns (Ma et al., 2016). Climate change can have an impact on how EDCs are able to move throughout aquatic food webs primarily by changing water-quality parameters such as temperature, salinity, and pH, which dictate the environmental fate of chemicals introduced into aquatic systems (Alava et al., 2017). These changes can result in an altered distribution of chemicals in aquatic systems, making them more bioavailable to be introduced into the food webs (Carere et al., 2011). In response to climate change many organisms in arctic regions have adjusted their diet to cope with lower sea ice. Species that have shifted to diets associated with lower sea ice have been found to have higher levels of contaminants, presumable because their dietary exposures to POPs have been increased (Mckinney et al., 2015). As temperatures warm, organisms that are not able to adapt to changing thermal regimes are expected to shift their ranges in a poleward direction (Chen et al., 2011). With this shift these organisms can introduce EDCs into less contaminated areas through their introduction in the food web through trophic transfer. Changes in water chemistry can affect the risk of bioaccumulation in exposed organisms (Maulvault et al., 2018b). One study found that as temperature increased, so did bioaccumulation, and acidification could either increase or bioaccumulation, depending on the chemical of interest (Maulvault et al., 2018b). This suggests that the interactions between the behavior of the chemical and the effects of abiotic factors are complex, illustrating the need for further research. Effects of both EDCs and climate change on species dynamics could pose an additional threat to communities in high latitudes that are notably affected by climate change.

10 CONCLUSIONS AND FUTURE RESEARCH DIRECTIONS

We have summarized recent findings on exposure to EDCs in a shifting climate, summarizing physiological and developmental effects, comparing across taxa and life stages, and highlighting current gaps in knowledge (Table 1). Exposures and responses of aquatic organisms to EDCs and mixtures in the context of GCC-induced alterations to temperature, hypoxia, altered salinity regimes, and acidification are wide-ranging and complex. GCC-related factors such as temperature and hypoxia can cause endocrine disruption independently, while changes in pH due to OA and altered salinity regimes may alter the bioavailability of EDCs due to requisite changes in osmoregulatory capacity on the part of exposed organisms, and the "salting-out" of more hydrophobic endocrine-active compounds into sediments and lipids. Ultimately, aquatic organisms are facing challenges to fundamental homeostatic processes as environmental conditions become more stressful, reducing their ability to cope with chemical exposure, while chemical exposure simultaneously interferes with their ability to maintain homeostasis. Many of these effects are dependent on life stage, with embryonic and larval stages being more sensitive to exposure than adults in many but not all cases. In effect, many organisms undergo physiological

10 CONCLUSIONS AND FUTURE RESEARCH DIRECTIONS

TABLE 1 Summary of Responses Across Life Stages, Species, Abiotic Factors, and EDCs

Species	Life stage	Abiotic factor(s)	EDC	Direction of response	Response	References
D. rerio	Embryo	Hypoxia	TCDD	↓	Reduced gene expression (heme oxygenase)	Prasch et al. (2004)
D. rerio	Embryo	Hypoxia	TCDD	↓	Reduced induction of TCDD response gene expression (zfcyp1a)	Prasch et al. (2004)
D. rerio	Embryo	Hypoxia	TCDD	↓	Reduced potency of TCDD-induced edema	Prasch et al. (2004)
D. rerio	Larvae	Hypoxia	PAH	↑	Increased pericardial effusion	Matson et al. (2008)
D. rerio	Larvae	Hypoxia	PAH	↑	Increased lordosis	Matson et al. (2008)
F. heteroclitus	Adult	Hypoxia	PCB	↓	Suppressed hypoxia-induced glycolytic enzyme production	Kraemer and Schulte (2004)
M. beryllina	Juvenile	Salinity	Diuron	↓	Decreased expression of thyroid receptor and growth hormone receptors	Moreira et al. (2018)
M. beryllina	Juvenile	Salinity	Diuron	↓	Decreased condition factor and body mass	Moreira et al. (2018)
H. azteca	Adult	Temperature	Pyrethroid	↓	Decreased survival	Hasenbein et al. (2018)
H. azteca	Adult	Salinity	Pyrethroid	↓	Decreased survival	Hasenbein et al. (2018)
C. sapidus	Juvenile	Salinity	Fipronil	↓	Reduced growth	Goff et al. (2017)
C. sapidus	Juvenile	Salinity	Fipronil	↓	Reduced gene expression of vitellogenin and ecdysone receptor	Goff et al. (2017)
C. gigas	Adult	Salinity	PFCs	↑	Increased bioaccumulation	Jeon et al. (2010a)

Continued

EVALUATING WATER QUALITY TO PREVENT FUTURE DISASTERS

2. COMPLEXITIES OF ENDOCRINE DISRUPTION AND CLIMATE CHANGE

TABLE 1 Summary of Responses Across Life Stages, Species, Abiotic Factors, and EDCs—cont'd

Species	Life stage	Abiotic factor(s)	EDC	Direction of response	Response	References
S. schlegeli	Adult	Salinity	PFCs	↑	Increased bioconcentration and elimination	Jeon et al. (2010b)
G. locusta	Embryo	Acidification	Levonorgestrel (synthetic androgen)	↓	Reduced growth rates	Loganimoce (2016)
A. regius	Juvenile	Acidification	Venlafaxine	↑	Increased bioaccumulation	Maulvault et al. (2018a)
A. regius	Juvenile	Acidification	Venlafaxine	↓	Decreased behavior	Maulvault et al. (2018a)
M. galloprovincialis	Adult	Acidification	Flame retardants (Dechloranes 602, 603, and 604, and TBBPA)	↑	Increased bioaccumulation	Maulvault et al. (2018b)
M. galloprovincialis	Adult	Acidification	Inorganic arsenic and PFCs (PFOA and PFOS)	↑	Increased elimination	Maulvault et al. (2018b)
G. morhua	Adult	Acidification	PFOS	↑	Increased gene expression for steroid hormone synthesis	Preus-Olsen et al. (2014)
M. beryllina	Larval	Temperature	EE2	↑	Increased skeletal deformities	DeCourten and Brander (2017)
M. beryllina	Adult	Temperature	EE2	↓	Decreased egg production	DeCourten and Brander (2017)
M. beryllina	Adult	Temperature	EE2	↑	Skewed sex ratio	DeCourten and Brander (2017)
D. rerio	Adult	Temperature	Levonorgestrel	↓	Decreased fecundity	Cardoso et al. (2017)
D. rerio	Embryonic	Temperature	Levonorgestrel	↓	Decreased hatch rate	Cardoso et al. (2017)
D. rerio	Adult	Temperature	Clotrimazole	↓	Population decline/ inbreeding depression	Brown et al. (2015)

EVALUATING WATER QUALITY TO PREVENT FUTURE DISASTERS

10 CONCLUSIONS AND FUTURE RESEARCH DIRECTIONS

TABLE 1 Summary of Responses Across Life Stages, Species, Abiotic Factors, and EDCs—cont'd

Species	Life stage	Abiotic factor(s)	EDC	Direction of response	Response	References
D. rerio	Adult	Temperature	Clotrimazole	↑	Skewed sex ratio	Brown et al. (2015)
M. beryllina	Juvenile	Temperature	Diuron	↓	Decreased expression of thyroid receptor and growth hormone receptors	Moreira et al. (2018)
M. beryllina	Juvenile	Temperature	Diuron	↓	Decreased condition factor and body mass	Moreira et al. (2018)
G. locusta	Adult	Temperature	Levonorgestrel	↓	Deceased growth	Cardoso et al. (2018)
G. locusta	Adult	Temperature	Levonorgestrel	↓	Reduced fecundity	Cardoso et al. (2018)
A. regius	Juvenile	Temperature	Venlafaxine	↓	Reduced shoaling behavior	Maulvault et al. (2018a)
A. regius	Juvenile	Temperature	Venlafaxine	↑	Increased risk behavior	Maulvault et al. (2018a)
A. regius	Juvenile	Temperature	Venlafaxine	↑	Increased bioaccumulation	Maulvault et al. (2018a)
M. galloprovincialis	Adult	Temperature	Dec 602, Dec 604, TBBPA	↑	Increased bioaccumulation	Maulvault et al. (2018b)
R. philippinarum	Adult	Temperature	Dec 602, Dec 604, TBBPA	↑	Increased bioaccumulation	Maulvault et al. (2018b)
P. promelas	Adult	Temperature	Estrone	↓	Decreased testis weight	Shappell et al. (2018)
P. promelas	Adult	Temperature	Estrone	↓	Decreased vitellogenin	Shappell et al. (2018)
F. heteroclitus	Adult	Temperature	EE2	↑	Increased EE2 uptake	Blewett et al. (2013)
D. rerio	Adult	Temperature	EE2/NP	↑	Increased transcription of vtg 1 gene	Jin et al. (2009)
P. promelas	Adult	Temperature	Estrogenic mixture	↑	Increased rate of vitellogenin production	Brian et al. (2008)
O. latipes	Adult	Temperature	Perchlorate	↓	Thyroid disruption; decreased T4	Lee and Choi (2014)

Continued

EVALUATING WATER QUALITY TO PREVENT FUTURE DISASTERS

2. COMPLEXITIES OF ENDOCRINE DISRUPTION AND CLIMATE CHANGE

TABLE 1 Summary of Responses Across Life Stages, Species, Abiotic Factors, and EDCs—cont'd

Species	Life stage	Abiotic factor(s)	EDC	Direction of response	Response	References
O. latipes	Adult	Temperature	Perchlorate	↓	Decreased egg production	Lee and Choi (2014)
D. rerio	Larval	Temperature	EE2	↑	Skewed sex ratios	Luzio et al. (2016)
D. rerio	Adult	Temperature	EE2	↓	Delayed testis maturation	Luzio et al. (2016)

trade-offs when exposed concurrently to GCC-related stressors and EDCs, sometimes only possessing the resources to deal with the most immediate threat. Furthermore, responses are dependent on environmental fate, which can be influenced by factors associated with GCC such as thawing and redistribution of POPs, increased frequency of storms and droughts, and bioavailability varying with changes in salinity, pH, and DO. The magnitude of EDCs and climate change impacts is also dependent on location, with higher latitudes being more severely affected, having had larger changes in environmental conditions thus far.

The most pressing issue facing the ecotoxicology research community is that these large fluxes in environmental conditions are not yet accounted for in environmental risk assessment, causing a greater mismatch between laboratory testing and the reality of present day aquatic ecosystems. As demonstrated in our review, both fish and invertebrates can have considerably different and varied responses depending on the environmental conditions under which they are exposed. This is inclusive of the molecular level (e.g., gene expression) as well as apical endpoints such as survival, growth, and reproduction that are more typically used for the assessment of risk. Consideration for adding varied abiotic conditions to standard testing practices, therefore accounting for differential responses across oxygen, salinity, temperature, and pH gradients, is of vital importance if we are to truly assess the magnitude of impacts to aquatic animals in the Anthropocene.

Abbreviation

DDT	dichlorodiphenyltrichloroethane
DO	dissolved oxygen
EDC	endocrine disrupting chemical
GCC	global climate change
IPCC	international panel on climate change
LC50	lethal concentration for 50% of organisms
OMZ	oxygen minimum zone
PBDE	polybrominated diphenyl ether
PCB	polychlorinated biphenyl
POP	persistent organic pollutant

ACKNOWLEDGMENTS

We acknowledge funding from the Environmental Protection Agency (#835799) and the California Department of Fish and Wildlife, Proposition 1 (#P1796002) that made writing this review possible. We also thank the Oregon State University Department of Environmental and Molecular Toxicology for their support.

AUTHOR CONTRIBUTIONS

DeCourten and Romney contributed ideas and wrote equal portions of the text, and thus are dual first authors of this manuscript. Brander conceptualized, organized, edited, and wrote portions of the review.

References

Al-Ansari, A.M., Atkinson, S.K., Doyle, J.R., Trudeau, V.L., Blais, J.M., 2013. Dynamics of uptake and elimination of 17α-ethinylestradiol in male goldfish (Carassius auratus). Aquat. Toxicol. 132, 134–140.

Alava, J.J., Cheung, W.W., Ross, P.S., Sumaila, U.R., 2017. Climate change–contaminant interactions in marine food webs: toward a conceptual framework. Glob. Chang. Biol. 23 (10), 3984–4001.

Altenritter, M.E., Cohuo, A., Walther, B.D., 2018. Proportions of demersal fish exposed to sublethal hypoxia revealed by otolith chemistry. Mar. Ecol. Prog. Ser. 589, 193–208.

Altieri, A.H., Gedan, K.B., 2015. Climate change and dead zones. Glob. Chang. Biol. 21 (4), 1395–1406.

Alvarado-Alvarez, R., Gould, M.C., Stephano, J.L., 1996. Spawning, in vitro maturation, and changes in oocyte electrophysiology induced by serotonin in *Tivela stultorum*. Biol. Bull. 190 (3), 322–328. Chicago.

Anderson, S.N., Podrabsky, J.E., 2014. The effects of hypoxia and temperature on metabolic aspects of embryonic development in the annual killifish *Austrofundulus limnaeus*. J. Comp. Physiol. B 184 (3), 355–370.

Armstrong, J.D., Nislow, K.H., 2006. Critical habitat during the transition from maternal provisioning in freshwater fish, with emphasis on Atlantic salmon (*Salmo salar*) and brown trout (*Salmo trutta*). J. Zool. 269 (4), 403–413.

Baishaw, S., Edwards, J., Daughtry, B., Ross, K., 2007. Mercury in seafood: mechanisms of accumulation and consequences for consumer health. Rev. Environ. Health 22 (2), 91–114.

Barrionuevo, W.R., Fernandes, M.N., Rocha, O., 2010. Aerobic and anaerobic metabolism for the zebrafish, *Danio rerio*, reared under normoxic and hypoxic conditions and exposed to acute hypoxia during development. Braz. J. Biol. 70 (2), 425–434.

Barth, J.A., Fram, J.P., Dever, E.P., Risien, C.M., Wingard, C.E., Collier, R.W., Kearney, T.D., 2018. Warm blobs, low-oxygen events, and an eclipse: the Ocean observatories initiative endurance array captures them all. Oceanography 31, 90–97.

Baumann, H., Talmage, S.C., Gobler, C.J., 2011. Reduced early life growth and survival in a fish in direct response to increased carbon dioxide. Nat. Clim. Change 2, 38–41.

Bell, A.M., 2004. An endocrine disrupter increases growth and risky behavior in threespined stickleback (*Gasterosteus aculeatus*). Horm. Behav. 45 (2), 108–114.

Bergeron, J.M., Crews, D., McLachlan, J.A., 1994. PCBs as environmental estrogens: turtle sex determination as a biomarker of environmental contamination. Environ. Health Perspect. 102 (9), 780.

Berkeley, S.A., Chapman, C., Sogard, S.M., 2004. Maternal age as a determinant of larval growth and survival in a marine fish, Sebastes melanops. Ecology 85 (5), 1258–1264.

Bhandari, R.K., 2016. Medaka as a model for studying environmentally induced epigenetic transgenerational inheritance of phenotypes. Environ. Epigenetics 2 (1).

Bibby, R., Widdicombe, S., Parry, H., Spicer, J., Pipe, R., 2008. Effects of ocean acidification on the immune response of the blue mussel *Mytilus edulis*. Aquat. Biol. 2 (1), 67–74.

Bickler, P.E., Buck, L.T., 2007. Hypoxia tolerance in reptiles, amphibians, and fishes: life with variable oxygen availability. Annu. Rev. Physiol. 69 (1), 145–170.

Blewett, T., MacLatchy, D.L., Wood, C.M., 2013. The effects of temperature and salinity on 17-α-ethynylestradiol uptake and its relationship to oxygen consumption in the model euryhaline teleost (*Fundulus heteroclitus*). Aquat. Toxicol. 127, 61–71.

Bonduriansky, R., Crean, A.J., Day, T., 2012. The implications of nongenetic inheritance for evolution in changing environments. Evol. Appl. 5 (2), 192–201. https://doi.org/10.1111/j.1752-4571.2011.00213.x.

Bonner, T.H., Brandt, T.M., Fries, J.N., Whiteside, B.G., 1998. Effects of temperature on egg production and early life stages of the fountain darter. Trans. Am. Fish. Soc. 127 (6), 971–978.

Borgå, K., Fisk, A.T., Hoekstra, P.F., Muir, D.C., 2004. Biological and chemical factors of importance in the bioaccumulation and trophic transfer of persistent organochlorine contaminants in arctic marine food webs. Environ. Toxicol. Chem. 23 (10), 2367–2385.

Borrirukwisitsak, S., Keenan, H.E., Gauchotte-Lindsay, C., 2012. Effects of salinity, pH and temperature on the octanol-water partition coefficient of bisphenol A. Int. J. Environ. Sci. Dev. 3 (5), 460.

Bosker, T., Santoro, G., Melvin, S.D., 2017. Salinity and sensitivity to endocrine disrupting chemicals: a comparison of reproductive endpoints in small-bodied fish exposed under different salinities. Chemosphere 183, 186–196.

Brander, S.M., 2013. Thinking outside the box: assessing endocrine disruption in aquatic life. In: Monitoring Water Quality: Pollution Assessment, Analysis, and Remediation. Elsevier, Waltham, MA, pp. 103–147.

Brander, S., Hecht, S., Incardona, J., Kidd, K., Kuivila, K., Kullman, S., 2015. The Challenge: "Bridging the gap" with fish: advances in assessing exposure and effects across biological scales. Environ. Toxicol. Chem. 34, 459.

Brander, S.M., Gabler, M.K., Fowler, N.L., Connon, R.E., Schlenk, D., 2016a. Pyrethroid pesticides as endocrine disruptors: molecular mechanisms in vertebrates with a focus on fishes. Environ. Sci. Technol. 50, 8977–8992.

Brander, S.M., Jeffries, K.M., Cole, B.J., DeCourten, B.M., White, J.W., Hasenbein, S., Fangue, N.A., Connon, R.E., 2016b. Transcriptomic changes underlie altered egg protein production and reduced fecundity in an estuarine model fish exposed to bifenthrin. Aquat. Toxicol. 174, 247–260.

Brander, S.M., Biales, A.D., Connon, R.E., 2017. The role of epigenomics in aquatic toxicology. Environ. Toxicol. Chem. 36, 2565–2573.

Breckels, R.D., Neff, B.D., 2013. The effects of elevated temperature on the sexual traits, immunology and survivorship of a tropical ectotherm. J. Exp. Biol. 216 (14), 2658–2664.

Breitburg, D., Levin, L.A., Oschlies, A., Grégoire, M., Chavez, F.P., Conley, D.J., …Jacinto, G.S., 2018. Declining oxygen in the global ocean and coastal waters. Science 359 (6371), eaam7240.

Brian, J.V., Harris, C.A., Runnalls, T.J., Fantinati, A., Pojana, G., Marcomini, A., …Sumpter, J.P., 2008. Evidence of temperature-dependent effects on the estrogenic response of fish: implications with regard to climate change. Sci. Total Environ. 397 (1-3), 72–81.

Brown, S.B., Fairchild, W.L., 2003. Evidence for a causal link between exposure to an insecticide formulation and declines in catch of Atlantic salmon. Hum. Ecol. Risk. Assess. 9 (1), 137–148.

Brown, A.R., Owen, S.F., Peters, J., Zhang, Y., Soffker, M., Paull, G.C., 2015. Climate change and pollution speed declines in zebrafish populations. PNAS 112 (11), E1237–E1246.

Byrne, M., 2012. Global change ecotoxicology: identification of early life history bottlenecks in marine invertebrates, variable species responses and variable experimental approaches. Mar. Environ. Res. 76, 3–15.

Byrne, M., Ho, M., Selvakumaraswamy, P., Nguyen, H.D., Dworjanyn, S.A., Davis, A.R., 2009. Temperature, but not pH, compromises sea urchin fertilization and early development under near-future climate change scenarios. Proc. R. Soc. Lond. B Biol. Sci. 276 (1663), 1883–1888.

Caldeira, K., Wickett, M.E., 2005. Ocean model predictions of chemistry changes from carbon dioxide emissions to the atmosphere and ocean. J. Geophys. Res. Oceans 110 (C9).

Cardoso, P.G., Rodrigues, D., Madureira, T.V., Oliveira, N., Rocha, M.J., Rocha, E., 2017. Warming modulates the effects of the endocrine disruptor progestin levonorgestrel on the zebrafish fitness, ovary maturation kinetics and reproduction success. Environ. Pollut. 229, 300–311.

Cardoso, P.G., Loganimoce, E.M., Neuparth, T., Rocha, M.J., Rocha, E., Arenas, F., 2018. Interactive effects of increased temperature, pCO2 and the synthetic progestin levonorgestrel on the fitness and breeding of the amphipod *Gammarus locusta*. Environ. Pollut. 236, 937–947.

Carere, M., Miniero, R., Cicero, M.R., 2011. Potential effects of climate change on the chemical quality of aquatic biota. TrAC Trends Anal. Chem. 30 (8), 1214–1221.

Chambers, R.C., Candelmo, A.C., Habeck, E.A., Poach, M.E., Wieczorek, D., Cooper, K.R., Greenfield, C.E., Phelan, B.A., 2014. Effects of elevated CO2 in the early life stages of summer flounder, *Paralichthys dentatus*, and potential consequences of ocean acidification. Biogeosciences 11, 1613–1626.

Charmantier, G.U.Y., 1998. Ontogeny of osmoregulation in crustaceans: a review. Invertebr. Reprod. Dev. 33 (2–3), 177–190.

Chen, I.C., Hill, J.K., Ohlemüller, R., Roy, D.B., Thomas, C.D., 2011. Rapid range shifts of species associated with high levels of climate warming. Science 333 (6045), 1024–1026.

Chou, C., Neelin, J.D., Chen, C.A., Tu, J.Y., 2009. Evaluating the "rich-get-richer" mechanism in tropical precipitation change under global warming. J. Climate 22 (8), 1982–2005.

Claiborne, J.B., Edwards, S.L., Morrison-Shetlar, A.I., 2002. Acid–base regulation in fishes: cellular and molecular mechanisms. J. Exp. Zool. A Ecol. Genet. Physiol. 293 (3), 302–319.

Clarke, A., Fraser, K.P.P., 2004. Why does metabolism scale with temperature? Funct. Ecol. 18 (2), 243–251.

Cloern, J.E., Knowles, N., Brown, L.R., Cayan, D., Dettinger, M.D., Morgan, T.L., …Jassby, A.D., 2011. Projected evolution of California's San Francisco bay-delta-river system in a century of climate change. PLoS One 6 (9).

Colborn, T., Thayer, K., 2000. Aquatic ecosystems: harbingers of endocrine disruption. Ecol. Appl. 10, 949–957.

Conover, D.O., 1984. Adaptive significance of temperature-dependent sex determination in a fish. Am. Nat. 123 (3), 297–313.

Conover, D.O., Kynard, B.E., 1981. Environmental sex determination: interaction of temperature and genotype in a fish. Science 213 (4507), 577–579.

Consuegra, S., Rodríguez López, C.M., 2016. Epigenetic-induced alterations in sex-ratios in response to climate change: an epigenetic trap? Bioessays 38, 950–958.

Corrales, J., Thornton, C., White, M., Willett, K.L., 2014. Multigenerational effects of benzo[a]pyrene exposure on survival and developmental deformities in zebrafish larvae. Aquat. Toxicol. 148, 16–26.

Coward, K., Bromage, N.R., Hibbitt, O., Parrington, J., 2002. Gamete physiology, fertilization and egg activation in teleost fish. Rev. Fish Biol. Fish. 12 (1), 33–58.

Cox, M.K., Peterson, K.N., Tan, D., Novak, P.J., Schoenfuss, H.L., Ward, J.L., 2017. Temperature modulates estrone degradation and biological effects of exposure in fathead minnows. Sci. Total Environ. 621, 1591–1600.

David, M., Marigoudar, S.R., Patil, V.K., Halappa, R., 2012. Behavioral, morphological deformities and biomarkers of oxidative damage as indicators of sublethal cypermethrin intoxication on the tadpoles of *D. melanostictus* (Schneider, 1799). Pestic. Biochem. Physiol. 103 (2), 127–134.

Day, T., Bonduriansky, R., 2011. A unified approach to the evolutionary consequences of genetic and nongenetic inheritance. Am. Nat. 178, E18–E36. Associate Editor: Maria RS, Editor: Mark AM.

REFERENCES

DeCourten, B.M., Brander, S.M., 2017. Combined effects of increased temperature and endocrine disrupting pollutants on sex determination, survival, and development across generations. Sci. Rep. 7 (1), 9310.

DeCourten, B.M., Connon, R.E., Brander, S.M., 2019. Direct and indirect parental exposure to endocrine disruptors and elevated temperature influences gene expression across generations in a euryhaline model fish. PeerJ 7, e6156.

DeGroot, B.C., Brander, S.M., 2014. The role of P450 metabolism in the estrogenic activity of bifenthrin in fish. Aquat. Toxicol. 156, 17–20.

De Schamphelaere, K.A.C., Stubblefield, W., Rodriguez, P., Vleminckx, K., Janssen, C.R., 2010. The chronic toxicity of molybdate to freshwater organisms. I. Generating reliable effects data. Sci. Total Environ. 408 (22), 5362–5371.

DeWitt, J.C., 2015. Toxicological Effects of Perfluoroalkyl and Polyfluoroalkyl Substances. Springer International Publishing, Switzerland.

Diaz, R.J., Breitburg, D.L., 2009. The hypoxic environment. In: Fish Physiology. vol. 27. Academic Press, pp. 1–23.

Diaz, R.J., Rosenberg, R., 2008. Spreading dead zones and consequences for marine ecosystems. Science 321, 926.

Doney, S.C., Mahowald, N., Lima, I., Feely, R.A., Mackenzie, F.T., Lamarque, J.F., Rasch, P.J., 2007. Impact of anthropogenic atmospheric nitrogen and sulfur deposition on ocean acidification and the inorganic carbon system. Proc. Natl. Acad.of Sci. 104 (37), 14580–14585.

Doong, R.A., Chang, S.M., Sun, Y.C., 2000. Solid-phase microextraction for determining the distribution of sixteen US environmental protection agency polycyclic aromatic hydrocarbons in water samples. J. Chromatogr. A 879 (2), 177–188.

Dore, J.E., Lukas, R., Sadler, D.W., Church, M.J., Karl, D.M., 2009. Physical and biogeochemical modulation of ocean acidification in the central North Pacific. Proc. Natl. Acad. Sci. U.S.A. 106 (30), 12235–12240.

Du, J., Shen, J., Park, K., Wang, Y.P., Yu, X., 2018. Worsened physical condition due to climate change contributes to the increasing hypoxia in Chesapeake Bay. Sci. Total Environ. 630, 707–717.

Duffy, T.A., McElroy, A.E., Conover, D.O., 2009. Variable susceptibility and response to estrogenic chemicals in *Menidia menidia*. Mar. Ecol. Prog. Ser. 380, 245–254.

Dunwoodie, S.L., 2009. The role of hypoxia in development of the mammalian embryo. Dev. Cell 17 (6), 755–773.

Durack, P.J., Wijffels, S.E., Matear, R.J., 2012. Ocean salinities reveal strong global water cycle intensification during 1950 to 2000. Science 336 (6080), 455–458.

Eganhouse, R.P., Calder, J.A., 1976. The solubility of medium molecular weight aromatic hydrocarbons and the effects of hydrocarbon co-solutes and salinity. Geochim. Cosmochim. Acta 40 (5), 555–561.

Evans, D.H., 1980. Osmotic and ionic regulation by freshwater and marine fishes. In: Environmental Physiology of Fishes. Springer, Boston, MA, pp. 93–122.

Fabry, V.J., Seibel, B.A., Feely, R.A., Orr, J.C., 2008. Impacts of ocean acidification on marine fauna and ecosystem processes. ICES J. Mar. Sci. 65 (3), 414–432.

Farmer, T.M., Marschall, E.A., Dabrowski, K., Ludsin, S.A., 2015. Short winters threaten temperate fish populations. Nat. Commun. 6, 1–10. https://doi.org/10.1038/ncomms8724.

Faust, D.R., Moore, M.T., Emison, G.A., Rush, S.A., 2016. Potential implications of approaches to climate change on the clean water rule definition of "Waters of the United States" Bull. Environ. Contam. Toxicol. 96, 565–572.

Finn, R.N., 2007. The physiology and toxicology of salmonid eggs and larvae in relation to water-quality criteria. Aquat. Toxicol. 81 (4), 337–354.

Forsgren, E., Dupont, S., Jutfelt, F., Amundsen, T., 2013. Elevated CO2 affects embryonic development and larval phototaxis in a temperate marine fish. Ecol. Evol. 3, 3637–3646.

Freitas, R., Almeida, Â., Calisto, V., Velez, C., Moreira, A., Schneider, R.J., … Figueira, E., 2015. How life history influences the responses of the clam *Scrobicularia plana* to the combined impacts of carbamazepine and pH decrease. Environ. Pollut. 202, 205–214.

Frommel, A.Y., Maneja, R., Lowe, D., Pascoe, C.K., Geffen, A.J., Folkvord, A., Piatkowski, U., Clemmensen, C., 2014. Organ damage in Atlantic herring larvae as a result of ocean acidification. Ecol. Appl. 24, 1131–1143.

Furlan, E., et al., 2017. Spatially explicit risk approach for multi-hazard assessment and management in marine environment: the case study of the Adriatic Sea. Sci. Total Environ. 618, 1008–1023.

Gazeau, F., Quiblier, C., Jansen, J.M., Gattuso, J.P., Middelburg, J.J., Heip, C.H., 2007. Impact of elevated CO2 on shellfish calcification. Geophys. Res. Lett. 34 (7), L07603.

Giaccia, A.J., Simon, M.C., Johnson, R., 2004. The biology of hypoxia: the role of oxygen sensing in development, normal function, and disease. Genes Dev. 18, 2183–2194.

Glinka, C.O., Frasca, S., Provatas, A.A., Lama, T., DeGuise, S., Bosker, T., 2015. The effects of model androgen 5α-dihydrotestosterone on mummichog (Fundulus heteroclitus) reproduction under different salinities. Aquat. Toxicol. 165, 266–276.

Goff, A.D., Saranjampour, P., Ryan, L.M., Hladik, M.L., Covi, J.A., Armbrust, K.L., Brander, S.M., 2017. The effects of fipronil and the photodegradation product fipronil desulfinyl on growth and gene expression in juvenile blue crabs, Callinectes sapidus, at different salinities. Aquat. Toxicol. 186, 96–104.

Green, C., Williams, R., Kanda, R., Churchley, J., He, Y., Thomas, S., Goonan, P., Kumar, A., Jobling, S., 2013. Modeling of steroid estrogen contamination in UK and

EVALUATING WATER QUALITY TO PREVENT FUTURE DISASTERS

South Australian rivers predicts modest increases in concentrations in the future. Environ. Sci. Technol. 47, 7224–7232.

Guillette Jr., L.J., 2006. Endocrine disrupting contaminants—beyond the dogma. Environ. Health Perspect. 114 (S1), 9–12.

Gupta, R.C. (Ed.), 2011. Reproductive and Developmental Toxicology. Academic Press.

Hallman, T.A., Brooks, M.L., 2015. The deal with diel: temperature fluctuations, asymmetrical warming, and ubiquitous metals contaminants. Environ. Pollut. 206, 88–94.

Halpern, B.S., Kappel, C.V., Selkoe, K.A., Micheli, F., Ebert, C.M., Kontgis, C., ...Teck, S.J., 2009. Mapping cumulative human impacts to California current marine ecosystems. Conserv. Lett. 2 (3), 138–148.

Hamdoun, A., Epel, D., 2007. Embryo stability and vulnerability in an always changing world. Proc. Natl. Acad. Sci. 104 (6), 1745–1750.

Hansen, J., Sato, M., Ruedy, R., 2012. Perception of climate change. Proc. Natl. Acad. Sci. 109.

Harris, C.A., Hamilton, P.B., Runnalls, T.J., Vinciotti, V., Henshaw, A., Hodgson, D., Coe, T.S., Jobling, S., Tyler, C.R., Sumpter, J.P., 2011. The consequences of feminization in breeding groups of wild fish. Environ. Health Perspect. 119, 306–311.

Hart, C.J., Kelly, R.P., Pearson, S.F., 2018. Will the California current lose its nesting tufted puffins? PeerJ 6.

Hasenbein, S., Poynton, H., Connon, R.E., 2018. Contaminant exposure effects in a changing climate: how multiple stressors can multiply exposure effects in the amphipod *Hyalella azteca*. Ecotoxicology, 1–15.

Hashimoto, Y., Tokura, K., Kishi, H., Strachan, W.M.J., 1984. Prediction of seawater solubility of aromatic compounds. Chemosphere 13 (8), 881–888.

Hassell, K.L., Coutin, P.C., Nugegoda, D., 2008. Hypoxia, low salinity and lowered temperature reduce embryo survival and hatch rates in black bream *Acanthopagrus butcheri* (Munro, 1949). J. Fish Biol. 72 (7), 1623–1636.

Hayes, T.B., Case, P., Chui, S., Chung, D., Haeffele, C., Haston, K., Lee, M., Mai, V.P., Marjuoa, Y., Parker, J., Tsui, M., 2006. Pesticide mixtures, endocrine disruption, and amphibian declines: are we underestimating the impact? *Environmental health perspectives, 114* (Suppl 1), 40–50.

Head, J.A., 2014. Patterns of DNA methylation in animals: an ecotoxicological perspective. Integr. Comp. Biol. 54 (1), 77–86.

Head, J.A., Dolinoy, D.C., Basu, N., 2012. Epigenetics for ecotoxicologists. Environ. Toxicol. Chem. 31, 221–227.

Heath, A.G., 1995. Environmental hypoxia. In: Water Pollution and Fish Physiology. CRC Press, pp. 29–60.

Held, I.M., Soden, B.J., 2006. Robust responses of the hydrological cycle to global warming. J. Climate 19 (21), 5686–5699.

Henry, T.R., Spitsbergen, J.M., Hornung, M.W., Abnet, C.C., Peterson, R.E., 1997. Early life stage toxicity of 2,3,7,8-tetrachlorodibenzo-p-dioxin in zebrafish (*Danio rerio*). Toxicol. Appl. Pharmacol. 142, 56–68.

Henry, R.P., Lucu, C., Onken, H., Weihrauch, D., 2012. Multiple functions of the crustacean gill: osmotic/ionic regulation, acid-base balance, ammonia excretion, and bioaccumulation of toxic metals. Front. Physiol. 3, 431.

Hilton, T.W., Najjar, R.G., Zhong, L., Li, M., 2008. Is there a signal of sea-level rise in Chesapeake Bay salinity? J. Geophys. Res. Oceans 113 (C9), C09002.

Hochachka, P., Buck, L., Doll, C., Land, S., 1996. Unifying theory of hypoxia tolerance: molecular/metabolic defense and rescue mechanisms for surviving oxygen lack. Proc. Natl. Acad. Sci. U.S.A. 93 (18), 9493–9498.

Hoegh-Guldberg, O., Mumby, P.J., Hooten, A.J., Steneck, R.S., Greenfield, P., Gomez, E., ... Knowlton, N., 2007. Coral reefs under rapid climate change and ocean acidification. Science 318 (5857), 1737–1742.

Holmstrup, M., Bindesbøl, A.M., Oostingh, G.J., Duschl, A., Scheil, V., Köhler, H.R., ...Gerhardt, A., 2010. Interactions between effects of environmental chemicals and natural stressors: a review. Sci. Total Environ. 408 (18), 3746–3762.

Hooper, M.J., Ankley, G.T., Cristol, D.A., Maryoung, L.A., Noyes, P.D., Pinkerton, K.E., 2013. Interactions between chemical and climate stressors: a role for mechanistic toxicology in assessing climate change risks. Environ. Toxicol. Chem. 32 (1), 32–48.

Horner, V.L., Wolfner, M.F., 2008. Transitioning from egg to embryo: triggers and mechanisms of egg activation. Dev. Dyn. 237 (3), 527–544.

Houde, E.D., 1989. Comparative growth, mortality, and energetics of marine fish larvae: temperature and implied latitudinal effects. Fish. Bull. 87 (3), 471–495.

Hughes, B.B., Levey, M.D., Fountain, M.C., Carlisle, A.B., Chavez, F.P., Gleason, M.G., 2015. Climate mediates hypoxic stress on fish diversity and nursery function at the land–sea interface. Proc. Natl. Acad. Sci. U.S.A. 112 (26), 8025–8030.

IPCC, 2013. Summary for policymakers. In: Stocker, T.F., Qin, D., Plattner, G.K., Tignor, M., Allen, S.K., Boschung, J., Nauels, A., Xia, Y., Bex, V., Midgley, P.M. (Eds.), Climate Change 2013: The Physical Science Basis. Contribution of Working Group I to the Fifth Assessment Report of the Intergovernmental Panel on Climate Change. Cambridge University Press, Cambridge, United Kingdom and New York, NY, USA pp. 1535. www.ipcc.ch/pdf/assessmentreport/ar5/wg1/WG1AR5_SPM_FINAL.pdf.

Ishimatsu, A., Kikkawa, T., Hayashi, M., Lee, K.S., Kita, J., 2004. Effects of CO 2 on marine fish: larvae and adults. J. Oceanogr. 60 (4), 731–741.

Ishimatsu, A., Hayashi, M., Lee, K.S., Kikkawa, T., Kita, J., 2005. Physiological effects on fishes in a high-CO2 world. J. Geophys. Res. Oceans 110 (C9).

REFERENCES

Ishimatsu, A., Hayashi, M., Kikkawa, T., 2008. Fishes in high-CO_2, acidified oceans. Mar. Ecol. Prog. Ser. 373, 295–302.

Ivy, C.M., Robertson, C.E., Bernier, N.J., 2017. Acute embryonic anoxia exposure favours the development of a dominant and aggressive phenotype in adult zebrafish. Proc. R. Soc. B 284 (1846), 20161868.

Jeon, J., Kannan, K., Lim, H.K., Moon, H.B., Ra, J.S., Kim, S.D., 2010a. Bioaccumulation of perfluorochemicals in pacific oyster under different salinity gradients. Environ. Sci. Technol. 44 (7), 2695–2701.

Jeon, J., Kannan, K., Lim, H.K., Moon, H.B., Kim, S.D., 2010b. Bioconcentration of perfluorinated compounds in blackrock fish, Sebastes schlegeli, at different salinity levels. Environ. Toxicol. Chem. 29 (11), 2529–2535.

Jeon, J., Kannan, K., Lim, B.J., An, K.G., Kim, S.D., 2011. Effects of salinity and organic matter on the partitioning of perfluoroalkyl acid (PFAs) to clay particles. J. Environ. Monit. 13 (6), 1803–1810.

Jin, M., Zhang, X., Wang, L., Huang, C., Zhang, Y., Zhao, M., 2009. Developmental toxicity of bifenthrin in embryo-larval stages of zebrafish. Aquat. Toxicol. 95 (4), 347–354.

Jobling, S., Nolan, M., Tyler, C.R., Brighty, G., Sumpter, J.P., 1998. Widespread sexual disruption in wild fish. Environ. Sci. Tech. 32, 2498–2506.

Kannan, K., 2011. Perfluoroalkyl and polyfluoroalkyl substances: current and future perspectives. Environ. Chem. 8 (4), 333–338.

Kessel, G.M., Phillips, N.E., 2018. Global change scenarios trigger carry-over effects across life stages and generations of the intertidal limpet, Siphonaria australis. PLoS One 13.

Kidd, K.A., Blanchfield, P.J., Mills, K.H., Palace, V.P., Evans, R.E., Lazorchak, J.M., Flick, R.W., 2007. Collapse of a fish population after exposure to a synthetic estrogen. Proc. Natl. Acad. Sci. 104 (21), 8897–8901.

Kiesecker, J.M., Blaustein, A.R., Belden, L.K., 2001. Complex causes of amphibian population declines. Nature 410 (6829), 681.

Knouft, J.H., Ficklin, D.L., 2017. The potential impacts of climate change on biodiversity in flowing freshwater systems. Annu. Rev. Ecol. Evol. Syst. 48, 111–133.

Komoroske, L.M., Connon, R.E., Lindberg, J., Cheng, B.S., Castillo, G., Hasenbein, M., Fangue, N.A., 2014. Ontogeny influences sensitivity to climate change stressors in an endangered fish. Conserv. Physiol. 2 (1), 1–13.

Kooijman, S.A.L.M., 1998. Process-oriented descriptions of toxic effects. In: Ecotoxicology: Ecological Fundamentals, Chemical Exposures, and Biological Effect. Wiley, NY, USA, pp. 483–520.

Kraemer, L.D., Schulte, P.M., 2004. Prior PCB exposure suppresses hypoxia-induced up-regulation of glycolytic enzymes in Fundulus heteroclitus. Comp. Biochem. Physiol., Part C: Toxicol. Pharmacol. 139 (1–3), 23–29.

Lai, K.P., Li, J.-W., Tse, A.C.-K., Chan, T.-F., Wu, R.S.-S., 2016. Hypoxia alters steroidogenesis in female marine medaka through miRNAs regulation. Aquat. Toxicol. 172, 1–8.

Laloë, J.O., Esteban, N., Berkel, J., Hays, G.C., 2016. Sand temperatures for nesting sea turtles in the Caribbean: implications for hatchling sex ratios in the face of climate change. J. Exp. Mar. Biol. Ecol. 474, 92–99.

Landry, C., Steele, S., Manning, S., Cheek, A., 2007. Long term hypoxia suppresses reproductive capacity in the estuarine fish, Fundulus grandis. Comp. Biochem. Physiol. A Mol. Integr. Physiol. 148 (2), 317–323.

Lange, R., Marshall, D., 2017. Ecologically relevant levels of multiple, common marine stressors suggest antagonistic effects. Sci. Rep. 7, 6281.

Lau, C., Anitole, K., Hodes, C., Lai, D., Pfahles-Hutchens, A., Seed, J., 2007. Perfluoroalkyl acids: a review of monitoring and toxicological findings. Toxicol. Sci. 99 (2), 366–394.

Lee, S., Ji, K., Choi, K., 2014. Effects of water temperature on perchlorate toxicity to the thyroid and reproductive system of Oryzias latipes. Ecotoxicology and environmental safety, 108, 311–317.

Linares-Casenave, J., Werner, I., Eenennaam, J.P., Doroshov, S.I., 2013. Temperature stress induces notochord abnormalities and heat shock proteins expression in larval green sturgeon (Acipenser medirostris Ayres 1854). J. Appl. Ichthyol. 29 (5), 958–967.

Liu, Y., Guan, Y., Gao, Q., Tam, N.F.Y., Zhu, W., 2010. Cellular responses, biodegradation and bioaccumulation of endocrine disrupting chemicals in marine diatom Navicula incerta. Chemosphere 80 (5), 592–599.

Loganimoce, E.M., 2016. Combined Effects of Warming, Acidification and the Synthetic Progestin Levonorgestrel on the Fitness of the Marine Amphipod Gammarus locusta (Crustacea). U. Porto Edições.

Luzio, A., Santos, D., Fontaínhas-Fernandes, A.A., Monteiro, S.M., Coimbra, A.M., 2016. Effects of 17α-ethinylestradiol at different water temperatures on zebrafish sex differentiation and gonad development. Aquat. Toxicol. 174, 22–35.

Ma, J., Hung, H., Macdonald, R.W., 2016. The influence of global climate change on the environmental fate of persistent organic pollutants: a review with emphasis on the Northern Hemisphere and the Arctic as a receptor. Global Planet. Change 146, 89–108.

Major, K.M., Weston Donald, P., Lydy Michael, J., Wellborn Gary, A., Poynton, H.C., 2017. Unintentional exposure to terrestrial pesticides drives widespread and predictable evolution of resistance in freshwater crustaceans. Evol. Appl. 11, 748–761.

Matson, C.W., Timme-Laragy, A.R., Di Giulio, R.T., 2008. Fluoranthene, but not benzo [a] pyrene, interacts with hypoxia resulting in pericardial effusion and lordosis in developing zebrafish. Chemosphere 74 (1), 149–154.

Matsumoto, Y., Crews, D., 2012. Molecular mechanisms of temperature-dependent sex determination in the context of ecological developmental biology. Mol. Cell. Endocrinol. 354 (1–2), 103–110.

Maulvault, A.L., Anacleto, P., Barbosa, V., Sloth, J.J., Rasmussen, R.R., Tediosi, A., … Marques, A., 2015. Toxic elements and speciation in seafood samples from different contaminated sites in Europe. Environ. Res. 143, 72–81.

Maulvault, A.L., Santos, L.H., Paula, J.R., Camacho, C., Pissarra, V., Fogaça, F., … Rodriguez-Mozaz, S., 2018a. Differential behavioural responses to venlafaxine exposure route, warming and acidification in juvenile fish (*Argyrosomus regius*). Sci. Total Environ. 634, 1136–1147.

Maulvault, A.L., Camacho, C., Barbosa, V., Alves, R., Anacleto, P., Fogaça, F., … Rasmussen, R.R., 2018b. Assessing the effects of seawater temperature and pH on the bioaccumulation of emerging chemical contaminants in marine bivalves. Environ. Res. 161, 236–247.

Mckinney, M.A., Pedro, S., Dietz, R., Sonne, C., Fisk, A.T., Roy, D., … Letcher, R.J., 2015. A review of ecological impacts of global climate change on persistent organic pollutant and mercury pathways and exposures in arctic marine ecosystems. Curr. Zool. 61 (4), 617–628.

Melzner, F., Gutowska, M.A., Langenbuch, M., Dupont, S., Lucassen, M., Thorndyke, M.C., Bleich, M., Portner, H.O., 2009. Physiological basis for high CO2 tolerance in marine ectothermic animals: pre-adaptation through lifestyle and ontogeny? Biogeosciences 6, 2313–2331.

Meyer, J.N., Nacci, D.E., Di Giulio, R.T., 2002. Cytochrome P4501A (CYP1A) in killifish (Fundulus heteroclitus): heritability of altered expression and relationship to survival in contaminated sediments. Toxicol. Sci. 68 (1), 69–81.

Millero, F.J., Woosley, R., Ditrolio, B., Waters, J., 2009. Effect of ocean acidification on the speciation of metals in seawater. Oceanography 22 (4), 72–85.

Moreira, L.B., Diamante, G., Giroux, M., Coffin, S., Xu, E.G., Moledo de Souza Abessa, D., & Schlenk, D., 2018. Impacts of salinity and temperature on the thyroidogenic effects of the biocide diuron in *Menidia beryllina*. Environ. Sci. Technol. 52 (5), 3146–3155.

Moyle, P.B., Hobbs, J.A., Durand, J.R., 2018. Delta smelt and water politics in California. Fisheries 43 (1), 42–50.

Mueller, C., Joss, J., Seymour, R.S., 2011. Effects of environmental oxygen on development and respiration of Australian lungfish (*Neoceratodus forsteri*) embryos. J. Comp. Physiol. B 181, 941–952.

Munday, P.L., Donelson, J.M., Dixson, D.L., Endo, G.G.K., 2009. Effects of ocean acidification on the early life history of a tropical marine fish. Proc. R. Soc. B 276 (1671), 3275–3283. https://doi.org/10.1098/rspb.2009.0784.

Munday, P.L., 2014. Transgenerational acclimation of fishes to climate change and ocean acidification. F1000Prime Rep. 6, 99.

Nash, J.P., Kime, D.E., Van der Ven, L.T., Wester, P.W., Brion, F., Maack, G., … Tyler, C.R., 2004. Long-term exposure to environmental concentrations of the pharmaceutical ethynylestradiol causes reproductive failure in fish. Environ. Health Perspect. 112 (17), 1725.

Noyes, P.D., Lema, S.C., 2015. Forecasting the impacts of chemical pollution and climate change interactions on the health of wildlife. Curr. Zool. 61 (4), 669–689.

Noyes, P.D., McElwee, M.K., Miller, H.D., Clark, B.W., Van Tiem, L.A., Walcott, K.C., … Levin, E.D., 2009. The toxicology of climate change: environmental contaminants in a warming world. Environ. Int. 35 (6), 971–986.

Ohtani, H., Miura, I., Ichikawa, Y., 2000. Effects of dibutyl phthalate as an environmental endocrine disruptor on gonadal sex differentiation of genetic males of the frog *Rana rugosa*. Environ. Health Perspect. 108 (12), 1189.

Ondricek, K., Thomas, P., 2018. Effects of hypoxia exposure on apoptosis and expression of membrane steroid receptors, ZIP9, mPR alpha, and GPER in Atlantic croaker ovaries. Comp. Biochem. Physiol. A Mol. Integr. Physiol. 224, 84–92. 2018.

Pait, A.S., Nelson, J.O., 2002. Endocrine Disruption in Fish: An Assessment of Recent Research and Results. 149 NOAA Technical Memorandum NOS NCCOS CCMA, p. 48.

Pepin, P., 1991. Effect of temperature and size on development, mortality, and survival rates of the pelagic early life history stages of marine fish. Can. J. Fish. Aquat. Sci. 48 (3), 503–518.

Pérez-Robles, J., Diaz, F., Ibarra-Castro, L., Giffard-Mena, I., Re, A.D., Ibarra, L.E.R., Soto, J.A.I., 2016. Effects of salinity on osmoregulation during the embryonic development of the bullseye puffer (*Sphoeroides annulatus* Jenyns 1842). Aquacult. Res. 47 (3), 838–846.

Pörtner, H.O., Berdal, B., Blust, R., Brix, O., Colosimo, A., De Wachter, B., … Lannig, G., 2001. Climate induced temperature effects on growth performance, fecundity and recruitment in marine fish: developing a hypothesis for cause and effect relationships in Atlantic cod (*Gadus morhua*) and common eelpout (Zoarces viviparus). Cont. Shelf Res. 21 (18–19), 1975–1997.

Pörtner, H.O., Langenbuch, M., Reipschläger, A., 2004. Biological impact of elevated ocean CO2 concentrations: lessons from animal physiology and earth history. J. Oceanogr. 60 (4), 705–718.

Pörtner, H.O., Langenbuch, M., Michaelidis, B., 2005. Synergistic effects of temperature extremes, hypoxia, and increases in CO2 on marine animals: from earth history to global change. J. Geophys. Res. Oceans 110 (C9), C09S10.

Pounds, J.A., Fogden, M.P., Campbell, J.H., 1999. Biological response to climate change on a tropical mountain. Nature 398 (6728), 611.

REFERENCES

Prasch, A.L., Andreasen, E.A., Peterson, R.E., Heideman, W., 2004. Interactions between 2, 3, 7, 8-tetrachlorodibenzo-p-dioxin (TCDD) and hypoxia signaling pathways in zebrafish: hypoxia decreases responses to TCDD in zebrafish embryos. Toxicol. Sci. 78 (1), 68–77.

Preus-Olsen, G., Olufsen, M.O., Pedersen, S.A., Letcher, R.J., Arukwe, A., 2014. Effects of elevated dissolved carbon dioxide and perfluorooctane sulfonic acid, given singly and in combination, on steroidogenic and biotransformation pathways of Atlantic cod. Aquat. Toxicol. 155, 222–235.

Ramach, S., Zachary Darnell, M., Avissar, N., Rittschof, D., 2009. Habitat use and population dynamics of blue crabs, Callinectes sapidus, in a high-salinity embayment. J. Shellfish. Res. 28 (3), 635–640.

Ramirez, J.M., Folkow, L.P., Blix, A.S., 2007. Hypoxia tolerance in mammals and birds: from the wilderness to the clinic. Annu. Rev. Physiol. 69 (1), 113–143.

Raven, J., Caldeira, K., Elderfield, H., Hoegh-Guldberg, O., Liss, P., Riebesell, U., … Watson, A., 2005. Ocean Acidification Due to Increasing Atmospheric Carbon Dioxide. The Royal Society.

Rodgers, G.G., Donelson, J.M., McCormick, M.I., Munday, P.L., 2018. In hot water: sustained ocean warming reduces survival of a low-latitude coral reef fish. Mar. Biol. 165 (4), 73.

Roegner, G.C., Needoba, J.A., Baptista, A.M., 2011. Coastal upwelling supplies oxygen-depleted water to the Columbia river estuary. PLoS One 6 (4).

Ross, P.S., Ellis, G.M., Ikonomou, M.G., Barrett-Lennard, L.G., Addison, R.F., 2000. High PCB concentrations in free-ranging Pacific killer whales, Orcinus orca: effects of age, sex and dietary preference. Mar. Pollut. Bull. 40 (6), 504–515.

Ruhí, A., Acuña, V., Barceló, D., Huerta, B., Mor, J.R., Rodríguez-Mozaz, S., Sabater, S., 2016. Bioaccumulation and trophic magnification of pharmaceuticals and endocrine disruptors in a Mediterranean river food web. Sci. Total Environ. 540, 250–259.

Russell, N.R., Fish, J.D., Wootton, R.J., 1996. Feeding and growth of juvenile sea bass: the effect of ration and temperature on growth rate and efficiency. J. Fish Biol. 49 (2), 206–220.

Salierno, J.D., Kane, A.S., 2009. 17α-Ethinylestradiol alters reproductive behaviors, circulating hormones, and sexual morphology in male fathead minnows (Pimephales promelas). Environ. Toxicol. Chem. 28 (5), 953–961.

Saranjampour, P., Vebrosky, E.N., Armbrust, K.L., 2017. Salinity impacts on water solubility and n-octanol/water partition coefficients of selected pesticides and oil constituents. Environ. Toxicol. Chem. 36 (9), 2274–2280.

Schiedek, D., Sundelin, B., Readman, J.W., Macdonald, R.W., 2007. Interactions between climate change and contaminants. Mar. Pollut. Bull. 54 (12), 1845–1856.

Schmidtko, S., Stramma, L., Visbeck, M., 2017. Decline in global oceanic oxygen content during the past five decades. Nature 542 (7641), 335.

Schug, T.T., Johnson, A.F., Birnbaum, L.S., Colborn, T., Guillette, L.J., Crews, D.P., Collins, T., Soto, A.M., vom Saal FS, McLachlan JA, Sonnenschein C, Heindel JJ, 2016. Minireview: endocrine disruptors: past lessons and future directions. Mol. Endocrinol. 30, 833–847.

Schulte, P.M., 2014. What is environmental stress? Insights from fish living in a variable environment. J. Exp. Biol. 217 (1), 23–34.

Semenza, G.L., 2000. Expression of hypoxia-inducible factor 1: mechanisms and consequences. Biochem. Pharmacol. 59 (1), 47–53.

Seymour, R.S., Roberts, J.D., Mitchell, N.J., Blaylock, A.J., 2000. Influence of environmental oxygen on development and hatching of aquatic eggs of the Australian frog, Crinia georgiana. Physiol. Biochem. Zool. 73 (4), 501–507.

Shama, L.N.S., Wegner, K.M., 2014. Grandparental effects in marine sticklebacks: transgenerational plasticity across multiple generations. J. Evol. Biol. 27 (11), 2297–2307. https://doi.org/10.1111/jeb.12490.

Shang, E.H., Yu, R.M., Wu, R.S., 2006. Hypoxia affects sex differentiation and development, leading to a male-dominated population in zebrafish (Danio rerio). Environ. Sci. Technol. 40 (9), 3118–3122.

Shappell, N.W., et al., 2018. Do environmental factors affect male fathead minnow (Pimephales promelas) response to estrone? Part 2. Temperature and food availability. Sci. Total Environ. 610, 32–43.

Shaw, J.L.A., Judy, J.D., Kumar, A., Bertsch, P., Wang, M.-B., Kirby, J.K., 2017. Incorporating transgenerational epigenetic inheritance into ecological risk assessment frameworks. Environ. Sci. Technol. 51, 9433–9445.

Smalling, K.L., Morgan, S., Kuivila, K.K., 2010. Accumulation of current-use and organochlorine pesticides in crab embryos from northern California, USA. Environ. Toxicol. Chem. 29 (11), 2593–2599.

Stoker, C., et al., 2003. Sex reversal effects on Caiman latirostris exposed to environmentally relevant doses of the xenoestrogen bisphenol A. Gen. Comp. Endocrinol. 133 (3), 287–296.

Stramma, L., Johnson, G.C., Sprintall, J., Mohrholz, V., 2008. Expanding oxygen-minimum zones in the tropical oceans. Science 320 (5876), 655–658.

Strüssmann, C.A., Conover, D.O., Somoza, G.M., Miranda, L.A., 2010. Implications of climate change for the reproductive capacity and survival of new world silversides (family Atherinopsidae). J. Fish Biol. 77 (8), 1818–1834.

Swanson, H.I., Chan, W.K., Bradfield, C.A., 1995. DNA binding specificities and pairing rules of the Ah receptor, ARNT, and SIM proteins. J. Biol. Chem. 270 (44), 26292–26302.

Takahashi, A., Higashitani, T., Yakou, Y., Saitou, M., Tamamoto, H., Tanaka, H., 2003. Evaluating bioaccumulation of suspected endocrine disruptors into periphytons and benthos in the Tama river. Water Sci. Technol. 47 (9), 71–76.

Tanguay, R.L., Andreasen, E.A., Walker, M.K., Peterson, R.E., 2003. Dioxin toxicity and aryl hydrocarbon receptor signaling in fish. In: Schecter, A., Gasiewicz, T.A. (Eds.), Dioxins and Health, second ed. John Wiley & Sons, New York, pp. 603–628.

Telemeco, R.S., Abbott, K.C., Janzen, F.J., 2013. Modeling the effects of climate change–induced shifts in reproductive phenology on temperature-dependent traits. Am. Nat. 181 (5), 637–648.

Thomas, P., 2017. Role of G-protein-coupled estrogen receptor (GPER/GPR30) in maintenance of meiotic arrest in fish oocytes. J. Steroid Biochem. Mol. Biol. 167, 153–161.

Thomas, P., Rahman, M.S., 2011. Extensive reproductive disruption, ovarian masculinization and aromatase suppression in Atlantic croaker in the northern Gulf of Mexico hypoxic zone. Proceedings of the Royal Society B, Biological Sciences.

Thomas, P., Rahman, M.S., Khan, I.A., Kummer, J.A., 2007. Widespread endocrine disruption and reproductive impairment in an estuarine fish population exposed to seasonal hypoxia. Proc. R. Soc. Lond. B Biol. Sci. 274 (1626), 2693–2702.

Thor, P., Dupont, S., 2015. Transgenerational effects alleviate severe fecundity loss during ocean acidification in a ubiquitous planktonic copepod. Glob. Chang. Biol. 21, 2261–2271.

Thorpe, K.L., Maack, G., Benstead, R., Tyler, C.R., 2009. Estrogenic wastewater treatment works effluents reduce egg production in fish. Environ. Sci. Technol. 43 (8), 2976–2982.

Todgham, A.E., Stillman, J.H., 2013. Physiological responses to shifts in multiple environmental stressors: relevance in a changing world. Integr. Comp. Biol. 53 (4), 539–544.

Tseng, Y.C., Hu, M.Y., Lin, L.Y., Melzner, F., Hwang, P.P., 2013. CO2-driven seawater acidification differentially affects development and molecular plasticity along life history of fish (Oryzias latipes). Comp. Biochem. Physiol. 165, 119–130.

Uller, T., Nakagawa, S., English, S., 2013. Weak evidence for anticipatory parental effects in plants and animals. J. Evol. Biol. 26, 2161–2170.

Varsamos, S., Nebel, C., Charmantier, G., 2005. Ontogeny of osmoregulation in postembryonic fish: a review. Comp. Biochem. Physiol. A Mol. Integr. Physiol. 141 (4), 401–429.

Vinogradov, G.A., 1985. Ion regulation in the perch, Perca fluviatilis, in connection with the problem of acidification of water bodies. J. Ichthyol. 25, 53–61.

Voisin, A.-S., Fellous, A., Earley, R.L., Silvestre, F., 2016. Delayed impacts of developmental exposure to 17-α-ethinylestradiol in the self-fertilizing fish Kryptolebias marmoratus. Aquat. Toxicol. 180, 247–257.

Vorrink, S.U., Domann, F.E., 2014. Regulatory crosstalk and interference between the xenobiotic and hypoxia sensing pathways at the AhR-ARNT-HIF1α signaling node. Chem. Biol. Interact. 218, 82–88.

Walker, M.K., Peterson, R.E., 1994. Aquatic toxicity of dioxins and related chemicals. In: Dioxins and Health. Springer, Boston, MA, pp. 347–387.

Wang, L.H., Tsai, C.L., 2000. Effects of temperature on the deformity and sex differentiation of tilapia, Oreochromis mossambicus. J. Exp. Zool. A Ecol. Genet. Physiol. 286 (5), 534–537.

Wang, G.L., Jiang, B.H., Semenza, G.L., 1995. Effect of protein kinase and phosphatase inhibitors on expression of hypoxia inducible factor 1. Biochem. Biophys. Res. Commun. 216 (2), 669–675.

Ward, T.J., Robinson, W.E., 2009. Evolution of cadmium resistance in Daphnia magna. Environ. Toxicol. Chem. 24 (9), 2341–2349. https://doi.org/10.1897/04-429R.1.

Warner, K.E., Jenkins, J.J., 2007. Effects of 17α-ethinylestradiol and bisphenol a on vertebral development in the fathead minnow (Pimephales Promelas). Environ. Toxicol. Chem. 26 (4), 732–737.

Wenger, R.H., 2000. Mammalian oxygen sensing, signalling and gene regulation. J. Exp. Biol. 203 (8), 1253–1263.

Wentz, F.J., Ricciardulli, L., Hilburn, K., Mears, C., 2007. How much more rain will global warming bring? Science 317 (5835), 233–235.

White, J.W., Cole, B.J., Cherr, G.N., Connon, R.E., Brander, S.M., 2017. Scaling up endocrine disruption effects from individuals to populations: outcomes depend on how many males a population needs. Environ. Sci. Technol. 51 (3), 1802–1810.

WHO, 2012. Biomonitoring-based Indicators of Exposure to Chemical Pollutants. Report of a Meeting, Catania, Italy, 19–20 April 2012, 41pp.

Windsor, F.M., Ormerod, S.J., Tyler, C.R., 2018. Endocrine disruption in aquatic systems: up-scaling research to address ecological consequences. Biol. Rev. 93, 626–641.

Wong, J.M., Johnson Kevin, M., Kelly Morgan, W., Hofmann, G.E., 2018. Transcriptomics reveal transgenerational effects in purple sea urchin embryos: adult acclimation to upwelling conditions alters the response of their progeny to differential pCO2 levels. Mol. Ecol. 27, 1120–1137.

Yang, L., Cheng, Q., Lin, L., Wang, X., Chen, B., Luan, T., Tam, N.F.Y., 2016. Partitions and vertical profiles of 9 endocrine disrupting chemicals in an estuarine environment: effect of tide, particle size and salinity. Environ. Pollut. 211, 58–66.

EVALUATING WATER QUALITY TO PREVENT FUTURE DISASTERS

Zha, J., Sun, L., Zhou, Y., Spear, P.A., Ma, M., Wang, Z., 2008. Assessment of 17α-ethinylestradiol effects and underlying mechanisms in a continuous, multigeneration exposure of the Chinese rare minnow (*Gobiocypris rarus*). Toxicol. Appl. Pharmacol. 226 (3), 298–308.

Zhang, J., Gilbert, D., Gooday, A., Levin, L., Naqvi, S.W.A., Middelburg, J.J., ... Oguz, T., 2010a. Natural and human-induced hypoxia and consequences for coastal areas: synthesis and future development. Biogeosciences 7, 1443–1467.

Zhang, G., Shi, Y., Zhu, Y., Liu, J., Zang, W., 2010b. Effects of salinity on embryos and larvae of tawny puffer *Takifugu flavidus*. Aquaculture 302 (1-2), 71–75.

Further Reading

Blewett, T.A., Simon, R.A., Turko, A.J., Wright, P.A., 2017. Copper alters hypoxia sensitivity and the behavioural emersion response in the amphibious fish *Kryptolebias marmoratus*. Aquat. Toxicol. 189, 25–30.

Brion, F., Le Page, Y., Piccini, B., Cardoso, O., Tong, S.-K., Chung, B.-c., Kah, O., 2012. Screening estrogenic activities of chemicals or mixtures in vivo using transgenic (cyp19a1b-GFP) zebrafish embryos. PLoS One 7 (5).

Chang, K., Xiao, D., Huang, X., Xue, Z., Yang, S., Longo, L.D., Zhang, L., 2010. Chronic hypoxia inhibits sex steroid hormone-mediated attenuation of ovine uterine arterial myogenic tone in pregnancy. Hypertension 56 (4), 750–757.

Cheung, C.H.Y., Chiu, J.M.Y., Wu, R.S.S., 2014. Hypoxia turns genotypic female medaka fish into phenotypic males. Ecotoxicology 23 (7), 1260–1269.

Deane, E.E., Woo, N.Y., 2009. Modulation of fish growth hormone levels by salinity, temperature, pollutants and aquaculture related stress: a review. Rev. Fish Biol. Fish. 19 (1), 97–120.

DeMicco, A., Cooper, K.R., Richardson, J.R., White, L.A., 2009. Developmental neurotoxicity of pyrethroid insecticides in zebrafish embryos. Toxicol. Sci. 113 (1), 177–186.

Deutsch, C., Ferrel, A., Seibel, B., Pörtner, H.-O., Huey, R.B., 2015. Climate change tightens a metabolic constraint on marine habitats. Science 348 (6239), 1132–1135.

Dinnel, P.A., Link, J.M., Stober, Q.J., 1987. Improved methodology for a sea urchin sperm cell bioassay for marine waters. Arch. Environ. Contam. Toxicol. 16 (1), 23–32.

Doney, S.C., 2010. The growing human footprint on coastal and open-ocean biogeochemistry. Science 328, 1512–1516.

Ferguson, E.M., Allinson, M., Allinson, G., Swearer, S.E., Hassell, K.L., 2013. Fluctuations in natural and synthetic estrogen concentrations in a tidal estuary in south-eastern Australia. Water Res. 47 (4), 1604–1615.

Fitzgerald, J.A., Jameson, H.M., Dewar Fowler, V.H., Bond, G.L., Bickley, L.K., Uren Webster, T.M., ... Santos, E.M., 2016. Hypoxia suppressed copper toxicity during early development in zebrafish embryos in a process mediated by the activation of the HIF signaling pathway. Environ. Sci. Technol. 50 (8), 4502–4512.

Fitzgerald, J.A., Katsiadaki, I., Santos, E.M., 2017. Contrasting effects of hypoxia on copper toxicity during development in the three-spined stickleback (*Gasterosteus aculeatus*). Environ. Pollut. 222, 433–443.

Gilles, R., Delpire, E., 2011. Variations in Salinity, Osmolarity, and Water Availability: Vertebrates and Invertebrates. Physiology, Comprehensive.

Grantham, B.A., Chan, F., Nielsen, K.J., Fox, D.S., Barth, J.A., Huyer, A., Lubchenco, J., Menge, B.A., 2004. Upwelling-driven nearshore hypoxia signals ecosystem and oceanographic changes in the northeast Pacific. Nature 429 (6993), 749.

Greulich, K., Pflugmacher, S., 2003. Differences in susceptibility of various life stages of amphibians to pesticide exposure. Aquat. Toxicol. 65 (3), 329–336.

Gupta, S.C., Sharma, A., Mishra, M., Mishra, R.K., Chowdhuri, D.K., 2010. Heat shock proteins in toxicology: how close and how far? Life Sci. 86 (11–12), 377–384.

Hallare, A., Schirling, M., Luckenbach, T., Köhler, H.-R., Triebskorn, R., 2005. Combined effects of temperature and cadmium on developmental parameters and biomarker responses in zebrafish (*Danio rerio*) embryos. J. Therm. Biol. 30 (1), 7–17.

Hu, M.Y., Michael, K., Kreiss, C.M., Stumpp, M., Dupont, S., Tseng, Y.-C., Lucassen, M., 2016. Temperature modulates the effects of ocean acidification on intestinal ion transport in Atlantic cod, *Gadus morhua*. Front. Physiol. 7, 198.

Hurst, T.P., Fernandez, E.R., Mathis, J.T., 2013. Effects of ocean acidification on hatch size and larval growth of walleye pollock (*Theragra chalcogramma*). ICES J. Mar. Sci. 70 (4), 812–822.

Isaksson, C., 2010. Pollution and its impact on wild animals: a meta-analysis on oxidative stress. Ecohealth 7 (3), 342–350.

Jin, Y., Pan, X., Cao, L., Ma, B., Fu, Z., 2013. Embryonic exposure to cis-bifenthrin enantioselectively induces the transcription of genes related to oxidative stress, apoptosis and immunotoxicity in zebrafish (*Danio rerio*). Fish Shellfish Immunol. 34 (2), 717–723.

Kinch, C.D., Ibhazehiebo, K., Jeong, J.-H., Habibi, H.R., Kurrasch, D.M., 2015. Low-dose exposure to bisphenol A and replacement bisphenol S induces precocious hypothalamic neurogenesis in embryonic zebrafish. Proc. Natl. Acad. Sci. 112 (5), 1475–1480.

Köprücü, K., Aydın, R., 2004. The toxic effects of pyrethroid deltamethrin on the common carp (*Cyprinus carpio* L.) embryos and larvae. Pestic. Biochem. Physiol. 80 (1), 47–53.

EVALUATING WATER QUALITY TO PREVENT FUTURE DISASTERS

Kubo, T., Maezawa, N., Osada, M., Katsumura, S., Funae, Y., Imaoka, S., 2004. Bisphenol A, an environmental endocrine-disrupting chemical, inhibits hypoxic response via degradation of hypoxia-inducible factor 1α (HIF-1α): structural requirement of bisphenol A for degradation of HIF-1α. Biochem. Biophys. Res. Commun. 318 (4), 1006–1011.

Kubota, A., Goldstone, J.V., Lemaire, B., Takata, M., Woodin, B.R., Stegeman, J.J., 2014. Role of pregnane X receptor and aryl hydrocarbon receptor in transcriptional regulation of pxr, CYP2, and CYP3 genes in developing zebrafish. Toxicol. Sci. 143 (2), 398–407.

Lee, S., Ji, K., Choi, K., 2014. Effects of water temperature on perchlorate toxicity to the thyroid and reproductive system of *Oryzias latipes*. Ecotoxicol. Environ. Saf. 108, 311–317.

Machado, B.E., Podrabsky, J.E., 2007. Salinity tolerance in diapausing embryos of the annual killifish *Austrofundulus limnaeus* is supported by exceptionally low water and ion permeability. J. Comp. Physiol. B 177 (7), 809–820.

Mandic, M., Todgham, A.E., Richards, J.G., 2009. Mechanisms and evolution of hypoxia tolerance in fish. Proc. R. Soc. Lond. B Biol. Sci. 276 (1657), 735–744.

McCormick, S.D., Bradshaw, D., 2006. Hormonal control of salt and water balance in vertebrates. Gen. Comp. Endocrinol. 147 (1), 3–8.

Miller, G.M., Kroon, F.J., Metcalfe, S., Munday, P., 2015. Temperature is the evil twin: effects of increased temperature and ocean acidification on reproduction in a reef fish. Ecol. Appl. 25 (3), 603–620.

Nielsen, D.L., Brock, M.A., 2009. Modified water regime and salinity as a consequence of climate change: prospects for wetlands of Southern Australia. Clim. Change 95, 523–533.

Oberdörster, E., Cheek, A.O., 2001. Gender benders at the beach: endocrine disruption in marine and estuarine organisms. Environ. Toxicol. Chem. 20 (1), 23–36.

Orłowski, G., Hałupka, L., Pokorny, P., Klimczuk, E., Sztwiertnia, H., Dobicki, W., 2016. The effect of embryonic development on metal and calcium content in eggs and eggshells in a small passerine. Ibis 158 (1), 144–154.

Osterberg, J.S., Darnell, K.M., Blickley, T.M., Romano, J.A., Rittschof, D., 2012. Acute toxicity and sub-lethal effects of common pesticides in post-larval and juvenile blue crabs, *Callinectes sapidus*. J. Exp. Mar. Biol. Ecol. 424, 5–14.

Pankhurst, N.W., Munday, P.L., 2011. Effects of climate change on fish reproduction and early life history stages. Mar. Freshw. Res. 62 (9), 1015–1026.

Pittman, K., Yúfera, M., Pavlidis, M., Geffen, A.J., Koven, W., Ribeiro, L., Zambonino-Infante, J.L., Tandler, A., 2013. Fantastically plastic: fish larvae equipped for a new world. Rev. Aquac. 5 (s1).

Pollock, M., Clarke, L., Dubé, M., 2007. The effects of hypoxia on fishes: from ecological relevance to physiological effects. Environ. Rev. 15, 1–14.

Räsänen, K., Laurila, A., Merilä, J., 2003. Geographic variation in acid stress tolerance of the moor frog, Rana arvalis. I. Local adaptation. Evolution 57 (2), 352–362.

Rocha, M.J., Ribeiro, M., Ribeiro, C., Couto, C., Cruzeiro, C., Rocha, E., 2012. Endocrine disruptors in the Leça river and nearby Porto coast (NW Portugal): presence of estrogenic compounds and hypoxic conditions. Toxicol. Environ. Chem. 94 (2), 262–274.

Roessig, J.M., Woodley, C.M., Cech, J.J., Hansen, L.J., 2004. Effects of global climate change on marine and estuarine fishes and fisheries. Rev. Fish Biol. Fish. 14 (2), 251–275.

Sabine, C.L., Feely, R.A., Gruber, N., Key, R.M., Lee, K., Bullister, J.L., … Millero, F.J., 2004. The oceanic sink for anthropogenic CO2. Science 305 (5682), 367–371.

Scholz, S., Mayer, I., 2008. Molecular biomarkers of endocrine disruption in small model fish. Mol. Cell. Endocrinol. 293 (1-2), 57–70.

Shang, E.H., Wu, R.S., 2004. Aquatic hypoxia is a teratogen and affects fish embryonic development. Environ. Sci. Technol. 38 (18), 4763–4767.

Shao, Y.T., Chang, F.Y., Fu, W.C., Yan, H.Y., 2016. Acidified seawater suppresses insulin-like growth factor I mRNA expression and reduces growth rate of juvenile orange-spotted groupers, *Epinephelus coioides* (Hamilton, 1822). Aquacult. Res. 47 (3), 721–731.

Sørensen, S.R., Butts, I.A.E., Munk, P., Tomkiewicz, J., 2016. Effects of salinity and sea salt type on egg activation, fertilization, buoyancy and early embryology of European eel, *Anguilla anguilla*. Zygote 24 (1), 121–138.

Suzawa, M., Ingraham, H.A., 2008. The herbicide atrazine activates endocrine gene networks via non-steroidal NR5A nuclear receptors in fish and mammalian cells. PLoS One 3 (5).

Tabb, M.M., Blumberg, B., 2006. New modes of action for endocrine-disrupting chemicals. Mol. Endocrinol. 20 (3), 475–482.

Tedeschi, J., Kennington, W., Berry, O., Whiting, S., Meekan, M., Mitchell, N., 2015. Increased expression of Hsp70 and Hsp90 mRNA as biomarkers of thermal stress in loggerhead turtle embryos (*Caretta caretta*). J. Therm. Biol. 47, 42–50.

Tseng, M.-C., Yang, D.-H., Yen, T.-B., 2017. Comparative study on hatching rate, survival rate, and feminization of Onychostoma barbatulum (Pellegrin, 1908) at different temperatures and examining sex change by Gonad and Karyotype analyses. Zool. Stud. 56, 16.

Tu, W., Xu, C., Lu, B., Lin, C., Wu, Y., Liu, W., 2016. Acute exposure to synthetic pyrethroids causes bioconcentration and disruption of the hypothalamus–pituitary–thyroid axis in zebrafish embryos. Sci. Total Environ. 542, 876–885.

EVALUATING WATER QUALITY TO PREVENT FUTURE DISASTERS

FURTHER READING

Vasseur, P., Cossu-Leguille, C., 2006. Linking molecular interactions to consequent effects of persistent organic pollutants (POPs) upon populations. Chemosphere 62 (7), 1033–1042.

Venturino, A., Rosenbaum, E., Caballero De Castro, A., Anguiano, O.L., Gauna, L., Fonovich De Schroeder, T., Pechen De D'Angelo, A.M., 2003. Biomarkers of effect in toads and frogs. Biomarkers 8 (3–4), 167–186.

Weber, S., Koch, A., Kankeleit, J., Schewe, J.C., Siekmann, U., Stüber, F., Hoeft, A., Schröder, S., 2009. Hyperbaric oxygen induces apoptosis via a mitochondrial mechanism. Apoptosis 14 (1), 97–107.

Wu, R.S., Zhou, B.S., Randall, D.J., Woo, N.Y., Lam, P.K., 2003. Aquatic hypoxia is an endocrine disruptor and impairs fish reproduction. Environ. Sci. Technol. 37 (6), 1137–1141.

Wu, Y., Zhou, Q., Li, H., Liu, W., Wang, T., Jiang, G., 2010. Effects of silver nanoparticles on the development and histopathology biomarkers of Japanese medaka (*Oryzias latipes*) using the partial-life test. Aquat. Toxicol. 100 (2), 160–167.

CHAPTER 3

Impact of Persistent Droughts on the Quality of the Middle East Water Resources

Y. Shevah*

TOHL, Civil and Environmental Engineering Inc., Atlanta, GA, United States
*Corresponding author: E-mail: ysheva@gmail.com

1 INTRODUCTION

1.1 The Middle East and the Regional Climate

The Middle East region stretches along a north–south axis between Turkey and the Arab Golf, through the ring enclosing the southern Turk, Syria, and Iraq and the Arabian Peninsula Bahrain, Kuwait, Oman, Qatar, Saudi Arabia, the United Arab Emirates, and Yemen, along the Red Sea. In between, the Levant Basin (the Eastern Mediterranean, comprising Lebanon, Israel, Palestine, Jordan, and Syria), connecting the three continents—Asia, Europe, and Africa. See Map (Fig. 1).

The geomorphology comprises sea coasts, mountains, deserts, and fertile plains, dissected by major rivers including the Tigris and Euphrates and the Jordan Rivers. The population in the levant countries amounts to about 120 million in 2015 and is projected to grow to about 225 million by 2050 (UN Department of Economic and Social Affairs, Population Division, 2015). In general, the region is endowed with natural wealth, although not evenly distributed and is oil dependent (International Monetary Fund, 2017). Sectarian divides and the entrenched rift between Sunnis and Shiites Muslims have immensely complicated the regional politics and stability (El-Tablawy, 2018).

The Levant climate is generally characterized by arid and semiarid conditions exhibiting warm temperatures, low rainfall. Because of drought, the daily maximum temperature has risen to over 44°C and the rainfall declined from the long-term average of 800–850 mm to the current level of 500–650 mm, and less. The drought which has began in 2014 caused the rainfall to drop, in the north east zone, bordering Israel, Lebanon, and Syria, by 20%–30% of the average (Cook et al., 2016) and below, as shown in Fig. 2 for the period: 1981–2018 (Israel Water Authority, 2018a,b,c).

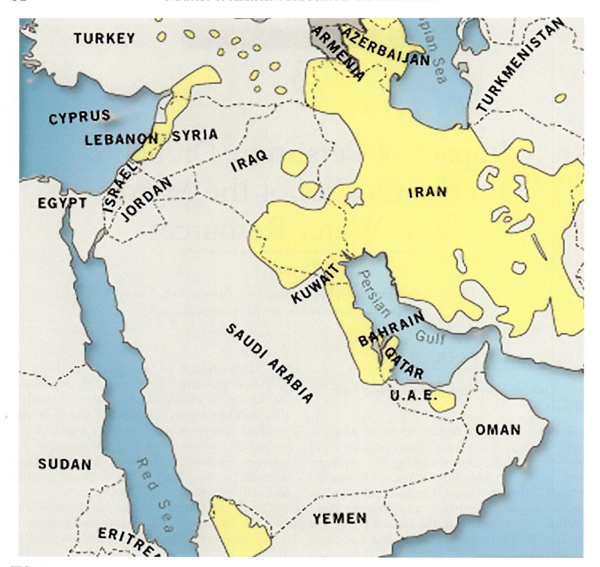

FIG. 1 Map of the Middle East. *Source: US CIA.*

The current and the previous drought during the period 2006–2009 show that the climatic conditions are worsening over time reflecting a decrease in precipitation and increase in temperature, evaporation, and rainfall. The climate change has led to increased variability and reduced availability of water, leading to severe water scarcity (Cook et al., 2016; Voss et al., 2013).

FIG. 2 Representative yearly rainfall 1985–2017, in % of the long-term average 1981–2010 (South Levant Area). *Source: Compiled from: Israel Water Authority 2018a. Rainfall Data. http://www.water.gov.il/Hebrew/ProfessionalInfoAndData/Pages/rainmap.aspx (accessed 16.04.18).*

1.2 Global Warming and Drought Impact

Globally, the recent years were generally warmer than the long-term average across most of the Earth's surface (IPCC, 2014). The record set in 2014 of 0.136–0.18°C was broken in 2015 in which the global surface temperature was higher by 0.426–0.466°C (0.76–0.83°F), above the 1980–2010 average (Blunden and Arndt, 2016). While 2016 sets a new warmer year, being the warmest in 150 years with 0.45–0.56°C above the 1981–2010 annual average (Blunden and Arndt, 2017). The impacts of climate change on natural and human systems are apparent on all continents and across the oceans (UN Department of Economic and Social Affairs and World Bank's Water Global Practice, 2018). Approximately 0.8 billion people, are assumed to be affected by the lack of safe drinking water, and the number may increase to more than half of the world's population by 2050 to around 4.5 billion people (UN and WB Report, 2018). The drought is also manifested in many of the world's largest cities, among them, Cape Town in South Africa in which the one in 100 years drought is threatening water supplies to its population, getting close to "day zero" (Desalination.biz, 2018). The world water crisis is ranked by the World Economic Forum (2018) high among the most world pressing problems such as weapons of mass destruction, cyberattacks, and involuntary migration. Articulated policies are essential to better management of the climate impacts on water resources worldwide, and in the most affected region, the Middle East.

1.3 The Drought in the Middle East

In the region, the record-breaking temperatures and other signs of climate change have greatly increased water stresses, depleting groundwater aquifers and reduced the flow of springs and streams, including among other, the major Tigris and Euphrates Rivers, the vital water sources for large portions of the Middle

East (Bozkurt and Sen, 2013). Out of 22 Arab countries, 18 are falling below the water poverty line of $1000\,m^3$ per capita in 2014 (WWAP, 2017). The drought and resulting water scarcity put more and more pressure on water resources leading to overexploitation, depletion, pollution, and degradation of water quality. Furthermore, the drought has led to reduced agricultural production and food shortage, which in turn caused abandonment of land and massive influx of rural migrants into already-crowded urban areas (Wrathall et al., 2018). This situation has increased the number of conflicts, followed by the erupted "Arab Spring" in 2011, and the civil war in Syrian, Iraq, and Yemen which provoked migration and dislocation with widespread consequences (Schuyler and Lauren, 2016), whose end, in early 2018, has yet to be seen.

2 DROUGHT IMPACT ON AQUATIC ECOSYSTEMS

2.1 Surface Water and Aquatic Ecosystems

The drought, which affect the hydrological cycle is also a factor in the deterioration of water quality, altering the physical and chemical composition of water. The reduced water flow and the drying of springs and other water bodies enhanced the impact of nutrient loads which derive from agricultural runoff, domestic and industrial effluents, and atmospheric deposition containing microbes, nutrients, heavy metals, and organic chemicals, substantially degrading the water quality (Sivan et al., 1998). The reduced flow decreased the dilution effects, causing an increase in the concentration of contaminants, including changes in DOC, pH, hardness, and DO and processes such as the redox potential, chelation, complexation, digestion, prey and grazing by zooplankton, protozoan and algae grazing influenced the assimilation of elements (Markel et al., 2014). As the result, the water bodies are exposed to excessive contents of suspended solids, turbidity, particulate matter entrainment, salinity, and even toxicity, which render the water unfit for human and animal consumption.

The increased concentrations of various pollutants impair the proper functioning of flora and fauna within the various aquatic ecosystems, greatly affecting important services of the ecosystems including biodiversity, flood and soil erosion control, food, carbon sequestration, fisheries, game, foraging, recreation, cultural, spiritual heritage, and research, among other (Gal, 2017; Rüdel et al., 2015). Mitigation measures are essential to restore the functioning of the ecosystems, seeing resilient ecosystems as a key priority for conservation and restoration of aquatic ecosystems, as discussed in the following (Willis et al., 2018).

River water quality. The rivers are part of the natural ecological systems, serving as a source of water for all purposes, as well as for recreation, moderating, and control flood events, in the surrounding watersheds and communities. However, as drought persists and the water flow is reduced, the rivers and streams become depleted and exposed to hazardous substances, heavily deteriorating the water quality (Abramson et al., 2010). In parallel, the excessive abstraction of water upstream and the turning of rivers into dumping sites for agricultural and other waste are severely damaging the quality of the remaining water, the ecosystem, and the wet habitats (Arnon et al., 2015; Shaofei et al., 2017).

These impacts, as already apparent, are greatly affecting the major rivers in the region, including, the Euphrates River, which until the 1990s, has discharged an annual quantity of about 32 billion m^3 of water to Syria and Iraq. However, currently because of drought and the development of the Turkey southeastern Anatolia Irrigation Project, only 17 billion m^3/year are available, shared between Syria and Iraq.

The reduced flow has greatly affected the water quality and the amount of water that is available to farmers and as the result it has restricted land cultivation downstream (Glass, 2017; Issa et al., 2014). Similarly, the reduced flow in the Litany River, in Lebanon, and the Lower Jordan River, shared by Israel, Jordan, and Palestine, has reduced the biodiversity and the natural functioning of the aquatic systems and the surrounding oases and wetlands, to become the recipient of untreated agricultural, industrial, and domestic wastewater (Boelee, 2011; David et al., 2014).

Lakes water quality. The persistent drought is affecting the state of water in lakes and reservoirs, drastically reducing the volume and the quality of water. The water level is approaching a historic low and expected to continue to decline, necessitating the ceasing of pumping of water from the lakes and the reservoirs.

In a representative lake in the north east of the Levant, the Sea of Galilee, draining a basin extending on an area of 2730 km^2, the water level has drastically dropped by almost 2 m below the lowest permissible water level, to −214.87 m bmsl. While in 1985–2008, the water recharge volume has dropped from an annual average of 630 million m^3, to about 500 million m^3 in 2009–2016, a decrease of an average 8 million m^3/year ($P=0.0001$) (Rimmer and Givati, 2014; Israel Water Authority, 2018a,b,c).

In parallel, the low water level and the slow turnover of water within the water body (the "water age") have caused a salinity increase (Zabel, 2016), from 210 mg/L, in 2010, to over 320 mg/L chloride in 2017 (Fig. 3).

The drastic drop in the water volume in lakes and reservoirs has had a far-reaching implication on the chemistry and biology of the water bodies, threatening the integrity of the natural ecosystems and their biodiversity (Rimmer and Givati, 2014). The reduced inflow leads to higher concentrations of biodegradable organic

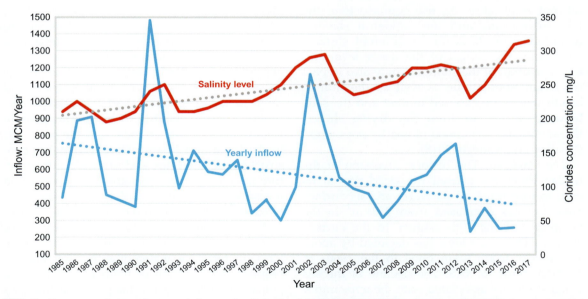

FIG. 3 Representative yearly water inflow and salinity level 1985–2016 (South Levant—Sea of Galilee). *Source: Compiled from: Rimmer, A., Givati, A., 2014. Hydrology. In: Zohary, T., Sukenik, A., Berman, T., Nishri, A. (Eds). Lake Kinneret: Ecology and Management. Heidelberg (Germany): Springer; pp. 97–112. Chapter 7; Israel Water Authority, 2018. Rainfall Data. http://www.water.gov.il/Hebrew/ProfessionalInfoAndData/Pages/rainmap.aspx (accessed 16.04.18); Israel Water Authority, 2018. Lake Kineret Monitoring Data. http://www.water.gov.il/Hebrew/ProfessionalInfoAndData/Kinneret-monitoring/Pages/default.aspx (accessed April, 2018); Israel Water Authority, 2018. Water Consumption Report, 2016. http://www.water.gov.il/Hebrew/ProfessionalInfoAndData/Allocation-Consumption-and-production/Pages/Consumer-survey.aspx (accessed 20.04.18).*

material, in terms of biological oxygen demand (BOD) and chemical oxygen demand (COD) followed by excessive release of nutrients to cause dense algal bloom and decreased dissolved oxygen (DO), resulting in hypoxia in the water layers, termed eutrophication (Kosten et al., 2012). The harmful algal blooms (Cyano-HABs)—blue green algae—emit foul odor and produce toxins which harm the ecological balance, indicating that the lake's natural "buffering capacity" may be in danger (Starkl et al., 2013). The deterioration process is normally gradual, but at a certain stage, it could become irreversible and difficult to restore the lake to the normal state.

The unprecedented record of the low level of water in lakes and the retreat of the shoreline also exposed the fish breeding littoral zone, which provide refuge habitats and feeding grounds for aquatic organisms and fish (Gaeta et al., 2014; Jeppesen et al., 2015) having a significant implication on fish reproduction, the lake food chain, and the ecological stability (Cummings et al., 2017). In different climatic conditions, the retreat of the water level and the exposition of the shoreline may also impact the avian populations as well as the human health which are exposed to dust and the salts blowing in the air from the drying shores, threatening the livelihood of nearby population (Bradley and Yanega, 2018).

The Dead Sea. The historic diversion of the Jordan River has caused the Dead Sea, at its end, to lose much of its water and its surface area that has decreased from $950 \, km^2$ in the beginning of the 20th century to the current $392 \, km^2$. The water level has declined by about 1 m a year from 394 m bmsl in the 1960s to 429.04 m bmsl in 2015. The decline of the sea water level was associated with the appearance of hundreds of sinkholes along the shoreline, ripping up buildings, beaches, and sections of the north–south artery along the seashore (Sim, 2015).

The Mediterranean Sea. The increase in the sea water temperature and evaporation have affected the salt balance and harmed the natural habitat such as coral, oysters and mussels, and other species that cannot tolerate and withstand the new heat conditions (Boero et al., 2008). The coral reefs, the iconic habitat of the sea ecosystems are sensitive to rising seawater temperature, causing coral bleaching and death due to the disappearance of the symbiotic algae (Hughes et al., 2018). The new conditions give room to invading species, from the Red Sea and the Indian Ocean, which are accustomed to heat and able to survive the warm temperature (Yeruham et al., 2015). In parallel, the sea water level has risen by 13 cm, between April 1992 and March 2014, almost double the global average (Church et al., 2013), showing a sharp surge over a short period (Hansen et al., 2016).

The higher seawater level as observed and predicted (Church et al., 2013; Hansen et al., 2016) is likely to cause sea water intrusion into the coastal zones, flooding on-shore installations, destroying roads, power, desalination plants, etc., and consequently, leading to salinity of the coastal aquifers, the lack of drinking water and rapid spread of epidemics and displaced people. Building coastal defenses such as the sea gates in Rotterdam, the Netherlands, and the sea walls in Australia and Miami, Florida, denying construction of buildings and strategic infrastructure in coastal areas are some of the protection measures that could avoid potential damages to coastal infrastructure (Dronkers et al., 1990).

Restoration and management of aquatic ecosystems. Avoiding the deterioration of the aquatic ecosystems, in terms of water quantity and quality, requires the adoption of national and regional policies, which recognize the principle "nature's right to water," and the need to secure adequate flow of water to maintain the aquatic habitats even during the drying water cycle (Adams, 2012). To conform with the RAMSAR Convention for the Rehabilitation and Conservation of Water, the authorities are required to apply mitigation measures including the demolishing of weirs and diversion structures, to circulate water from the downstream back to the upstream, and as necessary, to restore the water flow for the benefit of the public recreation and nature conservation (WWAP, 2018).

Under severe water scarcity, sustainable management of water bodies should be governed by the following principles:

- Conserving the streams, restoring, and maintaining environmental flow, flash floods, sediment flushing, re-meanders, and fish passages.
- Treatment of drainage water, Installing treatment facilities, green filters, buffer strips, and forest plantations, and diversion of wastewater and other phosphorus-rich waste, which influence the phosphorus and nitrogen balance, inducing the growth of poisonous blue algae (Markel et al., 2014).
- Preserving the shorelines. Maintaining the inundated vegetated littoral habitats and the fish breeding grounds, to avoid the depletion of fish stocks (Cummings et al., 2017; Zohary and Gasith, 2014) and applying wind abatement techniques (Bradley and Yanega, 2018).
- Involvement of water users. Educating the public about the benefits of healthy ecosystems, applying the "polluter pays" principle, to collect funds to invest in treatment, operation and maintenance of the water bodies, and ensuring that fees charged to water users are exclusively used for their purpose.
- Cost-effective investments, including monetary and nonmonetary incentives, for the communities to adopt sustainable cropping and other land use practices, to support conservation and restoration of the ecosystems.

- Studying and monitoring the limnology. Regularly collecting and analyzing meteorological, water balance, hydrology and hydrodynamic data, water quality, and other parameters that may have direct or indirect environmental effects (Hauck et al., 2015).
- Mediterranean Sea monitoring. Studying the impact of oil and gas exploration, power stations and desalination plants, maritime transportation and wastewater on the seawater quality, as well as carbon dioxide, particulates deposition, biological pump, and the food chain.

2.2 Groundwater Resources

State of the aquifers. Groundwater aquifers are encountered at variable depths below the ground and can extend over areas ranging from a few dozen km^2 to enormous stretches of land, ignoring administrative or political boundaries and many are of a transboundary nature. The groundwater quality is influenced by geogenic and external sources that can contaminate the water body as a result of dissolution of the geological formations and/or because of infiltration and movement of saline water as well as anthropogenic pollution from above the surface, including leachate of organic and inorganic substances, chloride-nitrate, metals, organic substances of industrial origin pesticides, fuel, perchlorate, and explosives from industrial sources. Groundwater aquifers are generally recharged by rainfall, but owing to the lasting drought of the recent years and consequently, the low recharge and excessive abstraction pattern, the state of the groundwater aquifers has greatly deteriorated. The low water level led to seawater intrusion, in coastal aquifers (Shevah, 2017) and/or to wastewater and chemicals contamination that may be too costly to treat or impossible to reverse (Grimmeisen et al., 2016).

In this context, the coastal aquifer, shared by Israel and Gaza Strip, is recording a significant drop in the water level, below optimal exploitation levels, and degraded water quality. The aquifer is subject to infiltration of sewage, run-off, fertilizers, pesticides, and landfill leachates, as well as from seawater intrusion, causing irreversible damage to the southern parts of the aquifer (Larsen, 2014). Weak regulations and overall low capacity to regulate and enforce regulations further exacerbates the issue endangering the health of Gazans and Israelis alike (OCHA, 2015). Low replenishment and excessive exploitation regime of the coastal aquifer caused a steady increase in the concentration of nitrates and chlorides, in 63% of the samples, much above the recommended threshold of 50 and 250 mg/L, respectively, rendering the water unsafe for drinking according to WHO standards. Other contaminants of industrial sources were found to be negligible. Rehabilitation of the contaminated wells is underway to restore the production and delivery of the water supply (Reshef, 2017).

In Jordan, the drilling of hundreds of wells contributes to the depletion of water and to excessive salinity of Jordan northern Sirhan, Hamad, and Azraq aquifers, while high levels of nitrates, phosphates and chlorides were observed in Mafraq, Agaba, and Ramtha aquifers, and in Jafr Dhulia Azraq and Agib aquifers, in the Amman-Zarka region (Alraggad et al., 2012). In Saudi Arabia, the depletion of aquifers and deterioration of groundwater quality are already apparent, in two-thirds of its groundwater supplies (Alrehaili and Tahir, 2012; Devlin, 2014), while in the United Arab Emirate, the ground water table level is dropping by 1 m per year (Al-Otaibi, 2015). In Yemen, many wells have dried up, and groundwater salinization and pollution are worsening as aquifers are exhausted, in a country, which is unable to invest in desalination in the same way as it is done in the wealthy neighbors (Mumssen and Abdulnour, 2015).

Groundwater management. To preserve the aquifers and water quality, a combination of activities can be applied, including inter alia, optimal operation, restricting drilling of wells and limiting water abstraction, as well as active recharge, infiltrating storm and runoff water, to preserve the aquifer and to establish strategic water storage for conjunctive use during period of scarcity, and regulation of the national water systems (Alrehaili and Tahir, 2012). Good management may also include dedicated monitoring systems, sending signals of any change in water quality, detecting pollutants, and treatment to avoid the spread of pollutant lenses. Survey of water wells in the vicinity of suspected contamination sites, to assess the extent and the spread of contamination near gas stations, tank farms, refineries, and oil pipe lines and major industrial parks.

Groundwater studies. Past trends and the current status of the aquifers are investigated, focusing on seawater intrusion and infiltration of anthropogenic pollutants. Novel methods (e.g., nontargeted analyses and bioanalytical tools) and integrated approached including in vitro, in vivo, and in situ investigations are conducted for a comprehensive assessment of negative ecological and human health effects. Risk assessment studies focusing on the reduction of the occurrence of organic micropollutants and emerging substances in water bodies are applied to quantify the level of contaminants and to evaluate mitigation processes, including mathematical models and employment of advanced treatment techniques.

Modeling tools integrating social, economic, and biophysical data are used, displaying management options for users, policy makers, researchers, and planners to assess the would be impacts of different management practices. Geophysics, graphical methods, numerical techniques, and finite difference modeling like MODFLOW and statistical analysis are used to simulate the groundwater conditions under different abstraction scenarios, making quantitative assessment, and hydraulic estimation as well as the evaluation of the chemical quality (Alfaifi et al., 2017).

Appropriate modeling tools can be adjusted to study:

- Water Balance/Hydrologic Models
- Hydraulic/Hydrodynamic Models
- Water-Quality Models
- Other Specialized Modeling
- Groundwater Systems Modeling

Groundwater monitoring. Data collection is the first and the most important step in acquiring an understanding of aquifer systems. Data is assembled using:

- Mechanical drilling to obtain stratigraphic information, geophysical soundings and exploration, and aerial or satellite photography to draw geological, geometrical, and structural contours;
- Drilling of (observation) wells, test pumping and tracer tests, and spring-flow and aquifer-level measurements to determine the hydrogeological parameters;
- Geochemical logging in wells and sampling for physicochemical, isotopic, and chlorofluorocarbon (CFC) analyses to establish the geochemical and mineralogical characteristics of the subsurface and of water quality.

Manual and automatic sampling stations are used to collect representative samples that are subjected to bio-physicochemical analyses. The measurements include the flow rates (natural and pumped) and static water levels, as well as parameters related to water quality (Pennequin and Foster, 2008), enabling the monitoring of

- the load of nutrients and other substances entering the water body,
- the changing trend in water quality due to natural processes or by varying operating conditions,
- the occurrence and movement of anthropogenic pollutants to predict

anticipated contamination of water pumped from the production wells, and
- the definition of water treatment requirement for specific pollutants to allow continuous operation of the boreholes and supply of water after treatment.

To control the quality of groundwater used for drinking water and to safeguard the public health, in compliance with the public health regulations. Groundwater sources are also monitored to ensure that the abstracted water is free from disease-causing microorganisms, as well as from chemicals that can affect human health or cause aesthetic taste and smell (WHO, 2017). As the case may be, point-of-use (at the tap) and point-of-entry (at the house) treatment systems, including reverse osmosis (RO), ion exchange, and distillation systems are used to reduce organic and inorganic contaminants (Hisham and Al-Najar, 2011).

3 IMPACT ON WATER RESOURCES AVAILABILITY AND QUALITY

3.1 Prevailing Water Scarcity

The annual volume of renewable water in the Levant region is $<200\,m^3$/capita, showing a down trend due to population growth and the economic development (8th World Water Forum, 2018) and the persistent drought which makes the situation worse, sparkling conflict within and between the countries that rely on the diminishing water resources for survival (Damania et al., 2017). As the result, the region is facing greater challenges to secure the demand for food, water, and energy, facing the climate change, desertification, and shrinking forests (FAO, 2011).

In Syria, the reduced flow of the Tigris and Euphrates Rivers, the depletion of the Khabur River in northeast Syria and the drying of the multipurpose Tishrin Dam, in northern Aleppo, among others, have caused the people in Eastern Syria to lose their livelihoods and to face extreme hardship (De Châtel, 2014). Further, inefficient water management and the civil war have enhanced the displacement and environmental migration of the rural population [UN Office for the Coordination of Humanitarian Affairs (2017), 2018].

Jordan, with just 92 m^3 of freshwater per capita, in 2014, is unable to meet the water demand of its 7 million population and close to 3 million refugees. Because of the intense and prolonged drought the available stock has dropped to about 20% of the storage capacity of 333 million m^3, adding to exhausted aquifers and bad water management (Hadadin et al., 2010). In Gaza Strip, the over exploited coastal aquifer, high above the safe pumping rate (50–60 million m^3 year^{-1}), is depleted and water quality has deteriorated to below the WHO standards for drinking water (Isaac and Rishmawi, 2015; UNOCHA, 2015).

In Israel, the hydraulic infrastructures built upstream by riparian states and the drying of rivers that once flow to the Mediterranean Sea, pose a serious threat to the ecology and the ability to secure regular water supply. The water shortage is alleviated by a large water reuse and seawater desalination programs which contribute close to 70% of the water demand, including the supply of agreed amounts of water to the Palestinians (72 million m^3) and Jordan (52 million m^3) in 2016 (International Water Association (IWA), 2018).

In Saudi Arabia, in Riyadh, the capital city and home to about 6 million people, natural groundwater resources are unable to meet the population water demand and most of the supply is met by water produced in desalination plants. With a disproportionate domestic consumption rate of up to 250 m^3/capita/year, Qatar suffers from wasteful and inefficient use of water, as well as excessive use of water for irrigation of up to 90% (Qatar National Development Strategy 2011~2016, 2011). In the neighboring country, Yemen's ablity to supply water is restricted by weak technical, administrative, and poor management of water resources <50%

served with piped water (UNOCHA, 2014). Yemen is also unable to invest in desalination plants in the same way as its wealthy neighbors (Mumssen and Abdulnour, 2015).

Against this background, unreliable service, deferred maintenance, and vulnerabilities of inadequate water supply as apparent, have contributed to humanitarian crisis, across the region and need to be altered. The Middle East nations have to put water security on top of the economic development agenda and to realize that growing regional water demand cannot be satisfied by intensified abstraction of the already overexploited sources, or by the harnessing additional, unattainable scarce resources.

The region, as the whole, have to understand the inherited linkages among water, energy, and food production and to adopt strategies and policies as required to incorporate new technologies, supported with educational and capacity building initiatives. Changes in attitude and new policies are required to be able to meet urban and rural water demand, as well as the water needs to preserve the aquatic ecosystems and the related social and natural services, as described in the following.

3.2 Water Management Principles

The existing water utilities within the region are diversified in terms of population size, geographic location, and economic status, while the changing climate and the unpredictable rainfall coupled with rapid urbanization are increasing the frequency of flooding and droughts and other water-related crises across the region. In accordance, the water authorities are required to adopt a water resilience framework for better preparing and responding to shocks and stresses to the respective water systems, to be able to identify and diagnose water challenges, and to monitor the technical, the financial, and the skills gain performance against prescribed targets, reaching planning, and investment decisions, under a good governance setting.

Proper planning and management decision-making, under uncertain and varying climate conditions and persistent drought, require the development and application of restructured management systems (Serrao-Neumann et al., 2018). This is crucial for effective short- and long-term drought adaptation strategies, to guide and organize the management to deal with the changing climatic conditions, forming a collective action taken by all actors involved in the management of water utility (Bressers et al., 2013). Good governance follows the transparency, accountability, and participation principles encouraging the involvement of civil societies and consumer groups, financiers, and policymakers in order to secure the ever-increasing demand for water as against the diminishing supplies.

This is in line with the new holistic management systems that were suggested to replace traditional fragmented management systems including IWRM—Integrated Water Resources Management (2009); Paris Agreement (2015); UN 2030 Water and Sanitation Development Agenda and Nature-Based Solutions (NBS) 2015. All represent major milestones for addressing sustainable management of water resources, as described below.

Integrated water resources management (IWRM). Under the IWRM, the development and management of water, land, and related resources are coordinated in order to maximize the resultant economic and social welfare in an equitable manner without compromising the sustainability of vital ecosystems (Integrated Water Resources Management, 1998). IWRM is a sound platform for the implementation of policies, plans, and laws for the control and protection of the water resources integrating investments, financial, economic, and institutional instruments in water management (Ingold et al., 2016). Water is seen as an economic good subjected to cost-benefits analyses, while securing the sustainable management of available surface and groundwater resources. Water reuse and materials recovery are also accounted to protect the environment and to sustain the quality and usability of aquatic ecosystems.

UN 2030 Sustainable Development Goals (SDGs). Of the 17 Sustainable Development Goals of the 2030 Agenda (UN Sustainable Development Goals (SDG), 2015), SDG6—Water and Sanitation—is dedicated to ensure availability of water and sanitation for all, providing effective management that would ensure access to safe and affordable drinking water and sanitation while protecting the environment and maintaining the integrity of the aquatic ecosystems. For the SDG 6 targets to be achieved in the region, coordinated efforts and close collaboration between governments and social and business communities at a regional level is required.

UN Nature-based solutions (NBS). Succeeding the SDG 2030 Agenda, the new NBS management principles (WWAP, 2018), in respect of water supply and sanitation services, emphasize also the conservation, rehabilitation, and/or the enhancement of natural processes in ecosystems in addition to the social, economic, and hydrological gains, based on natural processes, while reducing waste and environmental pollution, through reuse and recycling of waste. The NBS are central to achieving the 2030 SDG Agenda, playing an essential role toward the circular economy. The NBS address potential water-related disaster risks while preparing for a better response to the impacts of climate change and water management challenges. NBS enhance water availability (e.g., soil moisture retention, groundwater recharge), improve water quality (e.g., natural and constructed wetlands, riparian buffer strips), and reduce risks associated with water-related disasters (e.g., floodplain restoration and green roofs).

3.3 Water Management Tools

Decision support systems (DSS). To support water resources planning (e.g., watershed investment planning) or real-time systems operations (e.g., dams, flood management) and other needs for decision-making. DSS are useful for day-to-day operational and long-range strategic decision-making and in predicting the outcomes of alternative plans and programs, including developing different scenarios for policies and management plans, choice of priorities and optimal use of measures and available resources, setting financing options (Newman et al., 2017). In this context, together with expert systems and multicriteria analysis models, such as the Saaty's method (Saaty, 2008) DSS can help in the selection of appropriate and sustainable responses to water crisis.

Public participation. Under the current unprecedented water scarcity, engagement of local organizations and beneficiaries/stakeholders in the decision-making process and the day-to-day management is highly indispensable. The authorities are required to publicly recognize the communities and individuals who contribute to the common good of conserving and using water wisely, changing consumer behavior, and reduce water usage.

Public-private partnership (*PPP*). The water authorities are also required to encourage PPP agreements in addition to outsourcing of contracts for operation and maintenance of water supply and sanitation systems. Under the PPP, the private-sector invests in construction and management of water works, opening new sources for financing and costing systems, on commercial terms, facilitating an efficient operation of the water supply systems.

Assets management systems (*AMS*). Asset management systems are available to support the water utilities in optimizing the life cycle of the assets, the costs, and risks, balancing between operational and financial performance along the asset life cycle. AMS allow the analysis of risks as a basis for planning, installation and pricing of maintenance and upgrading of the various components (Water Finance Research Foundation, 2016). Advanced software for the assessment of the current state of the facilities is available, identifying the assets critical and in need for replacement, including:

- Capturing inventory and data/attributes
- Asset condition/performance and residual value
- Asset valuation (current value and replacement cost)/life cycle cost and depreciation
- Target level of service/failure mode analysis
- Operation and maintenance activities
- Investments prioritization according to business risk, and
- Funding strategy

Geographic information systems (*GIS*). GIS offer mapping of infrastructure/underground network, hydraulic modeling and decision analytics approach, enabling near real-time visibility into asset usage across multiple sites. GIS mapping applications create a complete, concise picture of a water network situation, illustrating relationships, connections, and patterns of spatial data, across organizational boundaries. Combined with sensors and devices, GIS provide warning signals reducing unplanned downtime and increasing operational efficiency, extending the useful life of equipment, improves return on assets and defers new purchases. GIS are widely used to organize geographically referenced data generating electronic maps available on the Internet. Visualization technologies that display the results for decision-support models are incorporated allowing users to search by land type, land use, management options, settlement patterns, land ownership, or planning zones.

3.4 Water Resources Data Management

Considering the growing challenges facing water resources availability, access to information concerning the status and evolution of the water resource and its uses is a crucial component of water resources management. Reliable, up-to-date, and pertinent information is essential to make sound decisions on regulation, planning, adaptation to drought, climate change, risk management, and public information at a local, national, and transboundary level. The data is also required to assess the state of the resource, and how they can impact the result of the use and development of resources, land use practices, and climate change, among other variables.

However, within the region, reliable datasets on water resources is scarce, especially in the less industrialized states, where monitoring systems and data generation capacity, the procedures, and tools for the production of information are lacking or collapsed as a result of social unrest (Voss et al., 2013). Where available, data are often produced and managed by different organizations, with little coordination among themselves, resulting with multiplicity and inconsistency of data and information. The data may be heterogeneous but not comparable and lacks standardization and insufficient for meaningful analysis. On the regional scale, the lack of timely, relevant, and credible information and indicators complicates the efforts to accurately pin down the volatility and the gravity of the drought and its impact on water resources. Because of lack of data, crucial planning decision are taken with partial, insufficient and imprecise data and information, leading the riparian parties to frame their position on ideological or political terms and not on actual facts, disregarding how one sided actions would affect the shared water quantity and quality.

Data generation. Against this background, there is an urgent need to take action, to produce, process, and validate, and finally share and disseminate data to better understand the hydrological cycle and impacts of climate- and human-induced change on water quality and quantity. Data generation is required to support water management and states' efforts in implementing IWRM and SDG principles as well as other goals and targets that are linked directly or indirectly to water management. It is also essential to facilitate dialogue and trust among riparian countries sharing transboundary water for which knowledge transfer and exchange of well-organized and harmonized data and information is required to support joint decisions and monitor their effects on transboundary water.

To generate reliable data, substantial capacities are required for producing, accessing, processing, and data sharing and dissemination, making good use of the water-related data and information. These capacities are required to ensure that data in quality and timeliness are accessible to the individuals who need them and are protected against misuse and abuse (Oracle White Paper, 2013), in line with expansion of applications of advanced and computerized communication systems including satellites, internet, remote sensing objects, crowd sourcing, etc., as highlighted in the following.

Big data. Nowadays, big data analytics and digital capabilities and intelligence, as available, would provide comprehensive and actionable information than previously possible (Ernest & Young, 2014). The big data that can be obtained from the United States National Air and Space Administration (NASA), the National Oceanic and Atmospheric Administration (NOAA), and US Geological Survey (USGS) can be integrated to enhance the knowledge base and access to information on a regional scale, to generate accurate and openly accessible water data from various sources. The new World Water Quality (IIWQ) Portal, launched by UNESCO's International Hydrological Program (IHP), provides time-series data on turbidity, sedimentation, chlorophyll-a, organic absorption, and surface temperature, among other.

Spatial data (mapping). A combination of GIS, remote sensing and modeling techniques could help to collate both dynamic and static characteristics of the prevailing climate, hydrological cycle, land use, vegetation, evapotranspiration, soil moisture, and other parameters to create easily accessible maps, figures, and tables. The spacial data would provide the users with near real-time data, allowing risks and vulnerability assessments before stresses and conflict arise. In this direction, an initiative was launched by the Arab Water Council, NASA, USAID, and the World Bank to provide the Middle East countries the ability to assimilate GIS and remote sensing systems under the GRACE (Gravity Recovery and Climate Experiment) Project (García et al., 2016).

Remote sensing data. Traditional data collection systems are increasingly supplemented by the new drone/aerial/satellite and communication tools, including means for remote sensing (RS) and crowd sourcing. Sensors and telemetry are revolutionizing the monitoring landscape, enabling the collection of high-frequency data from remote locations with high accuracy (Zhang et al., 2017). Remote sensors play an increasingly important role in providing complementary data needed to confront key water challenges, enabling, among others, early warning monitoring of many quantitative and qualitative parameters of surface water and groundwater, including the likelihood of algal blooms and other threats to the quality of water supply systems, land use practices and nonpoint-sources of pollution.

Crowd sourcing. Mobile phone technology creates great opportunities to increase the awareness of water quality in rural and urban communities, empowering citizens to collect information about the quality of water they use. The citizens may use their mobile phones to report the collected information to a control room, where it is processed and shared using messages, websites, and social media. The crowd provides a new way to inform the public and other stakeholders with information that is necessary for regular water management and delivery of safe water.

Blockchain for the water sector. The decentralized record and public ledger of information that are collected through the internet, allow the sharing of open source and open access data between users. Blockchain data is stored in "blocks" of information linked with other similar blocks of information, forming a chain. The chain of data blocks are considered reliable and trusted because the recorded data, after verification, cannot be altered without also changing all the blocks around it. The water sector may adopt the new technology to resolve the gaps in water data quality, quantity, and access, allowing data verification, communication and exchange, public participation, and transparency. Data collected and uploaded by institutions and monitoring systems is not kept in centralized database or ledger. Instead, many copies of the data are kept in different places on the network, to secure that no one piece of the record is changed without consensus from the rightful owners (Frøystad and Holm, 2015). New platforms such as Ethereum may increase scalability to enable the technology to potentially become part of nearly every digital process in the future.

Data Processing and access to data. Data and information that is generated by the various means as described above can be processed accessed through computers, tablets, and smartphones, using cloud computing, artificial intelligence, web services and internet of things (IOT). End users, even in the most remote locations, can have easy access to data and information that is understandable and adapted to their needs. The 3D visualization solutions can also enhance data interpretation and related impact.

3.5 Reorganization of Water Management Systems

Water demand management (WDM). WDM employs a set of complementary policy options, comprising regulatory, technical, public awareness, and economic incentives to optimize water allocation between the economic sectors and the environment. Water demand management aims to adjust demand with the available supply in an efficient manner, conserving and saving water, while considering socioeconomic, public health, financial, and environmental sustainability. Under water scarcity and persistent drought conditions,

WDM is considered to be far more important than trying to increase the supply-side which is finite and unattainable, getting harder to develop water resources to increase the supply side.

Main WDM principles that are, in particular, relevant for the arid Middle East are

- Developing drought early warning systems
- Adopting transboundary water resources allocation mechanisms, among the riparian states
- Reforming agricultural policies, food production, cropping, and irrigation systems
- Securing environmental and ecological sustainability, promoting water conservation and reuse
- Employing technical measures; metering, collecting data, monitoring and exchange of information
- Raising public awareness, capacity building, and meaningful stakeholder participation
- Expanding water services to vulnerable communities

3.5.1 WDM Supporting Measures

Water conservation campaigns. Effective water conservation campaigns and other communication and media measures make the water user aware of the need to conserve water, apprehending that conservation will reduce the risk of shortage. The resulting water saving expands availability of existing water supplies, while delaying or moderating the pace for new construction of expensive reservoirs and dams, dikes, pumping and water conveyance, and distribution systems.

Technical measures. Supporting conservation measures is the installation of water saving appliances, leakage and pressure control and other fixtures, including zonal and individual water automated meter reading (AMR) providing continuous and accurate real-time readings and water-quality alerts, making physical reading of water meters on-site, unnecessary. AMR detects burst and leaking pipes, alerting the consumer and or turning off the water flow, remotely transmitting household consumption and pressure readings to the data manager. New ultrasonic meters, IoT and cloud based, cellular-driven solutions are also being introduced to improve the efficiency of water transmission, distribution, and delivery systems.

Irrigation efficiency. In the region, irrigated agriculture, with more than 70% of the total use of water, is the largest water consumer, (Devlin, 2014). But, persistent drought and the lack of water for irrigation are affecting crop production, limiting food supply (Iizumi and Ramankutty, 2015) and consequently leading to desertification and destruction of rangelands in Syria, Jordan, and Iraq. To rationalize the use of water for irrigation, institutional and managerial reforms, considering the "true value" of water, are required to adjust irrigation water demand to the state of water scarcity, the competition with other users and the growing cost of obtaining additional water, (Briscoe, 1997; Zilberman and Schoengold, 2005).

Water charges. Market-based solutions, such as water's real value, cost recovery, and differential pricing that closely reflect the scarcity value of water shall be imposed as required to improve the efficiency of water use, eliminating free of charge or extremely cheap price which are responsible for the low water productivity rates and misuse of water (World Bank Group, 2016). Tiered tariff starting at a very low price, for a basic quantity of water, and increasing for a greater amount would modify water usage, reaching financial sustainability. Water is also considered as a "social good" that needs to be for low-income users, subsidizing or waiving the charge for very low-income groups.

Irrigation of farmland is also instrumental to ecosystem viability and environmental services, therefore, the price for water used for irrigation shall be attuned to the need for crop production to feed the growing population, alleviating food insecurity, and social instability (de Boer et al., 2016). Strong efforts are required to develop marginal water resources to substitute the lacking freshwater. Farmers are also in need for financial support to apply ultralow volume and least wasteful irrigation techniques and to avoid the irrigation of water-intensive crops, all in order to reduce water usage and increasing irrigation efficiency.

Virtual water. Virtual water trade in the form of food import to replace locally grown irrigated crops shall be considered to augment scarce water resources. Import of food commodities that require intensive irrigation such as rice, dairy, and beef represents a significant addition of virtual water to the meager water balance (Orlowsky et al., 2014).

3.6 Urban Storm Runoff Management

Cities are rapidly expanding and water resources are under increasing pressure, two challenges that need to be urgently addressed to ensure that cities are resilient to floods, droughts, and knowing of the growing water scarcity. By reason of the rapid urbanization process, the built-up areas are expanding to open fields, which become paved and asphalted grounds, preventing the infiltration of rainwater into the ground, increasing the risk of flooding before draining into the drainage systems, and eventually discharging into water bodies.

Generally, stormwater management is typically centered on flood protection and pollution mitigation without a focus on important hydrological aspects that can enrich the aquifers and augment water supplies, lessening the impacts of urban development on natural water cycles, seeing rainwater as an opportunity rather than a nuisance (Christopher et al., 2016). However, new approaches on urban design indicate that built-up areas can be structured to accommodate runoff management, adopting engineering measures that could support the preservation of runoff, allowing infiltration of the stormwater into the groundwater aquifers, transforming runoff from a nuisance to a resource that could benefit the water sector and the environment (Renouf et al., 2018).

Accordingly, water sensitive interventions able to improve the harvesting of rainwater and stormwater runoff control measures are suggested to detain rain events, delivering suitable flow in quality and quantity. Detention and vegetated infiltration basins, ponds, green walls, and roof gardens are constructed to reduce pollution (Blecken et al., 2009; Zinger et al., 2013). In addition, greening buildings, permeable paving, roadside green infrastructure, urban parks, and safe retention ponds were suggested by Carmon and Shamir (2010): based on the following design principles:

- Developing urban drainage modeling systems, addressing water quality and quantity, flood forecasting, risk analysis, and socioeconomic interactions to optimize the design of the drainage structures, considering the intensity of rainfall, the pace of the urbanization process, and the characteristics of the underline groundwater aquifers that could benefit from replenishment.
- Construction of stormwater structures able to attenuate the flood events and enrichment of aquifers and or direct use of water ("rain harvesting"), including wetlands and reforestation within urban environments to receive rain and stormwater, encourage landscaping and beautification.
- Revitalizing the rainwater harvesting concept, using innovative techniques and tools that can be applied in urban areas, helping to overcome current and future water scarcity challenges.
- Treating the runoff before reaching the receiving water to biodegrade or immobilize a range of emerging pollutants, before infiltration into the groundwater aquifers.

3.7 Engineered Solutions for Storm Runoff Management

Adequately designed engineered system can be incorporated in the urban design to preserve floodwater within the cities and in the

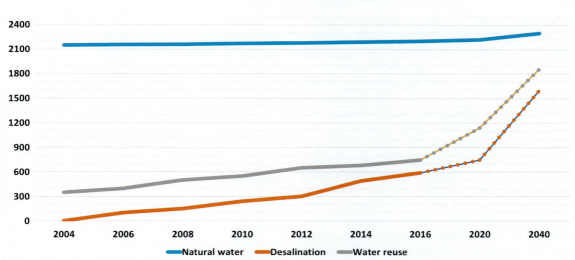

FIG. 4 Water reuse and sea water reverse osmosis desalination plan: 2004–2040 (Israel, Palestine & Jordan).

surrounding environment, reducing erosion and protecting the riverine. Such structures would attenuate peak flows, thereby reducing extreme risks from flash floods, providing storage capacity without occupying valuable open space above. New developments in permeable aggregate now make it possible to have water drain through porous, macadam surfaces of a parking lot, into purpose-designed underground storage space (USEPA, 2008). For the realization of efficient stormwater management, infrastructural, institutional, or governance systems are also required to encourage appropriate collective action based on the NBS guidelines, resulting with less impact on aquatic ecosystems, reduced loss of freshwater, and sustainable cities.

4 DEVELOPMENT OF NONCONVENTIONAL WATER RESOURCES

The continuous drought, and the increasing pressure on water resources magnify the impact of water scarcity that cannot be satisfied by the construction of additional infrastructure (e.g., dams, canals, water treatment plants, and other facilities), as it will not be economically feasible and will result in unsustainable systems and in further deterioration of the resources. Instead, the focus should turn into conservation, protection, and regulation of available resources, as discussed above, followed by the harnessing of nonconventional water resources (NCWR), including water reuse for irrigation and desalination of brackish and seawater to supply the urban and industrial sectors. The extension of water reuse and desalination capacity as is being pursued by Israel, Palestine, and Jordan (Fig. 4) would reduce the pressure and would conserve the natural water resources, while meeting current and future water needs of the expanding population, demonstrating a "new water culture," against the upward trend in water scarcity and persistent drought.

4.1 Wastewater Treatment and Reuse Systems

The Middle East region is ranked at the top of the arid and water scarce regions of the world, but, water reuse for irrigation purposes has been uneven and slow. In most of the region, wastewater is still discharged untreated into rivers, lakes, and oceans, creating health and environmental hazards (Koncagül et al., 2017). Connections to the sewerage networks and wastewater treatment facilities are generally limited (UNESCWA, 2014), and only 15% are reported to be reused in agriculture and industrial uses (World Bank and the Arab World Council, 2011). To safeguard the public health and to protect the environment, the region as a whole has to embark on a comprehensive wastewater treatment and recycling program, turning wastewater into a new resource, recovering water, energy, nutrients, and materials, toward building a circular economy. In parallel, the region is required to raise the public awareness and broader acceptance of water reuse, in line with the UN Paris Summit, 2015, aiming to accelerate the development of renewable energy and decarbonizing of the economies.

4.1.1 Centralized Wastewater Treatment Systems

Wastewater constituents: Municipal wastewater contains human waste combined with biological and chemical contaminants including nutrients, pathogens, and inorganic micropollutants, classified as:

- *Physical*: suspended solids.
- *Biological*: pathogens, microbial agents, and antibiotics compounds, including bacteria: *coli*, *Salmonella* and campylobacter; protozoa: *Cryptosporidium*, *Giardia* and various amoebae; viruses: enteroviruses, norovirus, rotavirus, and adenovirus; and other pathogens.
- *Chemicals*: pH, alkalinity, ions, metals, fats, oils, and grease (FOG), phosphorus, and organic and inorganic nitrogen in the forms of ammonia, amines, nitrate, nitrite, or more complex molecules. Other biologically active chemicals that are found are endocrine compounds (ECs) or pharmaceuticals and personal care products (PPCPs) and their

derivatives which have potential impact on aquatic and terrestrial ecosystems and human health (Shevah, 2014).

Physical and biological treatment. The wastewater treatment processes are generally categorized as preliminary and primary followed by secondary and tertiary treatment steps. The line of treatment includes a physical step intended for the separation and removal of suspended solids and particles, followed by a biological step for the decomposition and digestion of the residual solids. Activated sludge systems, combining biological and physical processes and forced aeration are used to induce suspended microbe cultures to degrade organic matter and convert the nutrients. The biological treatment stage is followed by a sedimentation stage designed to separate the sludge from the effluent using gravitational clarifiers (Metcalf and Eddy, 2014) or alternatively using immersed microfiltration or ultrafiltration membranes biological reactors (MBR) for further digestion of the bio-solids, to reduce the sludge fraction (Hai et al., 2013). The secondary effluent is subjected to a physicochemical step to remove remaining impurities and disinfection, to produce tertiary effluents suitable for unrestricted nonpotable applications. The results of the treatment process are harmless products, in the form of irrigation water, industrial process water, biogas, heat, electricity, nutrient-rich bio-solids for soil conditioning meet stringent quality standards, as well as public health regulations that can be reused and/or disposed to land and/or water bodies (WWAP, 2017).

Subsurface and surface storage of effluent. Until used, the treated effluent may be stored in surface or underground (artificial recharge) (Mor et al., 2015). The reservoirs, serve as large biological reactors, providing additional physical, biological, and chemical treatment, through chemical adsorption, ion exchange, precipitation of metals, and die-off of microorganisms, resulting in upgraded quality of the effluent

(Shevah, 2014). The tertiary effluent can be used indirectly or further upgraded for direct potable reuse (DPR).

Indirect reuse. Under indirect reuse tertiary treated effluent, can be reused for irrigation of crops, landscape or industrial uses before or after surface or underground storage. Major health, agronomic, and environmental benefits may include:

- Substituting freshwater that can be released for other potable uses
- Securing water for irrigation of crops and landscape
- Recovering of organic fertilizers, N, P, and K and other essential nutrients
- Contributing to the restoration of aquifers, rivers, wetlands, and other aquatic ecosystems.

Direct potable reuse (DPR). Wastewater effluent can be treated and upgraded to a quality that can be reused for human consumption (Mosher and Vartanian, 2015). DPR implies the use of advanced water treatment technologies (e.g., membrane filtration and advanced oxidation treatments) to remove viruses, bacteria, chemicals, and other contaminant residues that may be present in the effluent. DPR offers the opportunity to turn wastewater directly into drinking water. However, public acceptance and safety concerns may keep it from being a realistic technology. Further, in the case of Middle East where most of natural water is still used for irrigation (Droogers et al., 2012), indirect reuse is the most suitable option.

4.1.2 Decentralized Treatment Systems

Gray water systems. Properly installed systems for individually treating and reuse of black water and greywater systems near the point of generation reduce pressure on the use of potable water (Rozosa et al., 2017). Such systems minimize environmental footprints and provide opportunities for public education and awareness for a sustainable environment. Decentralized reuse systems, however, require separate plumbing and expose the public to health risks and could be costly, requiring additional infrastructure.

Wetland systems. Wetland systems are engineered ecosystems that can be used to treat wastewater, in urban and peri-urban settlements to produce effluent that can be used for the irrigation of green spaces within the neighborhoods (Zinger, 2015). Fundamental interactions among water, soil, plants, and bacteria inside the wetland treatment systems control, transform, and reduce the pollutants load (McNamaraa et al., 2017). Properly designed, constructed, and operated wetlands may perform well under certain environmental, social, and economic terms, providing sustainable and resilient solutions while providing added value to the landscape (Plevria et al., 2017), although, further research is required to fully realize the potential of wetland technology and its setting in densely populated areas.

4.1.3 Water Reuse Systems

Following collection and treatment of wastewater, reuse of reclaimed water is emerging as a well-established water management practice providing an attractive solution to climate change, contributing to the restoration of water resources, circumventing further depletion and degradation of water quality. However, the region policy-makers are yet to recognize the need for large-scale water reuse and build the lacking organizational and professional ability to collect waste and purify sewage and to establish the basic infrastructure to deal with environmental hazards (de Mes, 2014). The region has to embrace the experience that is already apparent in Israel, Jordan, Saudia Arabia, and others, setting the way for progress, globally (Fig. 5).

In Israel water reuse provides about 40% of the water allocated to irrigation and is projected to increase to 50% in 2025 and to 70% (900 million m^3) in 2050. Jordan water reuse forms about 10% of the total water supply, while

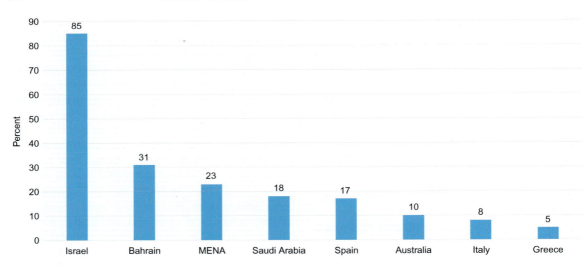

FIG. 5 Extent of water reuse in selected countries (2017) (in % of the generated wastewater).

in the Gulf Cooperation Countries (GCC) about 85% of wastewater is collected and 45% is used (WWAP, 2017, 2018). In Jordan, the Aqaba coastal city on the Red Sea was selected to apply Zero Discharge of effluent, taking proactive actions to improve water security (IWA, 2018).

Barriers to water reuse. The cost, the public opinion or "human perception" and the public fear about potential health risks are still a barrier to water reuse (Jiménez and Takashi, 2008), but the biggest barrier by far is for DPR. For indirect reuse, the mineral salts that may be found in excessive concentrations could harm the soil, crops, and the groundwater aquifers below the irrigation fields. To avoid any damage, the concentration of salts in the effluent shall be limited to 250 mg/L of chlorides and kept to a minimum possible.

Organic micropollutants. Similarly, residues of stable and biologically active micropollutants such as the personal care products (PPCPs) and the human and veterinary Pharmaceutical which are not adequately removed by the various treatment processes, may impact the aquatic ecosystems and the edible crops (Christou et al., 2017; Škrbić et al., 2018; Wang and Wang, 2016; Yin et al., 2017). However, the likely adverse impacts of the micropollutants on health and living organisms are inconclusive (Arnnok et al., 2017) and further research is suggested to verify evidence about their accumulation in soils and the ecosystems and especially their presence in food in particular (Larsen, 2014).

Regulation and quality standards. Wastewater treatment and reuse facilities are controlled to conform with health and environment regulatory requirements as needed for safe reuse and disposal of wastewater effluents, avoiding the pollution of streams, aquifers, and the sea. To clear obstacles to the use of the "new water" which could significantly reduce dependency on natural water resources, the effluent is required to meet the following quality guidelines (Ministry of the Environment (MOE), 2010; Mor et al., 2015):

- Biological oxygen demand (BOD5) < 10 mg/L
- Total suspended solids (TSS) < 10 mg/L
- Turbidity <2 Nephelometric Turbidity Units (NTU)
- E. coli <10 colony-forming units (cfu) per 100 mL
- Viruses/pathogens >5 log removal

4.2 Desalination of Sea and Brackish Water

With low rainfall and depleting water resources and the upward trend in water scarcity and persistent drought and in addition to conservation, protection, and regulation of available resources, as discussed above, the region has to turn to seawater desalination, as a strategic option. Water desalination has been practiced for more than 50 years in the region, accounting for most of the global desalination capacity and should be further expanded to become the primary response to water shortage, to satisfy current and future domestic water supply requirements of the expanding population.

The region desalination capacity. The region is already a world leader in desalination capacity producing 37.32 million m^3/day of desalinated water, out of the global capacity of 85.55 million m^3/day, or 44% of the world's capacity (Voutchkov, 2016), with 70% of the world's largest desalination plants located in Israel, Saudi Arabia, the United Arab Emirates, Kuwait, and Bahrain, (Desalination.biz, 2017), excluding the new Riyadh Ras Al Khair Desalination Plant (1.025 million m^3/day), in Saudi Arabia. The desalination capacity in the Gulf Cooperation Council (GCC) countries, having a total installed capacity of around 5000 million m^3 is expected to increase to around 9000 million m^3 by 2030 (Dawoud and Al Mulla, 2012). In Israel, desalination capacity of 1.5 million m^3/day in 2018, to supply about 80% of the domestic and industrial needs (Shevah, 2015 is to expand to about 3.5 million m^3/day by 2040 (Fig. 3).

Desalination technology. Two basic types of technologies are widely used to separate salts from ocean water, namely, distillation and thermal evaporation, including the multistage flash (MSF), multieffect distillation (MED), and vapor compression (VC), (Gebel, 2014) and the membrane separation technique—reverse osmosis (RO), in which sea water passes a semipermeable membrane under a pressure, typically 50–80 bar (RO Guide, 2015). Historically, cogeneration of water and power using distillation technologies have dominated the seawater desalination market, especially in the Arab Golf where thermal evaporation technology is of choice, due to access to low-cost fuel, but owing to technology development, RO technology is currently dominating the desalination markets. Improved membranes, reduced energy consumption, incorporation of energy recovery devices, to capture excess heat and pressure from the RO process, and reduced costs, enhanced the construction of flexible and modular brackish and seawater desalination plants, although, recently, forward osmosis (FO) is receiving much attention as potentially lower energy alternative to RO (Mazlan et al., 2016).

Environmental impact and mitigation measures. The desalination plants could produce the most needed water for the arid region, but despite the many benefits the technology has to offer, concerns are raised regarding the increasing dependency on a costly and intensive energy technology, instead of searching alternative solutions of water supply, having less social and environmental impacts. Key issues are related to the potential negative impacts on the environment from the discharge of concentrates and chemicals to the marine environment, the emission of pollutants and the energy demand of the processes, as described below.

Siting and marine environment. Desalination plants occupy coastal land, connected to underwater pipelines, potentially damaging the landscape, archeology, the marine and terrestrial ecosystems, and the environment, affecting ecologically sensitive coastal sites and aquatic creatures swimming near the deep seawater intake (Lokiec, 2013). Desalination plants also discharge concentrated brines, flocculants, and cleaning chemicals to the sea. A typical desalination plant producing 100 million m^3 year^{-1} of desalinated water discharges to the sea up to 535 tons per year of Fe and 40 tons per year of P and brine containing 73.5 g L^{-1} of salts, double the sea background level (Shevah, 2015). In the Gulf of Oman, during wintertime, HABs driven by climate change and growth-stimulating nutrients form a Red Tide present a serious threat to desalination plants, clogging filters, and fouling membranes, disrupting flow and reduce productivity (Anderson, 2014). The HABs also release harmful toxins that can be detected in the finished water. Desalination plants, like any industrial facility or infrastructure, may be subjected to technical failure and exposed to security incidents due to hostile and cyberattacks.

Energy consumption and emissions of greenhouse gases. The energy consumption of desalination plants is still relatively high, comprising 60% of the operation costs, generating greenhouse gases, emitting 4.7 g of sulfur dioxide and $286 g/m^3$ of carbon dioxide (Israel Electric Corporation, 2016), fueling the general water industry's perception that seawater desalination industry is inadequately viable (Schrotter et al., 2010). Although, in the recent years, as a result of the incorporation of advanced energy recovery devices and higher recovery rate, RO desalination plants have shown a significant drop from 4 to $5 kWh/m^3$ to about $3.5 kWh/m^3$.

Used membranes. In a typical $300,000 m^3/day$ capacity RO desalination plant, about 20,000 of 8″ membranes are installed. Because of fouling and the deterioration of the selective properties of membranes, their life cycle is short (7–10 years) and need to be replaced and disposed to landfills where they are grinded and buried. New membranes, having a longer life expectancy, made of carbon nanotubes and new cleaning methods, such as ultrasound radiation, are being investigated (Lawler et al., 2012).

Mitigation measures. To safeguard a sustainable use of desalination technology, the impacts of desalination project should be investigated employing environmental impact assessment (EIA) study and mitigated, avoiding some of the adverse impacts to the environment. In this context, the off-shore seawater intake is built as a submerged structure with slow intake velocity to avoid interference with the aquatic life and where possible, built close to power plants, where the brine and the backwash water are blended with the power plant cooling water and/or using diffusers laid on the sea bed to achieve a rapid mixing with the entire water column, resulting in lower salinity and lower temperature. New desalination plants are required to treat the filtration leachates and remove the chemical sludge to certified landfills. Long-term monitoring systems are employed to monitor any adverse impact on the sea environment.

Research and future desalination development. Sustainable desalination technologies capable of meeting future demand for freshwater are being investigated aiming to improve the efficiency and reduced environmental impact including reduced carbon emissions. 16″ membranes in a vertical array, nano-membranes, FO, and ion concentrate polarization and positive displacement pistons to increase energy recovery to 90%–95% are being introduced (Elimelech and Phillip, 2011; Namboodiri and Rajagopalan, 2014). In addition, ultrathin graphene-oxide membranes for desalination applications are being investigated as well as techniques to reduce mineral and organic fouling, biofouling, scaling, and energy consumption.

5 REGIONAL COOPERATION ON TRANSBOUNDARY WATER

A regional hydro-diplomacy approach coupled with scientific and innovative action could avert water crisis preventing cross border conflicts while successfully navigating the effects of climate change over increased water shortage. The applicability of the conclusions of the Conference of the Parties (COP 21), Paris 2015, adjusted to the "Water-Energy-Food Nexus" and the IWRM and NBS principles shall be evaluated, for the regional hydro-geographic basins, motivating the disadvantaged countries

5 REGIONAL COOPERATION ON TRANSBOUNDARY WATER

to benefit from regional cooperation to safeguard their water and food security. Challenging water allocation issues between upstream and downstream, water pollution, and water overexploitation, under severe water scarcity. Challenges that cannot be adequately addressed by the individual states, only joint regional efforts are the best to achieve sound transboundary and watershed management. Further, water can be a bridge between communities and countries as a means of building trust and peaceful dialogue between the neighboring states (Cosgrove, 2003). The cross-boundary nature of water resources can be a powerful entry point for peaceful cooperation rather than a cause of conflict, allowing water sharing between riparian states (Williams, 2018).

In the context of the Middle East, transboundary cooperation on water resources, agriculture, environment, and energy need to be institutionalized to reduce insecurity, minimizing the risks of water scarcity, food poverty, and energy shortage. The states within the region are required to move from being adversaries to the position of supporters in which the shared water resources are dealt with applicable instruments, regarding.

- Transboundary water management policies.
- Quantitative and qualitative baseline levels of transboundary water resources.
- Structured mechanisms for confidence building, sharing know-how and data exchange.
- Integration of aquatic ecosystems into sustainable water management.
- Applying cost-effective engineering solutions, offering benefits to the riparian parties.
- Setting a functional R&D on transboundary water resources.

On-going regional cooperation. The case of Israel, Jordan, and Palestine cooperation. In line with the above cooperation principles, instances of positive cooperation on the management of

transboundary water are already apparent in the region and significant initiatives have been made, developing trade-off mechanisms for water sharing under bilateral and multilateral agreements. However, much remains to be accomplished, leveraging on on-going cooperation, seeing water as an element for cooperation rather than a cause for conflict and confrontation.

In the case of the Israel, Palestine, and Jordan, the Oslo Agreement 1993 and the 1994 Jordan and Israel peace treaty (Israel Ministry of Foreign Affairs, 1994, 1995) specified water sharing and transboundary water agreements which allow the estimation of a unified water balance for the three states (Fig. 6) and the initiation of bi- and triparties transboundary water projects.

Restoration of the Lower Jordan River. The river and tributaries which criss-cross along the borders of Lebanon, Syria, Israel, Palestine, and Jordan, whose natural flow was fully diverted by the riparian states, the remaining flow which ends in the Dead Sea consists mainly of excess stormwater, sewage and fish pond and agricultural drainage. The lower part of the river, being rich of natural ecology, the route of migrant birds, pilgrimage and of economic potential, is now the subject for restoration by the three riparian states who has embarked on a comprehensive restoration and development plan aiming to creat thousands of tourism jobs, boosting agricultural output (EcoPeace Middle East, 2015).

East Jerusalem wastewater treatment plant. Israel and Palestine reached an agreement for the treatment of wastewater, draining the East of Jerusalem which is currently discharged to the Kidron River, polluting the river and the underlined aquifer. A wastewater treatment plant will be built to generate effluent suitable for use for irrigation by Palestinian farmers, while facilitating the development of scenic and touristic corridors along the river's historic sites (Bryant, 2013).

The Red Sea-Dead Sea Project. The sponsored project by Israel, Jordan, and Palestine aims to produce desalinated water at the Red Sea coast

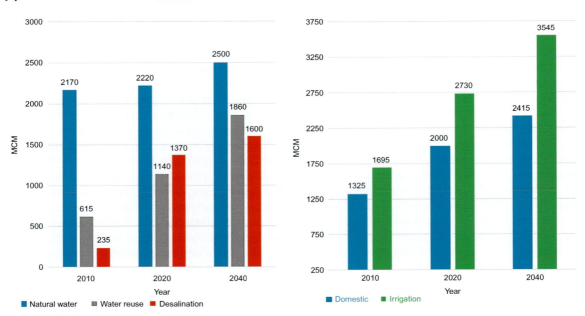

FIG. 6 Water demand and sources of supply 2010–2040 (Israel, Palestine & Jordan). *Source: author's estimates.*

and pumping the brine to the Dead Sea to halt the declining of seawater level in which because of diversion of the Jordan River water upstream and high evaporation, the seawater level is dropping at an accelerated rate of about 1 m/year to the current 428 m below sea level (Al Rawashdeh et al., 2013; Neugebauer et al., 2014). At its final stage, the project will transfer, annually, about 1 billion m^3 of brine to the Dead Sea while the desalinated water will be used locally to supply Jordanian and Israeli consumers in the south of the two countries, in exchange of water to be supplied by Israel to Jordanian and Palestinian consumers in the north (World Bank Group, 2012; Jordan Valley Authority, 2014).

Jordan-Palestine-Israel Solar Energy Water Nexus Project. Water, energy, and food sectors, which are inextricably linked, and any action in any one of them which usually impact the others, can be used to strengthen transboundary collaboration, to yield significant economic advantages (Jägerskog et al., 2013). In this context, the reliance on expensive energy for water production in desalination plants, built on the Mediterranean coast, in Israel, can be paired with generation of relatively cheap solar energy in the Jordanian desert (Bkayrat, 2015). The maturity of solar technology confidently determines its usage to supply the energy-intensive water production, to save costs. Such transaction would allow the region to enhance renewable energy economy, contributing to stability and long-lasting peace between the three parties while being socially, economically, and environmentally sustainable, significantly cutting down the emission of greenhouse gases.

6 IMPROVING THE KNOWLEDGE BASE

6.1 Capacity Building

Education, capacity building, and technology transfer are an essential element, as much as other system components, in order to work toward sustainable conservation and investment in advanced water and energy technologies, aiming to deal with water scarcity in the region. Training, research and development (R&D), and scientific efforts are required to better understand the ecological impacts and on how to mitigate the adverse effects of the persistent drought on the state of water resources as well as the natural ecosystems and the biodiversity in the region and beyond.

Effective capacity building is also essential to culture consultation and conflict resolution mechanisms of the relevant officials, within the riparian states, to be able to support transboundary water resources management, taking into account hydrological events and societal values, as well as avoiding causing harm to the neighboring states. Well-trained staff will be able to integrate advanced solutions for the management of the hydrological cycle without creating adverse negative consequences. International governments and agencies including Malta Conferences Foundation (MCF), International Union of Pure and Applied Chemistry (IUPAC), EcoPeace Middle East, and the Euro-Mediterranean Information System (EMWIS, 2018) (http://www.emwis.net) and other environmental activists and experts are active across the region, contributing to effectively develop workable institutions and trained staff to manage water resources. EMWIS gathers all available information with open and easy access to promote information exchange and dissemination, among the Mediterranean countries.

6.2 Research and Development

Significant knowledge gaps in understanding of biophysical processes and natural resources management in basins need are being rectified to fill the gaps in areas where research is missing analyzing processes and linkages within and between ecosystems, developing models to predict ecological and hydrological changes, as summarized below.

Rain enhancement. Rain enhancement studies provide a platform for the development of innovating techniques, aiming to improve the efficiency and predictive capabilities of targeted and effective cloud seeding operations intended for rain enhancement in dry regions (Delene and Shamrukh, 2016).

Water leakage detection and repair. Excessive water losses through leaky pipes, broken water mains and faulty meters can be prevented using sensors that monitor the utilities' water network. A sponge that moves through the pipe remove scale and slime, while fixing any holes detected in the pipe (Aquarius Spectrum, 2009). Supported with a video inspection equipment and acoustic measurements, the leak location detector sends SMS alerts, allowing repair without massive digging of roads and trenches.

Graphene. Single-atom thick carbon nanosheets of the innovative graphene nanomaterials may play a role in water treatment. Adsorption, filtration, and photodegradation processes are being investigated, exploring the full potential of graphene materials along with nano-toxicity and hazards potential (Machado et al., 2015).

Detection of microbial pollution. Greater awareness of the human value to the safety of drinking water, novel molecular methods enabling rapid identification and characterization of microorganisms (bacteria, viruses, fungi, and parasites) are being developed, moving from traditional culture-based methods that are widely used to determine the presence of disease-causing bacteria. Continuous and online monitoring methods based on biochemical, physiochemical, and enzymatic methods (ATP, DNA, and PCR) are being developed to support regulations that drive for advanced disinfection and oxidation technologies to eliminate micropollutants (Deshmukh et al., 2016).

Resources recovery. Wastewater treatment facilities that traditionally focused on the removal of pollutants are being seen as resources recovery systems aiming to cut treatment costs and reduce energy consumption, employing microbial systems to recover chemicals and nutrients to maximize the recovery of valuable resources (Matsakas et al., 2017; Puyol et al., 2017).

De-ammonification of the sludge return flow. Removal of nitrogen from the supernatant deriving from the sludge dewatering system before it is sent back for treatment at the biological reactor, may significantly reduce the nitrogen load by 20%–30% in the aerobic reactor (Jardin and Hennerkes, 2012).

Disintegration of the biomass. Vibration of the sludge flocs using high amplitude ultrasonic was found to enhance the breakage of the digestion of the flocs, at less retention time and less residual amount, while improving the dewatering and the handling properties of the digested sludge (Luo et al., 2014; Zawieja and Wolny, 2013).

Smart artificial intelligence, Internet of Things, and predictive tools. To capitalize on the changing market dynamics the use of smart technologies to support the operation of the water systems to control energy, chemical use, labor, to increase reliability and to safeguard public health and environmental quality are attracting the interest of the water industry (Water Environment Federation, 2017). Smart water networks and cloud-based control systems, based on the internet, the smartphone and the cloud, provide online platforms for continuous online information, through sensors, which automatically manage the water supply (Talari et al., 2017).

Cybersecurity. Cyber security is taking a center stage to prepare and protect the water systems from global threats. Critical assets of water supply could be a target for cyber hackers and other malicious adversaries requiring cyber security and means for preparedness, restoration of the service immediately after disruptive events, at the water utility level (Tehan, 2017).

7 CONCLUSIONS AND THE WAY FORWARD

Water scarcity. The inherited water scarcity in the Middle East, lacking sufficient water resources is being aggravated by the alarming impact of the global warming on the hydrological cycles, causing the climate of the Eastern Mediterranean, to become drier and hotter (Kelley et al., 2015). The global warming is pushing water scarcity to an all-time high, leading to irreversible depletion of the water resources and degradation of water quality. Strong intervention measures are required to alleviate the predicted adverse impacts on water resources in the years to come.

Regional challenges. The growing demand for water and sanitation, driven by urbanization and population growth, at a time of less predictable hydrological patterns, adds new challenges, hazards, and complexity to the water utilities, which are required to adapt new and innovative models of governance regarding policies, institutions, and financing mechanisms. Progressing from the MDGs to SDGs and beyond, avoiding the "as usual scenario," in which the demand for water is satisfied by the no longer feasible option of construction of costly and environmentally undesirable dams, conveyance systems, and other infrastructure.

Administrative solutions. Innovative water management approaches as highlighted under the UN IWRM, NBS, and SDG 2030 agenda, shall be adopted, to ensure sustainable high-quality water and sanitation availability for all. Emerging technologies and business models, adequately adjusted to the low-income-emerging economies and higher income countries in the region, alike, shall be adopted to reduce or shift demand for water, reducing losses, and increasing water use efficiency to reflect the true cost of water under scarcity. Reliable, timely, and accessible data is also required, for follow-up, review and report at the local and regional levels.

Nonconventional sources. In addition to effective use of available resources, the use of nonconventional sources shall be expanded comprising water reuse and desalination of brackish water and sea water which emerge as the optimal solutions to fill in the widening gap between demand and supply, representing a "game changer" and a method of choice to scale up water supply. Desalination is the most promising sector to supply high-quality water for drinking purposes, while water reuse and recovery of water, energy, nutrients, and materials which are technically feasible and practical on a growing scale, allow the inherited exchange of waste for material to provide the ground for synergies of the "water-energy-waste nexus" to enabling the development of the circular economy and healthy environment. Further efforts are, however, needed to raise public awareness and broader acceptance of using recovered resources, with due consideration to adversary environmental and public health impacts, changing perceptions and attitudes toward water reuse and desalination.

Regional conflicts and cooperation. The water scarcity is paired with a transboundary nature of rivers and aquifers, which have suffered from the violent conflicts, often with disastrous results for the resources, the environment and the society, as strongly observed in Yemen, Syria, and Iraq. To avoid further forced mass displacements and environmental migration, the region has to promote peace and stability as a key for human security. In this context, cooperation on shared water resources would play a vital role in initiating the essential conflict resolution mechanisms, as required to minimize disputes among riparian states, in general, and on shared and cross border water, in particular. To overcome political barriers and to regulate water sharing between the riparian countries, and securing ecosystems integrity, while implementing cross border water projects, involving the local communities and the private sector. Regional public-private-people partnerships, based on economic terms would yield mutual benefits for all, while exploring the "water-Energy Nexus" advanatges among riparian parties. Assisted by international science and technology transfer the region would be to upscale cooperation on knowledge hubs and capacity building to bridge across the existing disparity in economic and technical capability among the region, sharing experiences and knowledge, looking at the whole value chain to achieve water security for all.

Acknowledgments

This work was supported by the Division of the Chemistry and Environment, International Union of Pure and Applied Chemistry (IUPAC), and Malta Conferences Foundation (MCF). Their support is gratefully acknowledged.

References

Abramson, A., Tal, A., Becker, N., et al., 2010. Stream restoration as a basis for Israeli-Palestinian cooperation—a comparative analysis of two transboundary rivers. Int. J. River Basin Management 8 (1), 39–53.

Adams, J., 2012. Determination and implementation of environmental water requirements for estuaries. Ramsar Technical Report No. 9/CBD Technical Series No. 69. Ramsar Convention Secretariat, Gland, Switzerland.

Al Rawashdeh, S., Ruzouq, R., Al-Fugara, A., et al., 2013. Monitoring of Dead Sea water surface variation using multi-temporal satellite data and GIS. Arab. J. Geosci. 6 (9), 3241–3248.

Alfaifi, H.J., Abdelfatah, M.S., Abdelrahman, K., et al., 2017. Groundwater management scenarios for the Biyadh-Wasia aquifer systems in the eastern part of Riyadh region. Saudi Arab. J Geol. Soc. India 89, 669–674. https://doi.org/10.1007/s12594-017-0676-x.

Al-Otaibi, G., 2015. By the Numbers: Facts About Water Crisis in the Arab World. World Bank. Available at http://blogs.worldbank.org/arabvoices/numbers-facts-about-water-crisis-arab-world. [(Accessed 21 March 2018)].

Alraggad, M., Al-Saleh, S., Al-Amoush, H., et al., 2012. Vulnerability of groundwater system in Central Jordan Valley/pollution indicators and decontamination process. J. Water Resour. Prot. 4, 133–139.

Alrehaili, A.M., Tahir, H.M., 2012. Use of remote sensing, GIS and groundwater monitoring to estimate artificial groundwater recharge in Riyadh, Saudi Arabia. Arab J. Geosci. 5, 1367. https://doi.org/10.1007/s12517-011-0306-7.

Anderson, D.M., 2014. Harmful algal blooms (HABs) in a changing world: a perspective on HABs, their impacts, and research and management in a dynamic era of climactic and environmental change. In: Kim, H.G., Reguera, B., Hallegraeff, G., Lee, C.K., Han, M.S., Choi, J.K. (Eds.), Proceedings of the 15th International Conference on Harmful Algae. International Society for the Study of Harmful Algae, pp. 3–25.

Aquarius Spectrum, 2009. Multi-layer Leak Detection in Urban Water Systems. http://www.aquarius-spectrum.com/index.html#what-we-do. [(Accessed 21 March 2018)].

Arnnok, P., Randolph, R., Singh, R.R., et al., 2017. Selective uptake and bioaccumulation of antidepressants in fish from effluent-impacted Niagara River. Environ. Sci. Tech. 51 (18), 10652–10662.

Arnon, S., Avni, N., Gafny, S., 2015. Nutrient uptake and macro-invertebrate community structure in a highly regulated Mediterranean stream receiving treated wastewater. Aquat. Sci. 77, 623–637.

Bkayrat, R., 2015. Desalination is a thirsty business, but does solar provide an answer? December 20, 2015 Updated: December 21, 2015. http://www.thenational.ae/opinion/desalination-is-a-thirsty-business-but-does-solar-provide-an-answer. [(Accessed 20 February 2018)].

Blecken, G., Zinger, Y., Deletic, A., Viklander, M., 2009. Impact of a submerged zone and carbon source on heavy metal removal in stormwater biofilters. Ecol. Eng. 35 (5), 769–778. https://doi.org/10.1016/j.ecoleng.2008.12.009.

Blunden, J., Arndt, D.S., 2016. State of the climate in 2015. Bull. Amer. Meteor. Soc. 97 (8), S1–S275. https://doi.org/10.1175/2016BAMSStateoftheClimate.

Blunden, J., Arndt, D.S., 2017. State of the climate in 2016. Bull. Amer. Meteor. Soc. 98 (8), S1–S277. https://doi.org/10.1175/2017BAMSStateoftheClimate.

Boelee, E. (Ed.), 2011. Ecosystems for Water and Food Security. Nairobi: United Nations Environment Program. International Water Management Institute, Colombo, p. 194.

Boero, F., Féral, J.P., Azzurro, E., et al., 2008. Climate Warming and Related Changes in Mediterranean Marine Biota. Helgoland CIESM Workshop Monographs, CIESM the Mediterranean Science Commission, Monaco, pp. 1–17. http://www.ciesm.org/online/monographs/Helgoland08_ExecSum.pdf. [(Accessed 30 March 2018)].

Bozkurt, D., Sen, L., 2013. Climate change impacts in the Euphrates-Tigris basin based on different model and scenario simulations. J. Hydrol. 480, 149–161. https://doi.org/10.1016/j.jhydrol.2012.12.021.

Bradley, T.J., Yanega, G.M., 2018. Salton Sea: ecosystem in transition. Science 359 (6377), 754. https://isramar.ocean.org.il/isramar2009/Downloads/ISRAMAR_Report-H24-2016.pdf. [(Accessed 30 March 2018)].

Bressers, H., de Boer, C., Lordkipanidze, M., et al., 2013. Water Governance Assessment Tool With an Elaboration for Drought Resilience. Vechtstromen Water Board, Almelo, the Netherland INTERREG IVb DROP Project.

Briscoe, J., 1997. Managing water as an economic good: rules for reformers. Water Supply. 15 (4), 153–112.

Bryant, C.C., 2013. How Sewage is Bridging the Israeli-Palestinian Divide in Jerusalem. The Christian Science Monitor, Boston, MA. https://www.csmonitor.com/World/Middle-East/2013/1206/How-sewage-is-bridging-the-Israeli-Palestinian-divide-in-Jerusalem. [(Accessed 20 February 2018)].

Carmon, N., Shamir, U., 2010. Water-sensitive planning (WSP): integrating water considerations into urban and regional planning. Water Environ. J. 24 (3), 181–191.

Christopher, J.W., Derek, B., Booth, M., et al., 2016. Principles for urban storm-water management to protect stream ecosystems. Freshwater Sci. 35 (1), 398–411.

Christou, A., Agüera, A., Bayona, J.M., et al., 2017. The potential implications of reclaimed wastewater reuse for irrigation on the agricultural environment. The knowns and unknowns of the fate of antibiotics and antibiotic resistant bacteria and resistance genes—a review. Water Res. 123, 448–467.

Church, J.A., Clark, P.U., Cazenave, A., et al., 2013. Sea level change. In: Stocker, T.F., Qin, D., Plattner, G.-K. et al., (Eds.), Climate Change 2013: The Physical Science Basis. Contribution of Working Group I to the Fifth Assessment Report of the Intergovernmental Panel on Climate Change. Cambridge University Press, Cambridge, NY, pp. 1137–1216.

Cook, B.I., Anchukaitis, K.J., Touchan, R., et al., 2016. Spatiotemporal drought variability in the Mediterranean over the last 900 years. J. Geophys. Res. Atmos. 121 (5), 2060–2074. https://doi.org/10.1002/2015JD023929.

Cosgrove, W.J., 2003. Water Security and Peace. A Synthesis of Studies Prepared Under the PCCP-Water for Peace Process (An UNESCO–Green Cross International Initiative). Published as UNESCO–IHP technical documents—PCCP series. http://unesdoc.unesco.org/images/0013/001333/133318e.pdf. [(Accessed 15 April 2018)].

Cummings, D., Goren, M., Gasith, A., Zohary, T., 2017. Inundated shore vegetation as habitat for cichlids breeding in a lake subjected to extreme water level fluctuations. Inlands Waters 7 (4), 449–460. https://doi.org/10.1080/20442041.2017.1388984.

Damania, R., Desbureaux, S., Hyland, M., et al., 2017. Uncharted waters: the new economics of water scarcity and variability. In: License: creative commons attribution CC BY 3.0 IGO. World Bank, Washington, DC. https://doi.org/10.1596/978-1-4648-1179-1.

David, G.A., Craig, R.A., Hannah, E.B., et al., 2014. Assessing and managing freshwater ecosystems vulnerable to environmental change. Ambio 43 (Suppl. 1), 113–125. https://doi.org/10.1007/s13280-014-0566-z.

Dawoud, A.M., Al Mulla, M.M., 2012. Environmental impacts of seawater desalination: Arabian gulf case study Mohamed A. Int. J. Environ. Sustain. 1 (3), 22–37.

de Boer, C., Vinke-de Kruijf, J., Ozerol, G., Bressers, H., 2016. Collaborative water resource management: what makes up a supportive governance system?: Supportive governance for collaborative WRM. Environ. Policy and Governance 26 (4), 229–241. https://doi.org/10.1002/eet.1714.

REFERENCES

De Châtel, F., 2014. The role of drought and climate change in the Syrian uprising: untangling the triggers of the revolution. J. Middle Easter Stud. 50 (4), 521–535.

de Mes, T., 2014. Arcadis Middle East Aquifer Recharge Report. Arcadis NV, Amsterdam, the Netherlands. www.nww.ir/DorsaPax/Userfiles/file/.../AWW-January. [(Accessed 25 January 2018)].

Delene, D., Shamrukh, M., 2016. Investing in rainfall enhancement: an innovative plan for arid regions Qatar Foundation Annual Research Conference Proceedings. Hamad Bin Khalifa University Press, Qatar. https://doi.org/10.5339/qfarc.2016.EEPP1916. (accessed 10.03.18).

Desalination.biz, (2017) https://www.desalination.biz/news/0/Saudi-Arabia-and-China-to-develop-nuclear-desalination-projects/8827/ 30 August 2017 (accessed 12.04.18).

Desalination.biz. (2018). South Africa Declares Drought a National Disaster, Weighs Desalination Options, Water: Desalination + Reuse; West Sussex, UK. (accessed 14.02.18).

Deshmukh, R.A., Joshi, K., Bhand, S., Roy, U., 2016. Recent developments in detection and enumeration of waterborne bacteria: a retrospective mini review. Microbiol. Open 5 (6), 901–922. https://doi.org/10.1002/mbo3.383.

Devlin, J., 2014. Is Water Scarcity Dampening Growth Prospects in the Middle East and North Africa? Brookings. Available at: https://www.brookings.edu/opinions/is-water-scarcity-dampening-growth-prospects-in-the-middle-east-and-north-africa/. [(Accessed 15 April 2018)].

Dronkers, J., Gilbert, J.T., Butler, L.W., et al., 1990. Strategies for adaptation to sea level rise. In: Coastal Zone Management Subgroup Intergovernmental Panel on Climate Change 1990. Report of the IPCC Coastal Zone Management Subgroup: Intergovernmental Panel on Climate Change, Geneva: Intergovernmental Panel on Climate Change, pp. 148. http://papers.risingsea.net/IPCC-1990-Strategies-for-Adaption-to-Sea-Level-Rise.html. [(Accessed 18 March 2018)].

Droogers, P., Immerzeel, W., Terink, W., et al., 2012. Water resources trends in Middle East and North Africa towards 2050. Hydrol. Earth Syst. Sci. 16, 3101–3114. www.hydrol-earth-syst-sci.net/16/3101/2012/. https://doi.org/10.5194/hess-16-3101-2012.

EcoPeace Middle East, 2015. Regional Master Plan for Sustainable Development and Regional Investment Strategy for the Jordan River Valley, Final Report. June 2015, p. 190.

Elimelech, M., Phillip, W.A., 2011. The future of seawater desalination: energy, technology and the environment. Science 333 (6043), 712–717.

El-Tablawy, T. (2018). Sunni-Shiite Rift feeds Iranian-Saudi tension. Bloomberg, January 19, 2018. The Washington Post. https://www.washingtonpost.com/business/sunni-shiite-rift-feeds-iranian-saudi-tension-quicktake/2018/01/19/a879e886-fd0a-11e7-9b5d-bbf0da31214d_story.html?utm_term=.19f8d318cc8d (accessed 16.02.18).

EMWIS, 2018. Euro-Mediterranean Information System on Know-How in the Water Sector. U.T. SEMIDE/EMWIS T.U., Sophia Antipolis, France. http://www.emwis.org/overview. [(Accessed 28 April 2018)].

Ernest &Young, 2014. Big data. In: Changing the Way Businesses Compete and Operate, Ernst & Young Global Limited ("EYG"), p. 32. www.ey.com/Publication/vwLUAssets/EY_-_Big_data:_changing_the_way_businesses_operate/%24FILE/EY-Insights-on-GRC-Big-data.pdf. [(Accessed 15 April 2018)].

FAO, 2011. The state of the world's land and water resources for food and agriculture (SOLAW)—managing systems at risk. In: Food and Agriculture Organization of the United Nations. Rome and Earth scan, London, p. 286.

Frøystad, P., Holm, J., 2015. Whitepaper Evry: Blockchain: Powering the Internet of Value. Available at: https://www.evry.com/globalassets/insight/bank2020/bank-2020-blockchain-powering-the-internet-of-value-whitepaper.pdf. [(Accessed 22 April 2018)].

Gaeta, J.W., Sass, G.G., Carpenter, S.R., 2014. Drought-driven lake level decline: effects on coarse woody habitat and shells. Can. J. Fish. Aquat. Sci. 71, 315–325.

Gal, G., 2017. Annual Activity Report of the Kinneret Limnological Laboratory, 2016. Israel Oceanographic and Limnological Research (IOLR) Ltd Publication.

García, L.E., Rodríguez, D.J., Wijnen, M., Pakulski, I., 2016. Earth observation for water resources management: Current Use and Future Opportunities for the Water Sector. World Bank Group, Washington, DC. https://doi.org/10.1596/978-1-4648-0475-5. License: Creative Commons Attribution CC BY 3.0 IGO.

Gebel, J., 2014. Thermal desalination processes. In: Kucera, J. (Ed.), Desalination: Water from Water. John Wiley & Sons, Inc., Hoboken, NJ, USA. https://doi.org/10.1002/9781118904855.ch2

Glass, S., 2017. Twisting the Tap: Water scarcity and conflict in the Euphrates-Tigris River Basin. Independent Study Project (ISP) Collection, p. 2594. http://digitalcollections.sit.edu/isp_collection/2594.

Grimmeisen, F., Zemann, M., Goeppert, N., Goldscheider, N., 2016. Weekly variations of discharge and groundwater quality caused by intermittent water supply in an urbanized karst catchment. J. Hydrol. 537, 157–171.

Hadadin, N., Qaqish, M., Akawwi, E., Bdour, A., 2010. Water shortage in Jordan sustainable solutions. Desalination 250, 197–202. www.elsevier.com/locate/desal.

Hai, F.I., Yamamoto, K., Lee, C.H., 2013. Membrane biological reactors (MBR), theory, modeling, design, management and applications to wastewater reuse. IWS Publishing, p. 504.

Hansen, J., Sato, M., Hearty, P., Ruedy, R., et al., (2016). Ice melt, sea level rise and superstorms: evidence from paleoclimate data, climate modeling, and modern observations that 2°C global warming could be dangerous,

Atmos. Chem. Phys., 16: 3761–3812. doi: https://doi.org/10.5194/acp-16-3761-2016, (accessed 15.01.18).

Hauck, J., Winkler, K.J., Priess, J. A. (2015). Reviewing drivers of ecosystem change as input for environmental and ecosystem services modelling. Sustainability Water Qual. Ecol. 5, 9–30 doi: https://doi.org/10.1016/j.swaqe.2015.01.003 (accessed 10.01.18).

Hisham, A.H., Al-Najar, H., 2011. Brackish water desalination is the merely potable water potential in the Gaza strip: prospective and limitations. J. Environ. Sci. Tech. 4, 158–171. https://doi.org/10.3923/jest.2011.158.171.

Hughes, T.P., Anderson, K.D., Connolly, S.R., et al., 2018. Spatial and temporal patterns of mass bleaching of corals in the Anthropocene. Science 359 (6371), 80–83. https://doi.org/10.1126/science.aan8048.

Iizumi, T., Ramankutty, N., 2015. How do weather and climate influence cropping area and intensity? Glob. Food Sec. 4, 46–50. https://doi.org/10.1016/j.gfs.2014.11.003.

Ingold, K., Fischer, M., Boer, C., Mollinga, P., 2016. Water management across borders, scales and sectors: recent developments and future challenges in water policy analysis. Environ. Policy and Gov. 26 (4), 223–228.

Integrated Water Resources Management, 1998. Global Water Partnership Technical Advisory Committee (TAC). Stockholm, Sweden, GWP Global Secretariat, Stockholm, Sweden, p. 71. https://www.gwp.org/globalassets/global/toolbox/publications/background-papers/04-integrated-water-resources-management-2000-english.pdf. [(Accessed 30 March 2018)].

International Monetary Fund, 2017. Regional Economic Outlook. Middle East and Central Asia. Washington, DC, International Monetary Fund, Washington, D.C., p. 124. ISBN 9781484312520 (paper).

International Water Association (IWA), 2018. The Reuse Opportunity, Cities Seizing the Reuse Opportunity in a Circular Economy. IWA Publishing, London, UK. http://www.iwa-network.org/wp-content/uploads/2018/02/v7-OFID-Wastewater-report-2017-screen-1.pdf. [(Accessed 15 April 2018)].

IPCC, 2014. In: Team, C.W., Pachauri, R.K., Meyer, L.A. (Eds.), Climate Change Synthesis Report (2014). Contribution of Working Groups I, II and III to the Fifth Assessment Report of the Intergovernmental Panel on Climate Change. IPCC, Geneva, Switzerland, p. 151.

Isaac, J., Rishmawi, K., 2015. Status of the Environment in the State of Palestine. Publications of the Applied Research Institute—Jerusalem (ARIJ).

Israel Electric Corporation, 2016. Environmental Report for 2014-2015. Israel Electric Corp. Publication, Haifa, Israel.

Israel Ministry of Foreign Affairs, 1994. Treaty of Peace Between the State of Israel and the Hashemite Kingdom of Jordan. October 26, 1994, Article 6, Water, Israel Foreign Affairs Publishers, Jerusalem, Israel. http://www.mfa.gov.il/mfa/foreignpolicy/peace/guide/pages/israel-jordan%20peace%20treaty.aspx. [(Accessed 15 February 2018)].

Israel Ministry of Foreign Affairs, 1995. Israeli-Palestinian Interim Agreement on the West Bank and the Gaza Strip, Washington, D.C., Anne III, Article 40. Israel Foreign Affairs Publishers, Jerusalem, Israel. http://www.mfa.gov.il/mfa/foreignpolicy/peace/guide/pages/the%20israeli-palestinian%20interim%20agreement.aspx. [(Accessed 15 February 2018)].

Israel Water Authority, 2018a. Rainfall Data. http://www.water.gov.il/Hebrew/ProfessionalInfoAndData/Pages/rainmap.aspx. [(Accessed 16 April 2018)].

Israel Water Authority, 2018b. Lake Kineret Monitoring Data. Israel Water Authority Publications, Tel-Aviv, Israel. http://www.water.gov.il/Hebrew/ProfessionalInfoAndData/Kinneret-monitoring/Pages/default.aspx. [(Accessed April 2018)].

Israel Water Authority, 2018c. Water Consumption Report, 2016. Israel Water Authority Publications, Tel-Aviv, Israel. http://www.water.gov.il/Hebrew/ProfessionalInfoAndData/Allocation-Consumption-and-production/Pages/Consumer-survey.aspx. [(Accessed 20 April 2018)].

Issa, I.E., Al-Ansari, N.A., Sherwany, G., Knutsson, S., 2014. Expected future of water resources within Tigris-Euphrates Rivers Basin, Iraq. J. Water Res. Prot. 6, 421–432. https://doi.org/10.4236/jwarp.2014.65042.

Jägerskog, A., Clausen, T.J., Lexén, K., Holmgren, T. (Eds.), 2013. Cooperation for a Water Wise world—Partnerships for Sustainable Development. SIWI, Stockholm. Report Nr. 32.

Jardin, N., Hennerkes, J., 2012. Full-scale experience with the deammonification process to treat high strength sludge water—a case study. Water Sci. Technol. 65 (3), 447–455. https://doi.org/10.2166/wst.2012.867.

Jeppesen, E., Brucet, S., Naselli-Flores, L., et al., 2015. Ecological impacts of global warming and water abstraction on lakes and reservoirs due to changes in water level and related changes in salinity. Hydrobiologia 750, 201–227.

Jiménez, B., Takashi, A., 2008. Water Reuse. An International Survey of Current Practice, Issues and Needs. IWA Publishing, London.https://www.scribd.com/doc/262954539/Water-Reuse-An-International-Survey-of-Current-Practice-Issues-and-Needs-pdf. [(Accessed 15 April 2018)].

Jordan Valley Authority, 2014. Project Concept for Red Sea-Dead Sea Project, Initial Phase. Red Sea Desalination Project, Aqaba.http://www.jva.gov.jo/sites/en-us/RSDS/SiteAssets/rsds%20phase1.aspx?PageView=/ Shared. [(Accessed 15 April 2018)].

Kelley, C.P., Mohtadi, S., Crane, M.A., et al., 2015. Climate change in the Fertile Crescent and implications of the recent Syrian drought. Proc. Natl. Acad. Sci. U.S.A. 112 (11), 3241–3246. https://doi.org/10.1073/pnas.1421533112.

REFERENCES

Koncagül, E., Tran, M., Connor, R., et al., 2017. Wastewate—the untapped resource, facts and figures. In: UN World Water Assessment Programme Office for Global Water Assessment, Division of Water Sciences. UNESCO. http://www.unesco.org/new/en/natural-sciences/environment/water/wwap/wwdr/2017-wastewater-the-untapped-resource/unesco.org/water/wwap. [(Accessed 15 April 2018)].

Kosten, S., Huszar, V.L.M., et al., 2012. Warmer climate boosts cyanobacterial dominance in lakes. Glob. Chang. Biol. 18 (1), 118–126.

Larsen, M.C., 2014. Global change and water availability and quality, challenges ahead. In: Ahuja, S. (Ed.), Comprehensive Water Quality and Purification. vol. 1. Elsevier, Amsterdam, pp. 11–20.

Lawler, W., Bradford-Hartke, Z., Cran, M.J., et al., 2012. Towards new opportunities for reuse, recycling and disposal of used reverse osmosis membranes. Desalination 299, 103–112.

Lokiec, F., 2013. Sustainable desalination- environmental approaches. In: The Proceedings of the International Desalination Association World Congress on Desalination and Water Reuse 2013/Tianjin, China, pp. 20–25.

Luo, J., Fang, Z., Smith, R.L., 2014. Ultrasound-enhanced conversion of biomass to biofuels. Progress in energy and combustion. Science 41 (1), 56–93.

Machado, F.M., Fagan, S.B., da Silva, I.Z., de Andrade, M.J., 2015. Carbon Nanoadsorbents chap. 2. In: Bergmann, C.P., Machado, F.M. (Eds.), Carbon Nanomaterials as Adsorbents for Environmental and Biological Applications, Carbon Nanostructures. Springer International Publishing Switzerland. https://doi.org/10.1007/978-3-319-18875-1_2.

Markel, D., Shamir, U., Green, P., 2014. Operational management of Lake Kinneret and its watershed. In: Zohary, T., Sukenik, A., Berman, T., Nishri, A. (Eds.), Lake Kinneret: Ecology and Management. Springer, Heidelberg (Germany), pp. 541–560. Chapter 31.

Matsakas, L., Gao, Q., Jansson, S., et al., 2017. Green conversion of municipal solid wastes into fuels and chemicals. Electron. J. Biotechnol. 26, 69–83.

Mazlan, N.M., Peshev, D., Livingston, A.G., 2016. Energy consumption for desalination—a comparison of forward osmosis with reverse osmosis, and the potential for perfect membranes. Desalination 377, 138–151. https://doi.org/10.1016/j.desal.2015.08.011.

McNamaraa, L., Fitzsimonsa, L., Doherty, E., et al., 2017. The evaluation of technologies for small, new design wastewater treatment systems. Desalin. Water Treat. 91, 12–22. https://doi.org/10.5004/dwt. 2017.21247.

Metcalf and Eddy, 2014. Wastewater Engineering, Treatment and Reuse, fifth ed. McGraw Hill, New York, NY, p. 1966. ISBN 0-07-11225—8. 10121. https://courses.edx.org/c4x/DelftX/CTB3365STx/asset/Wastewater_Lecture_Note.pdf.

Ministry of the Environment (MOE), 2010. Public Health Standards for Sewage Treatment and Reuse of Effluent, 201. Israel Ministry of the Environment Publications, Jerusalem, Israel. http://www.sviva.gov.il/subjectsEnv/Streams/SewageStandards/Pages/Milestones.aspx. [(Accessed 25 January 2018)].

Mor, R., Kraitzer, T., Michail, M., et al., 2015. Soreq Mechanical Biological Wastewater Treatment Plant Operation Report. Mekorot National Water Co., for Mey Ezor Dan. Agricultural Cooperative Water Society Ltd, Tel-Aviv.

Mosher, J.J., Vartanian, G.M. (Eds.), 2015. Framework for direct potable reuse. In: Water Reuse Research Foundation. WEF, AWWA, NWRI, Alexandria, VA.

Mumssen, Y.U., Abdulnour, R., 2015. How Can Growing Cities Achieve Water Security for All in a World of Scarcity? The Water Blog, The World Bank Group, Washington, D.C. http://blogs.worldbank.org/water/how-can-growing-cities-achieve-water-security-all-world-scarcity. [(Accessed 25 January 2018)].

Namboodiri, V., Rajagopalan, N., 2014. Desalination. In: Ahuja, S. (Ed.), Comprehensive Water Quality and Purification. In: vol. 2. Elsevier, Amsterdam, pp. 98–119.

Neugebauer, I., Brauer, A., Schwab, M.J., et al., 2014. Lithology of the long sediment record recovered by the ICDP Dead Sea deep drilling project (DSDDP). Quat. Sci. Rev. 102, 149–165.

Newman, J.P., Maier, H.R., Riddel, G.A., et al., 2017. Review of literature on decision support systems for natural hazard risk reduction: current status and future research directions. Environ. Model. Software 96, 378–409. https://doi.org/10.1016/j.envsoft.2017.06.042.

OCHA, 2015. Humanitarian bulletin. In: Monthly Report—May. UNRWA Gaza situation report. Issue No. 93.

Oracle White Paper, 2013. Information management and big data. A Reference Architecture, p. 29. http://www.oracle.com/technetwork/topics/entarch/articles/info-mgmt-big-data-ref-arch-1902853.pdf. [(Accessed 25 January 2018)].

Orlowsky, B., Hoekstra, A.Y., Gudmundsson, L., Seneviratne, S.I., 2014. Today's virtual water consumption and trade under future water scarcity. Environ. Res. Lett. 9 (7), 074007. http://iopscience.iop.org/article/10.1088/1748-9326/9/7/074007/meta. [(Accessed 25 January 2018)].

Pennequin, D., Foster, S., 2008. Groundwater quality monitoring: the overriding importance of hydrogeologic typology (and need for 4D thinking), Chapter 5.1. In: The Water Framework Directive: Ecological and Chemical Status Monitoring. John Wiley and Sons Ed, Hoboken.

Plevria, A., Mamaisa, D., Noutsopoulos, C., et al., 2017. Promoting on-site urban wastewater reuse through MBR–RO treatment. Desalin. Water Treat. 91, 2–11. https://doi.org/10.5004/dwt.2017.20804.

Puyol, D., Batstone, D.J., Hülsen, T., et al., 2017. Resource recovery from wastewater by biological technologies: opportunities, challenges, and prospects. Front. Microbiol. 7, 2106. https://doi.org/10.3389/fmicb.2016.02106.

Qatar National Development Strategy, 2011. 2011~2016. Towards Qatar national vision 2030. Qatar General Secretariat for Development Planning. Gulf Publishing and Printing Company, Doha, p. 286. www.gsdp.gov.qa.

Renouf, M.A., Kenway, S., Leung Lam, K.A., et al., 2018. Understanding urban water performance at the city-region scale using an urban water metabolism evaluation framework. Water Res. 137, 395–406. https://doi.org/10.1016/j.watres.2018.01.070.

Reshef, G., 2017. Coastal Aquifer Groundwater Report, 2015. Israel Water Authority. http://www.water.gov.il/Hebrew/ProfessionalInfoAndData/Water-Quality/DocLib5/water_quality_report_2015.pdf. [(Accessed 25 January 2018)].

Rimmer, A., Givati, A., 2014. Hydrology. In: Zohary, T., Sukenik, A., Berman, T., Nishri, A. (Eds.), Lake Kinneret: Ecology and Management. Springer, Heidelberg (Germany), pp. 97–112. Chapter 7.

RO Guide, 2015. In: Burn, S., Gray, S. (Eds.), Efficient Desalination by Reverse Osmosis. A Guide to RO Practice. IWA publishing, p. 272.

Rozosa, E., Tsoukalasa, I., Ripis, K., et al. (2017). Turning black into green: ecosystem services from treated wastewater. Desalin. Water Treat. 91: 198–205. www.deswater.com doi: https://doi.org/10.5004/dwt.2017.20926.

Rüdel, H., Muñiz, C.D., Garelick, H., et al., 2015. Consideration of the bioavailability of metal/metalloid species in freshwaters: experiences regarding the implementation of biotic ligand model-based approaches in risk assessment frameworks. Environ. Sci. Pollution Res. 22, 7405–7421. https://doi.org/10.1007/s11356-015-4257-5.

Saaty, T.L., 2008. Decision making with the analytic hierarchy process. Int. J. Serv. Sci. 1 (1), 83–98.

Schrotter, J.C., Rapenne, S., Leparc, J., Remize, P.J., Casas, S., 2010. Comprehensive Membrane Science and Engineering. vol. 2. 35–65.

Schuyler, N. & Lauren, H. R. (2016). Navigating complexity: climate, migration and conflict in a changing world USAID's Office of Conflict Management and Mitigation, Wilson Center, Nov 2016, 44 pages. www.usaid.gov, www.wilsoncenter.org.

Serrao-Neumann, S., Darryl, C., Choy, L., 2018. Uncertainty and future planning: the use of scenario planning for climate change, adaptation planning and decision. In: Serrao-Neumann, S., Coudrain, A., Coulter, L. (Eds.), Communicating Climate Change Information for Decision-Making. Springer International Publishing. https://doi.org/10.1007/978–3–319-74669-2_6.

Shaofei, W., Xiang, Z., Dunxian, S., 2017. Joint occurrence of water quality indexes in relation to river streamflow in the heavily polluted Huai River Basin, China. Water Sci. Technol. Water Supply 17 (6), 1602–1615. https://doi.org/10.2166/ws.2017.046.

Shevah, Y., 2014. Water scarcity, water reuse and environmental safety. Pure Appl. Chem. 86 (7), 1205–1214.

Shevah, Y., 2015. Water resources, water scarcity challenges, and perspectives. In: Ahuja, S., de Andrade, J.B., Dionysiou, D.D., Hristovski, K.D., Loganathan, B.G. (Eds.), Water Challenges and Solutions on a Global Scale. 2015. In: ACS symposium series, vol. 1206. pp. 185–219. https://doi.org/10.1021/bk-2015-1206.ch010. Chapter 10.

Shevah, Y., 2017. Challenges and solutions to water problems in the Middle East. In: Ahuja, S. (Ed.), Chemistry and Water. The Science Behind Sustaining the World's Most Crucial Resource. Elsevier Inc, pp. 207–258.

Sim, D., 2015. Israel: The Dead Sea is Srinking and Hundreds of Sinkholes are Opening up on its Shores. Science, International Business Times, New York, NY. https://www.ibtimes.co.uk/israel-dead-sea-shrinking-hundreds-sinkholes-are-opening-its-shores-photos-1513207. (accessed 25.01.18).

Sivan, O., Erel, Y., Mandler, D., Nishri, A., 1998. The dynamic redox chemistry of iron in the epilimnion of Lake Kinneret. Geochim. Cosmochim. Acta 62 (4), 565–576.

Škrbić, B.D., Kadokami, K., Antić, I., et al., 2018. Micropollutants in sediment samples in the middle Danube region, Serbia: occurrence and risk assessment. Environ. Sci. Pollut. Res. 25, 260. https://doi.org/10.1007/s11356-017-0406-3.

Starkl, M., Brunner, N., Lopez, E., Martinez-Ruiz, J.L., 2013. A planning-oriented sustainability assessment framework for peri-urban water management in developing countries. Water Res. 47, 7175–7183.

Talari, S., Shafie-khah, M., Siano, P., et al., 2017. A review of smart cities based on the internet of things concept. Energies 10 (4), 421. https://doi.org/10.3390/en10040421. www.mdpi.com/journal/energies.

Tehan, R., 2017. Cybersecurity: Critical Infrastructure Authoritative Reports and Resources. US Congressional Research Service, p. 40. https://fas.org/sgp/crs/misc/R44410.pdf. [(Accessed 20 March 2018)].

UN Department of Economic and Social Affairs and World Bank's Water Global Practice, 2018. Making every drop count. In: An Agenda for Water Action High-Level Panel On Water Outcome Document. High Level Panel on Water Outcome Report, UN Division for Sustainable Development Goals; UN-DESA, New York, p. 34. https://sustainabledevelopment.un.org/content/documents/17825HLPW_Outcome.pdf. [(Accessed 25 January 2018)].

UN Department of Economic and Social Affairs, Population Division, 2015. World Population Prospects: the 2015 Revision, Key Findings and Advance Tables. Working Paper No. ESA/P/WP.241.

REFERENCES

UN Office for the Coordination of Humanitarian Affairs (2017), 2018. Humanitarian Needs Overview: Syrian Arab Republic Report. UN, Strategic Steering Group (SSG) and Humanitarian Partners, the Whole of Syria (WoS) Framework Report, New York, NY,pp. 77. https://reliefweb.int/report/syrian-arab-republic/2018-humanitarian-needs-overview-syrian-arab-republic-enar. [(Accessed 20 January 2018)].

UN Sustainable Development Goals (SDG), 2015. SDG 2030 Agenda and its 17 Sustainable Development Goals. https://www.un.org/sustainabledevelopment/development-agenda/. [(Accessed 15 January 2018)].

UNESCWA, 2014. The Economic and Social Commission for Western Asia Annual Report 2013. 40 years with the Arab World. E/ESCWA/OES/2014/1. United Nations Publication. 14–00091 - May 2014–1000.

UNOCHA, 2014. Humanitarian Response Plan 2014, Yemen. OCHA on behalf of the Humanitarian Country Team. Pp.108. http://www.unocha.org/sites/dms/CAP/HRP_2014_Yemen.pdf. [(Accessed 15 January 2018)].

UNOCHA, 2015. Humanitarian bulletin monthly report—May. UNRWA (2015). In: Gaza Situation Report. Issue No. 196. https://www.unrwa.org/newsroom/emergency-reports/gaza-situation-report-196. [(Accessed 20 January 2018)].

United States Environmental Protection Agency (USEPA). 2008. June 13, 2008 Memo. L. Boornaizian and S. Heare. Clarification on which stormwater infiltration practices/technologies have the potential to be regulated as "Class V" wells by the Underground Injection Control Program. Water Permits Division and Drinking Water Protection Division. Washington, D.C.

Voss, K.A., Famiglietti, J.S., de Linage, L.M., et al., 2013. Groundwater depletion in the Middle East from GRACE with implications for transboundary water management in the Tigris-Euphrates-Western Iran region. Water Resour. Res. 49 (2), 904–914.

Voutchkov, N., 2016. Desalination, Past, Present and Future. The International Water Association Organization, London, UK. http://www.iwa-network.org/desalination-past-present-future/. [(Accessed 15 January 2018)].

Wang, J., Wang, S., 2016. Removal of pharmaceuticals and personal care products (PPCPs) from wastewater: a review. J. Environ. Manage. 182, 620–640.

Water Environment Federation (2017). Intelligent Water Systems: The Path to a Smart Utility. The Water Research Foundation, Denver, CO., pp. 20. https://www.wef.org/globalassets/assets-wef/directdownload-library/public/03—resources/wsec-2016-tr-002-iws-smart-start—final.pdf. (accessed 20.01.18).

Water Finance Research Foundation, 2016. Municipal maintenance and infrastructure asset management systems. In: The 2016 Comparative Review. Water Finance Research Foundation Publications, Washington, D.C., p. 41.

http://www.waterfinancerf.org/studies-and-reports.html. [(Accessed 20 January 2018)].

WHO, 2017. Guidelines for Drinking-Water Quality: Fourth Edition Incorporating the First Addendum. World Health Organization, Geneva. License: CC BY-NC-SA 3.0 IGO.

Williams, P.A., 2018. Peace like a river: institutionalizing cooperation over water resources in the Jordan River Basin. Colo. Nat. Resour. Energy Environ. Law Rev. 28 (2), 315–351. https://www.colorado.edu/law/research/journals/colorado-natural-resources-energy-environmental-law-review.

Willis, K.J., Jeffers, E.S., Tovar, C., 2018. What makes a terrestrial ecosystem resilient? Science 359 (6379), 988–989. https://doi.org/10.1126/science.aar5439.

World Bank and the Arab World Council, 2011. Water reuse in the Arab world, from principle to practice. In: A Summary of Proceedings. Expert Consultation Wastewater Management in the Arab World, Egypt, Dubai-UAW.

World Bank Group (2012). Red Sea—Dead Sea Water Conveyance Study Program: Study of Alternatives (Preliminary Draft Report), The World Bank Group; Washington, D.C. http://siteresources.worldbank.org/INTREDSEADEADSEA/Resources/Study_of_Alternatives_Report_EN.pdf.

World Bank Group (2016). High and Dry: Climate Change, Water, and the Economy. World Bank, Washington, D.C. © World Bank. https://openknowledge.worldbank.org/handle/10986/23665 License: CC BY 3.0 IGO.

World Economic Forum, 2018. The Global Risks Report, Fractures, Fears and Failures, 13th ed. p.80. http://reports.weforum.org/global-risks-2018/. [(Accessed April 2018)].

World Water Forum, 2018. Regional Process Commission. League of Arab States (LAS), Arab Region, Coordinator. Pre-forum version March 2018, pp.93. www.worldwaterforum8.org. [(accessed 25.01.18)].

Wrathall, D.J., Van Den Hoek, J., Walters, A., Devenish, A., 2018. Water Stress and Human Migration: A Global, Georeferenced Review of Empirical Research. UN FAO, Rome. 2018. http://www.fao.org/3/i8867en/I8867EN.pdf. [(Accessed 25 January 2018)].

WWAP (United Nations World Water Assessment Program), 2017. The United Nations World Water Development Report 2017. Wastewater: The Untapped Resource. UNESCO, Paris. p. 198.

WWAP (United Nations World Water Assessment Program)/UN-Water, 2018. The United Nations World Water Development Report 2018: Nature-Based Solutions for Water. UNESCO, Paris.

Yeruham, E., Rilov, G., Shpigel, M., Abelson, A., 2015. Collapse of the echinoid Paracentrotus lividus populations in the Eastern Mediterranean—result of climate change? Sci. Rep. 5, 13479. https://doi.org/10.1038/srep13479.

Yin, L., Wang, B., Yuan, H., et al., 2017. Pay special attention to the transformation products of PPPs in environment. Emerg. Contam. 3 (2), 69–75.

Zabel, F., 2016. Impact of climate change on water availability. In: Mauser, W., Prasch, M. (Eds.), Regional Assessment of Global Change Impacts. Springer, Cham (Switzerland), pp. 463–469.

Zawieja, I., Wolny, L., 2013. Ultrasonic disintegration of sewage sludge to increase biogas generation. Chem. Biochem. Eng. Q. 27 (4), 491–497.

Zhang, Y., Giardino, C., Linhai, L., 2017. Water Optics and Water colour Remote Sensing. MDPI Books, China. https://doi.org/10.3390/books978-3-03842-509-0. under CC BY-NC-ND license. Pages 425.

Zilberman, D., Schoengold, K., 2005. The use of pricing and markets for water allocation. Can. Water Res. J. 30 (1), 47–54.

Zinger, Y. (2015). Kfar Saba Biofilter http://www.kkl-jnf.org/about-kkl-jnf/green-israel-news/october-2015/kenya-delegation-kfar-saba-biofilter (accessed 5.08.18).

Zinger, Y., Godecke, B., Fletcher, T.D., Deletic, A., 2013. Optimizing zing nitrogen removal in existing stormwater biofilters: benefits and tradeoffs of a retrofitted saturated zone. Ecol. Eng. 51, 75–82. https://doi.org/10.1016/j.ecoleng.2012.12.007.

Zohary, T., Gasith, A., 2014. The littoral zone. Chap. 29. In: Zohary, T., Sukenik, A., Berman, T., Nishri, A. (Eds.), Lake Kinneret: Ecology and Management. Springer, Heidelberg (Germany), pp. 517–532.

Further Reading

Marcus, F., Spivy-Weber, F., Doduc, M., et al., 2015. Water quality control plan for Ocean waters of California. Addressing desalination facility intakes, brine discharges, and the incorporation of other non-substantive changes. Division of Water Quality State Water Resources Control Board, California Environmental Protection Agency, p. 242. https://www.waterboards.ca.gov/board_decisions/adopted_orders/resolutions/2015/rs2015_0033_sr_apx.pdf. [(Accessed 15 April 2018)].

Nidal, H., 2015. Dams in Jordan. Current and future perspective. Can. J. Pure Appl. Sci. 9 (1), 3279–3290.

Sixth Environment Action Program of the European Community, 2002–2012. Final assessment of the 6th Environment Action Program. http://ec.europa.eu/environment/archives/newprg/final.htm. [(Accessed April 2018)].

Tapscott, D., Tapscott, A., 2016. Blockchain Revolution: How the technology behind Bitcoin is Changing Money Business and the World. Penguin Random House Publishing Group, p. 368.

VA DEQ Stormwater Design Specification No. 7 Permeable Pavement Version 2.0, January 1, 2013 pp. 33. https://chesapeakestormwater.net/2012/03/design-specification-no-7-permeable-pavement/.

CHAPTER

4

Present and Potential Water-Quality Challenges in India

Sanjay Bajpai*, Neelima Alam, Piyalee Biswas

Technology Mission Division, Department of Science and Technology, Technology Bhawan, New Delhi, India
*Corresponding author: E-mail: sbajpai.dst@gmail.com

1 INTRODUCTION

India is rich in water resources, being gifted with network of rivers and snow covered Himalayan ranges that can meet a variety of water requirements of the country. However, as a result of the speedy rise in already large population of the country and the need to meet the growing demands of irrigation and industrial consumption, the existing water resources in many parts of the country are getting depleted and the quality of water has deteriorated.

As per the NITI Aayog report, India will be water-stressed country from 2020 onward, which implies $<1000 \, m^3$ of water availability per person per annum (http://niti.gov.in/writereaddata/files/document_publication/2018-05-18-Water-Index-Report_vS8-compressed.pdf). This calls for an urgent action. First, India needs a lot more water infrastructure. Compared to other semi-arid countries, India can store comparatively small amount of its rainfall. Whereas, India's dams can accumulate $200 \, m^3$ of water per person, other middle-income countries like China,

South Africa, and Mexico can store about $1000 \, m^3$ per capita. Furthermore, India has exploited merely about 20% of its cost-effective hydropower, compared to 80% in developed countries. The critical water–energy symbiotic relationship needs to be explored in a more sustainable mode for their mutual cogeneration. India needs to spend in water infrastructure at all levels from large multipurpose water dams to small neighborhood watershed management and rainwater harvesting projects. India also needs to achieve water use efficiency across all sectors including recycling and reuse.

2 QUANTITY AND QUALITY OF WATER

India is encircled by Bay of Bengal, Arabian Sea, and Indian Ocean and has a coastline of more than 7500 km. Precipitation in the form of rain and snowfall provide over 4000 trillion liters of freshwater to India. On an average the rainfall received in India is 1200 mm, with maximum of 11,000 mm in Meghalaya and the minimum in West Rajasthan of about 250–300 mm. The total traversable measurement lengthwise of inland waterways is 14,500 km. With view to this, the country now requires well-organized and competent water resource planning and management (Fig. 1).

As per the Ministry of Water Resources estimates, per capita water availability in 2025 and 2050 is estimated to come down by almost 36% and 60%, respectively of the 2001 levels. With the population already crossing the 1.3 billion mark, the annual per capita water accessibility of the country is just enough to satisfy per capita needs. The population of India is believed to become stable around the year 2050 at 1.64 billion. If the populace is not contained, the requirement of water may go up (Fig. 2). The percentage use of water by 2050 as forecasted by National Commission for Integrated Water Resources Development is irrigation (68%), drinking (9%), industry (7%), power (6%), inland navigation (1%), environment (2%), and rest is loss because of evaporation (7%).

One of the goals of India's Sustainable Development, related to drinking water, is to improve the quality of water by 2030. Poor water quality and water pollution are the two most crucial challenges India is facing. Pollution caused by pollutants and improper treatment of wastewater discharged into lakes, ponds, and rivers leads to unavailability of freshwater. This in turn shifts the focus to groundwater. In India, about 85% of drinking water comes from aquifers. Excessive consumption of groundwater has declined the level of water table causing water scarcity due to inadequate replenishment of water in aquifers. The water-quality problem can be Physical (Turbidity, Color, Odor), Organic [Microorganism, Bacterial (Total coliform, fecal coliform), Virus (associated with particulate/

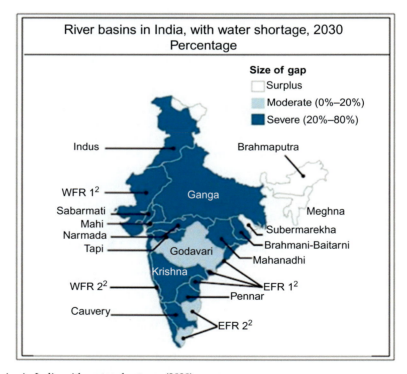

FIG. 1 River basins in India with water shortages (2030).

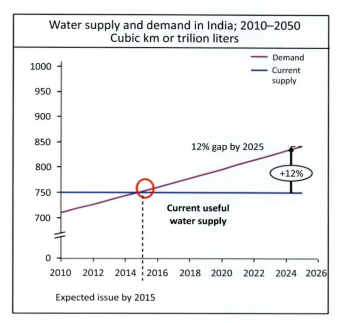

FIG. 2   Water supply and demand in India (2010–2050).

colloidal impurities and adsorbed to surface), Protozoa (Gyptosporodia, Giardia-cysts, Oocysts), Dissolved organics such as pesticides, etc.] and Chemical (Iron, Arsenic, Fluoride, Magnesium, Sodium, etc., which are responsible for hardness, alkalinity and salinity).

3 ENLISTMENT OF EXISTING WATER QUALITY CHALLENGES

A nation wide exercise for identification of major water issues amenable to technology solution in water challenged hot spots, spread across the country led to identification of 26 water related challenges in India (http://www.dst.gov.in/water-technology-initiative-programme-wti). These water related challenges were classified into three major categories, viz., water shortage, water quality, and site-specific water use management (Table 1).

The water-quality issues related to biological contamination, excessive presence of salinity, arsenic, fluoride, iron, silica, strontium besides problems caused by inappropriate reject management of residues from water treatment technologies.

4 MAJOR WATER-QUALITY PROBLEMS PREVALENT IN INDIA

Freshwater ecosystems are finite and globally threatened by increasing environmental degradation caused by destructive land-use, ill water management practices, and increasing industrialization. For a decent quality of living, an individual requires about 100 L of water per day, out of which, about 10 L are requisite for drinking, cooking, and washing. The remaining is necessary for washing of clothes, toilet flushing, etc. Central Public Health Environment and Engineering Organisation (CPHEEO) determine the standards of water quantity required in rural and urban areas. The dependence on unsafe drinking water sources,

TABLE 1 Water Challenges Facing the Country

Water shortage	Water quality	Site specific water use management
Low per capita availability	Quality deficit of available water for specified uses	Storage capacity for seasonally available water
Evaporation loss from water bodies	Geological contamination through arsenic	Surface run-off on account of nature of geological terrain
Nonsustainable water cycle management	Contamination through fluoride	Water body disuse
Water winning and mining in water starved areas	Contamination through iron	Mismatched rates withdrawal and recharging capacity
	Contamination through multiple species	Nonoptimal use of water in agriculture
	Biological contamination	Nonoptimal use of water in industrial sector
	Alkali metal ion salinity	Unplanned water use and demand
	Alkaline earth metal salt salinity and hardness	River flood management
	Contamination of water on account of pesticides and other water derived residues	Wetland management
	Deficit of assurance for drinking water quality	
	Sea water intrusion in coastal areas	
	Reject management from water related technologies	
	Any other including Silica and Strontium contamination	

can lead to serious health effects. In India, quality of drinking water is regulated by Bureau of Indian Standard (BIS).

The major problems that constrain the availability of safe drinking water are:

- Excess salinity because of geological circumstances.
- Contamination of water resource by microorganisms, Personal and Pharmaceutical Care Products (PPCPs), and other emerging contaminants (ECs).
- Contamination of groundwater due to arsenic, iron, nitrate, chromium, fluoride, silica and from the geological and anthropogenic cause.

- Contaminants like Uranium have multiple sources (geological and anthropogenic or co-effect of both).

In order to use water efficiently, pioneering approaches for adopting best practices, utilizing existing technologies smartly and promoting innovation to augment quality reduce usage are necessary. For example, operating the water treatment plants at small recoveries would lead to less energy utilization, less maintenance, and enhanced permeate quality. This combined with partial recirculation of reject, depending on the feed salinity can lead to less misuse of the ground and surface water.

4.1 Contamination of Water By Fluoride

Groundwater is used by 85% of rural population of India for drinking and domestic use. This is the most widely prevalent contaminant with nearly three-fourths of Indian provinces affected by fluoride contamination. High fluoride concentration in groundwater past the permissible limit of 1.5 mg/L causes health problems. Considering its spread, extent, and health impacts, it is a serious concern for public health (Fig. 3).

The presence of fluoride in low concentrations has been proven to be advantageous in the prevention of enamel demineralization and during osteoporosis treatment. However, in excess quantity, the fluoride can cause metabolic disorders, skeletal disease, and mottled teeth. Fluoride also acts as neurotoxin and also leads to nerve damage upon amassing in nerve tissues. In severe manifestations of the disease, the bones become deformed, severe joints pain, and people find it difficult to even walk. Excessive fluoride is detected in body through urine samples.

The Nayagarh district in the state of Odisha came into limelight when the people of the area were affected by a mysterious disease way back in 1987–88. Later on it was established that the disease in question was fluorosis (http://www.dst.gov.in/water-technology-initiative-programme-wti). Analysis of the groundwater in the area by State Government agencies revealed that fluoride level in the water in the village tube wells as well as in the water in the open wells and ponds was much higher than the permissible levels. The alarming effect of fluoride such as dental and skeletal fluorosis has been found to be evident in some districts of Andhra Pradesh and Rajasthan (Barathi et al., 2015). Excessive fluoride in groundwater is observed in 224 districts of 19 provinces in India affecting around 65 million people in India, including 6 million children (Chatterjee et al., 2018).

Fluorosis has resulted in drastic fall in the productivity of the affected population. It increases the morbidity index and has an overall socioeconomic affect on the impacted community. Fluorosis not only affects the human population but the cattle also are found to have lesser ability to procreate and to give milk. Agricultural fields are also affected adversely by fluoride contaminated water.

Because of lack of fluoride-safe wells, awareness on fluoride toxicity and its consequences, rural population often end up consuming fluoride contaminated water (Pipileima and Kumar, 2016). Studies have underlined the need for continued monitoring of fluoride contamination and development of suitable mitigation technologies.

A research supported by Department of Science and Technology studied fluoride contamination in Bansathi village, near Kanpur to observe seasonal variation in fluoride concentration. Geochemical analysis was carried out to identify possible sources of fluoride contamination (Mohapatra et al., 2017) (Fig. 4).

Several mitigation technologies from adsorption to membrane separation are being used for fluoride mitigation. Activated alumina remains a preferred choice, though reverse osmosis (RO) systems are penetrating in larger numbers as broad spectrum technology capable of removing multiple contaminants.

In a DST supported project at IIT Kanpur, a calcium-based treatment technology for fluoride removal has been developed at the lab-scale which works in situ at the contaminated sites. Fluoride uptake on calcite in the presence of phosphate, either through the formation of hydroxyapatite-coated calcite or through other calcium phosphate solids, has been developed.

A team of researchers at IIT Kharagpur developed carbonized bone meal (CBM) and chemically treated carbonized bone meal (CTBM) as novel adsorbent for fluoride removal. Maximum fluoride adsorption capacity of CBM and CTBM are measured to be 14 and 150 mg/g, respectively.

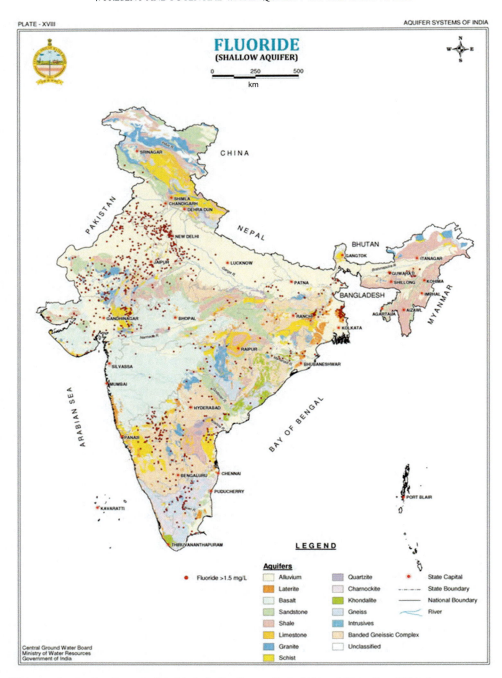

FIG. 3 Water contamination through fluoride in India. *Source: Central Ground Water Board Website.*

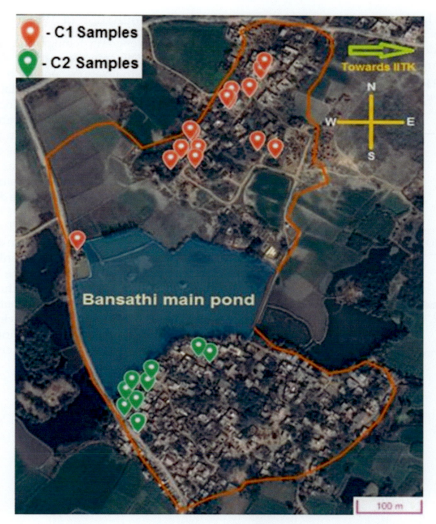

FIG. 4 Sampling locations in the Bansathi village Kanpur Nagar were shown in the map. *Red points* indicate fluoride concentrations above drinking water limit (DWL; 1.5 mg/L). *Green points* indicate fluoride concentration below DWL.

A composite bed was designed to cater to domestic drinking water requirement at the rate of 100 L/day using the technology to remove bacteria and iron along with fluoride. After the fluoride removal capacity is exhausted, the material can be mixed with coal and used as fuel (Fig. 5).

At Veer Surendra Sai University, a portable low-maintenance and eco-friendly indigenous water filter using laterite, is being developed for fluoride detection. The device will be able to detect fluoride as low as 1 ppm. The removal efficiency of the device was found to be in the range of 70%–80% for 10 mg/L using 200–500 μm particle size. Since it was found that the locally available laterite soil of Nayagarh district in the state of Odisha is able to absorb the fluoride, the process of developing a filter is going on which would dispense potable water at a low cost.

A RO unit has been functioning at Yellampalli village near Bagepalli town, in the state of Andhra Pradesh for over 2 years. In order to avoid discharge of reject water discharged via a drain into a pond, the reject water is being treated to remove excess fluoride (Fig. 6).

4.2 Contamination of Water by ECs

Owing to ongoing demographic shifts, urbanization and changing life styles supported by rapid industrialization, pollution by ECs is becoming a serious environmental and public health concern globally. PPCP, endocrine disrupting chemicals (EDCs) pesticides and industrial compounds are collectively known as ECs. Concentration of PPCPs increases in the environment due to individual human activity, as well as, residues from drug industries, agribusiness, and biotechnological sectors, irresponsible disposal from hospital and community uses, etc. Human excretion and bathing incorporate PPCPs in the environment. In addition, direct disposal of waste/unused products in trash, septic tanks, or sewers increase their concentration in the domestic or municipal sewage water which gradually accumulate in the water and soil bodies (Bhati et al., 2013).

There is lack of adequate knowledge of their impact in long-term effect on human health, environment, and aquatic environments.

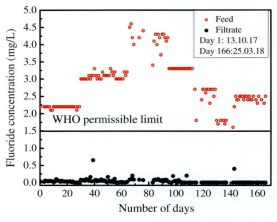

FIG. 5 Performance of domestic fluoride removal unit.

FIG. 6 Field-scale defluoridation unit at Yellampalli, Andhra Pradesh.

4 MAJOR WATER-QUALITY PROBLEMS PREVALENT IN INDIA

Antimicrobial resistance (AMR) represents a major threat to modern healthcare. Since India has one of the highest per capita uses of antibiotics it faces very significant challenges. It is also now known that antibiotic-resistant microbes found in the environment (e.g., water) contribute to the transfer of resistance and in recognition of this the World Health Organization (WHO) Global Action recognizes the important role played by water.

An advanced treatment for PPCPs is generally not employed by the on-site treatment plants which discharge effluent into the native soil. As a result, different pharmaceutical and other trace compounds of personal care products reach groundwater systems by the process of on-site system wastewater infiltration (Carrara et al., 2008). These compounds interfere with the secretion, synthesis, binding, transport, and indigenous action of natural hormones which are responsible for maintenance of various body functions (Oudejans, 2017).

Of late, the presence of ECs in the environment has got greater attention in India. The extent of contamination of Indian water sources is dependent on the consumption pattern of pharmaceuticals, effluent quality of Wastewater Treatment Plants (WWTPs), hospitals and pharmaceutical industry effluents, and discharge through other point and nonpoint sources.

Pesticides comprise a broad range of chemicals with different physicochemical characteristics and are used in agricultural applications to limit or prevent the growth of harmful insects, weeds, and microorganisms such as bacteria and fungi.

India accounts for approximately 3% of total pesticides consumption in the world. Though the level of consumption is less, the main concern is about the toxicity and life of class of pesticides predominant in India. The Indian consumption pattern includes 76% of insecticides followed by fungicides (13%) and herbicides (10%) and the consumption is extreme particularly in trade crops such as cotton, bananas, coffee, vegetables, and flowers (Mukherjee et al., 2018).

Despite their potential threat, the distribution, scale, and levels of this category of newly emerging water contaminants are largely unknown. The WHO report determined that low and ultra low concentration levels of these contaminants in drinking water fall outside the extent and understanding of the analytical methodologies that are approved for complete analysis of drinking water. Hence, a concerted effort to detect and treat these compounds with a viable and cost-effective method is necessary.

There are also very few systematic monitoring programs or comprehensive, methodical studies on the occurrence of these contaminants in drinking water. However, Indian Institute of Technology has mapped ECs of Delhi city by collecting samples from diverse environmental settings (major drains, hospitals, STPs, and the river Yamuna) in summer and winter season and different chemical and biochemical parameters of the samples have been analyzed. Antibiotic resistance bacteria (ARB) were found to be associated with the fecal matter and higher abundance of ARB and antibiotic resistance genes was observed in winter season.

In an Indian-United Kingdom collaborative program, Yamuna river in the north (in the most polluted stretch, contributing to 70% of Delhi's water supply needs) and the Cauvery river in the south (the most abstracted river in India) are being studied for prevalence of ECs. Investigations will also be made on the fate of ECs and use of bio-solids during wastewater and sludge treatment line at 10 WWTPs in India. The attempt is to develop evidence based wastewater discharge standards and guidance for safe use of contaminated sludge.

Another bilateral research program attempts to develop novel sensors to detect and monitor pollutants focusing on studying AMR co-selectors to improve the understanding of how these pollutants mediate and impact AMR. A bacterial sensor, using bio-reporter strains, for the detection of EDCs in discharges and freshwaters is also envisaged.

EVALUATING WATER QUALITY TO PREVENT FUTURE DISASTERS

4.3 Contamination of Water By Bacteria

The bacteriological contamination of drinking water is a key concern for public health authorities in almost all developing countries including India. Monitoring of bacteriological water quality of drinking water sources mainly depends on analysis of indicator organisms, e.g., coliform bacteria. *E. coli* is a member of fecal coliform group, which is a more explicit indicator of fecal contamination than any other fecal coliform.

The presence of coliform bacteria reveals that water is being polluted with fecal matter of humans, live stocks, and other domestic as well as wild animals. The coliform bacteria in water should be absent/100 mL, which is considered to be safe for general human use. Total coliform and fecal coliform concentration in the rivers of the country are recorded to be ranging between 500 and 100,000 MPN/100 mL (Martone et al., 2014).

The occurrence of fecal coliform is likely to affect human beings more than aquatic organisms. The presence of these organisms in drinking water in general is a result of improper water treatment or the seepage of the pipes, which distribute the water. Few water-borne pathogenic diseases include hepatitis-A, typhoid fever, ear infections, dysentery, viral, and bacterial gastroenteritis. The diseases and effects mentioned, even when they are not fatal, are still debilitating to the afflicted, increase general morbidity, lower performance in the workplace, cause loss of revenue, and lower quality of life.

Control of bacteriological contamination, which is of much serious concern, being capable of creating mass health hazard, requires more attention of the policy makers and drinking water supply departments to reduce and fully check the increasing biological pollution level and also maintain safety of drinking water sources before supply and consumption. The current water-quality evaluation methods are time consuming and by then people have already consumed the infected water. The need is to develop simple and reliable evaluation method that can predict water quality almost in real time so that consumption of contaminated water can be prevented beforehand.

The bacteriological contamination is the most wide spread in India. For example, In Uttarakhand, the surface water is the main source of drinking water in the hilly region, whereas, groundwater sources are used mainly in plain regions of the state such as parts of Dehradun, Pauri, Nainital, entire Haridwar, and entire Udham Singh Nagar districts. Most of the drinking water sources of Uttarakhand are contaminated by biological parameters namely Total Coliform, Fecal Coliform, and *E. coli*, wherein open defecation and slope factor of hills contribute significantly.

An accredited State Level Water-Quality Testing and Monitoring Laboratory assess the quality of water sources being used for drinking purpose in entire Uttarakhand. This laboratory acts as referral laboratory having analysis facility of 45 drinking water-quality parameters including biological parameters. A field effort in the state for the development of Water-Quality Monitoring & Surveillance Model in selected rural areas of Uttarakhand has also been implemented.

A bilateral project, between Department of Science and Technology with United Kingdom on microbiological pollution in the Vembanad Lake in southern state of Kerala aims to understand the conditions that allow pathogens to proliferate in the lake. Improving the water quality in Vembanad Lake will provide a solid foundation for augmentation of public health and welfare, an essential step in improving quality of life and securing a sustainable economic development of Kerala so that economic growth can proceed hand in hand with maintenance of traditional life styles of fishing and farming communities.

DEVELOPMENT OF NOVEL DISINFECTION TECHNOLOGY

With the support of DST's Water Technology Initiative, Central Electronics Engineering Research Institute (CEERI) Pilani has developed technology for dielectric barrier discharge (DBD)-based plasma system for disinfection. The technology has been successfully transferred for commercialization to Turners Pvt. Ltd. Jaipur. The prototype is an 8 in. tube mercury free plasma UV lamp, which produces ultraviolet radiations to disinfect the impure water and costs approximately Rs. 250 (Figs. 7 and 8).

FIG. 8 A potable water purifier system developed by CEERI, Pilani.

FIG. 7 Potable water purification system based on dielectric barrier discharge (DBD) developed by Central Electronics Engineering Research Institute (CEERI) Pilani.

4.4 Contamination of Water By Iron

Iron is one of the most plentiful elements found in our planet. However, its concentration in water is usually very less. The chemical behavior of iron and its solubility in water depends primarily on the oxidation intensity and pH in the system in which it occurs.

Iron is crucial for the metabolism of animals and plants. If it is present in water in large quantity, it forms red oxyhydroxide precipitates that stains cloths and rusts plumbing fittings, makes the water taste astringent and, therefore, is an undesirable element in domestic and industrial water supplies (http://cgwb.gov.in/WQ/Geogenic%20Final). The Environment Protection Agency (EPA) cautions that even though iron in drinking water is not dangerous to consume, yet the iron sediments may contain trace impurities or harbor microorganism that can be harmful. Constantly consuming excess amounts of iron can lead to a condition known as iron overload. This condition usually results in a gene mutation. If left untreated, iron overload can lead to hemochromatosis, a brutal ailment that can harm the body's organs. The symptoms of hemochromatosis consist of weight loss, joint pain, and fatigue, and if not treated, it can lead to diabetes, liver problems, and heart disease.

In India excess concentration of iron (>1.0 mg/L) in groundwater has been detected in more than 1.1 million villages of 24 provinces namely Andhra Pradesh, Assam, Bihar, Chhattisgarh, Goa, Gujarat, Haryana, J&K, Jharkhand, Karnataka, Kerala, Madhya Pradesh, Maharashtra, Manipur, Meghalaya, Orissa, Punjab, Rajasthan, Tamil Nadu, Tripura, Uttar Pradesh, West Bengal, and Union Territory of Andaman & Nicobar (http://cgwb.gov.in/WQ/Geogenic%20Final).

Technological Solution has been demonstrated for removing iron from sources affected by excess iron (>1 ppm) from drinking water in West Bengal and North Eastern States of India using Ceramic Membrane Size-Based Exclusion (Micro Filtration) system by Central Glass and Ceramics Research Institute. The

IRON REMOVAL TECHNOLOGY DEVELOPED BY CGCRI

Department of Science and Technology supported iron removal plant has been set up in different parts of North East India. The technology of Ceramic Membrane Systems is developed by Central Glass & Ceramics Research Institute (CGCRI) for removal of iron from drinking water sources affected by excess iron (>1 ppm) (Fig. 9).

FIG. 9 Iron removal plant developed by CGCRI, demonstrated at Guwahati, Assam.

technology developed removes bacteria and turbidity also. Ceramic micro-filtration membranes have high potential for separation of fine particulates, suspended solids, microorganism, and colloids from water. Thus product water is of the potable quality and meets standards.

No chemical additive is required for water clarification, which also minimizes sludge disposal problem. Technology is faster, more reliable, and maintenance free compared to traditional techniques based on slow sand filtration.

4.5 Contamination of Water By Silica

Silica (Si) in groundwater can be found as dissolved silica and particulate matter. Silica is released as a result of chemical weathering of silicate minerals in rocks and sediments acquired by flowing water and thus the cause of silica (SiO_2) in groundwater is largely from water-rock contact. Silica is known to cause kidney damage on long-term inhalational exposure and in drinking water it poses as a risk factor for Balkan Nephropathy. As high as 15%–20% of the population in Uddanam region, in north coastal Andhra Pradesh was found to be suffering from kidney diseases leading to fatal mortality (Pradeep et al., 2016).

Six mandals of Srikakulam district, viz., Vajrapukothuru, Palasa, Mandasa, Itchapuram, Kaviti, Sompeta, and Kanchili and the two surrounding mandals comprising of around 120 villages in Southern State of Andhra Pradesh in India were identified to be predominantly affected by Renal Parenchymal diseases. Four mandals of Srikakulam, Andhra Pradesh have been studied, covering 11 villages to assess and classify the renal and related diseases prevalent in the region. Initial studies indicated the presence of higher traces of Silica and other trace elements in the various water sources. Also a thorough observation and analysis of the water quality of shallow aquifers, deep aquifers, surface waters, and soil samples collected from the study area indicated higher concentrations of elements like Cobalt and radioactive elements like Gallium, Strontium, Lithium, and Rubidium, etc. in addition to Silica.

In Tindivanam Taluk, of Villupuram district, the northern part of Tamil Nadu, near the capital of Chennai, the higher silica from high land in archean formations demonstrate that there is dissolution of silica because of farming activities and the weathering enhances the HCO_3 concentration along with escalating silica concentration (Pradeep et al., 2016).

4.6 Contamination of Water By Nitrate

Nitrates are chiefly produced to be used as fertilizers in agriculture because they have high solubility and biodegradability. The main nitrate fertilizers are ammonium, sodium, potassium, and calcium salts. Nitrate is highly leachable and can easily travel with water from soil if there is excessive rainfall and can reach groundwater. It is found that the concentration of nitrates in water varies with distance from the surface of earth, shallow water sources have nitrate concentration as high as 100–400 mg/L and deep water sources have less concentration of nitrate ions in the range of 50–100 mg/L (Rao et al., 2013; Townsend and Young, 1995). The suggested acceptable upper limit for nitrate in water according to EPA is $\leq$45 mg/L.

The distribution of NO_3 in groundwater is controlled by many factors. They include source precipitation, groundwater flow, availability, and composition of the vadose zone, irrigation, aquifer heterogeneity, dissolved oxygen concentrations, and electron donor availability and dispersion. Few examples of nitrogen compounds found naturally in waters are ammonia (NH_3), nitrite (NO_2^-), nitrate (NO_3^-), and the ammonium ion (NH_4^+) (Hudak, 1999).

Excess amount of nitrate in drinking water promulgates methemoglobinemia, or "blue baby syndrome," to which infants are especially susceptible. Methemoglobinemia reduces ability of the blood to carry oxygen because of reduced hemoglobin, which leads to blueness around the mouth, hands, and feet in baby. In extreme cases, there is noticeable fatigue, increased saliva production, unconsciousness, seizures, and in some cases it may be fatal. Severe methemoglobinemia may cause brain damage and death. Prolonged consumption of high levels of nitrate is related to gastric troubles caused by the creation of nitrosamines in grown-up. Nitrosamines are also known to cause cancer in test animals.

In India, high concentration, more than 45 mg/L of nitrate, has been found in many districts of Andhra Pradesh, Bihar, Delhi, Haryana, Himachal Pradesh, Karnataka, Kerala, Madhya Pradesh, Maharashtra, Orissa, Punjab, Tamil Nadu, Rajasthan, West Bengal, and Uttar Pradesh (Suthar et al., 2009). The highest concentration of nitrate in the range of 3080 mg/L is found in Bikaner, Rajasthan.

Research efforts have been made for detection and removal of nitrates from drinking water. A handheld sensor measurement system has been developed for measuring nitrate ions in water, which is based on molecular modeling and computational simulation. The model also optimizes the molecular architecture, which is suitable for sensing nitrate and nitrite ions. In addition, android-based application is under development for measuring change in resistance of the sensor in the presence of nitrate ions.

In another initiative, N,N,N-triethyl ammonium functionalized cross-linked chitosan beads has been developed for the removal of nitrate from water. Its nitrate removal efficiency from brackish water system has been demonstrated at a laboratory scale.

4.7 Contamination of Water by Chromium

The most prevalent forms of chromium that is found naturally in the environment are Trivalent chromium (chromium-III) and Hexavalent chromium (chromium-VI). Chromium-III is found in many vegetables, fruits, meats, grains, and yeast. Chromium-VI is found in nature because of the erosion of natural chromium deposits and it can also be produced by industrial processes. It has been confirmed that instances of chromium being released to the nature are by poor storage, leakage, or poor industrial waste disposal practices. The EPA has a drinking water standard of 0.1 mg/L or 100 ppb for chromium, which includes all forms of chromium, including chromium-VI (https://www.epa.gov/dwstandardsregulations/chromium-drinking-water).

In major parts of India, trace concentrations of heavy metals (Cd, Cr, Fe, Ni) in water have been found. Plants and sediments of river Yamuna flowing through Haryana and Delhi have reported chromium occurrence due to major industrial complexes (Kaushik et al., 2009). Some of the largest Chromate producing industries are in Tamil Nadu Industrial Estate, Ranipet, India, with an annual production of 2500 metric tons of Sodium dichromate and 3300 metric tons of Basic Chromium Sulfate (Tanning powders) partly supplying to 440 leather tanning industries in the Tamil Nadu state (Khasim et al., 1989). The factory area receives an average annual rainfall of 1000 mm. About 1.5 lakh tons of solid wastes rich in hexavalent chromium (Cr VI), having an area of $14,000 \, m^2$ (about 3.5 acres), of about 4 m thickness, is collected in an open area within the factory grounds (Sankaran et al., 2010). Chromium has also been found to be high in and around the tannery belt in Kanpur area in the northern state of Uttar Pradesh.

USE OF NANOMATERIALS FOR DETECTION OF CHROMIUM VI

A method has been developed for the colorimetric determination of Cr VI in water with the use of nanomaterial. The device consists of a source, sample holder, detector, and a microcontroller-processing unit with a display unit and detects and quantifies the color complex for chromium levels detection. The method does not interfere with any ions present in the water. The color complex formation takes 1–2 min compared to the conventional method of using diphenyl-carbazide that takes 10–15 min. The device can be used for chromium detection up to ppb level for on-field monitoring (Fig. 10).

FIG. 10 Colorimetric determination of Cr VI in water with the use of nanomaterial.

Surface plasmon resonance (SPR)-based colorimetric sensing of mercury and chromium species in the contaminated water and a porous silicon-based biosensor for chromium detection have been developed.

4.8 Contamination of Water Through Uranium

Uranium is known to cause both chemical and radiological toxicity and known to affect organs like kidney and lungs. The WHO has set a provisional safety standard of 30 μg of Uranium per liter. Uranium is not, however, included in the list of contaminants monitored under the BIS's Drinking Water Specifications IS 10500-2012 although Bhabha Atomic Research Centre has set the drinking water limit for uranium as 60 ppb. Although health effects of uranium contamination in drinking water are yet to be fully studied, it is hard to deny that uranium may pose threat as a contaminant (Chakraborti et al., 2011). Several studies have linked uranium in drinking water to kidney diseases.

The usual concentration of Uranium in our planet is 2.7 ppm (Siegel and Bryan, 2004). In nature, Uranium is normally present in tetravalent state as insoluble and in hexavalent state as highly soluble. The Uranium concentration in groundwater depends on lithology, geomorphology, and other geological conditions of the region (Kumar et al., 2011; Michel, 1991; Ortega et al., 1996).

There are several factors, such as uranium content in source, leachability, proximity and hydraulic isolation of water, climatic effects and their seasonal variability, pH, and oxidation state of the water, concentrations of carbonate, nitrates, phosphate and other species, overexploitation of groundwater, which can influence the levels of uranium dissolved in water. These factors coexist in many parts of the country and thus results in high uranium concentrations in groundwater. Agriculture and industries are the main sources of uranium presence in groundwater of Punjab and Tamil Nadu.

A recent research study conducted by Duke University found uranium contamination in the groundwater from aquifers in 16 states of India, highlighting the need for attention in the states to prevent collateral damage. Nearly a third of water wells tested in Rajasthan, contained uranium levels that exceed the WHO and the US Environmental Protection Agency's safe drinking water standards. By analyzing previous water-quality reports, aquifers polluted with high levels of Uranium in 26 other districts of north-west India and 9 districts in south or south-east India were also identified. The report states that human activities, especially the over-exploitation of groundwater for agricultural irrigation, have contributed to the problem.

4.9 Contamination of Water Through Salinity and Hardness

Water salinity is a well-known problem around the world with undesirable consequences on health, soil quality, and overall eco-systems. Salinity in groundwater is largely divided into Inland and Coastal salinity. A large part of India is underlain with saline aquifers particularly in areas of Rajasthan, Delhi, Haryana, Punjab, Uttar Pradesh, and Gujarat besides the coastal aquifers of Kerala and West Bengal. In many aquifers in Rajasthan and Gujarat exceeds 10,000 ppm. The salinity level in Haryana and Punjab are also rising with Salinity ranging from 2500 to 5000 ppm.

4.9.1 Inland salinity

Groundwater salinity is common in the regions of Rajasthan, Haryana, Punjab, Gujarat, Uttar Pradesh, and to a lesser extent, in Madhya Pradesh, Maharashtra, Karnataka, Bihar, and Tamil Nadu. In some parts of Gujarat and Rajasthan, the salinity in groundwater is so high that salt is manufactured directly from these wells by solar evaporation (http://cgwb.gov.in/wqoverview.html).

The saline aquifer areas are enlarging because of the intensive development of fresh groundwater resources in Punjab, Haryana, and Uttar Pradesh, thereby enlarging the area covered with saline groundwater. Thus, there is a need to check the ingress of saline groundwater in fresh groundwater areas by intensive development of the saline groundwater and treating the same with updated technologies. Besides geogenic reasons, salinity of aquifer is also caused because of the practice of irrigation by surface water without taking into consideration the status of groundwater. The rise in groundwater levels with time has resulted in water logging and intense water evaporation in semiarid regions leading to salinity in the affected areas.

4.9.2 Coastal Salinity

Coastal Salinity is found across the country's coastal region, which has a vibrant shoreline of about 7500 km in length extending from Gulf of Khambhat in Gujarat to Konkan and Malabar Coast to Cape Comorin (Kanyakumari) in the south to northward along the Coromandel coast to Sunderbans in Bay of Bengal (http://cgwb.gov.in/wqoverview.html).

Saline groundwater is not being developed as a resource since it cannot be used directly without treatment. Therefore, the nondevelopment of saline groundwater is giving rise to water logging and reduction in the fertility of soil.

Women and young girls have to travel long distances to obtain water resulting in drudgery, increased school dropout rates, and wastage of productive time. The socioeconomic conditions of the community affected by saline water can considerably improve if the beneficiaries are provided good drinking water.

The procedure of removing excess salts from sea or brackish water to make the water fit for human use is called desalination. Different technologies are used for water desalination such as thermal and membrane processes. The main thermal desalination processes consist of multistage flash distillation, multi-effect distillation (MED), ocean-based low-temperature thermal desalination, and mechanical/thermal vapor compression. The common membrane desalination processes include RO, electrodialysis, and membrane distillation. In addition to these, ion-exchange methods such as capacitive deionization, multi-effect humidification, and biological method of desalination are also identified as potential methods as there is rising necessity for unconventional energy-based cost-efficient methods of water desalination. Another approach is to use combination of technologies for optimizing the energy efficiency/cost effectiveness and make the customized hybrid solution more sustainable and viable. In coastal areas, salt-loaded brine is getting discharged back into the sea after desalination which can damage the biodiversity and environment. On the other hand, removal of salts in inland salinity affected areas results in highly saline reject water flowing into soil or adjacent freshwater aquifer adversely affecting the water and soil quality. This issue needs to be adequately and urgently addressed. It is therefore of utmost importance that the saline groundwater is treated with new technological options where the discharge of reject does not create a problem and is dissipated in the deep aquifer in the same tube well.

HOLISTIC WATER SOLUTION FOR DRINKING, DOMESTIC AND IRRIGATION NEEDS

Drinking water in the Bhuja Bhuja village in Nellore District of coastal Andhra Pradesh was affected with salinity due to seawater intrusion, hardness, iron, and silica. Drinking water was being supplied in the village at the rate of 3 L per person once in 3 days by Nellore Corporation through water tankers. The water for other domestic needs and general purpose water was drawn from two bore wells in Chemudugunta lake located about 1000 m away from the village.

The poor quality of water was causing enormous health issues such as high incidence of gastrointestinal complaints, jaundice, dysentery, kidney stones, viral infections, and dental fluorosis. A water treatment plant of 80 kL/day using RO membrane technology has been set up in the Bhuja Bhuja village. The water treatment plant caters to a beneficiary population of around 12,000 people for their drinking water needs and addresses domestic needs of 3000 families. The reject wastewater from the RO system is collected and treated along with village wastewater thereby reducing the incidence of secondary pollution. The treated waste and reject water are used for irrigation and groundwater recharge which has high-quality nutrients for enhancing the soil fertility and yield per crop. Also, fish are reared in the wastewater ponds to further utilize the improved quality treated wastewater. The biological sludge is dried and reused as manure in village green belt development.

With the improvement of the quality of drinking water, the living conditions of the villagers have also improved. The absenteeism of children in school because of sickness has reduced and increased productivity level of villager's considerably. The villagers are paying for quality water to sustain the intervention. The drinking water treatment plant and WWTP are being monitored by the Public Health Department, Nellore Municipal Corporation regularly (Figs. 11 and 12).

FIG. 11 Reverse osmosis plant installed at Bhuja Bhuja village in Nellore District of Andhra Pradesh.

FIG. 12 Drinking water distribution to villagers at Bhuja Bhuja Village in Nellore District of Andhra Pradesh.

SOLAR POWER DESALINATION AT NARIPPAIYUR VILLAGE, RAMANATHAPURAM, TAMIL NADU

A Solar-Biomass-based multi-effect desalination (MED) system has been set up at Narippaiyur, a coastal village in a rain shadow area, for providing potable water in arid rural areas from desalination of sea water in the Bay of Bengal utilizing solar thermal energy.

The plant delivers potable water at a rate of 6000 L/h and could run for 24 h a day. This plant produces high-purity desalinated water for specific industrial units and also by appropriate simple and inexpensive remineralization, produces desalinated potable water meeting requirements of the WHO. This is a first-of-its-kind demonstration project which can be replicated along the coastal and arid coastal regions of India where there is acute water shortage. Brine reject from the desalination system can be directly discharged into the sea without affecting the local system and no harmful chemicals are used. The system is maintenance-friendly (Fig. 13).

FIG. 13 Solar-Biomass based multi-effect desalination (MED) system at Narippaiyur, Ramanathapuram, Tamil Nadu.

WATER SCARCITY IN LATUR, MARATHWADA REGION (AURANGABAD DIVISION) MAHARASHTRA

Latur is one of the districts of Maharashtra in Western India affected by drought in this region. The problem is aggravated because of cultivation of water intensive crops such as sugarcane which has further lowered the water table. In 2016, the severity of shortage of drinking water in the area was very acute. Based on a survey of water requirement, a high recovery mobile Reverse Osmosis (RO) unit was deployed in the area during the severe drought period in peak summers (May–June, 2016) at critical points in the Latur area with open and unused water sources.

This movable unit was equipped with indigenously developed membrane technology capable of purifying and cleaning turbid/suspended particles as well as removing unwanted dissolved salts (like fluoride, nitrate, and arsenic) that could be harmful for human health. Through this deployment about 50,000 L of pure and treated water was made available for drinking to the people of Latur. The attempt is now to make the intervention holistic by treating wastewater and making it good for specific use (Fig. 14).

FIG. 14 Deployment of a mobile Reverse Osmosis unit in Latur Marathwada for providing potable water.

4.10 Contamination of Water Through Arsenic

Arsenic occurs naturally in the environment as an element. Arsenic as a contaminant is important in terms of its poisonous nature with exceptionally diverse manifestations of poisoning. Drinking water is the major pathway for intake of arsenic by a human being as well as eating food that has been contaminated with arsenic.

As per BIS 2012 (IS 10500:2012), the acceptable limit of Arsenic is 0.01 mg/L and the permissible limit, in absence of alternate source, is 0.05 mg/L. In groundwater inorganic arsenic is present, commonly as arsenate (As V) and arsenite (As III) (http://cgwb.gov.in/wqoverview.html). Organic forms of arsenic like monomethylarsonic acid and its analogs are less toxic than the inorganics. Compounds like arsenobetaine and arsenocholine are relatively nontoxic to humans (Maliyekkal et al., 2009) (Fig. 15).

In India, Assam, Bihar, Jharkhand, Manipur, Uttar Pradesh, West Bengal, and some part of Chhattisgarh and Karnataka are reported having arsenic contamination affecting more than 50 million people. In the Gangetic River Basin, which supports a population of over 500 million people, and people are exposed to groundwater arsenic, resulting in enhanced morbidity and reduced economic productivity.

In affected areas, arsenic concentration is as high as 600 μg/L, which is much higher than WHO allowable limit of 10 μg/L (Sankar et al., 2013). Excessive arsenic in groundwater is observed in 86 districts of 10 states in India.

West Bengal has attracted much attention for extensive occurrence of arsenic since it was first reported in 1976. According to the West Bengal Public Health Engineering Department Integrated Management Information System database (2016–2017), the highest arsenic contamination of >1.0 mg/L has been recorded in Sarangabad gram panchayat of Canning Block in South 24 Parganas district. Arsenic contamination is also a big threat to a huge population of Bihar. Though several parts of Bihar have been severely affected by arsenic toxicity, there are limited reports on arsenic bio-geochemical cycle in this region.

Symptoms of arsenic poisoning are headaches, severe diarrhea, confusion, and drowsiness. As the poisoning grows spasm and pigmentation in nails of fingers may take place (Biswas and Sarkar, 2019). When the severity

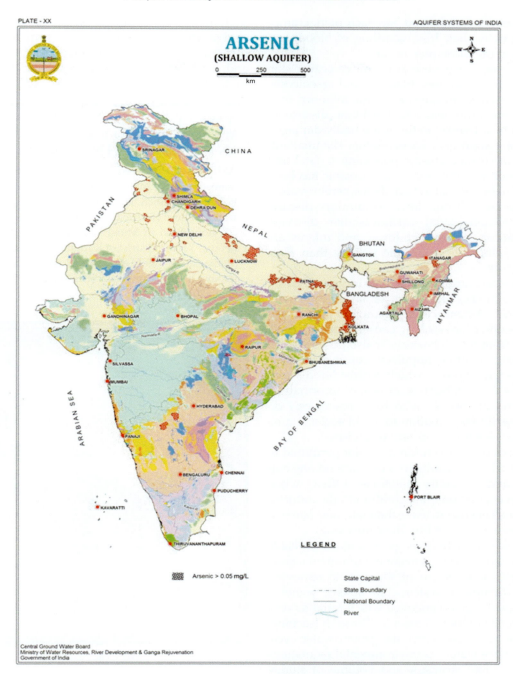

FIG. 15 Water Contamination through Arsenic in India. *Source: Central Ground Water Board Website.*

of toxicity is heightened, stomach pain, cramping muscles, hair loss, vomiting of blood, and blood in the urine may occur.

Long-term exposure to arsenic can cause skin cancer, lungs, kidney, bladder, and arsenicosis. Regular consumption of arsenic at more than allowable limit leads to blackfoot disease in which blood vessel in limbs are badly damaged resulting in progressive gangrene. The final outcome of chronic arsenic poisoning is fatal and can result in coma and death. Arsenic has been found to be correlated to diabetes, cerebrovascular diseases, chronic lower respiratory disease, hypertension-related cardiovascular diseases, and cancer. Women, children, and infants are more vulnerable to arsenic toxicity than male and adults.

The concentration of arsenic in surface freshwater sources, like rivers or lakes, is usually <10 µg/L. Arsenic contamination in groundwater has other far-reaching consequences as well. Apart from health hazards, it also poses a social problem. Illiterate populace, confuse the skin lesions with leprosy, which is considered a contagious disease, because of lack of information. As a result, people who have arsenicosis do not reveal their situation to avoid social stigma associated with it and face ostracism.

Affected school children are not permitted to attend schools. Adults try not to attend any cultural and religious functions. Often, when employers find out their conditions, the unfortunate workers had to leave their jobs and married women are deserted by their husbands.

While several technology options are available to remediate arsenic, a major environmental issue is reject management of the arsenic removing water treatment systems. In arsenic removal plants, a sludge is generated which contains very high concentration of arsenic. Through leaching, Arsenic can again enter the groundwater cycle and surface water. As environmental regulations become more stringent and volume of sludge generated continues to increase due to increase in numbers of treatment systems, traditional sludge disposal methods are coming under increasing pressure to change. Fast, reliable, and cost-effective detection of arsenic presence in drinking water is another major challenge.

AMRIT-ARSENIC AND METAL REMOVAL BY INDIAN TECHNOLOGY

AMRIT is a gravity-fed, compact, engineered, and nanostructured materials-based water purification system which make available clean drinking water without electricity or piped water supply to 200 household in Murshidabad district in West Bengal, Yadgir District of Karnataka and in Bihar. Media can be safely disposed or reused for brick making as it is composed of iron oxides (Sankar et al., 2013) (Fig. 16).

FIG. 16 Domestic AMRIT Filter for arsenic removal.

LATERITE-BASED WATER FILTER DEVELOPED BY IIT KHARAGPUR

The Laterite-based Arsenic filter developed by IIT Kharagpur uses naturally plentiful raw laterite customized for higher adsorption capacity by appropriate chemical treatment (acid-alkali treatment). It requires no power for functioning. Arsenic being adsorbed chemically on the adsorbent surface, forms highly stable structure and does not leach out in due course. Leaching does not occur from the spent laterite (adsorbent meets TCLP protocol) and can be used for road laying purpose after its adsorption capacity is exhausted (Figs. 17 and 18).

FIG. 17 500 L/day Arsenic removal units installed in a primary school in Kashinathpur, West Bengal.

FIG. 18   2000 L/h arsenic removal unit installed at Duttapukur, West Bengal.

ARSENIC FILTER BY INDIAN INSTITUTE OF TECHNOLOGY—BOMBAY

These community scale hand pump attached arsenic removal filter use zero-valent iron (ZVI). The method is based on corrosion of ZVI and generation of hydrous ferric oxides (HFOs) (adsorbent for arsenic) and subsequent filtration. The process is so designed that oxidation of As (III) to As (V) is attained and also the As (V) formed is adsorbed on HFO. These filters do not need energy or electricity to function. Now, totally 58 units of 600 Lph flow rate are operating in various parts of Uttar Pradesh, Bihar, West Bengal, and Assam (Fig. 19).

FIG. 19 Arsenic removal units developed by IIT Bombay operating in different parts of West Bengal.

5 EVALUATING WATER QUALITY TO PREVENT DISASTERS

India has robust water-quality monitoring network. However, as several new contaminants of concern emerge and gain importance, capacity and capability of this network needs to be augmented with sophisticated equipment, and skilled scientific and technical personnel. This is especially important as many of these contaminants, though present in trace and ultralow level can pose huge risk to human health. An innovative model for evaluating water quality involving water utility, academics and research institution ensures monitoring water quality accurately, tracking spatial and temporal variations and capacity development on continuous basis has been developed. Three such analytical facilities operating on this networked model are operational in north, north-eastern, and southern India.

ACCREDITED WATER-QUALITY LABORATORY IN UTTARAKHAND

This referral Water-Quality Analytical Laboratory in the north Indian hilly state of Uttarakhand in Dehradun analyzes water samples from different sources like drinking, effluent/wastewater and agriculture for testing physical and chemical parameters and the presence of metal ions.

The laboratory covers beneficiary population of 95 blocks in 13 districts covering the entire population of 17.5 million population of Uttarakhand. The parameters being evaluated on site with laboratory support are pH, turbidity, and total hardness (mg/L), while parameters such as Color, Odor, Taste, Residual Free Chlorine, Total dissolved solids, Alkalinity, Chloride, Copper, Sulfate, Calcium, Fluoride, Nitrate, Sulfate, Magnesium, Manganese, Cadmium, Lead, Zinc,

Chromium, Iron, Aluminum, Arsenic, Sodium, Potassium, Total and Fecal Coliform, and Cobalt are tested in laboratory.

The GIS Coordinates (longitude and latitudes) and height of water source (above mean sea level) of each source of water sample is recorded. Both raw and supply waters types are being analyzed and the effect of seasonal variations on water quality (especially pre- and post-monsoon seasons) is also determined.

NAGALAND STATE REFERRAL LABORATORY FOR WATER-QUALITY TESTING AND MONITORING

This Referral Water Laboratory in the remote hilly state is fully equipped with the sophisticated analytical equipment.

The laboratory has a wide scope of analyzing water samples from different sources like drinking, effluents, and agriculture for testing physical, chemical parameters, pesticides, and metal ions. The laboratory provides support for on-site monitoring of select parameters also. The laboratory covers a beneficiary population of 1.5 lakhs directly and 19 lakhs indirectly within the state covering almost all the districts of the state.

A total of 15 parameters being analyzed on-site with Water-Quality laboratory assistance include appearance, odor, turbidity, pH, alkalinity, hardness, chloride, fluoride, iron, ammonia, nitrite, nitrate, phosphate, residual chlorine, and total dissolved solids.

A total of 36 parameters being analyzed in Water-Quality laboratory include all the above and magnesium, calcium, copper, manganese, sulfate, nitrates, fluoride, phenolic compounds, mercury, cadmium, selenium, arsenic, cyanide, lead, zinc, anionic detergents, chromium, poly nuclear aromatic hydrocarbons, mineral oil, pesticides, salinity, alkalinity, aluminum, boron, total and fecal coliform.

FACILITY FOR DRINKING WATER-QUALITY ANALYSIS AND MONITORING IN NORTH COASTAL DISTRICTS OF ANDHRA PRADESH

A continuous water-quality analysis and monitoring facility has been established at Vizag, southern coastal state of Andhra Pradesh, especially for rural and tribal population. The facility is creating awareness about drinking water-quality deviations in the area and providing a source of an equipped water-quality analytical facility.

There are 25 parameters being analyzed in the laboratory. They are dissolved oxygen, pH, conductivity, chloride, fluoride, total hardness, calcium, magnesium, total alkalinity, total dissolved solids, nitrates, sulfates, iron, sodium, potassium, anionic detergents, phenolic compounds, mercury, sulfides as H_2S, metals, pesticides, E. coli, and total and fecal coliforms.

The facility analyzes drinking water-quality parameters as per BIS 10500. The laboratory covers a total of 3 districts, 20 mandals, and 86 villages in the state of Andhra Pradesh. Awareness campaigns for school children and women in the villages are being taken up on water-quality issues.

6 CONCLUSIONS

Because of rapid population growth, industrialization, climate change, etc., there is an increase in water demand, stress on water resources, and increased quantity of wastewater generation. In addition to this, the per capita water availability is reducing day by day in India. The availability of water would be greatly affected because of declining water table and increasing runoff affecting the supply side, on the one hand, and increasing population, industrialization, and urbanization, on the other hand. The water demand has surged dramatically as a result of improvements in living standards, more nuclear families, and changing lifestyle. In addition to the measures of water use efficiency, augmentation of water quality, water reuse, recycling, and wastewater treatment are important for arresting declining per capita water availability. Since water quality is a dynamic phenomenon affected by several anthropogenic and geogenic factors, evaluation and near real-time reliable monitoring of water quality, for existing as well as ECs is critical for preventing future disasters. India is developing its research capacities and capabilities for timely detection and mitigation of contaminants and current research endeavors will provide important policy inputs for formulating robust monitoring strategy.

Acknowledgments

The authors are thankful to Prof. (Dr.) Ashutosh Sharma, Secretary, Department of Science & Technology for encouraging research initiatives on water quality. The authors would also like to place on record grateful thanks to esteemed water researchers, both in India and abroad and other stakeholders' for their interest in Indian water-quality issues and worthy scientific contributions. We would also like to acknowledge information obtained from websites of Ministry of Water Resources and its constituent organizations. The authors are also thankful to their colleagues in the Technology Mission Division, especially Mrs. Indira Srihari, Mr. Rajender Kumar, and Mr. Neeraj Gouhar for their unstinted support.

References

Barathi, M., Kumar, A.S.K., Rajesh, N., 2015. Aluminium hydroxide impregnated macroreticular aromatic polymeric resin as a sustainable option for defluoridation. J. Environ. Chem. Eng. 3, 630–641.

Bhati, I., Dhawan, N.G., Maheshwari, R.K., 2013. Greener route to prevent pharmaceutical pollution. Int. J. Pharm. Chem. Sci. 2 (4), 1781–1787.

Biswas, R., Sarkar, A., 2019. Characterization of arsenite-oxidizing bacteria to decipher their role in arsenic bioremediation. Prep. Biochem. Biotechnol. 49 (1), 30–37. https://doi.org/10.1080/10826068.2018.1476883.

Carrara, C., Ptacek, C.J., Robertson, W.D., Blowes, D.W., Moncur, M.C., Sverko, E., et al., 2008. Fate of pharmaceuticals and trace organic compounds in three septic system plumes, Ontario, Canada. Env. Sci. Tech. 42 (8), 2805–2811.

Chakraborti, D., Das, B., Murrill, M.T., 2011. Examining India's groundwater-quality management. Environ. Sci. Technol. 45 (1), 27–33.

Chatterjee, S., Jha, S., De, S., 2018. Novel carbonized bone meal for defluoridation of groundwater: batch and column study. J. Environ. Sci. Health A 53 (9), 832–846. https://doi.org/10.1080/10934529.2018.1455378.

Hudak, P.F., 1999. Regional trends in nitrate content of Texas groundwater. J. Hydrol. 228 (1–2), 37–47.

Kaushik, A., Kansal, A., Kumari, S., Kaushik, C.P., 2009. Heavy metal contamination of river Yamuna, Haryana, India: assessment by metal enrichment factor of the sediments. J. Hazard. Mater. 164 (1), 265–270.

Khasim, I., Kumar, N.V., Hussain, R.C., 1989. Environmental contamination of chromium in agricultural and animal products near a chromate industry. Bull. Environ. Contam. Toxicol. 43 (5), pp. 742.

Kumar, A., Rout, S., Narayanan, U., Mishra, M.K., Tripathi, R.M., Singh, J., Kumar, S., Kushwaha, H.S., 2011. Geochemical modelling of uranium speciation in the subsurface aquatic environment of Punjab State, India. J. Geol. Min. Res. 3 (5), 137–146.

Maliyekkal, S.M., Philip, L., Pradeep, T., 2009. As(III) removal from drinking water using manganese oxide-coated-alumina: performance evaluation and mechanistic details of surface binding. Chem. Eng. J. 153, 101–107.

Martone, R.S., Martins, B.E., Peternella, S.F., Razzolini, M., Tereza, P., 2014. Assessment of the bacteriological quality of the drinking water consumed in a condominium of students. Sch. Acad. J. Biosci. 2 (9), 564–596.

Michel, J., 1991. Relationship of radium and radon with geological formations. In: Cothern, C.R., Robers, P.A. (Eds.), Radon, Radium, and Uranium in Drinking Water. Lewis Publishers, Michigan, pp. 83–95.

Mohapatra, A.K., Ekamparam, A.S.S., Adla, S., Singh, A., 2017. Mechanism of fluoride mobilization in groundwater of Bansathi, Kanpur Nagar, India. In: 7th International Ground Water Conference (IGWC-2017), New Delhi, India (Dec 11–13, 2017).

Mukherjee, D., Bhattacharya, P., Jana, A., Bhattacharya, S., Sarkar, S., Ghosh, S., Majumdar, S., Swarnakar, S., 2018. Synthesis of Ceramic Ultrafiltration Membrane and Application in Membrane Bioreactor Process for Pesticide Remediation From Wastewater Process Safety and Environmental Protection. vol. 116 Elsevier, pp. 22–33.

Ortega, X., Valles, I., Serrano, I., 1996. Natural radioactivity in drinking water in Catalonia (Spain). Environ. Int. 22, 347–354.

Oudejans, L., 2017. Report on the 2016 U.S. Environmental Protection Agency (EPA) International Decontamination Research and Development Conference. U.S. Environmental Protection Agency, Washington, DC, EPA/600/R-17/174.

Pipileima, S., Kumar, P.A., 2016. Comparative studies between conventional and polymer based adsorption for defluoridization for drinking water. J. Energy Res. Environ. Technol. 3 (1), 34–37.

K., Pradeep et al., 2016. IOP Conference Series Materials Science and Engineering. 121, 012008, (2016).

Rao, S.M., Sekhar, M., Rao, P.R., 2013. Impact of pit-toilet leachate on groundwater chemistry and role of vadose zone in removal of nitrate and E. coli pollutants in Kolar District, Karnataka, India. Environ. Earth Sci. 68, 927–938.

Sankar, M.U., Aigal, S., Chaudhary, A., Anshup, Maliyekkal, S.M., Kumar, A.A., Chaudhari, K., Pradeep, T., 2013. Biopolymer reinforced synthetic granular nano composites for affordable point-of-use water purification. Proc. Natl. Acad. Sci. U. S. A. 110, 8459–8464.

Sankaran, S., Rangarajan, R., Krishna Kumar, K., Saheb Rao, S., Humbarde, S.V., 2010. Geophysical and tracer studies to detect subsurface chromium contamination and suitable site for waste disposal in Ranipet, Vellore District, Tamil Nadu, India. Environ. Earth Sci. 60 (4), 757.

Siegel, M.D., Bryan, C.R., 2004. Environmental geochemistry of radioactive contamination. In: Treatise on Geochemistry, p. 262.

Suthar, S., Bishnoi, P., Singh, S., Mutiyar, P.K., Nema, A.K., Patil, N.S., 2009. J. Hazard. Mater. 171, 189–199.

Townsend, M.A., Young, D.P., 1995. Factors affecting nitrate concentrations in ground water in Stafford County, Kansas. In: Kansas Geological Survey, Current Research in Earth Sciences, Bulletin 238.

Further Reading

WHO (World Health Organization), 2011. Guidelines for Drinking Water-Quality Recommendations, Geneva, fourth ed. vol. 1, p. 515.

CHAPTER

5

Arsenic Contamination in South Asian Regions: The Difficulties, Challenges and Vision for the Future

Sukanchan Palit[a,b,], Kshipra Misra[c], Jigni Mishra[c]*

[a]Department of Chemical Engineering, University of Petroleum and Energy Studies, Dehradun, India
[b]43, Judges Bagan, Post-Office-Haridevpur, Kolkata, India [c]Department of Biochemical Sciences, Defence Institute of Physiology and Allied Sciences, Delhi, India
*Corresponding author: E-mails: sukanchan68@gmail.com; sukanchan92@gmail.com; sukanchanp@rediffmail.com

1 INTRODUCTION

Arsenic and heavy metal groundwater contamination are vexing issues in many developing and developed nations around the world. The South Asian countries of Afghanistan, Bangladesh, India, Nepal, Pakistan and Sri Lanka face the severest degree of arsenic contamination in groundwater (Luby et al., 2008) containing high levels of arsenous acid, arsenic acid, and their derivatives that lead to both acute and chronic health concerns. Surveys reveal that population in rural and suburban regions of South and Southeast Asian countries like Bangladesh, Cambodia, India, Nepal, Pakistan, and Vietnam, exposed to groundwater arsenic contamination is much above 100 million (Mukherjee et al., 2006). Consequently, consumption of untreated and arsenic-laden groundwater by them is alarmingly increasing mortality and risk of high-grade illnesses. This alarming picture of hazards caused by arsenic in South Asia calls for remedial approaches to generate arsenic-free water and make it accessible, specially to people residing in rural and remote areas (Fig. 1).

2 HEALTH PROBLEMS DUE TO ARSENIC CONTAMINATION

The primary mode of exposure to arsenic poisoning is ingestion of arsenic-contaminated water which instigates several illnesses, viz., arsenicosis, cancer, cognitive disorders, etc. (Kapaj et al., 2006). The risks associated with these particular health disorders are discussed in brief in the succeeding paragraphs.

2.1 Arsenicosis

Reports have shown that consumption of arsenic-contaminated water for more than 6 months inflict damages on urine, hair, and nails sometimes causing serious melanosis and

---

Separation Science and Technology, Volume 11
https://doi.org/10.1016/B978-0-12-815730-5.00005-3

Copyright © 2019 Elsevier Inc. All rights reserved.

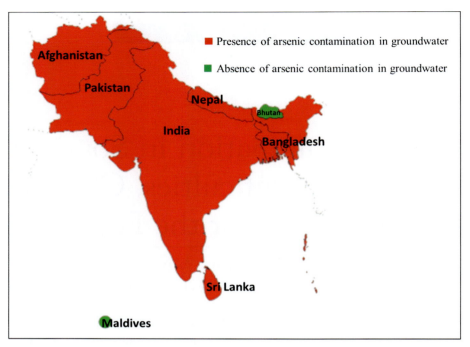

FIG. 1   Arsenic contamination in South Asia.

keratosis (Das and Sengupta, 2008). Also known as arsenic poisoning, arsenicosis encompasses a vast array of maladies, e.g., hyperpigmentation, brittle nails, skin lesions, muscle atrophy, respiratory, and renal disorders.

2.2 Cancer

The International Agency for Research on Cancer (IARC, 2004) has laid down arsenic as a potent carcinogen causing cancer of mainly the skin, lung, and gut. It has been reported that arsenic brings about alteration in genetic disposition that in turn, triggers incidence of various cancers (Mahata et al., 2004).

(i) Skin cancer: Arsenic exposure is known to induce biotransformation and increase in methylation capacity, engendering skin cancers (Yu et al., 2000). Recent studies have found the association of p62 protein upregulation and increase in Nrf2 levels as the underlying mechanisms for arsenic-induced cancer in keratinocytes (Shah et al., 2017).

(ii) Lung cancer: Hopenhayn-Rich et al. (1998) established a direct dose-dependence incidence of lung cancers with increasing arsenic concentration in drinking water. Lamm et al. (2015) inferred that 200 µg/L of arsenic is enough to cause high-level lung carcinoma.

2.3 Neuropathy

Chronic arsenic toxicity has been known to cause neuropathy, where the person affected shows symptoms of muscle weakness, atrophy, and nerve degeneration (Guha Mazumder, 2003). In an elaborate study carried out on residents exposed to arsenic-contaminated groundwater, it

2.4 Respiratory Disorders

Consumption of water containing moderate to high levels of arsenic reportedly disrupts normal breathing, leading to infection of the respiratory tract in general (Kapaj et al., 2006).

These detrimental effects stated above caused due to arsenic poisoning have prodded government and other regulatory boards to amend the existing guidelines in order to bring down the permissible limits of arsenic below their current values, for instance the United States and Canada have brought down the permissible arsenic levels in potable water from 50 to $10\,\mu g/L$.

3 COUNTRY-WISE STATUS OF ARSENIC CONTAMINATION IN SOUTH ASIA

The succeeding paragraphs shed light on the repercussions of arsenic contamination in groundwater in the affected countries of South Asia.

3.1 Afghanistan

An earlier study (Saltori, 2004) reported that Ghazni and Logar regions of Afghanistan were the most affected parts in Afghanistan. A survey comprised testing of water quality in drinking wells and it was seen that arsenic levels were much above the permitted limit of $10\,\mu g/L$, and ranged from 10 to $500\,\mu g/L$.

3.2 Bangladesh

Groundwater in the deltaic plains formed near Padma–Meghna rivers in Bangladesh are the most severely affected areas with respect to arsenic contamination. There are aplenty tube wells in this area which the residents use for various household purposes, including drinking (Mukherjee et al., 2006). Numerous tube wells are dug up to depths of $\sim 350\,m$ or more, which results in exposure of the water to high deposition of arsenic (Fendorf et al., 2010). Alarmingly, studies revealed that around 30% of the tube wells had serious grade arsenic contamination. Arsenic concentrations in groundwater exceeded 50 ppb in 61 out of the 64 districts that had participated in the survey. Health repercussions involved skin lesions and arsenic poisoning among one-fourth population in Bangladesh (Brinkel et al., 2009; Tareq et al., 2010).

3.3 India

Several Indian states face risk of arsenic poisoning in groundwater, for instance, Assam, Bihar, Chhattisgarh, Jharkhand, and Uttar Pradesh have arsenic levels ranging from 50 to 100 ppb or more in water being consumed for household uses. However, West Bengal happens to be the worst affected state of all. In fact, the first incidence of arsenicosis in India was reported in West Bengal. In a survey spanning about 20 years, it was observed that almost half of the tube wells installed had arsenic concentration of 10 ppb and almost a fourth of the total number of tube wells contained arsenic levels more than 50 ppb. Thousands of people in these affected areas suffer from serious cases of arsenicosis, respiratory diseases, skin lesions and skin cancer, caused due to contamination of arsenic-laden water (Mukherjee et al., 2006). The School of Environmental Studies, Jadavpur University, West Bengal, has segregated the arsenic-affected districts of West Bengal into three zones, namely, (i) "highly affected" districts (Bardhaman, Howrah, Hooghly, Kolkata, Malda, Murshidabad, Nadia, North 24-Parganas, and South 24-Parganas), with arsenic levels exceeding 50 ppb, (ii) "mildly affected" districts with arsenic levels ranging between 10 and 50 ppb, and (iii) "arsenic-safe" districts where arsenic concentrations are around 2–3 ppb in groundwater (Adhikary and Mandal, 2017; Dey et al., 2014). Similar to the situation as mentioned in Bangladesh, the major cause of arsenic poisoning in the aforementioned states is ascribed to their location on the Ganga–Brahmaputra basin which itself is subjected to grave arsenic contamination (Shewale et al., 2017).

3.4 Nepal

Arsenic contamination in groundwater is a gripping issue in the Terai region of Nepal. The WHO-financed survey carried out by the Department of Water Supply and Sewage discovered that 90% of the groundwater in Terai had high levels of arsenic. Arsenic blanket test conducted by the National Sanitation Steering Committee in 25 districts of Nepal analyzed 737,009 groundwater samples. The outcomes revealed that many of the test cases contained 10–50 ppb of arsenic (Kayastha, 2015; Ngai et al., 2006). Use of this arsenic-contaminated water in irrigation and agricultural purposes hampers the quality of crops, subsequently causing ill health effects (Dahal et al., 2008). Maharjan et al. (2005) reported a community-based, dose response study on arsenic contamination in the lowland areas of Nepal where almost 7% of the inhabitants suffered from severe arsenicosis.

3.5 Pakistan

Public health in the Sindh and Punjab provinces of Pakistan is highly affected by the diabolical effects of arsenic. Alarmingly, more than 50 million people in Indus valley consume arsenic-contaminated water that exceeds the safety level as prescribed by WHO (Sanhrani et al., 2017). Water in Dadu city was found to contain arsenic of concentrations as high as $100\,\mu g/L$. Manchar lake in Sindh had an average arsenic concentration of $97.5\,\mu g/L$. In the same province, water in Jamshoro city had arsenic

levels much above $50\,\mu g/L$. Similar situation looms over various other major Pakistani cities like Sukkur, Nawabshah, Hyderabad, Karachi, and Thar. Such disagreeable observations regarding arsenic contamination in groundwater of Pakistan is recently being correlated to a general lack of awareness about water quality and purification schemes.

3.6 Sri Lanka

Arsenic has recently been discovered in groundwater in a few selected localities in southern Sri Lanka. Although investigative monitoring surveys like the ones being carried out in Bangladesh or India (Hossain et al., 2007) are in a nascent state, however the limited reports published describe the presence of arsenic concentrations above $0.35\,\mu g/L$. Mineralogical conditions suggest sulfide (pyrite) oxidation as a process for the release of arsenic into groundwater (Rajasooriyar, 2003; Villholth and Rajasooriyar, 2010).

4 DIFFICULTIES AND CHALLENGES ASSOCIATED WITH ARSENIC CONTAMINATION IN SOUTH ASIA

Although many remediation technologies and awareness programs are currently in action to counter the issue of arsenic contamination in South Asia; however, there are certain problems which need to be addressed in order to facilitate supply of arsenic free water in different countries of South Asia. These problems are discussed in the succeeding paragraphs.

4.1 Lack of Awareness

Surveys state that people inhabiting rural and remote regions of Bangladesh, Pakistan, and West Bengal, India are mostly ignorant about arsenic contamination in groundwater and their repercussions on health. For example, residents in Moribund delta regions of Bangladesh, who happen to be the immediate beneficiaries of curative activities far from aware of the arsenic contamination and its aftermath. Considering that most of the countries in South Asia fall under the categories of either underdeveloped or developing countries, acceptance for social activities and awareness programs pertaining to arsenic exposure has not been able to reach full audience. However, in the past decade social measures in order to spread awareness about arsenic poisoning and ways to overcome its ill health effects are gaining pace (Paul, 2004).

4.2 Absence of Sophisticated Remediation Technology

Limited technological advancement, owing to the developing and underdeveloped status of the South Asian countries is yet another hindrance to groundwater treatment for arsenic removal. Absence of modern equipments and sophisticated materials lead to unfavorable circumstances in the regions experiencing arsenic related hazards. Economy of the South Asian countries is yet another issue which makes it less likely for these provinces to initiate and implement high-end arsenic remediation technologies to household levels. This concern can be overcome by encouraging the use of indigenous small-scale technologies which will reach out to even remote areas, for facilitating supply of arsenic-free water.

4.3 Accessibility

Arsenic contamination in South Asia prevails mostly in the low-lying, topographically flat plains of rivers (Fendorf et al., 2010). Regions lying south of Ganga River in Bangladesh are more prone to elevated arsenic levels than other parts of the country. Due to higher elevation, these plains generally witness lesser flow of water and also reduced rainfall. Consequently, arsenic

concentration increases in these areas. Moreover, due to higher elevation, these particular regions are less accessible than rest of the country, thus posing a challenge in approachability of remedial programs to these areas (Paul, 2004).

5 REMEDIAL MEASURES FOR REMOVAL OF ARSENIC IN GROUNDWATER

Research is being conducted all over the world to overcome the serious problem of arsenic contamination in groundwater. Technologies, applicable in household scale as well as in industrial scale have been conceptualized. Out of these techniques, some are indigenous to South Asia whereas some other technologies have been developed abroad and have found use in Bangladesh and India. Some of the most widely used arsenic remediation technologies have been compiled in Table 1, along with the countries that have adopted a particular technology. Additionally, the following paragraphs describe certain such remediation technologies, categorized into "domestic" and "industrial" measures.

5.1 Domestic Measures

5.1.1 Rapid Small Scale Column Test (RSSCT)

The RSSCT method is based mainly on the principle of adsorption. Fixed bed adsorbers comprising iron-impregnated granular activated carbon are used here for arsenic removal. The advantages of this method include (i) low capital and operating cost and (ii) good mechanical stability for use in household taps and drinking water systems (Assaf-Anid et al., 2003).

5.1.2 Combination Treatment

Litter et al. (2012) have elaborately described the efficacy of arsenic removal techniques developed in Latin America that collectively

118

5. ARSENIC CONTAMINATION IN SOUTH ASIAN REGIONS

TABLE 1 List of Some Arsenic Remediation Technologies

Sl. No.	Name of arsenic remediation technology	Countries where the technology is in use	References
1.	Rapid Small-Scale Column Test (New York)	Bangladesh	Assaf-Anid et al. (2003)
2.	Alufloc (Peru)	Argentina, Mexico, Peru	Castro de Esparza et al. (2005)
3.	Natural geological materials (Latin America)	Argentina, Mexico	Litter et al. (2012)
4.	RO desalinator (Germany)	Bangladesh	Geucke et al. (2009)
5.	Passive sedimentation (Bangladesh)	Bangladesh	
6.	In situ oxidation (DPHE-Danida laboratory)	Bangladesh	
7.	SORAS (Switzerland)	Bangladesh, India	Wegelin et al. (2000)
8.	Bucket treatment units (DPHE-Danida laboratory)	Bangladesh	Bangladesh
9.	Sono filter (Bangladesh)	Bangladesh	Hussam (2009)
10.	Subterranean arsenic removal technology (Queen's University, Belfast)	Bangladesh, India	Sen Gupta et al. (2006)
11.	DRDO arsenic removal technology (DRDO, India)	India	www.drdo.gov.in
12.	Neosepta AX 01 membrane (Portugal)	Bangladesh, Cambodia	Pessoa-Lopes et al. (2016)
13.	Multiwall carbon nanotubes (Bangladesh)	Bangladesh	Roy et al. (2013)
14.	Magnetite-reduced graphene oxide (Korea)	Bangladesh, India	Chandra et al. (2010)
15.	Kanchan arsenic filter (MIT, ENPHO & RWSSSP)	Nepal	Ngai et al. (2006)

employ oxidation, adsorption, coagulation, and flocculation. A low cost household methodology (Castro de Esparza, 2010; Castro de Esparza et al., 2005) using Alufloc was used in Peru that consisted of chlorine as oxidant, activated clay as adsorbent, and alum as coagulant. Alufloc reportedly brought down arsenic levels in water by 98%. In Argentina, a hydrogel made up of aluminum sulfate, powdered calcium hypochlorite, ammonium hydroxide, and demineralized water was developed. This hydrogel reduced arsenic concentrations to below 10 ppb (Luján, 2001).

5.1.3 *Use of Natural Geological Material*

Iron and aluminum rich minerals, e.g., goethite, gibbsite, and hematite act as economical adsorbents for arsenic removal (Litter et al., 2012). Indigenous limestone has also been found to be quite effective in removing arsenic (Romero et al., 2004).

5.1.4 *Desalinator*

A cost effective, small-scale desalinator based on reverse osmosis (RO) and nanofiltration has been described for efficient arsenic removal especially from drinking water (Geucke et al., 2009). In this, a RO desalinator-type power survivor 160E (designed in Switzerland) is used to bring down arsenic removals to the WHO-prescribed limit of 10 ppb or below. The advantages of the RO desalinator include high efficiency, easy operation, higher effluent quality, and durability, by virtue of which the use of this technology has been suggested for prospective arsenic removal in South Asian countries.

5.1.5 *Oxidation*

Some research endeavor gives a detailed account of oxidation-based arsenic removal methodologies. Here, oxidation happens to be

EVALUATING WATER QUALITY TO PREVENT FUTURE DISASTERS

the pretreatment step to convert arsenite into arsenate. Arsenite can be oxidized by oxygen, ozone, chlorine, hydrogen peroxide, etc. However, in developing countries, oxidation for arsenic remediation is facilitated by cost-effective options like atmospheric oxygen, hypochlorite, and permanganate. Some examples of usage of oxidation for arsenic removal are given below.

(i) Passive sedimentation

Passive sedimentation functions as yet another option for arsenic removal. In passive sedimentation, oxidation of water takes place during collection and storage. Ahmed (2003) reported that more than 50% of arsenic is reduced, possibly by sedimentation, in water containing around 400 mg/L of alkalinity. Calcium carbonate and iron are said to play a major role in this.

(ii) In situ oxidation

This helps in minimization of arsenic levels in water. A project entitled DPHE-Danida Arsenic mitigation project worked toward lessening of arsenic levels in tube well water. The steps include aeration of tube well water and storage in reservoirs, followed by its release back into aquifers. In this way, the dissolved oxygen oxidizes the arsenate to the less diabolical arsenite form (Ahmed, 2001).

(iii) Solar oxidation

Water stored in transparent bottles is exposed to long periods of sunlight by which the ultraviolet radiations catalyze the process of oxidation (Wegelin et al., 2000). This method reportedly reduces arsenic concentration by a third.

5.1.6 Coprecipitation

Bangladesh is the worst-stricken country in South Asia in terms of arsenic contamination in groundwater. The country has adopted certain indigenous small-scale and cost-effective technologies based on coprecipitation to attenuate the problem. Some examples are stated as follows:

(i) Iron-based chemical coprecipitation

Iron-based chemical coprecipitation followed by sand filtration has been developed by the Bangladesh Council of Scientific and Industrial research.

(ii) Bucket treatment units

Bucket treatment units or BTU have been conceptualized by DPHE-Danida arsenic mitigation project. In BTU, aluminum sulfate and potassium permanganate are utilized to coprecipitate out arsenic. Better results are obtained on addition of ferric chloride to the aforementioned chemicals.

5.1.7 Sono Filter

It is a simple device made up of composite iron matrix. Two buckets are used here; the first bucket is placed atop and is filled with coarse sand and composite iron matrix. The sand filters coarse particles and controls the flow of the water, while the iron removes inorganic arsenic. Then, the water flows into a second bucket where it again travels through a sand layer, followed by migrating through wood charcoal. In the final step, water again seeps through a bed of finer sand and wet brick chips to remove fine particles and stabilize water flow (Hussam, 2009).

5.1.8 Subterranean Arsenic Removal Technology

Subterranean arsenic removal technology (SART) developed by TIPOT is based on pumping of water into well and its aeration and reintroduction to accentuate arsenic absorption capacity. An oxidation zone is created which entraps arsenic on the soil particles through adsorption process. One advantage of this technique is that it is completely free from inclusion of chemicals. Thus, risks associated with waste disposal are completely prevented. This technique has been widely employed for arsenic removal in West Bengal, India, and many regions of Bangladesh. The first SART water treatment plant was established at Kashimpur,

West Bengal, India in 2004. Many such systems have been installed since then. On an average, one SART system supplies 3000 L of arsenic-free water per day (Sen Gupta et al., 2006).

5.1.9 DRDO Arsenic Removal Technology

DRDO arsenic removal technology (DART) is based on coprecipitation and adsorption. Field trials have shown that the device brings down arsenic levels in water from >1000 ppb to the EPA-prescribed permissible limits of <10 ppb. Owing to its major advantages like cost-effectiveness, nondependence on electricity, durability, and easy maintenance, over 5000 filters have been successfully installed in selected villages of Bihar, Chhattisgarh, North-Eastern India, Uttar Pradesh, and West Bengal for providing arsenic-free water. Waste generated is safely disposed of as concrete blocks, thus making it a green technology.

5.1.10 Kanchan Arsenic Filter

The Massachusetts Institute of Technology along with two local teams, Environment and Public Health Organization (ENPHO), and Rural Water Supply and Sanitation Support Programme (RWSSSP) in Nepal have conceptualized an indigenous arsenic removal filter, i.e., the Kanchan arsenic filter (KAF). The KAF consists of numerous iron nails that are deliberately subjected to rusting in order to produce ferric hydroxide, which in turn happens to be an efficient adsorbent for arsenic. Upon passage of contaminated water through the KAF, the arsenic present gets adsorbed onto the rust material. After this, the arsenic-laden rust seeps down to a sand layer present beneath and is easily removed. Reports suggest that use of KAF facilitates arsenic removal up to 90% (Ngai et al., 2006).

5.2 Industrial Measures

5.2.1 Ion Exchange

Conventional anion exchange resins are quite efficient in removing arsenic from groundwater. An instance is the Neosepta AX 01 anion exchange membrane. This membrane operates on the principle of Donnan dialysis, with chlorine being used as counter-ion. This type of membrane offers simultaneous precipitation of iron and arsenic from water. Its obvious advantages are that it is an environment friendly technique and also, requires minimalistic maintenance (Pessoa-Lopes et al., 2016).

5.2.2 Hybrid Technology Based on Ion Exchange and Electrodialysis

Recently, Ortega et al. (2017) have described a hybrid technology where electro-regenerated anion exchange has been associated with electrodialysis. This technology, abbreviated as "IXED" relies on a membrane-resin-membrane arrangement. As(V) gets exchanged with hydroxyl anions, accompanied with regeneration of resin. In this method, arsenic concentrations in water tested reportedly reduce to less than 10 ppb or below.

5.2.3 Use of Nanoparticles

Nicomel et al. (2015) have compiled a list of nanoparticles that have been developed in recent times for reducing arsenic levels in water. Nanoparticles are particularly preferred because of their higher surface area, hence more adsorption. Some instances are mentioned in the following:

(i) Carbon nanotubes

Carbon nanotubes (CNTs) have been observed to be fairly efficient in reducing arsenic concentrations by half, provided the initial concentration is restricted to less than 1 mg/L (Roy et al., 2013).

(ii) Titanium-based nanoparticles

In contrast to CNTs, titanium-based nanoparticles prove to be much more efficient in almost completely converting arsenite to arsenate, in the presence of sunlight and dissolved oxygen (Pena et al., 2005).

(iii) Iron-based nanoparticles

In case of iron-based nanoparticles, zerovalent iron nanoparticles and iron oxide nanoparticles

6 CONCLUSIONS AND VISION FOR THE FUTURE

display considerable efficacy in treatment of arsenic-contaminated water. The principles here involve mainly, adsorption and coprecipitation (Tang and Lo, 2013).

(iv) Zirconium oxide nanoparticles

Zirconium oxide nanoparticles are favorable for removal of arsenic from water, because of their stability and nonreactivity. Unlike ceria nanoparticles which are pH dependent, zirconium oxide nanoparticles operate irrespective of the prevailing pH conditions (Cui et al., 2012).

5.2.4 Magnetite–Graphene Hybrids

Developed in Korea, the magnetite-reduced graphene oxide (M-RGO) hybrids display much better arsenic removal efficiency than magnetite alone. On account of more number of adsorption sites, the M-RGO hybrids exhibit high binding capacity for As(III) and As(V), and therefore are able to bring down arsenic levels to as low as 1 ppb (Chandra et al., 2010).

Apart from the aforementioned technologies, there are reports of several Indian technologies developed for arsenic remediation that have been elaborated elsewhere in this book.

6 CONCLUSIONS AND VISION FOR THE FUTURE

The detrimental effects inflicted by high arsenic concentrations in groundwater are continuously exceeding the prescribed limits of 10 ppb set by the WHO, sometimes increasing far beyond 10 ppb. Industrial operations such as anthropogenic emissions and mining are some unprecedented reasons that are accountable for this continual rise of arsenic contamination. As per recent reports, in Bangladesh alone, 80 million people are suffering due to the unavoidable dependence on arsenic-contaminated water, and incidences of arsenic poisoning and associated cancers are on the rise. In view of these dire consequences, it is essential that this deleterious

issue of arsenic contamination must be addressed and curative steps be taken at the earliest in order to circumvent the negative impacts of arsenic on health. Based on this information, it could be suggested that the technologies discussed in this chapter may be adopted by the South Asian countries to ensure an economical way of averting the difficulties of arsenic contamination.

References

Adhikary, R., Mandal, V., 2017. Status of arsenic toxicity in ground water in West Bengal. MOJ Toxicology 3 (5), 00063.

Ahmed, M.F., 2001. An overview of arsenic removal technologies in Bangladesh and India. In: Proceedings of BUET-UNU International Workshop on Technologies for Arsenic Removal From Drinking Water. Dhaka, pp. 5–7.

Ahmed, M.F., 2003. Treatment of arsenic contaminated water. In: Arsenic Contamination: Bangladesh Perspective, pp. 354–403.

Assaf-Anid, N.M., Duby, P.F., 2003. A Novel Adsorption Technology for Small-Scale Treatment of Arsenic. United States Environmental Protection Agency.

Brinkel, J., Khan, M.H., Kraemer, A., 2009. A systematic review of arsenic exposure and its social and mental health effects with special reference to Bangladesh. Int. J. Environ. Res. Public Health 6 (5), 1609–1619.

Castro de Esparza, M.L., 2010. Mejoramiento de la Calidaddel Agua de Pozosen Zonas Rurales de Puno. In: Litter, M.I., Sancha, A.M., Ingallinella, A.M. (Eds.), Tecnologíaseconómicas para el abatimiento de arsénicoenaguasBuenos Aires: Editorial ProgramaIberoamericano de Cienciay Tecnología para el Desarrollo, pp. 243–255. Available from, http://www.cnea.gov.ar/xxi/ambiental/iberoarsen/docs/libro%20remocion%20As%20Marta%20Litter.pdf.

Castro de Esparza, M.L., Noriega, P.R., Wong, N.M., Inamine, A.A., 2005. Estudio para el Mejoramiento de la Calidaddel Agua de Pozosen Zonas Rurales de Puno. CEPIS, OPS, Lima. Available from, http://www.bvsde.paho.org/bvsacd/cd53/punoinforme/puno-informe.html.

Chandra, V., Park, J., Chun, Y., Lee, J.W., Hwang, I.C., Kim, K.S., 2010. Water-dispersible magnetite-reduced graphene oxide composites for arsenic removal. ACS Nano 4 (7), 3979–3986.

Cui, H., Li, Q., Gao, S., Shang, J.K., 2012. Strong adsorption of arsenic species by amorphous zirconium oxide nanoparticles. J. Ind. Eng. Chem. 18 (4), 1418–1427.

Dahal, B.M., Fuerhacker, M., Mentler, A., Karki, K.B., Shrestha, R.R., Blum, W.E., 2008. Arsenic contamination of soils and agricultural plants through irrigation water in Nepal. Environ. Pollut. 155 (1), 157–163.

Das, N.K., Sengupta, S.R., 2008. Arsenicosis: diagnosis and treatment. Indian J. Dermatol. Venereol. Leprol. 74 (6), 571.

Dey, T.K., Banerjee, P., Bakshi, M., Kar, A., Ghosh, S., 2014. Groundwater arsenic contamination in West Bengal: current scenario, effects and probable ways of mitigation. Int. Lett. Nat. Sci. 8 (1), 45–58.

Fendorf, S., Michael, H.A., van Green, A., 2010. Spatial and temporal variations of groundwater arsenic in South and Southeast Asia. Science 328 (5982), 1123–1127.

Geucke, T., Deowan, S.A., Hoinkis, J., Pätzold, C., 2009. Performance of a small-scale RO desalinator for arsenic removal. Desalination 239 (1–3), 198–206.

Guha Mazumder, D.N., 2003. Chronic arsenic toxicity: clinical features, epidemiology and treatment: experience in West Bengal. J. Environ. Sci. Health A 38 (1), 141–163.

Hopenhayn-Rich, C., Biggs, M.L., Smith, A.H., 1998. Lung and kidney cancer mortality associated with arsenic in drinking water in Cordoba, Argentina. Int. J. Epidemiol. 27 (4), 561–569.

Hossain, F., Hill, J., Bagtzoglou, A.C., 2007. Geostatistically based management of arsenic contaminated ground water in shallow wells of Bangladesh. Water Resour. Manag. 21 (7), 1245–1261.

Hussam, A., 2009. Contending with a development disaster: SONO filters remove arsenic from well water in Bangladesh (innovations case discussion: SONO filters). Innovations Technol. Gov. Glob. 4 (3), 89–102.

International Agency for Research on Cancer Working Group on the Evaluation of Carcinogenic Risks to Human, World Health Organization and International Agency for Research on Cancer, 2004. Some Drinking Water Disinfectants and Contaminants Including Arsenic, p. 84.

Kapaj, S., Peterson, H., Liber, K., Bhattacharya, P., 2006. Human health effects from chronic arsenic poisoning—a review. J. Environ. Sci. Health A 41 (10), 2399–2428.

Kayastha, S.P., 2015. Arsenic contamination in the deep and shallow groundwater of Bara district, Nepal. Cross. Border Int. J. Interdiscip. Stud. 3 (1), 65–70.

Lamm, S.H., Ferdosi, H., Dissen, E.K., Li, J., Ahn, J., 2015. A systematic review and meta-regression analysis of lung cancer risk and inorganic arsenic in drinking water. Int. J. Environ. Res. Public Health 12 (12), 15498–15515.

Litter, M.I., Alarcón-Herrera, M.T., Arenas, M.J., Armienta, M.A., Avilés, M., Cáceres, R.E., Cipriani, H.N., Cornejo, L., Dias, L.E., Cirelli, A.F., Farfán, E.M., 2012. Small-scale and household methods to remove arsenic from water for drinking purposes in Latin America. Sci. Total Environ. 429, 107–122.

Luby, S.P., Gupta, S.K., Sheikh, M.A., Johnston, R.B., Ram, P.K., Islam, M.A., 2008. Tubewell water quality and predictors of contamination in three flood-prone areas in Bangladesh. J. Appl. Microbiol. 105 (4), 1002–1008.

Luján, J.C., 2001. Aluminum hydroxide hydrogel to remove arsenic from water. Pan Am. J. Public Health 9 (5), 302–305.

Maharjan, M., Watanabe, C., Ahmad, S.A., Ohtsuka, R., 2005. Arsenic contamination in drinking water and skin manifestations in lowland Nepal: the first community-based survey. Am. J. Trop. Med. Hyg. 73 (2), 477–479.

Mahata, J., Chaki, M., Ghosh, P., Das, L.K., Baidya, K., Ray, K., Natarajan, A.T., Giri, A.K., 2004. Chromosomal aberrations in arsenic-exposed human populations: a review with special reference to a comprehensive study in West Bengal, India. Cytogenet. Genome Res. 104 (1–4), 359–364.

Mukherjee, A., Sengupta, M.K., Hossain, M.A., Ahamed, S., Das, B., Nayak, B., Lodh, D., Rahman, M.M., Chakraborti, D., 2006. Arsenic contamination in groundwater: a global perspective with emphasis on the Asian scenario. J. Health Popul. Nutr. 24, 142–163.

Mukherjee, A., Fryar, A.E., Scanlon, B.R., Bhattacharya, P., Bhattacharya, A., 2011. Elevated arsenic in deeper groundwater of the western Bengal basin, India: extent and controls from regional to local scale. Appl. Geochem. 26 (4), 600–613.

Ngai, T.K., Murcott, S., Shrestha, R.R., Dangol, B., Maharjan, M., 2006. Development and dissemination of Kanchan™ arsenic filter in rural Nepal. Water Sci. Technol. Water Supply 6 (3), 137–146.

Nicomel, N.R., Leus, K., Folens, K., Van Der Voort, P., Du Laing, G., 2015. Technologies for arsenic removal from water: current status and future perspectives. Int. J. Environ. Res. Public Health 13 (1), 62.

Ortega, A., Oliva, I., Contreras, K.E., González, I., Cruz-Díaz, M.R., Rivero, E.P., 2017. Arsenic removal from water by hybrid electro-regenerated anion exchange resin/electrodialysis process. Sep. Purif. Technol. 184, 319–326.

Paul, B.K., 2004. Arsenic contamination awareness among the rural residents in Bangladesh. Soc. Sci. Med. 59 (8), 1741–1755.

Pena, M.E., Korfiatis, G.P., Patel, M., Lippincott, L., Meng, X., 2005. Adsorption of As (V) and As (III) by nanocrystalline titanium dioxide. Water Res. 39 (11), 2327–2337.

Pessoa-Lopes, M., Crespo, J.G., Velizarov, S., 2016. Arsenate removal from sulphate-containing water streams by an ion-exchange membrane process. Sep. Purif. Technol. 166, 125–134.

Rajasooriyar, L.D., 2003. A Study of the Hydrochemistry of the Uda Walawe Basin, Sri Lanka, and the Factors That Influence Groundwater Quality. Doctoral Dissertation, University of East Anglia.

Romero, F.M., Armienta, M.A., Carrillo-Chavez, A., 2004. Arsenic sorption by carbonate-rich aquifer material, a

REFERENCES

control on arsenic mobility at Zimapan, Mexico. Arch. Environ. Contam. Toxicol. 47 (1), 1–13.

Roy, P., Choudhury, M., Ali, M., 2013. As(III) and As(V) adsorption on magnetite nanoparticles: adsorption isotherms, effect of pH and phosphate and adsorption kinetics. Int. J. Chem. Environ. Eng. 4, 55–63.

Saltori, R., 2004. Arsenic Contamination in Afghanistan: Preliminary Findings. DACCAR, Kabul, pp. 1–5.

Sanhrani, M.A., Teshome, M., Sanjrani, N.D., Leghari, S.J., Moryani, H.T., Shabnam, A.B., 2017. Current situation of aqueous arsenic contamination in Pakistan, focused on Sindh and Punjab province, Pakistan: a review. J. Pollut. Eff. 5 (4), 207.

Sen Gupta, B., Bandopadhyay, A., Mukhopadhyay, S., 2006. Subterranean Arsenic Removal (SAR) Technology for Groundwater Remediation. Available from, www.insituarsenic.org.

Shah, P., Trinh, E., Qiang, L., Xie, L., Hu, W.Y., Prins, G.S., Pi, J., He, Y.Y., 2017. Arsenic induces p62 expression to form a positive feedback loop with Nrf2 in human epidermal keratinocytes: implications for preventing arsenic-induced skin cancer. Molecules 22 (2), 194.

Shewale, M., Bhandari, D., Garg, R.K., 2017. A review of arsenic in drinking water: Indian scenario. Int. J. Sci. Res. Sci. Eng. Technol. 3 (5), 300–304.

Tang, S.C., Lo, I.M., 2013. Magnetic nanoparticles: essential factors for sustainable environmental applications. Water Res. 47 (8), 2613–2632.

Tareq, S.M., Islam, S.N., Rahmam, M.M., Chowdhury, D.A., 2010. Arsenic pollution in groundwater of Southeast Asia: an overview on mobilization process and health effects. Bangladesh J. Environ. Res. 8, 47–67.

Villholth, K.G., Rajasooriyar, L.D., 2010. Groundwater resources and management challenges in Sri Lanka—an overview. Water Resour. Manag. 24 (8), 1489–1513.

Wegelin, M., Gechter, D., Hug, S., Mahmud, A., Motaleb, A., 2000. SORAS—a simple arsenic removal process. In: WEDC Conference 26, pp. 255–258.

Yu, R.C., Hsu, K.H., Chen, C.J., Froines, J.R., 2000. Arsenic methylation capacity and skin cancer. Cancer Epidemiol. Biomarkers Prev. 9 (11), 1259–1262.

CHAPTER

6

Cyanobacteria and Their Toxins

J.S. Metcalf*, N.R. Souza

Brain Chemistry Labs, Institute for Ethnomedicine, Jackson, WY, United States
*Corresponding author: E-mail: james@ethnomedicine.org

1 INTRODUCTION

Freshwater resources are finite with many demands placed upon them. The action of global climate change and human population growth are resulting in increased abstraction and the need to prepare drinking water from water sources, such as seawater, through desalination (Sheppard et al., 2010). This often results in a requirement for some form of physical and/or chemical treatment. As a result of, e.g. sanitation or from washing, such wastewater and gray water also pose a burden on facilities with concomitant energy requirements. Within fresh, brackish, and marine ecosystems, many food chains and food webs exist, comprised of plankton, both phyto- and zoo-plankton, invertebrates, fish, birds, and mammals. All of these organisms require good quality water in order to allow them to survive and reproduce. Adversely affecting such waters, whether through pollution or changes to the chemistry of the water, can result in shifts in the natural ecosystem dynamics. Such changes can alter the dominance of organisms present in the waters and can be noticed by increased growth of plants and algae, largely attributed to

increases in nutrients such as nitrogen and phosphorous in the water (O'Hare et al., 2018). Concerning the various algal groups in freshwater, often in temperate climates during the spring, diatoms are the first algal group to bloom and once the silica present in the waterbody has been used up, this algal group is no longer able to bloom and subsequently cyanobacteria become dominant where they can form mass populations within waterbodies (Xu et al., 2009).

2 WHAT ARE CYANOBACTERIA?

Cyanobacteria, previously known as blue-green algae, are Gram-negative photosynthetic bacteria. Members of true bacteria, lacking a nucleus, they are also able to grow heterotrophically (Fogg et al., 1973) and have a number of adaptations that allow them to thrive in water, which are also key characteristics for traditional taxonomic identification. Some members of the cyanobacteria have gas vesicles, which collapse when the cells sink. Often such buoyant cyanobacteria will be found at the surface of waterbodies early in the morning and once exposed to

Separation Science and Technology, Volume 11
https://doi.org/10.1016/B978-0-12-815730-5.00006-5

Copyright © 2019 Elsevier Inc. All rights reserved.

sunlight can begin to photosynthesize. During this time, carbon fixation occurs and sugars are produced within the cell. As a result of their increased density they begin to sink, collapsing the gas vesicles. At depth, the sugars are used for cellular respiration and gas vesicle proteins are produced resulting in a decrease in density and an increase in buoyancy such that the cyanobacteria are subsequently able to float toward the surface where they can again perform photosynthesis (Ibelings et al., 1991). This ability of cyanobacteria to be present at various depths can also affect water treatment plant intakes as subsurface maxima of cyanobacteria, such as *Planktothrix*, are able to survive under the low light conditions below the thermocline in thermally stratified waterbodies (Halstvedt et al., 2007). In addition to chlorophyll *a*, cyanobacteria potentially contain many pigments which allow these organisms to appear in a variety of colors, from green to blue green and black to red. The pigments include the blue-colored phycocyanin and allophycocyanin which exist in the phycobilisome, the red pigment phycoerythrin and the carotenoid pigments, ultimately allowing them to harvest energy over a range of light wavelengths (Fogg et al., 1973; Komarek and Johansen, 2015). Monitoring phycocyanin also allows for the detection of cyanobacteria and can be useful for assessing waters, including lakes and rivers (Bertone et al., 2018; Brient et al., 2008).

Other adaptations include the ability of some cyanobacterial genera to fix atmospheric diazotrophic nitrogen. In genera such as *Dolichospermum*, *Aphanizomenon*, and *Cylindrospermopsis*, cell differentiation can occur to produce specialized cells within the filament termed heterocysts (Fogg et al., 1973). By excluding oxygen, heterocysts allow enzymatic nitrogen fixation to occur, producing organic nitrogen sources for use by the cyanobacterial cell (Stewart, 1972, 1980). Within a waterbody, when nitrogen rather than phosphorous is the limiting nutrient for cyanobacterial and algal growth, nitrogen-fixing cyanobacteria may become dominant as there is now no limiting requirement concerning nitrogen (Smith, 1983). Akinetes, thick-walled cyanobacterial spores produced by many filamentous strains of cyanobacteria (e.g., *Dolichospermum*, *Gloeotrichia*, *Aphanizomenon*) are resistant to adverse conditions, working as perennating structures, allowing for germination under favorable conditions (e.g., light, temperature; Cardemil and Wolk, 1979; Yamamoto, 1976). Other structural modifications among the cyanobacteria include branching, both true and false branching and the ability to form aggregates of cells and filaments. Such aggregations may be observed by the naked eye in waters and can appear as green clumps such as with *Microcystis* or green straw-like tufts as found with *Aphanizomenon* (Fogg et al., 1973). Not all cyanobacteria form aggregates but mass accumulations can occur and individual cells and filaments can produce many thousands of cells per mL, resulting in a green "pea-soup" appearance (Plate 1). Under certain conditions such as calm, warm days that allow cyanobacteria to proliferate, the potential to form accumulations on the surface of waters such as through positive buoyancy from gas vesicles and other physical factors can lead to large concentration factors acting on the bloom. This largely occurs through the action of winds which ultimately push surface cyanobacterial accumulations across lakes to the leeward shore where they can become stranded on shorelines and in embayments (Fogg et al., 1973). The ultimate movement of cyanobacterial blooms across lakes can result in very large concentration factors acting upon cell numbers such that consistencies of thick green paint can result from these cells on or near shores. The connection between the deaths of wild and domestic animals and the occurrence of cyanobacteria (blue-green algae) have been known since at least the 19th century (Codd et al., 2005). Although blooms were known to be harmful, it was not until the advancements in analytical chemistry that the nature of the risks posed by cyanobacteria began to be understood.

EVALUATING WATER QUALITY TO PREVENT FUTURE DISASTERS

PLATE 1 Representative blooms of cyanobacteria: (A) representative bloom comprising *Dolichospermum* sp., (B) from an Idaho freshwater lake, (C) representative bloom of *Microcystis* spp., and (D) from St. Lucie canal, Florida.

3 WHAT ARE CYANOBACTERIAL TOXINS?

Meticulous testing with cyanobacterial strains and bloom samples, in addition to samples from animal poisoning incidents, led to an understanding of the association of cyanobacteria with adverse health outcomes. Through preparative chemistry and bioactivity guided fractionation, "toxic" fractions were investigated by analytical chemistry. As a result, low molecular weight alkaloids and peptides were identified. Based on their chemical structures and modes of action, cyanotoxins were classified into different groups.

3.1 Hepatotoxins

Assessment of extracts of the cyanobacterium *Microcystis* with test animals resulted in their deaths, accompanied by swollen livers. Subsequently, potent hepatotoxins comprising cyclic seven amino acid peptides, were identified and named microcystins (Fig. 1; Catherine et al., 2017). Multiple microcystins have been identified, due to variation in the structure at a number of sites leading to the naming of individual microcystins, for example, with microcystin-LR, which is one of the most common and toxic microcystins, possessing amino acids L-leucine (L) and L-arginine (R) at positions 2 and 4 of the microcystin ring. Based on substitutions at these sites and chemical modifications at other positions in the molecule over 200 microcystin molecules have been identified (Spoof and Catherine, 2017). Analysis of *Nodularia spumigena* extracts also showed hepatotoxicity and although microcystins were not identified, a related class of compounds, the nodularins (Fig. 2), were extracted, purified, and characterized (Rinehart et al., 1988). Similar in terms of

6. CYANOBACTERIA AND THEIR TOXINS

128

FIG. 1 Generic structure of the microcystins where X and Y represent variable amino acids.

FIG. 2 The structure of nodularin.

their acute toxicity, these cyclic peptides were found to differ from microcystins by having fewer (only five) amino acids in the peptide structure. As with microcystins, some modifications to the structure have led to the recent identification of 10 variants (Spoof and Catherine, 2017). Although nodularins have only been documented in the genus *Nodularia*, microcystin production has been documented in many genera and species of cyanobacteria including *Microcystis*, *Dolichospermum*, *Planktothrix*, *Arthrospira*, *Rivularia*, and *Hapalosiphon* (Metcalf and Codd, 2012). As *Nodularia* is a brackish water species, nodularins are not commonly found in freshwaters, therefore, microcystins are the most likely encountered hepatotoxic peptides in these environments. As their classification suggests, acute doses of microcystins and nodularins result in hepatotoxicity to mammals. This largely occurs through the destruction of the liver cytoskeleton resulting in pooling of the blood in this organ such that animals die of hypovolemic shock and there is insufficient blood for the remaining organs to function in the body (Carmichael, 1994). Although an acute dose usually results in death within 1–2 h, exposure to lower doses has been implicated in a higher incidence of primary liver cancer, as evidenced from China and Eastern Europe (Svircev et al., 2013; Ueno et al., 1996). Both microcystins and nodularins are capable of inhibiting protein phosphatases, essential mammalian cellular enzymes (Metcalf et al., 2001). These enzymes regulate various aspects of cell division and due to structural differences, microcystins and nodularins interact slightly differently with this

EVALUATING WATER QUALITY TO PREVENT FUTURE DISASTERS

3 WHAT ARE CYANOBACTERIAL TOXINS?

enzyme. In microcystins, the *N*-methyldehydroalanine (Mdha) moiety is a common component of the peptide structure which is able to chemically react with thiols (Metcalf et al., 2000a). Conversely, nodularins contain dehydrobutyric acid (dhb; Catherine et al., 2017) which is unable to undergo the same chemical reaction. In the case of microcystins, the Mdha moiety can covalently bind to cysteine 273 of the protein phosphatase active site whereas the dhb moiety of nodularin is unable to do this (Metcalf et al., 2017). This simple difference results in the ability of microcystin to act as a tumor promoter, whereas nodularin is considered a carcinogen (Nishiwaki-Matsushima et al., 1992; Ohta et al., 1994).

3.2 Cytotoxic and Genotoxic Alkaloids

After an outbreak of hepatic enteritis among aboriginal people in Australia, the source of the outbreak was identified as the local drinking water supply. Subsequently, from the reservoir a strain of *Cylindrospermopsis raciborskii* was isolated (Hawkins et al., 1985). Toxicity testing with this strain showed that it was hepatotoxic to mice and a guanidine alkaloid toxin, cylindrospermopsin (Fig. 3), was isolated (Hawkins et al., 1985; Terao et al., 1994). Although originally considered to be a tropical/subtropical toxin, as it was produced by strains of *C. raciborskii* in Australia, analysis of cyanobacterial strains and bloom material has increased the known

geographic and phylogenetic range of this toxin (Kokocinski et al., 2017). The screening of cyanobacterial strains has shown that cylindrospermopsin can be produced by genera of cyanobacteria including *Aphanizomenon*, *Umezakia*, and *Raphidiopsis* (Metcalf and Codd, 2012). After the discovery of cylindrospermopsin, two further variants, 7-epicylindrospermopsin and deoxycylindrospermopsin, were identified with the latter considered nontoxic (Banker et al., 1997, 2001; Norris et al., 1999). More recent analysis of Thai material has identified further variants of cylindrospermopsin, the 7 deoxy-desulfocylindrospermopsin and 7-deoxy-desulfo-12-acetylcylindrospermopsin (Wimmer et al., 2014).

Due to the planar nature of the molecule, cylindrospermopsin is considered to interact with the double helix of DNA, which indicates that this alkaloid toxin may be carcinogenic (Falconer and Humpage, 2001). Cylindrospermopsin has also been shown to inhibit mammalian protein translation (Ohtani et al., 1992; Terao et al., 1994) with protein synthesis also inhibited in plants (Metcalf et al., 2004).

3.3 Alkaloid Neurotoxins

The alkaloid neurotoxins are highly toxic, low molecular weight compounds. There are a number of groups that comprise this class of compounds, based on their structures and modes of action.

FIG. 3 The structure of cylindrospermopsin.

EVALUATING WATER QUALITY TO PREVENT FUTURE DISASTERS

3.3.1 Anatoxin-a

A toxin originally isolated from *Anabaena*, anatoxin-a (Fig. 4; Carmichael et al., 1975), has also been shown to be produced by, for example, *Aphanizomenon, Arthrospira, Cylindrospermum,* and *Phormidium* (Metcalf and Codd, 2012). Anatoxin-a is known to have three structural variants, homoanatoxin-a, 4-hydroxyhomoanatoxin-a, and 11-carboxyl anatoxin-a, and several breakdown products have been documented (Bruno et al., 2017; Carmichael, 1994; Carmichael et al., 1979; Devlin et al., 1977; James et al., 1998). Anatoxin-a works by acting as an acetylcholine mimic binding to nicotinic acetylcholine receptors, continually stimulating the receptors (Carmichael, 1994; Carmichael et al., 1979; Devlin et al., 1977). In sufficient doses, such stimulation can result in paralysis and death, largely as a result of the paralysis of the muscles controlling breathing. If the acute dose is survived then currently it is believed that there are no known long-term adverse health effects from exposure to anatoxin-a (Bruno et al., 2017).

3.3.2 Anatoxin-a(S)

Although similar in name, but not in structure or mode of action to anatoxin-a, anatoxin-a(*S*) (Fig. 5) is a naturally occurring organophosphate, similar to synthetic insecticides and pesticides (Mahmood and Carmichael, 1986). Mainly found in aquatic environments, where *Dolichospermum* is often the major producer, this toxic molecule has also been suspected of being produced by *Microcoleus* in desert crusts from Qatar (Metcalf et al., 2012). Anatoxin-a(*S*), like synthetic insecticides and pesticides, are able to inhibit acetylcholinesterases and sufficient doses can lead to paralysis and death (Carmichael, 1994; Cook et al., 1989; Henriksen et al., 1997; Mahmood and Carmichael, 1986).

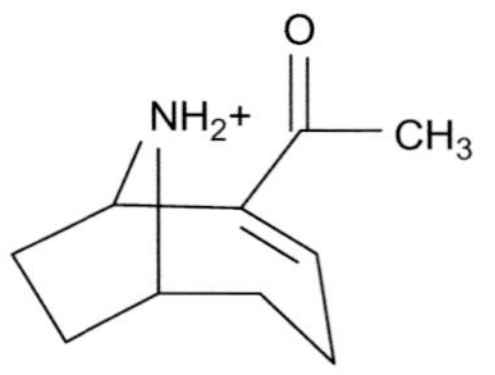

3.3.3 Saxitoxins

More commonly associated with eukaryotic algae, the saxitoxins (Fig. 6) are a group of highly potent toxic alkaloids (Bordner et al., 1975; Schantz et al., 1957), with around 25 structural variants known (Ballot et al., 2017). They can be produced by a number of genera of freshwater cyanobacteria including *Dolichospermum, Aphanizomenon, Cylindrospermopsis, Lyngbya,* and *Planktothrix* (Metcalf and Codd, 2012). In high doses, death can occur rapidly, as a result of their ability to block voltage-gated sodium channels leading to respiratory arrest (Ballot et al., 2017; Carmichael, 1994).

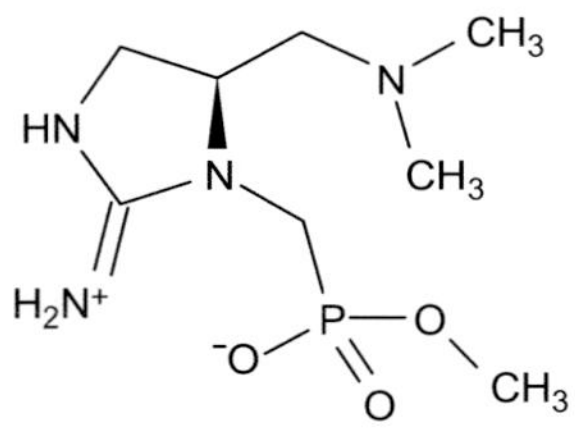

FIG. 5   Anatoxin-a(*S*).

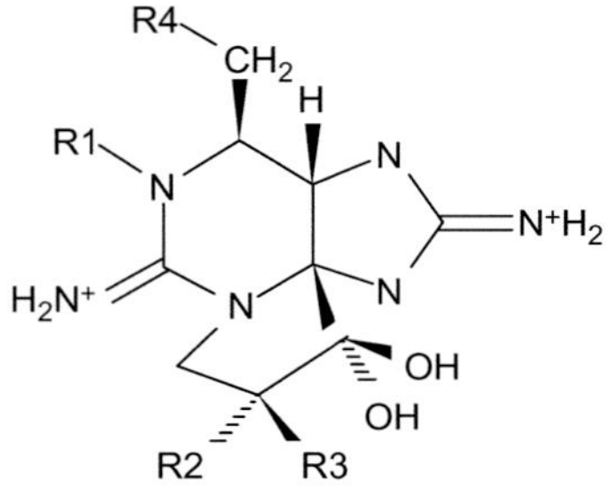

FIG. 4   Anatoxin-a.

FIG. 6   The generic structure of the saxitoxin molecule.

EVALUATING WATER QUALITY TO PREVENT FUTURE DISASTERS

3.4 Lipopolysaccharide

As members of the bacteria, cyanobacteria also possess cell wall structures consistent with them being Gram-negative bacteria (Drews and Weckesser, 1982). The lipopolysaccharide (LPS) part of the outer cell wall is antigenic when present in mammals and can also pose potential health problems causing gastrointestinal (GI) upset (Codd et al., 2005). From the information currently known, cyanobacterial LPS is not considered to be as toxic as other Gram-negative heterotrophic bacterial LPS such as that from *E. coli* (Monteiro et al., 2017). However, cyanobacteria are known to act as support structures for other heterotrophic bacteria, for example, a correlation was shown between outbreaks of cholera, caused by *Vibrio cholerae*, and cyanobacterial blooms in Bangladesh (Islam et al., 1994). Therefore, during such outbreaks LPS from a range of bacteria, including cyanobacteria, may also contribute to adverse human health effects.

3.5 Neurotoxic Amino Acids

Although many cyanobacterial toxins have acute toxic effects, increasing evidence suggests that long-term exposure to cyanobacterial toxins can have adverse human health effects including Alzheimer, ALS, and other neurodegenerative diseases (Murch et al., 2004; Pablo et al., 2009). Of the compounds potentially present within cyanobacteria that may cause neurological disease, β-N-methylamino-L-alanine (BMAA; Fig. 7) has been shown to cause neuropathologies, such as neurofibrillary tangles and amyloid plaques in experimental animals and a single prenatal dose of BMAA has been shown to cause neurological deficits in experimental animals, consistent with neurological disease (Cox et al., 2016; Scott and Downing, 2018). The analysis of cyanobacterial strains and bloom samples indicated that BMAA production is potentially widespread and although no variants are known for this small amino acid, isomers, namely 2,4-diaminobutyric acid and N-(2-aminoethyl)glycine, have been documented as occurring in cyanobacterial blooms and strains (Banack et al., 2010, 2012; Metcalf et al., 2008). BMAA is known to act as a glutamate agonist (Weiss et al., 1989) affecting nerve transmission and has been shown to be misincorporated into proteins in the place of L-serine, potentially allowing misfolded proteins to function incorrectly within the cell (Dunlop et al., 2013).

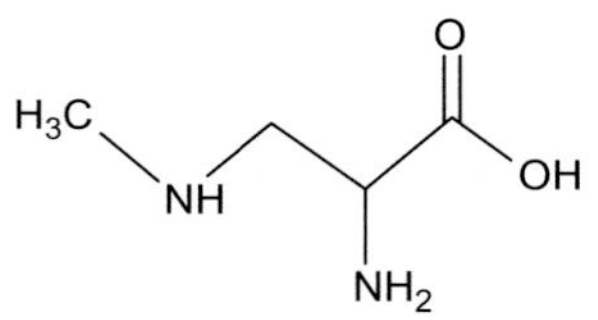

3.6 Dermatotoxins

Although there are a plethora of low molecular weight compounds in cyanobacteria, there are some that can inflict dermatotoxic effects in mammals. They are also considered to be tumor promoters in the skin and the main toxins comprising lyngbyatoxins (variants A, B, and C), aplysiatoxin, and debromoaplysiatoxin (Cardellina et al., 1979; Fujiki et al., 1981, 1983; Mynderse et al., 1977).

4 WHAT CONDITIONS MAKE CYANOBACTERIA BLOOM AND UNDER WHAT CIRCUMSTANCES?

Cyanobacteria are important and common components of freshwater ecosystems. However, when they become dominant and form large blooms, this may result in ecological and environmental health issues. The dominance of cyanobacteria is largely due to the presence of high concentrations of nutrients in waters (Fogg et al., 1973; Paerl, 2014). The major

FIG. 7   β-N-methylamino-L-alanine (BMAA).

nutrients that cyanobacteria require to grow are nitrogen and phosphorous, which can then be used for making nitrogen- and phosphorous-containing components such as proteins and phospholipids (Paerl, 2014). With respect to carbon, as these are photosynthetic organisms, they are able to remove carbon directly from water, often present as carbonic acid, and then lead to subsequent production of complex sugars through photosynthesis within the cell which can result in large changes in the pH of the waterbody, with pH values of 9–10 being observed in waters with extensive cyanobacterial blooms due to the removal of carbonic acid (Gao et al., 2012). Like carbon, nitrogen and phosphorous will be used according to the Redfield ratio until one of the nutrients becomes limiting (Paerl, 2014). If the bloom is phosphorous limited, then no further growth will occur until more phosphorous enters the system, regardless of the amount of excess nitrogen in the water. Similarly, if a bloom is nitrogen limited then there will be no further cyanobacterial growth, regardless of the excess phosphorous concentration within the water until more nitrogen enters the system. Under such scenarios, population changes may shift with a move away from non-nitrogen-fixing cyanobacteria to nitrogen-fixing cyanobacteria (Fogg et al., 1973), negating the need for new nitrogen salts to enter the freshwater system as the atmospheric nitrogen present in the water will be sufficient to allow organic forms of nitrogen to be produced. The only difference that might be observed is microscopic changes to cyanobacterial species and genera that may occur as nitrogen-fixing cyanobacteria become dominant.

Besides nutrients, others physical factors can influence cyanobacterial growth. Cyanobacteria dominate and succeed in waters with a long residence time and such waters may undergo physical changes during the seasons of the year. When calm, still warm weather impacts a waterbody, stratification can result. During such periods, a thermocline may become established in the waterbody such that nutrient-poor warm water sits above nutrient-rich cold water (e.g., Halstvedt et al., 2007). Although the thermocline is a characteristic of summer, storms, such as those in the autumn or fall, have the ability to break the thermocline and "mix" the waterbody. As a result, this will most likely disturb any cyanobacterial bloom that is present at the surface and re-introduce nutrient-rich waters into these surface waters. In addition to nutrients, akinetes and settled colonies can also be resuspended from sediments and enter the water column with the potential to form blooms under appropriate conditions.

High temperature is another factor that can favor cyanobacterial growth and competition with other algal groups. With global warming/climate change, one may expect a higher frequency and intensity of cyanobacterial blooms along with geographical expansion of certain cyanobacterial species (Paerl and Huisman, 2008, 2009; Paerl and Paul, 2012; Sinha et al., 2012).

5 HOW DO CYANOBACTERIA AFFECT WATERBODIES?

Reports from the late 19th century documented a number of animals deaths associated with cyanobacterial blooms (Codd et al., 2005). Since that time, regular human and animal illnesses attributed to cyanobacterial blooms have been documented, often with analysis occurring retrospectively as cyanobacteria have often been overlooked as the initial cause of animal deaths (Metcalf and Codd, 2012; Saker et al., 1999). One of the first scientifically documented experiments examining the association of cyanobacteria in the cases of livestock deaths was conducted by Francis in 1878, where sheep were experimentally dosed with *Nodularia* scum from Lake Alexandrina reproducing the poisoning symptoms and deaths (Francis, 1878).

The production of large-scale cyanobacterial blooms and scums are considered to be aesthetically unpleasant. Taste and odor compounds, mainly geosmin and methylisoborneol, are nontoxic secondary cyanobacterial metabolites that can give water and scum material a pungent earthy smell. These can produce an off-flavor in tap water (Du Preez et al., 2017) and in fish (Smith et al., 2008).

Such smells can be attractive to animals, in particular dogs, leading to exposure and possibly death as a result of cyanobacterial intoxications (Cadel-Six et al., 2007; Codd et al., 1992; Edwards et al., 1992; Wood et al., 2007). This often occurs due to exposure to accumulations of scum material or cyanobacterial mats on the shorelines of lakes, rivers, and other waterbodies that can be attractive to dogs and the animals then play in the water and self-clean their fur afterward, resulting in primary and secondary intoxications (Codd et al., 1992; Gugger et al., 2005; Puschner et al., 2008). In the United States, from the late 1920s to mid-2012, 260 dogs were poisoned due to cyanotoxin (anatoxin-a and microcystin) exposure, in which 83% of the dogs died (Backer et al., 2013). In Berlin, Germany, cases of dog deaths due to anatoxin-a intoxication have been reported with some caused by scums, in addition to suspended populations and small amounts of cyanobacteria present in plant material (Fastner et al., 2018). Incidents of dog poisoning due to cyanobacteria have also been reported from desert crust, in Qatar, with analysis of the crust showing the presence of hepatotoxins and neurotoxins (Chatziefthimiou et al., 2014).

A die-off of an estimated 1000 bats and 24 ducks (Mallards and American wigeons) was reported from Cross Lake Provincial Park (Alberta, Canada) that supported a bloom of *Anabaena*. Analysis performed on the slime of some carcasses revealed a toxic alkaloid (Pybus et al., 1986). Mortalities of Greater Flamingo chicks (*Phoenicopterus ruber*) in Spain, Chilean Flamingos (*Phoenicopterus chilensis*) at Sea World (USA), Lesser Flamingos (*Phoeniconaias minor*) in African Rift Valley lakes (Codd et al., 2003),

and Dalmatian pelicans (*Pelecanus crispus*) from a Greek Reservoir have been attributed to ingestion of cyanotoxins (Codd et al., 2003; Papadimitriou et al., 2018).

In humans, cases of poisoning due to cyanobacterial toxins have been reported ranging from skin rashes to fatalities (Aimi et al., 1990; Azevedo et al., 2002). Exposures can occur during water recreation, through aerosols/inhalation, ingestion of water, food and dietary supplements, and even through the use of water in medical procedures (Codd et al., 2005). Pilotto et al. (1997) and Osborne and Shaw (2008) have reported the cases of dermatological reactions and allergy/flu-like symptoms in people recreating in cyanobacterially contaminated waters. Other recreational exposures with more severe outcomes have also been reported in Finland and Sweden in 1989 and 1994, respectively, where acute illnesses were reported and attributed to the cyanobacterium *Planktothrix* (Annadotter et al., 2001). After recreational exposure to a bloom of *Microcystis aeruginosa*, a patient showing symptoms of flu, hepatotoxicosis, and pneumonia was hospitalized (Giannuzzi et al., 2011). Vidal et al. (2017) reported a case of a family in Uruguay in 2015, who after recreating at a lake containing a cyanobacterial bloom suffered GI illness. One of the family members, a 20-month old baby, presented more severe symptoms leading to liver failure and subsequently required a liver transplant (Vidal et al., 2017).

Exposure to cyanotoxins via aerosols has not been extensively studied but has been shown to occur, as microcystins have been detected in aerosols generated in the laboratory and lakeside (Backer et al., 2008, 2010; Cheng et al., 2007). Caller et al. (2009) found clusters of ALS patients living around lakes with cyanobacterial blooms in New Hampshire. BMAA and isomers have been detected in air samples taken at lakes with cyanobacterial blooms (Henegan et al., 2017) and aerosols from dust storms, especially in desert environments such as Qatar where 87% of the soil crust is composed of cyanobacteria (Richer et al., 2012), which can pose as another source of material for airborne exposures (Cox et al., 2009; Metcalf et al., 2012).

Cyanotoxins, such as microcystins, cylindrospermopsins, saxitoxins, anatoxin, and BMAA have been found to exist in various trophic levels, including aquatic and crop plants, molluscs, amphibians, fish, birds, and mammals (Al-Sammak et al., 2014; Berry et al., 2011; Codd et al., 1999; Kittler et al., 2012; Magalhaes et al., 2001; Sipia et al., 2007). The accumulation of cyanotoxins in food, including dietary supplements (Bautista et al., 2015; Bruno et al., 2006; Roy-Lachapelle et al., 2017), can also pose a potential risk (reviewed by Testai et al., 2016).

Contaminated drinking water poses another potential route of exposure as seen on Palm Island (Australia) in 1979 after drinking water prepared from Solomon Dam resulted in more than 100 people presenting symptoms of bloody diarrhea and vomiting (Byth, 1980). Cyanobacterial blooms in waterbodies designated as sources for drinking water have been reported all over the world (Chatziefthimiou et al., 2016; Gaget et al., 2017; Mhlanga et al., 2006; Molica et al., 2005; Pelley, 2016). Furthermore, correlations have been shown between water sources containing cyanobacteria and the incidence of primary liver cancer, presumed to be due to exposure to microcystins (e.g., Ueno et al., 1996).

Water is required for medical hemodialysis procedures and is another possible route for intoxication. One of the worst cases of human cyanotoxin fatalities happened in 1996 in Brazil, when water from a local reservoir in Caruaru, supporting a cyanobacterial bloom, was used at the hemodialysis facility. Lysis of the cyanobacterial cells and insufficient treatment at the clinic did not remove the cyanotoxins from the water before being used in the medical procedure. The outcome was the deaths of 100 patients with liver failure, with 52 of them showing an association with cyanotoxins (Azevedo et al., 2002; Carmichael et al., 2001; Jochimsen et al., 1998; Pouria et al., 1998).

In addition to the toxins, cyanobacterial blooms can also adversely affect waterbodies, particularly in the case where cyanobacterial blooms are beginning to senesce. Although cyanobacterial blooms are largely seasonal in temperate climates, blooms will break down due to lack of nutrients and changes in the physical conditions in which the bloom persists. As cyanobacterial blooms can cover a large area and potentially have a high density of cells, when the bloom senesces there can be a large amount of decaying matter present in the water. During this decay process, heterotrophic bacteria undergo respiration to break down this organic material. As a result, oxygen is consumed in the process and the oxygen concentration of waters can decrease significantly, potentially causing deaths of aquatic animals (Townsend et al., 1992).

6 HOW CAN WE DETECT CYANOBACTERIAL TOXINS?

Although the modes of action and the chemical structures of cyanotoxins differ widely, a large amount of effort has been placed on detecting cyanobacterial toxins with great precision and sensitivity. The variety of methods available reflects the differing needs of various water users, managers, and scientists and as such are able to assess the concentration or the potential toxicity of cyanobacterial material, in animals and in clinical materials, while also accommodating financial constraints to suit a range of applications and budgets. Furthermore, advancements in the understanding of cyanobacterial genetics has allowed polymerase chain reaction (PCR)-based tests to be developed to determine the potential for cyanotoxin production. The following outlines the procedures used from physicochemical to biochemical methods, in addition to assessment of genetic potential.

6.1 Physicochemical Methods

6.1.1 High-Performance Liquid Chromatography

High-performance liquid chromatography (HPLC) was, in many cases, the original analytical method used to develop and determine the presence of a number of cyanotoxins in a wide range of environmental and clinical materials (Codd et al., 2001). The use of HPLC allows the separation of metabolites and compounds within complex samples. Coupled to HPLC is a requirement for detectors, the majority of which are based on fluorescence or ultraviolet (UV) absorbance. Fluorescence methods have proven useful to detect cyanobacterial toxins such as the saxitoxins and BMAA (Metcalf and Pereira, 2017). This is made possible by chemical modification of the compound of interest which is then able to fluoresce after chemical introduction of a known fluorophore (Metcalf and Pereira, 2017). The UV detectors make use of the natural ability of a compound to absorb light of a specific wavelength without chemical modification. At first using only one wavelength, a chromatogram could be created from a known sample and the retention time of peaks compared with known cyanobacterial toxin standards. Using such UV and fluorescence methods, the only comparison was retention time, which, although can be used when assessing experimental material with well characterized cyanotoxins, is not suitable for, e.g., understanding poisoning incidents where greater specificity is required. Improvements in UV detectors resulted in the introduction and use of photodiode array detection (PDA), which provides greater specificity when assessing UV-absorbing compounds, including cyanobacterial toxins. This is because the detector, rather than scanning at one wavelength, scans a range of wavelengths, often 200–300 nm. This provides a spectra of the molecule which, along with retention time, provides greater specificity as to the nature of the compound present. Methods employing PDA have been developed for microcystins, anatoxin-a, nodularins, and cylindrospermopsin (Metcalf and Codd, 2012). Although a relatively old technology, HPLC has been superseded by ultra high-performance liquid chromatography (UHPLC). These methods have allowed analysis times to be drastically reduced from, e.g., 45 min (HPLC) to <10 min due to the ability to run samples at very high pressures (>10,000 PSI). Methods employing UHPLC have now largely replaced HPLC methods for cyanotoxins and are allowing increased throughput of samples (e.g., Spoof et al., 2009). The versatility of HPLC and UHPLC is now achieved by the various phases and solvents which are amenable to the analysis of hydrophobic compounds using C18 (Spoof et al., 2009) to very hydrophilic compounds using HILIC methods (Dörr et al., 2010), permitting LC methods to detect an extremely wide range of compounds.

6.1.2 Mass Spectrometry

Mass spectrometric methods for cyanotoxins provide greater specificity, and often sensitivity when compared with UV and fluorescence methods. By identifying the mass of the compound, all other compounds that may be present at a different mass are excluded. Therefore, high precision can be attained with mass spectrometry, even though such equipment may require a much greater capital expenditure. Methods such as matrix-assisted laser desorption ionization time of flight (Erhard et al., 1997) and surface-enhanced laser desorption ionization time of flight (Yuan and Carmichael, 2004) mass spectrometry do not require any traditional separation of compounds from the extracts due to the ability of using a chemical matrix to hold, and a laser to ablate the sample, before detection in the mass spectrometer (MS), based on mass to charge ratio as determined by the time of flight through the MS.

In order to permit good detection and quantification by mass spectrometry, some liquid chromatography is, however, generally necessary.

This is often achieved using HPLC or UHPLC, although with some MS's nano-flow systems, which use orders of magnitude lower flow rates, are becoming more common. Once the cyanobacterial toxin has been characterized, the mass and the mass to charge ratios when ionized can then be determined. This information can then be used to develop methods necessary to detect cyanobacterial toxins, either as an ionized parent molecule (single quadrupole; Banack et al., 2010) or by using more advanced methods such as triple quadrupole and tandem mass spectrometry. Daughter ions and mass fragments can be analyzed to help construct the molecule (Banack et al., 2012) and can also help with identifying variants and chemical adducts. Mass spectrometric methods have been developed for the majority of the cyanobacterial toxins, including microcystins, anatoxin-a, anatoxin-a(S), cylindrospermopsin, saxitoxins, and BMAA (Caixach et al., 2017; Codd et al., 2001), either as a native compound (Spoof et al., 2009) or after chemical modification (Banack et al., 2012). This allows samples to be separated using conventional C18 phases, although with anatoxin-a(S), HILIC methods work well for the assessment of this very hydrophilic toxin (Dörr et al., 2010). Due to the specificity, sensitivity, and rapidity of the analysis, mass spectrometry is now the physicochemical method of choice for the analysis of most cyanobacterial toxins in complex matrices.

6.1.3 Nuclear Magnetic Resonance

Although not particularly useful for routine analysis of cyanobacterial toxins, nuclear magnetic resonance (NMR) methods have proven useful for the determination of their structures. The most specific analytical method, NMR can be used to measure interactions of cyanobacterial toxins with atoms and chemicals. The most likely use of NMR will be the determination of the structures of newly discovered cyanobacterial toxins, and in the case of toxin classes such as the microcystins and nodularins (Trogen et al., 1996), the elucidation of new variants.

6.2 Biochemical Methods

In addition to the identification and quantification provided by physicochemical methods, biological and biochemical methods are important complementary methods. Rather than having a basis in optics or mass spectrometry, these methods use toxicity, enzyme inhibition, and antigen–antibody interactions to determine the presence and amount of a compound.

6.2.1 Biological Assays

The original toxicity method employed biological assay or bioassay to determine the presence of cyanobacterial toxins, either as components of blooms and water samples or as purified toxins. Since the work of Francis (1878), whereby sheep were dosed with *Nodularia* scum and died, the ability to use whole organisms to test the toxicity of cyanobacterial material has become common. Originally, small animal bioassays such as those with mice were used to assess aqueous extracts of cyanobacterial material and purified toxins, extending to, e.g., the use of pigs for the assessment of long-term exposure to microcystins in *Microcystis* bloom material. By using bioassays various organs that may be affected are identified, such as liver intoxication with microcystins (Metcalf et al., 2000b). Due to ethical considerations, in addition to the sensitivity of assays such as mouse bioassay which potentially require large amounts of material, other alternative bioassays have been developed. These include many early life stages of small organisms such as zebrafish (*Danio rerio*), brine shrimp (*Artemia salina*), *Thamnocephalus,* and others (Bláha et al., 2017). Using such assays generally allows a greater sensitivity as, in general, the smaller the organisms, the less a dose is required to produce a toxic outcome. Furthermore, with small aquatic organisms such as *A. salina*, statistical power and dose–response curves can be developed due to the use of multiple organisms per test and multiple tests run at the same time. Plants can also be used to assess the toxicity of cyanobacterial toxins as has been shown with pollen tubes (Metcalf et al., 2004) and mustard when assessing cylindrospermopsin (Vasas et al., 2002).

6.2.2 Biochemical Assays

Extensive testing of purified toxins within whole organisms has led to an understanding of the molecular targets in mammalian systems. Studies of the interaction of cyanobacterial toxins with these receptors have often resulted in their implementation as in vitro tests. For the assessment of microcystins, protein phosphatase inhibition assays have been developed with colorimetric substrates that, when the enzyme is inhibited, result in less color development and these assays work well with extracts from cyanobacterial strains and blooms when compared with quantitative HPLC-PDA analysis of these materials (Ward et al., 1997). Such bioassays are useful for a number of reasons. Firstly, in the case of microcystins there are currently over 200 known variants, each with differing toxicities (Spoof and Catherine, 2017). If multiple microcystins are present within bloom samples, then determining the toxicity of the bloom material may be difficult using HPLC-PDA or mass spectrometry. With a bioassay, how toxic a bloom sample might actually be could be determined through understanding cumulative effects of, e.g., variants in a biological system (Bláha et al., 2017). Secondly, if other compounds are present within the bloom then they may cause synergistic or antagonistic effects which may be observed and understood when bioassays are carried out. If bioassays are used for determining the risks of exposure to humans, mammalian bioassays do provide some degree of complementarity with humans and have been used as the basis for deriving drinking water Guidelines, largely through the use of orally-dosed pigs in the case of microcystins (Falconer et al., 1994). Ultimately, bioassays show a toxic response that may end up as being translatable in understanding the toxic effect on humans.

Biochemical assays have been developed for other cyanobacterial toxins, including saxitoxins (saxiphilin binding; sodium channels; Negri and Llewellyn, 1998), cylindrospermopsin (rabbit reticulocyte lysate assay; Froscio et al., 2001), LPS (*Limulus* amoebocyte lysate assay; Lindsay et al., 2009), anatoxin-a (acetylcholine binding; Aráoz et al., 2005), and anatoxin-a(S) (acetylcholinesterase inhibition assay; Mahmood and Carmichael, 1986). As other molecular targets of cyanobacterial toxins are identified, further biochemical tests are likely to be developed and implemented.

6.2.3 Immunoassays

As most cyanobacterial toxins are too small to be recognized by the mammalian immune system, they are referred to as haptens. In order to produce antibodies against the majority of cyanobacterial toxins, they require chemical linkage to larger, more immunogenic proteins. This complex which is immunogenic is then used to generate antibodies. Of all the antibodies produced, a small pool should be specific to the hapten of interest which is bound to the surface of the protein. Ultimately, these antibodies can then be used to detect cyanobacterial toxins in a wide range of matrices, e.g., from environmental to clinical sources and have been used in the investigations of poisoning incidents in animals and contamination of plant material. Antibodies have been generated against microcystins (-LR, -RR, -LA, Adda), nodularins, saxitoxins (saxitoxin, neosaxitoxin) and commercial ELISAs are available for anatoxin-a and cylindrospermopsin (Metcalf and Codd, 2017).

One advantage of ELISAs over traditional analytical techniques such as HPLC and mass spectrometry is that they are amenable to field analysis of cyanobacteria and their toxins. Rather than collecting and then transporting cyanobacterial samples to the laboratory, scientists can take ELISAs into the field, and combined with field microscopy can perform early assessment of cyanobacterial bloom material.

The ELISAs are also useful screening tools for cyanobacterial toxins and can be used as a preliminary detection method prior to confirmatory techniques such as mass spectrometry to verify the presence of these toxins during, e.g., poisoning events (Metcalf and Codd, 2017).

6.3 Genetic Potential

Although microcystins and nodularins are peptide toxins, they are not produced as direct products of transcription and translation. Instead, the genetic material of the cyanobacterial cell encodes for enzymes which are able to synthesize the peptide and alkaloid toxins biochemically. Genes have been identified for the production of enzymatic components of microcystins, saxitoxins, cylindrospermopsin, and anatoxin-a. Understanding these gene clusters and the diversity of cyanotoxin production among the cyanobacteria has led to PCR methods to detect these genes in individual genera of cyanobacteria or for broad application to a range of cyanobacteria. Although the use of PCR methods does not always translate into toxin concentrations within the organisms or in a bloom, when mixed cyanobacterial assemblages are present in blooms, individual filaments and/or colonies can be assessed genetically. Subsequently, microscopy can then be used to measure the proportions of the colonies and filaments present in the bloom, and ultimately determine whether the risk of exposure to cyanobacterial toxins is likely to increase or decrease, depending on what happens to the mixed cyanobacterial population (Pacheco et al., 2016).

Recently, a whole cell tyramide signal amplification-fluorescent in situ hybridization (TSA-FISH) assay was developed targeting mcyA mRNA transcription as a proxy of MC-synthetase production (Zeller et al., 2016). The mcyA-mRNA TSA-FISH assay can label microcystin producing cells from natural environments and identify microcystin-producing species using a specific MICR3-16S rDNA probe. This technique can be used as an effective tool for environmental monitoring (Brient et al., 2017; Metcalf et al., 2009).

6.4 Intracellular vs Extracellular

During cyanobacterial blooms, cellular material can be taken and processed to determine the concentration of various compounds, including toxins contained within. The majority of cyanobacterial toxins remain within the cell during conditions when the blooms are healthy, but once the bloom decays then the toxins contained within the cell can be released. Some cyanobacterial toxins, such as cylindrospermopsin, are often found outside of the cell in the water (extracellular) phase, even when the filaments of cylindrospermopsis are healthy (Metcalf et al., 2002). If toxins are in the extracellular phase then this will have implications for understanding the efficacy of water treatment. If analysis of extracellular water is required, then methods are available to perform solid-phase extraction (SPE) using various commercial sorbents to concentrate cyanobacterial toxins and remove potential interfering compounds from the analysis. SPE methods have been developed for microcystins, anatoxin-a, BMAA, and cylindrospermopsin (Codd et al., 2001; Metcalf et al., 2002, 2013). Depending on the analytical method and the sensitivity required, SPE methods can frequently allow detection at concentrations covered by Guideline Values, either as legislation or alerts for risk assessment and human health protection. For the assessment of cyanobacterial bloom material, extraction methods have been tailored to the particular cyanotoxin class that requires analysis, in addition to the suitability of the solvent for the subsequent analytical method. For methods such as ELISA, simple methods such as boiling in water can release the toxins for analysis (Metcalf and Codd, 2000). However, when physicochemical methods such as mass spectrometry are used, organic solvents can allow better detection of cyanotoxins by only extracting small molecular weight compounds and leaving more aqueous-soluble compounds in the pelleted material with the further potential to allow concentration of the cyanotoxins contained within through evaporation (Codd et al., 2001). Ultimately, the final purpose of the water, the information required, types of samples and the various compartments that need to be analyzed will help in determining what sample preparation methods are to be carried out.

7 HOW CAN WE REMOVE CYANOBACTERIAL TOXINS?

Cyanobacterial toxins when present within the environment can have a number of fates. Biologically, the toxins themselves have biochemical targets which may act as fates for the toxins, such as mammalian protein phosphatase enzymes in the case of microcystins. These molecular targets allow the toxins to be removed and may or may not result in intoxication and adverse health events. In the case of microcystins, when they possess methyldehydroalanine, covalent binding to protein phosphatases can occur (Metcalf et al., 2017). With nodularins that possess methyldehydrobutyrine, no covalent attachment can occur and the toxin can inhibit the enzyme but still be available to undergo other biochemical effects (Metcalf et al., 2017). Similarly, with cylindrospermopsin this toxin can inhibit protein translation and therefore the protein translation machinery may act as an endpoint for the toxin (Froscio et al., 2001). Although organs, enzymes, and other cellular structures can act as targets for cyanobacterial toxins, when the dose of cyanobacterial toxin is not sufficient to cause a fatal intoxication then metabolism of the cyanobacterial toxin has been documented. Detoxication products have been observed for microcystins acting through glutathione-S-transferases (Metcalf et al., 2000b) and the cytochrome P450 enzyme system with cylindrospermopsin (Norris et al., 2002). These detoxication products are often of lower toxicity than the parent compound and may reduce the burden of the cyanobacterial toxin on the animal. The deposition of cyanobacterial toxins has also been reported for microcystins, BMAA, and anatoxin-a in bird feathers of Lesser Flamingo (Metcalf et al., 2006, 2013) and nodularin in eider duck feathers (Sipia et al., 2008). The ability to deposit cyanobacterial toxins in keratinous tissues removes them from the blood stream. When an organism, such as a healthy mammal is exposed to low doses, then the ability to displace cyanobacterial toxins either through deposition into tissues or into detoxication pathways may result in less likelihood of adverse health outcomes. Toxins can also be taken up by aquatic plants (e.g., Vasas et al., 2002). Furthermore, bacteria are known to degrade a number of cyanobacterial toxins through enzymatic reactions and these enzymes may prove useful for degrading cyanobacterial toxins, either naturally or commercially (e.g., Bourne et al., 1996). Many sediments contain minerals such as pumice (Gurbuz and Codd, 2008) that can bind cyanobacterial toxins further removing the pool of cyanobacterial toxins that may be present within a waterbody.

With regards to water treatment, generally, the processes may involve coagulation, flocculation, sedimentation, filtration, and disinfection. For the removal of cyanobacterial cells, the first four steps can usually be effective (Westrick, 2008). Nevertheless, cyanotoxins released into the water by cell lysis during the treatment process, decaying blooms, or the biological nature of the cyanobacteria (such as *Cylindrospermopsis* that has intra and extracellular toxin pools; Metcalf et al., 2002) needs to be taken into account when planning for treatment as cyanotoxins released into the water are more difficult to remove than the cells. Extracellular toxins can reach the disinfection step and may not be completely removed, requiring additional and/or advanced treatment technologies, such as the combination of chlorine and adsorption with activated carbon, to provide safe drinking water (e.g., Westrick, 2008). According to Westrick et al. (2010), the implementation of removal of intact cyanobacterial cells and auxiliary treatment barriers is empirical and that knowledge concerning cyanobacteria and their toxins should be taking into consideration when designing new or retrofitting old water treatment plants.

Other techniques involve the management of water supplies, whether through ultrasonication or destratification to minimize the ability of cyanobacteria to bloom using physical processes (Peczula, 2012; Simmons, 1998) and/or catchment management which requires reduction of the nutrient load entering waters and preventing mass proliferation of cyanobacteria and their potential to cause problems to human populations (Paerl, 2014).

8 HOW DO WE REDUCE THE LIKELIHOOD OF EXPOSURE TO CYANOBACTERIA AND THEIR TOXINS?

According to the fossil record, cyanobacteria are some of the oldest known living organisms (Schopf, 2000). Cyanobacteria are considered to have caused the Great Oxidation Event millions of years ago, ultimately changing atmospheric conditions to those rich in oxygen (~20%), which led to the evolution of many complex organisms that we see today, including humans (Knoll, 2003). Humans have also altered the atmosphere through the introduction of compounds including chlorofluoro-carbons and carbon dioxide. Global climate change often referred to as global warming is increasing the carbon dioxide

concentration of the atmosphere, which in addition to other compounds, is causing a general trend of increasing temperature. With increasing population pressures and reliance on finite water resources, exposure to cyanobacteria and their toxins is likely to increase. This is largely due to the fact that warmer temperatures due to climate change and longer residence times of water as a result of abstraction will encourage cyanobacterial blooms (Paerl and Huisman, 2008, 2009; Paerl and Paul, 2012; Sinha et al., 2012). Such changes can lead to sea-level rise, droughts, animal deaths, and food shortages, leading to changes in the way we live our lives. In most cases, these changes may actually provide "perfect" environmental conditions for cyanobacteria to thrive (Paerl and Huisman, 2008).

In dealing with cyanobacteria in the 21st century, we ultimately need to live with cyanobacteria. These common (and normal) components of waters do not need to be removed or destroyed. By employing water treatment techniques, toxins and the organisms can be removed at source. Furthermore, another necessity is guidelines and legislation for drinking and bathing waters in order to protect human and animal health (e.g., Metcalf et al., 2018).

In terms of the strategies needed to reduce public health risk, source water quality, environmental conditions, the dominance of cyanobacterial species, and the water treatment technologies available are just some of the factors that should be taken into consideration (Cheung et al., 2013). Monitoring of waterbodies, especially the ones that are more likely to result in human exposure to toxins, appropriate warning notices and information about blooms should also be provided to alert the local human population. The use of satellite imagery maps can also assist managers in implementing mitigation protocols for blooms, such as choosing or closing certain drinking water intakes, assistance in sampling strategies, and also helping the public to avoid recreational areas that may have blooms, ultimately reducing the risk of exposure (Recknagel et al., 2018; Wynne and Stumpf, 2015). Rapid communication between the public and appropriate authorities concerning animal illnesses and/or deaths, information on the potential human health effects related to algal blooms, and the participation of multidisciplinary professionals can aid in efficient approaches to managing environmental risks as required (Hilborn and Beasley, 2015; Metcalf et al., 2018). The awareness of medical staff with respect to cyanotoxin-related hepatotoxicoses, especially for areas where cyanobacteria blooms are known to occur, is also needed (Vidal et al., 2017).

Ultimately, we have to appreciate that cyanobacteria are a fundamental part of our ecosystem and perform very important functions such as the production of oxygen, supporting aquatic food chains and fixing atmospheric nitrogen into organic sources (Fogg et al., 1973). By curtailing our pollution and with responsible stewardship of our water supplies, the negative effects of cyanobacteria, whether through unsightly blooms, odors, or through the production of toxins, can be reduced to an extent whereby human and animal populations can use waters without fear of intoxication or disease.

References

Aimi, N., Odaka, H., Sakai, S., Fujiki, H., Suganuma, M., Moorre, R.E., Patterson, G.M., 1990. Lyngbyatoxins B and C, two new irritants from *Lyngbya majuscula*. J. Nat. Prod. 53, 1593–1596.

Al-Sammak, M.A., Hoagland, K.D., Cassada, D., Snow, D.D., 2014. Co-occurrence of the cyanotoxins BMAA, DABA and anatoxin-*a* in Nebraska reservoirs, fish, and aquatic plants. Toxins 6, 488–508.

Annadotter, H., Cronberg, G., Lawton, L.A., Hansson, H.B., Gothe, U., Skulberg, O.M., 2001. An extensive outbreak of gastroenteritis associated with the toxic cyanobacterium *Planktothrix agardhii* (Oscillatoriales, Cyanophyceae) in Scania, South Sweden. In: Chorus, I. (Ed.), Cyanotoxins Occurrence, Causes, Consequences. Springer, Berlin, pp. 200–208.

Aráoz, R., Nghiêm, H.-O., Rippka, R., Palibroda, N., de Marsac, N.T., Herdman, M., 2005. Neurotoxins in axenic oscillatorian cyanobacteria: coexistence of anatoxin-a and homoanatoxin-a determined by ligand-binding assay and GC/MS. Microbiology 151, 1263–1273.

Azevedo, S.M.F.O., Carmichael, W.W., Jochimsen, E.M., Rinehard, K.L., Lau, S., Shaw, G.R., Eaglesham, G.K., 2002. Human intoxication by microcystin during renal dialysis treatment in Caruaru—Brazil. Toxicology 181–182, 441–446.

Backer, L.C., Carmichael, W., Kirkpatrick, B., Williams, C., Irvin, M., Zhou, Y., Johnson, T.B., Nierenberg, K., Hill, V.R., Kieszak, S.M., Cheng, Y.-S., 2008. Recreational exposure to low concentrations of microcystins during an algal bloom in a small lake. Mar. Drugs 6, 389–406.

Backer, L.C., McNeel, S.V., Barber, T., Kirkpatrick, B., Williams, C., Irvin, M., Zhou, Y., Johnson, T.B., Nierenberg, K., Aubel, M., LePrell, R., Chapman, A., Foss, A., Corum, S., Hill, V.R., Kieszak, S.M., Chen, Y.-S., 2010. Recreational exposure to microcystins during algal blooms in two California lakes. Toxicon 55, 909–921.

Backer, L.C., Landsberg, J.H., Miller, M., Keel, K., Taylor, T.K., 2013. Canine cyanotoxin poisonings in the United States (1920s-2012): review of suspected and confirmed cases from three data sources. Toxins 5, 1597–1628.

Ballot, A., Bernard, C., Fastner, J., 2017. Saxitoxins and analogues. In: Meriuoto, J., Spoof, L., Codd, G.A. (Eds.), Handbook of Cyanobacterial Monitoring and Cyanotoxin Analysis. John Wiley & Sons, Chichester, UK, pp. 148–154.

Banack, S.A., Downing, T.G., Spácil, Z., Purdie, E.L., Metcalf, J.S., Searle, S., Esterhuizen, M., Codd, G.A., Cox, P.A., 2010. Distinguishing the cyanobacterial neurotoxin β-N-methylamino-L-alanine (BMAA) from its structural isomer 2,4-diaminobutyric acid (2,4-DAB). Toxicon 56, 868–879.

Banack, S.A., Metcalf, J.S., Jiang, L., Craighead, D., Ilag, L., Cox, P.A., 2012. Cyanobacteria produce N-(2-aminoethyl)glycine, a backbone for peptide nucleic acids which may have been the first genetic molecules for life on earth. PLoS One 7, e49043. https://doi.org/10.371/journal.pone.0049043.

Banker, R., Carmeli, S., Hadas, O., Teltsch, B., Porat, R., Sukenik, A., 1997. Identification of cylindrospermopsin in *Aphanizomenon ovalisporum* (cyanophyceae) isolated from Lake Kinneret. Israel. J. Phycol. 33, 613–616.

Banker, R., Carmeli, S., Werman, M., Teltsch, B., Porat, R., Sukenik, A., 2001. Uracil moiety is required for toxicity of the cyanobacterial hepatotoxin cylindrospermopsin. J. Toxicol. Environ. Health A 62, 281–288.

Bautista, A.C., Moore, C.E., Lin, Y., Cline, M.G., Benitah, N., Puschner, B., 2015. Hepatopathy following consumption of a commercially available blue-green algae dietary supplement in a dog. BMC Vet. Res. 11, 136.

Berry, J.P., Lee, E., Walton, K., Wilson, A.E., Bernal-Brooks, F., 2011. Bioaccumulation of microcystins by fish associated with a persistent cyanobacterial bloom in Lago de Patzcuaro (Michoacan, Mexico). Environ. Toxicol. Chem. 30, 1621–1628.

Bertone, E., Burford, M.A., Hamilton, D.P., 2018. Fluorescence probes for real-time remote cyanobacteria monitoring: a review of challenges and opportunities. Water Res. 141, 152–162.

Bláha, L., Cameán, A.M., Fessard, V., Gutiérrez-Praena, D., Jos, A., Marie, B., Metcalf, J.S., Pichardo, S., Puerto, M., Torokne, A., Vasas, G., Zegura, B., 2017. Bioassay use in the field of toxic cyanobacteria. In: Meriluoto, J., Spoof, L., Codd, G.A. (Eds.), Handbook of Cyanobacterial Monitoring and Cyanotoxin Analysis. John Wiley & Sons, Chichester, UK, pp. 272–279.

Bordner, J., Thiessen, W.E., Bates, H.A., Rapoport, H., 1975. Structure of a crystalline derivative of saxitoxin. Structure of saxitoxin. J. Am. Chem. Soc. 97, 6008–6012.

Bourne, D.G., Jones, G.J., Blakeley, R.L., Jones, A., Negri, A.P., Riddles, P., 1996. Enzymatic pathway for the bacterial degradation of the cyanobacterial cyclic peptide toxin microcystin LR. Appl. Environ. Microbiol. 62 (11), 4086–4094.

Brient, L., Lengronne, M., Bertrand, E., Rolland, D., Sipel, A., Steinmann, D., Baudin, I., Legeas, M., Le Rouzic, B., Bormans, M., 2008. A phycocyanin probe as a tool for monitoring cyanobacteria in freshwater bodies. J. Environ. Monit. 10 (2), 248–255.

Brient, L., Gamra, N.B., Periot, M., Roumagnac, M., Zeller, P., Bormans, M., Méjean, A., Ploux, O., Biegala, I.C., 2017. Rapid Characterization of microcystin-producing cyanobacteria in freshwater lakes by TSA-FISH (tyramid signal amplification-fluorescent in situ hybridization). Front. Environ. Sci. 5. https://doi.org/10.3389/fenvs.2017.00043.

Bruno, M., Fiori, M., Mattei, D., Melchiorre, S., Messineo, V., Volpi, F., Bogialli, S., Nazzari, M., 2006. ELISA and LC-MS/MS methods for determining cyanobacterial toxins in blue-green algae food supplements. Nat. Prod. Res. 20 (9), 827–834. https://doi.org/10.1080/14786410500410859.

Bruno, M., Ploux, O., Metcalf, J.S., Mejean, A., Pawlik-Skowronska, B., Furey, A., 2017. Anatoxin-a, homoanatoxin-a, and natural analogues. In: Meriluoto, J., Spoof, L., Codd, G.A. (Eds.), Handbook of Cyanobacterial Monitoring and Cyanotoxin Analysis. John Wiley & Sons, Chichester, UK, pp. 138–147.

Byth, S., 1980. Palm Island mystery disease. Med. J. Aust. 2, 40–42.

Cadel-Six, S., Peyraud-Thomas, C., Brient, L., de Marsac, N.T., Rippka, R., Méjean, A., 2007. Different genotypes of anatoxin-producing cyanobacteria coexist in the Tarn River, France. Appl. Environ. Microbiol. 73 (23), 7605–7614. https://doi.org/10.1128/AEM.01225-0.

Caixach, J., Flores, C., Spoof, L., Meriluotp, J., Schimidt, W., Mazur-Marzec, H., Hiskia, A., Kaloudis, T., Furey, A., 2017. Liquid chromatography—mass spectrometry. In: Meriluoto, J., Spoof, L., Codd, G.A. (Eds.), Handbook of Cyanobacterial Monitoring and Cyanotoxin Analysis. John Wiley & Sons, Chichester, UK, pp. 218–257.

Caller, T.A., Doolin, J.W., Haney, J.F., Murby, A.J., West, K.G., Farrar, H.E., Ball, A., Harris, B.T., Stommel, E.W., 2009. A cluster of amyotrophic lateral sclerosis in New Hampshire: a possible role for toxic cyanobacteria blooms. Amyotroph. Lateral Scler. 10, 101–108. https://doi.org/10.3109/174829609.

Cardellina, J.H., Marner, F.-J., Moore, R.E., 1979. Seaweed dermatitis: structure of lyngbyatoxin A. Science 204, 193–195.

Cardemil, L., Wolk, C.P., 1979. The polysaccharides from heterocyst and spore envelopes of a blue-green alga—structure of the basic repeating unit. J. Biol. Chem. 254 (3), 736–741.

Carmichael, W.W., 1994. The toxins of cyanobacteria. Sci. Am. 270, 78–86.

Carmichael, W.W., Biggs, D.F., Gorham, P.R., 1975. Toxicology and pharmacological action of *Anabaena flos-aquae* toxin. Science 187, 542–544.

Carmichael, W.W., Biggs, D.F., Peterson, M.A., 1979. Pharmacology of anatoxin-a, produced by the freshwater cyanophyte *Anabaena flos-aquae* NRC-44-1. Toxicon 17, 229–236.

Carmichael, W.W., Azevedo, S.M.F.O., An, J.S., Molica, R.J.R., Jochimsen, E.M., Lau, S., Rinehart, K.L., Shaw, G.R., Eaglesham, G.K., 2001. Human fatalities from cyanobacteria: chemical and biological evidence for cyanotoxins. Environ. Health Perspect. 109 (7), 663–668.

Catherine, A., Bernhardt, C., Spoof, L., Bruno, M., 2017. Microcystins and nodularins. In: Meriluoto, J., Spoof, L., Codd, G.A. (Eds.), Handbook of Cyanobacterial Monitoring and Cyanotoxin Analysis. John Wiley & Sons, Chichester, UK, pp. 109–126.

Chatziefthimiou, A.D., Richer, R., Rowles, H., Powell, J.T., Metcalf, J.S., 2014. Cyanotoxins as a potential cause of dog poisonings in desert environments. Vet. Rec. 174 (19), 484–485.

Chatziefthimiou, A.D., Metcalf, J.S., Glover, W.B., Banack, A.A., Dargham, S.R., Richer, R.A., 2016. Cyanobacteria and cyanotoxins are present in drinking water impoundments and groundwater wells in desert environments. Toxicon 114, 75–84.

Cheng, Y.S., Zhou, Y., Irvin, C.M., Kirkpatrick, B., Backer, L.C., 2007. Characterization of aerosols containing microcystin. Mar. Drugs 5, 136–150.

Cheung, M.Y., Liang, S., Lee, J., 2013. Toxin producing cyanobacteria in freshwater: a review of the problems, impact on drinking water safety, and efforts for protecting public health. J. Microbiol. 51 (1), 1–10.

Codd, G.A., Edwards, C., Beattie, K.A., 1992. Fatal attraction to cyanobacteria? Nature 359, 110–111.

REFERENCES

Codd, G.A., Metcalf, J.S., Beattie, K.A., 1999. Retention of Microcystis aeruginosa and microcystin by salad lettuce (Lactuca sativa) after spray irrigation with water containing cyanobacteria. Toxicon 37, 1181–1185.

Codd, G.A., Metcalf, J.S., Ward, C.J., Beattie, K.A., Kaya, K., Poon, G.K., 2001. Analysis of cyanobacterial toxins by physicochemical and biochemical methods. J. AOAC Int. 84, 1626–1635.

Codd, G.A., Metcalf, J.S., Morrison, L.F., Krienitz, L., Ballot, A., Pflugmacher, S., Wiegand, C., Kotut, K., 2003. Susceptibility of flamingos to cyanobacterial toxins via feeding. Vet. Rec. 152, 722–723.

Codd, G.A., Lindsay, J., Young, F.M., Morrison, L.F., Metcalf, J.S., 2005. Harmful cyanobacteria: from mass mortalities to management measures. In: Huisman, J., Matthijs, H.C.P., Visser, P.M. (Eds.), Harmful Cyanobacteria. Springer, pp. 1–23. © 2005. Printed in the Netherlands.

Cook, W.O., Beasley, V.R., Lovell, R.A., Dahlem, A.M., Hooser, S.B., Mahmood, N.A., Carmichael, W.W., 1989. Consistent inhibition of peripheral cholinesterases by neurotoxins from the freshwater cyanobacterium Anabaena flos–aquae: studies of ducks, swine, mice, and a steer. Environ. Toxicol. Chem. 8 (10), 915–922.

Cox, P.A., Richer, R., Metcalf, J.S., Banack, S.A., Codd, G.A., Bradley, W.G., 2009. Cyanobacteria and BMAA exposure from desert dust: a possible link to sporadic ALS among Gulf War veterans. Amyotroph. Lateral Scler. 10 (Suppl. 2), 109–117.

Cox, P.A., Davis, D.A., Mash, D.C., Metcalf, J.S., Banack, S.A., 2016. Dietary exposure to an environmental toxin triggers neurofibrillary tangles and amyloid deposits in the brain. Proc. R. Soc. B. 283. pii:20152397.

Devlin, J.P., Edwards, O.E., Gorham, P.R., Hunter, N.R., Pike, R.K., Stavric, B., 1977. Anatoxin-a, a toxic alkaloid from Anabaena flos-aquae NRC-44H. Can. J. Chem. 55F, 1367–1371.

Dörr, F.A., Rodrıguez, V., Molica, R., Henriksen, P., Krock, B., Pinto, E., 2010. Methods for detection of anatoxin-a(s) by liquid chromatography coupled to electrospray ionization-tandem mass spectrometry. Toxicon 55, 92–99.

Drews, G., Weckesser, J., 1982. Function, structure and composition of cell wall and external layers. In: Carr, N.G., Whitton, B.A. (Eds.), The Biology of Cyanobacteria. Blackwell Scientific Publications, Oxford, UK, pp. 33–357.

Dunlop, R.A., Cox, P.A., Banack, S.A., Rodgers, K.J., 2013. The non-protein amino acid BMAA is misincorporated into human proteins in place of L-serine causing protein misfolding and aggregation. PLoS One 8(9). https://doi.org/10.1371/journal.pone.0075376.

Du Preez, H.H., Swanepoel, A., Cloete, N., 2017. The occurrence and removal of algae (including cyanobacteria) and their related organic compounds from source water in Vaalkop Dam with conventional and advanced drinking water treatment processes. Water SA 43, 67–80.

Edwards, C., Beattie, K.A., Scrimgeour, C.M., Codd, G.A., 1992. Identification of anatoxin-a in benthic cyanobacteria (blue-green-algae) and in associated dog poisonings at Loch Insh, Scotland. Toxicon 30, 1165–1175.

Erhard, M., von Döhren, H., Jungblut, P., 1997. Rapid typing and elucidation of new secondary metabolites of intact cyanobacteria using MALDI-TOF mass spectrometry. Nat. Biotechnol. 15, 906–909.

Falconer, I.R., Humpage, A.R., 2001. Preliminary evidence for in vivo tumour initiation by oral administration of extracts of the blue-green alga Cylindrospermopsis raciborskii containing the toxin cylindrospermopsin. Environ. Toxicol. 16, 192–195.

Falconer, I.R., Burch, M.D., Steffensen, D.A., Choice, M., Coverdale, O.R., 1994. Toxicity of the blue-green alga (cyanobacterium) Microcystis aeruginosa in drinking water to growing pigs, as an animal model for human injury and risk assessment. Environ. Toxicol. 9 (2), 131–139.

Fastner, J., Beulker, C., Geiser, B., Hoffmann, A., Kröger, R., Teske, K., Hoppe, J., Mundhenk, L., Neurath, H., Sagebiel, D., Chorus, I., 2018. Fatal neurotoxicosis in dogs associated with tychoplanktic, anatoxin-a producing Tychonema sp. in mesotrophic Lake Tegel, Berlin. Toxins 10, 60. https://doi.org/10.3390/toxins10020060.

Fogg, G., Stewart, W.D.P., Fay, P., Walsby, A.E., 1973. The Blue-Green Algae. Academic Press, London.

Francis, G., 1878. Poisonous Australian lake. Nature 18, 11–12.

Froscio, S.M., Humpage, A.R., Burcham, P.C., Falconer, I.R., 2001. Cell-free protein synthesis inhibition assay for the cyanobacterial toxin cylindrospermopsin. Environ. Toxicol. 16, 408–412.

Fujiki, H., Mori, M., Nakayasu, M., Terada, M., Sugimura, T., Moore, R.E., 1981. Indole alkaloids: dihydroteleocidin B, teleocidin, and lyngbyatoxin A as members of a new class of tumor promoters. Proc. Natl. Acad. Sci. U. S. A. 78, 3872–3876.

Fujiki, H., Mori Sugimura, T., Moore, R.E., 1983. New classes of environmental tumor promoters: indole alkaloids and polyacetates. Environ. Health Perspect. 50, 85–90.

Gaget, V., Humpage, A.R., Huang, Q., Monis, P., Brookes, J.D., 2017. Benthic cyanobacteria: a source of cylindrospermopsin and microcystin in Australian drinking water reservoirs. Water Res. 124, 454–464.

Gao, Y., Cornwell, J.C., Stoecker, D.K., Owens, M.S., 2012. Effects of cyanobacterial-driven pH increases on sediment nutrient fluxes and coupled nitrification-denitrification in a shallow fresh water estuary. Biogeosciences 9, 2697–2710.

Giannuzzi, L., Sedan, D., Echenique, R., Andrinolo, D., 2011. An acute case of intoxication with cyanobacteria and cyanotoxins in recreational water in Salto Grande Dam, Argentina. Mar. Drugs 9, 2164–2175.

EVALUATING WATER QUALITY TO PREVENT FUTURE DISASTERS

Gugger, M., Lenoira, S., Bergera, C., Ledreuxa, A., Druartc, J., Humbertc, J., Guettea, C., Bernarda, C., 2005. First report in a river in France of the benthic cyanobacterium Phormidium favosum producing anatoxin-a associated with dog neurotoxicosis. Toxicon 45 (7), 919–928.

Gurbuz, F., Codd, G.A., 2008. Microcystin removal by naturally-occurring substance: pumice. Bull. Environ. Contam. Toxicol. 81 (3), 323–327. https://doi.org/10.1007/s00128-008-9458-x.

Halstvedt, C.B., Rohrlack, T., Andersen, T., Skulberg, O., Edvardsen, B., 2007. Seasonal dynamics and depth distribution of *Planktothrix* spp. in Lake Steinsfjorden (Norway) related to environmental factors. J. Plankton Res. 29, 471–482.

Hawkins, P., Runnegar, M.C., Jackson, A.B., Falconer, I.R., 1985. Severe hepatotoxicity caused by the tropical cyanobacterium (blue-green alga) *Cylindrospermopsis raciborskii* (Woloszynska) Seenaya and Subba Raju isolated from a domestic water supply reservoir. Appl. Environ. Microbiol. 50 (5), 1292–1295.

Henegan, P., Andrew, A., Kuczmarski, T., Michaelson, N., Storm, J., Atkinson, A., Wters, B., Crothers, J., Gallagher, T., Tsongalis, G., Bradley, W., Stommel, E., 2017. Aerosol exposure to cyanobacteria as a potential risk factor for neurological disease (P5.086). Neurology 88 (16 Suppl). P5.086.

Henriksen, P., Carmichael, W.W., An, J.S., Moestrup, O., 1997. Detection of an anatoxin-a(s)-like anticholinesterase in natural blooms and cultures of cyanobacteria/blue–green algae from Danish lakes and in the stomach contents of poisoned birds. Toxicon 35 (6), 901–913.

Hilborn, E.D., Beasley, V.R., 2015. One health and cyanobacteria in freshwater systems: animal illnesses and deaths are sentinel events for human health risks. Toxins 7, 1374–1395.

Ibelings, B.W., Mur, L.R., Waslby, A.E., 1991. Diurnal changes in buoyancy and vertical distribution in populations of *Microcystis* in two shallow lakes. J. Plankton Res. 13, 419–436.

Islam, M.S., Drasar, B.S., Sack, B., 1994. Probable role of blue-green algae in maintaining endemicity and seasonality of cholera in Bangladesh: a hypothesis. J. Diarrhoeal Dis. Res. 12 (4), 245–256.

James, K.J., Furey, A., Sherlock, I.R., Stack, M.A., Twohig, M., Caudwell, F.B., Skulberg, O.M., 1998. Sensitive determination of anatoxin-a, homoanatoxin-a and their degradation products by liquid chromatography with fluorimetric detection. J. Chromatogr. 798 (1–2), 147–157.

Jochimsen, E.M., Carmichael, W.W., An, J.S., Cardo, D.M., 1998. Liver failure and death after exposure to microcystins at a hemodialysis center in Brazil. N. Engl. J. Med. 338, 873–878.

Kittler, K., Schreiner, M., Krumnain, A., Manzei, S., Koch, M., Rohn, S., Maul, R., 2012. Uptake of the cyanobacterial toxin cylindrospermopsin in Brassica vegetables. Food Chem. 133, 875–879. https://doi.org/10.1016/j.foodchem.2012.01.107.

Knoll, A.H., 2003. Life on a Young Planet: The First Three Billion Years of Evolution on Earth. Princeton University Press, Princeton, NJ, USA.

Kokocinski, M., Camean, A.M., Carmeli, S., Guzman-Guillen, R., Jos, A., Mankiewicz-Boczek, J., Metcalf, J.S., Moreno, I.M., Prieto, A.I., Sukenik, A., 2017. Cylindrospermopsin and congeners. In: Meriluoto, J., Spoof, L., Codd, G.A. (Eds.), Handbook of Cyanobacterial Monitoring and Cyanotoxin Analysis. John Wiley & Sons, Chichester, UK, pp. 127–137.

Komarek, J., Johansen, J.R., 2015. Coccoid cyanobacteria. In: Wehr, J.D., Sheath, R.G., Kociolek, J.P. (Eds.), Freshwater Algae of North America, second ed Elsevier, London, UK, pp. 75–133.

Lindsay, J., Metcalf, J.S., Codd, G.A., 2009. Comparison of four methods for the extraction of lipopolysaccharide from cyanobacteria. Toxicol. Environ. Chem. 91, 1253–1262.

Magalhaes, V.F., Azevedo, S.M.O., Soares, R.M., 2001. Microcystin contamination in fish from the Jacarepagua Lagoon (Rio de Janeiro, Brazil): ecological implication and human health risk. Toxicon 39 (7), 1077–1085.

Mahmood, N.A., Carmichael, W.W., 1986. The pharmacology of anatoxin-a(s), a neurotoxin produced by the freshwater cyanobacterium *Anabaena flos-aquae*. Toxicon 24, 425–434.

Metcalf, J.S., Codd, G.A., 2000. Microwave oven and boiling waterbath extraction of hepatotoxins from cyanobacterial cells. FEMS Microbiol. Lett. 184, 241–246.

Metcalf, J.S., Codd, G.A., 2012. Cyanotoxins. In: Whitton, B.A. (Ed.), Ecology of Cyanobacteria II: Their Diversity in Space and Time. Spring Science Business Media B.V, pp. 651–676. https://doi.org/10.1007/978-94-007-3855_3224.

Metcalf, J.S., Codd, G.A., 2017. Protein phosphatase inhibition assay. In: Meriluoto, J., Spoof, L., Codd, G.A. (Eds.), Handbook of Cyanobacterial Monitoring and Cyanotoxin Analysis. John Wiley & Sons, Chichester, UK, pp. 263–266.

Metcalf, J.S., Pereira, P.B., 2017. Determination of cyanotoxins by high-performance liquid chromatography with fluorescence derivatization. In: Meriluoto, J., Spoof, L., Codd, G.A. (Eds.), Handbook of Cyanobacterial Monitoring and Cyanotoxin Analysis. John Wiley & Sons, Chichester, UK, pp. 212–217.

Metcalf, J.S., Bell, S.G., Codd, G.A., 2000a. Production of novel polyclonal antibodies against the cyanobacterial toxin microcystin-LR and their application for the detection and quantification of microcystins and nodularin. Water Res. 34, 2761–2769.

Metcalf, J.S., Beattie, K.A., Pflugmacher, S., Codd, G.A., 2000b. Immuno-cross reactivity and toxicity assessment of conjugation products of the cyanobacterial toxin, microcystin-LR. FEMS Microbiol. Lett. 189, 155–158.

REFERENCES

Metcalf, J.S., Bell, S.G., Codd, G.A., 2001. Colorimetric immuno-protein phosphatase inhibition assay for specific detection of microcystins and nodularins of cyanobacteria. Appl. Environ. Microbiol. 67, 904–909.

Metcalf, J.S., Beattie, K.A., Saker, M.L., Codd, G.A., 2002. Effects of organic solvents on the high performance liquid chromatographic analysis of the cyanobacterial toxin cylindrospermopsin and its recovery from environmental eutrophic waters by solid phase extraction. FEMS Microbiol. Lett. 216, 159–164.

Metcalf, J.S., Barakate, A., Codd, G.A., 2004. Inhibition of plant protein synthesis by the cyanobacterial hepatotoxin cylindrospermopsin. FEMS Microbiol. Lett. 235, 125–129.

Metcalf, J.S., Morrison, L.F., Krienitz, L., Ballot, A., Krause, E., Kotut, K., Putz, S., Wiegand, C., Pflugmacher, S., Codd, G.A., 2006. Analysis of the cyanotoxins anatoxin-a and microcystins in Lesser Flamingo feathers. Toxicol. Environ. Chem. 88, 159–167.

Metcalf, J.S., Banack, S.A., Lindsay, J., Morrison, L.F., Cox, P.A., 2008. Co-occurrence of beta-N-methylamino-L-alanine, a neurotoxic amino acid with other cyanobacterial toxins in British waterbodies, 1990-2004. Environ. Microbiol. 10, 702–708.

Metcalf, J.S., Reilly, M., Young, F.M., Codd, G.A., 2009. Localization of microcystin synthetase genes in colonies of the cyanobacterium *Microcystis* using fluorescence *in situ* hybridization. J. Phycol. 45, 1400–1404.

Metcalf, J.S., Richer, R., Cox, P.A., Codd, G.A., 2012. Cyanotoxins in desert environments may present a risk to human health. Sci. Total Environ. 421–422, 118–123.

Metcalf, J.S., Banack, S.A., Kotut, K., Krienitz, L., Codd, G.A., 2013. Amino acid neurotoxins in feathers of the Lesser Flamingo, *Phoeniconaias minor*. Chemosphere 90 (2), 835–839. https://doi.org/10.1016/j.chemosphere.2012.09.094.

Metcalf, J.S., Hiskia, A., Kaloudis, T., 2017. Protein phosphatase inhibition assays. In: Meriluoto, J., Spoof, L., Codd, G.A. (Eds.), Handbook of Cyanobacterial Monitoring and Cyanotoxin Analysis. John Wiley & Sons, Chichester, UK, pp. 267–271.

Metcalf, J.S., Banack, S.A., Powell, J.T., Tymm, F.J.M., Murch, S.J., Brand, L.E., Cox, P.A., 2018. Public health responses to toxic cyanobacterial blooms: perspectives from the 2016 Florida event. Water Policy 20, 919–932. wp2018012, https://doi.org/10.2166/wp.2018.012.

Mhlanga, L., Day, J., Cronberg, G., Chimbari, M., Siziba, N., Annadotter, H., 2006. Cyanobacteria and cyanotoxins in the source water from Lake Chivero, Harare, Zimbabwe, and the presence of cyanotoxins in drinking water. Afr. J. Aquat. Sci. 31 (2), 165–173.

Molica, R.J.R., Oliveira, E.J.A., Carvalho, P.V.C., Costa, A.N.S.F., Cunha, C.C., Melo, G., Azevedo, S.M., 2005. Occurrence of saxitoxins and an anatoxin-a(s)-like anticholinesterase in a Brazilian drinking water supply. Harmful Algae 4, 743–753.

Monteiro, S., Santos, R., Blaha, L., Codd, G.A., 2017. Lipopolysaccharide endotoxins. In: Meriuoto, J., Spoof, L., Codd, G.A. (Eds.), Handbook of Cyanobacterial Monitoring and Cyanotoxin Analysis. John Wiley & Sons, Chichester, UK, pp. 109–126.

Murch, S.J., Cox, P.A., Banack, S.A., 2004. A mechanism for slow release of biomagnified cyanobacterial neurotoxin and neurodegenerative disease in Guam. Proc. Natl. Acad. Sci. U. S. A. 101, 12228–12231.

Mynderse, J.S., Moore, R.E., Kashiwagi, M., Norton, T.R., 1977. Antileukemia activity in the Oscillatoriaceae: isolation of Debromoaplysiatoxin from *Lyngbya*. Science 196, 538–540.

Negri, A., Llewellyn, L., 1998. Comparative analyses by HPLC and the sodium channel and saxiphilin 3H-saxitoxin receptor assays for paralytic shellfish toxins in crustaceans and molluscs from tropical North West Australia. Toxicon 36 (2), 283–298.

Nishiwaki-Matsushima, R., Ohta, T., Nishiwaki, S., Suganuma, M., Kohyama, K., Ishikawa, T., Carmichael, W.W., Fujiki, H., 1992. Liver tumor protomotion by the cyanobacterial cyclic peptide toxin microcystin-LR. J. Cancer Res. Clin. Oncol. 118, 420–424.

Norris, R.L., Eagelsham, G.K., Pierans, G., Shaw, G.R., Smith, M.J., Chiswell, R.K., Seawright, A.A., Moore, M.R., 1999. Deoxycylindrospermopsin, an analog of cylindrospermopsin from *Cylindrospermopsis raciborskii*. Environ. Toxicol. 14, 163–165.

Norris, R.L., Seawright, A.A., Shaw, G.R., Senogles, P., Eaglesham, G.K., Smith, M.J., Chiswell, R.K., Moore, M.R., 2002. Hepatic xenobiotic metabolism of cylindrospermopsin in vivo in the mouse. Toxicon 40, 471–476.

O'Hare, M.T., Baattrup-pedersen, A., Baumgarte, I., Freeman, A., Gunn, I.D.M., Lazar, A.N., Sinclair, R., Wade, A.J., Bowes, M.J., 2018. Responses of aquatic plants to eutrophication in rivers: a revised conceptual model. Front. Plant Sci. 9, 451.

Ohta, T., Sueoka, E., Lida, N., Komori, A., Suganuma, M., Nishiwaki, R., Tatematsu, M., Kim, S.J., Carmichael, W.W., Fujiki, H., 1994. Nodularin, a potent inhibitor of protein phosphatases 1 and 2A, is a new environmental carcinogen in male F344 rat liver. Cancer Res. 54, 6402–6406.

Ohtani, I., Moore, R.E., Runnegar, M.T.C., 1992. Cylindrospermopsin: a potent hepatotoxin from the blue-green alga *Cylindrospermopsis raciborskii*. J. Am. Chem. Soc. 114, 7941–7942.

Osborne, N.J., Shaw, G.R., 2008. Dermatitis associated with exposure to a marine cyanobacterium during recreational water exposure. BMC Dermatol. 8, 5.

Pablo, J., Banack, S.A., Cox, P.A., Johnson, T.E., Papapetropoulos, S., Bradley, W.G., Buck, A., Mash, D.C., 2009. Cyanobacterial neurotoxin BMAA in ALS and Alzheimer's disease. Acta Neurol. Scand. 120 (4), 216–225.

Pacheco, A.B.F., Guedes, I.A., Azevedo, S.M.F.O., 2016. Is qPCR a reliable indicator of cyanotoxin risk in freshwater? Toxins 8 (6), 172. https://doi.org/10.3390/toxins8060172.

Paerl, H.W., 2014. Mitigating harmful cyanobacterial blooms in a human- and climatically-impacted world. Life 4, 988–1012.

Paerl, H.W., Huisman, J., 2008. Blooms like it hot. Science 320, 57–58.

Paerl, H.W., Huisman, J., 2009. Mini review: climate change: a catalyst for global expansion of harmful cyanobacterial blooms. Environ. Microbiol. Rep. 1 (1), 27–37.

Paerl, H.W., Paul, V.J., 2012. Climate change: links to global expansion of harmful cyanobacteria. Water Res. 46, 1349–1363.

Papadimitriou, T., Katsiapi, M., Vlachopoulos, K., Christopoulos, A., Laspidou, C., Moustaka-Gouni, M., Kormas, K., 2018. Cyanotoxins as the "common suspects" for the Dalmatian pelican (*Pelecanus crispus*) deaths in a Mediterranean reconstructed reservoir. Environ. Pollut. 234, 779–787.

Peczula, W., 2012. Methods applied in cyanobacterial bloom control in shallow lakes and reservoirs. Ecol. Chem. Eng. A. 19, 795–806.

Pelley, J., 2016. Taming toxic algae blooms. ACS Cent. Sci. 2, 270–273.

Pilotto, L., Douglas, R., Burch, M., Cameron, S., Beers, M., Rouch, G., Robinson, P., Kirk, M., Cowie, C., Hardiman, S., Moore, C., Attewell, R., 1997. Health effects of exposure to cyanobacteria (blue-green algae) during recreational water-related activities. Aust. N. Z. J. Public Health 21, 562–566.

Pouria, S., de Andrade, A., Barbosa, J., Cavalcanti, R.L., Barreto, V.T.S., Ward, C.J., Preiser, W., Poon, G.K., Neild, G.H., Codd, G.A., 1998. Fatal microcystin intoxication in haemodialysis unit in Caruaru, Brazil. Lancet 352 (9121), 21–26.

Puschner, B., Hoff, B., Tor, E.R., 2008. Diagnosis of anatoxin-a poisoning in dogs from North America. J. Vet. Diagn. Invest. 20, 89–92.

Pybus, M.J., Hobron, D.P., Onderka, D.K., 1986. Mass mortality of bats due to probable blue-green algal toxicity. J. Wildl. Dis. 22 (3), 449–450.

Recknagel, F., Orr, P., Swanepoel, A., Joehnk, K., Anstee, J., 2018. Operational forecasting in ecology by inferential models and remote sensing. chapter 15. In: Recknagel, F., Michener, W.K. (Eds.), Ecological Informatics. ©Springer International Publishing AG 2018. https://doi.org/10.1007/978-3-319-59928-1_15.

Richer, R., Anchassi, D., El-Assaad, I., El-Matbouly, M., Ali, F., Makki, I., Metcalf, J.S., 2012. Variation in the coverage of biological soil crusts in the state of Qatar. J. Arid Environ. 78, 187–190.

Rinehart, K.L., Harada, K.-I., Namikoshi, M., Chen, C., Harvis, C.A., Munro, M.H.G., Blunt, J.W., Mulligan, P.E., Beasley, V.R., Dahlem, A.M., Carmichael, W.W., 1988. Nodularin, microcystin and the configuration of Adda. J. Am. Chem. Soc. 110, 8557–8558.

Roy-Lachapelle, A., Solliec, M., Bouchard, M.F., Sauvé, S., 2017. Detection of cyanotoxins in algae dietary supplements. Toxins 9, 76.

Saker, M.L., Thomas, A.D., Norton, J.H., 1999. Cattle mortality attributed to the toxic cyanobacterium *Cylindrospermopsis raciborskii* in an outback region of North Queensland. Environ. Toxicol. 14 (1), 179–182.

Schantz, E.J., Mold, J., Stanger, D., Shavel, J., Riel, F., Bowden, J., Lynch, J., Wyler, R., Riegel, B., Sommer, H., 1957. Paralytic shellfish poison VI. A procedure for the isolation and purification of the poison from toxic clams and mussel tissues. J. Am. Chem. Soc. 79, 5230–5235.

Schopf, J.W., 2000. The fossil record: tracing the roots of the cyanobacterial lineage. In: Whitton, B.A., Potts, M. (Eds.), The Ecology of Cyanobacteria. Kluwer Academic Publishers, Dordrecht, The Netherlands, pp. 13–35.

Scott, L.L., Downing, T.G., 2018. A single neonatal exposure to BMAA in a rat model produces neuropathology consistent with neurodegenerative diseases. Toxins 10 (1), 22.

Sheppard, C., Al-Husiani, M., Al-Jamali, F., Al-Yamani, F., Baldwin, R., Bishop, J., Benzoni, F., Dutrieux, E., Dulvy, N.K., Durvasula, S.R., Jones, D.A., Loughland, R., Medio, D., Nithyanandan, M., Pilling, G.M., Polikarpov, I., Price, A.R., Purkis, S., Riegl, B., Saburova, M., Namin, K.S., Taylor, O., Wilson, S., Zainal, K., 2010. The Gulf: a young sea in decline. Mar. Pollut. Bull. 60, 13–38.

Simmons, J., 1998. Algal control and destratification at Hanningfield reservoir. Water Sci. Technol. 37, 309–316.

Sinha, R., Pearson, L.A., Davis, T.W., Burford, M.A., Orr, P.T., Neilan, B.A., 2012. Increased incidence of *Cylindrospermopsis raciborskii* in temperate zones—is climate change responsible? Water Res. 46, 1408–1419.

Sipia, V., Kankaanpaa, H., Peltonen, H., Vinni, M., Meriluoto, J., 2007. Transfer of nodularin to three-spined stickleback (*Gasterosteus aculeatus* L.), herring (*Clupea harengus* L.) and salmon (*Salmo salar* L.) in the northern Baltic Sea. Ecotoxicol. Environ. Saf. 66 (3), 421–425.

Sipia, V.O., Neffling, M.-R., Metcalf, J.S., Nybom, S.M.K., Meriluoto, J.A.O., Codd, G.A., 2008. Nodularin in feathers and livers of eiders (Somateria mollissima) caught from the western Gulf of Finland in June-September 2005. Harmful Algae 7, 99–105.

REFERENCES

Smith, V.H., 1983. Low nitrogen to phosphorus ratios favor dominance by blue-green algae in lake phytoplankton. Science 221, 669–671.

Smith, J.L., Boyer, G.L., Zimba, P.V., 2008. A review of cyanobacteria odorous and bioactive metabolites: impacts and management alternatives in aquaculture. Aquaculture 280, 5–20.

Spoof, L., Catherine, A., 2017. Appendix 3, tables of microcystins and nodularins. In: Meriuoto, J., Spoof, L., Codd, G.A. (Eds.), Handbook of Cyanobacterial Monitoring and Cyanotoxin Analysis. John Wiley & Sons, Chichester, UK, pp. 526–537.

Spoof, L., Neffing, M.R., Meriuoto, J., 2009. Separation of microcystins and nodularins by ultra performance liquid chromatography. J. Chromatogr. B Analyt. Technol. Biomed. Life Sci. 877 (30), 3822–3830. https://doi.org/10.1016/j.jchromb.2009.09.028.

Stewart, W.D.P., 1972. Heterocysts of blue-green algae. In: Desikachary, T.V. (Ed.), Taxonomy and Biology of Blue-Green Algae. University of Madras, pp. 227–235.

Stewart, W.D.P., 1980. Some aspects in structure and functions in N2-fixing cyanobacteria. Annu. Rev. Microbiol. 34, 497–536.

Svircev, Z., Drobac, D., Tokodi, N., Vidovic, M., Simeunovic, J., Miladinov-MIkov, M., Baltic, V., 2013. Epidemiology of primary liver cancer in Serbia and possible connection with cyanobacterial blooms. J. Environ. Sci. Health C Environ. Carcinog. Ecotoxicol. Rev. 31, 181–200.

Terao, K., Ohmori, S., Igarashi, K., Ohtani, I., Watanabe, M.F., Harada, K.I., Ito, E., Watanabe, M., 1994. Electron microscopic studies on experimental poisoning in mice induced by cylindrospermopsin isolated from blue-green alga *Umezakia natans*. Toxicon 32, 833–843.

Testai, E., Buratti, F.M., Funari, E., Manganelli, M., Vichi, S., Arnich, N., Biré, R., Fessard, V., Sialehaamoa, A., 2016. Review and Analysis of Occurrence, Exposure and Toxicity of Cyanobacteria Toxins in Food. EFSA Supporting Publication. EN-998. 309 pp.

Townsend, S.A., Boland, K.T., Wrigley, T.J., 1992. Factors contributing to a fish kill in the Australian wet/dry tropics. Water Res. 26 (8), 1039–1044.

Trogen, G.B., Annila, A., Eriksson, J., Kontteli, M., Meriuoto, J., Sethson, I., Zdunek, J., Edlund, U., 1996. Conformational studies of microcystin-LR using NMR spectroscopy and molecular dynamics calculations. Biochemistry 35 (10), 3197–3205.

Ueno, Y., Nagata, S., Tsutsumi, T., Hasegawa, A., Watanabe, M.F., Park, H.D., Chen, G.C., Chen, G., Yu, S.Z., 1996. Detection of microcystins, a blue-green algal hepatotoxin, in drinking water sampled in Haimen and Fusui, endemic areas of primary liver cancer in China, by highly sensitive immunoassay. Carcinogenesis 17, 1317–1321.

Vasas, G., Gáspár, A., Surányi, G., Batta, G., Gyémánt, G., M-Hamvas, M., Máthé, C., Grigorsky, I., Molnár, E., Borbély, G., 2002. Capillary electrophoretic assay and purification of cylindrospermopsin, a cyanobacterial toxin from *Aphanizomenon ovalisporum*, by plant test (blue-green *Sinapis* test). Anal. Biochem. 302 (1), 95–103.

Vidal, F., Sedan, D., D'Agostino, D., Cavalieri, M.L., Mullen, E., Varela, M.M.P., Flores, C., Caixach, J., Andrinolo, D., 2017. Recreational exposure during algal bloom in Carrasco beach, Uruguay: a liver failure case report. Toxins 9, 267. https://doi.org/10.3390/toxins9090267.

Ward, C.J., Beattie, K.A., Lee, E.Y., Codd, G.A., 1997. Colorimetric protein phosphatase inhibition assay of laboratory strains and natural blooms of cyanobacteria: comparisons with high-performance liquid chromatographic analysis for microcystins. FEMS Microbiol. Lett. 153 (2), 465–473.

Weiss, J.H., Christine, C.W., Choi, D.W., 1989. Bicarbonate dependence of glutamate receptor activation by beta-N-methylamino-L-alamine: channel recording and study with related compounds. Neuron 3 (3), 321–326.

Westrick, J.A., 2008. Cyanobacterial toxin removal in drinking water treatment processes and recreational waters. Adv. Exp. Med. Biol. 619, 275–290. https://doi.org/10.1007/978-0-387-75865-7_13.

Westrick, J.A., Szlag, D.C., Southwell, B.J., Sinclair, J., 2010. A review of cyanobacteria and cyanotoxins removal/inactivation in drinking water treatment. Anal. Bioanal. Chem. 397, 1705–1714. https://doi.org/10.1007/s00216-010-3709-5.

Wimmer, K.M., Strangman, W.K., Wright, J.L.C., 2014. 7-deoxy-desulfo-cylindrospermopsin and 7-deoxy-desulfo-12-acetylcylindrospermopsin: two new cylindrospermopsin isolated from a Thai strain of *Cylindrospermmopsis raciborskii*. Harmful Algae 37, 203–206.

Wood, S.A., Selwood, A.I., Rueckert, A., Holland, P.T., Milne, J.R., Smith, K.F., Smits, B., Watts, L.F., Cary, C.S., 2007. First report of homoanatoxin-a and associated dog neurotoxicosis in New Zealand. Toxicon 50, 292–301.

Wynne, T.T., Stumpf, R.P., 2015. Spatial and temporal patterns in the seasonal distribution of toxic cyanobacteria in Western Lake Erie from 2002–2014. Toxins 7, 1649–1663.

Xu, Y., Cai, Q., Ye, L., Zhou, S., Han, X., 2009. Spring diatom blooming phases in a representative eutrophic bay of the Three-Gorges Reservoir, China. J. Freshw. Ecol. 24, 191–198.

Yamamoto, Y., 1976. Effects of some physical and chemical factors on the germination of akinetes of *Anabaena cylindrica*. J. Gen. Appl. Microbiol. 22, 311–323.

EVALUATING WATER QUALITY TO PREVENT FUTURE DISASTERS

Yuan, M., Carmichael, W.W., 2004. Detection and analysis of the cyanobacterial peptide hepatotoxins microcystin and nodular using SELDI-TOF mass spectrometry. Toxicon 44 (5), 561–570.

Zeller, P., Méjean, A., Biegala, I., Contremoulins, V., Ploux, O., 2016. Fluorescence *in situ* hybridization (FISH) of Microcystis strains producing microcystin using specific mRNA probes. Lett. Appl. Microbiol. 63, 376–384.

Further Reading

Zilliges, Y., Kehr, J.-C., Meissner, S., Ishida, K., Mikkat, S., Hagemann, M., Kaplan, A., Borner, T., Dittmann, E., 2011. The cyanobacterial hepatotoxin microcystin binds to proteins and increases the fitness of microcystis under oxidative stress conditions. PLoS One 6, e17615. https://doi.org/10.1371/journal.pone.0017615.

CHAPTER 7

Educational Partnerships Combined With Research on Emerging Pollutants for Long-Term Water-Quality Monitoring

Julie R. Peller[a,*], Laurie Eberhardt[b], Erin Argyilan[c], Carrie Sanidas[d]

[a]Department of Chemistry, Valparaiso University, Valparaiso, IN, United States [b]Department of Biology, Valparaiso University, Valparaiso, IN, United States [c]Department of Geosciences, Indiana University Northwest, Gary, IN, United States [d]Willowcreek Middle School, Portage, IN, United States
*Corresponding author: E-mail: julie.peller@valpo.edu

1 INTRODUCTION

Surface water quality varies from location to location and over time, as changes in land use, society, and climate take place within a watershed (Arnell and Lloyd-Hughes, 2014). An effective water-quality monitoring plan requires an understanding of these changes and the prominent pollution sources, which can be complicated by the rising number of pollutants affecting most surface waters (Gotz et al., 2010). In some geographical areas, the choice of measured parameters is fairly straightforward, and reflects known contaminant sources. This is the case for surface waters surrounded by agricultural land, which are mostly influenced by runoff of fertilizers and pesticides from farm fields; dissolved substances such as phosphates,

nitrates, chlorides, and select pesticides need to be routinely monitored. In areas where underground storage tanks or septic tanks reside in shallow groundwater systems, the impact of leaking infrastructure on water quantity and quality need to be determined. For urban areas, surface waters receive a myriad of point and nonpoint contaminants from urban pollution, such as road runoff, aerial deposition, and pollution from businesses and homes (Mansour et al., 2016). Industries in urban or rural areas may discharge pollutants into the waste stream, which can negatively affect water quality.

Northwest Indiana is a highly industrialized area, where the growth of manufacturing along Lake Michigan began in the late 1800s and early 1900s in recognition of the area's natural resources. Prior to industrialization, the land in Northwest Indiana was covered with forest and known for its white pine (Schoon, 2003). The area experienced a change in land use that included deforestation, the loss of unique dune and swale landscapes, wetland loss and degradation, accumulation of industrial waste including, and most notably, steel slag and coal ash, among others connected to industrialization. The pollution load on surface waters from these changes is difficult to quantify. Major industries developed and remain in Northwest Indiana, including oil refineries, steel producers, railroads, chemical producers, materials manufacturing, and others. Clearly, there is a need for scientifically sound water-quality monitoring over the long term.

Over the years, major industrial contamination from ineffective regulatory practices and mistakes/contaminant spills in Northwest Indiana and surrounding regions of southern Lake Michigan have compromised ground and surface water quality and aquatic ecosystems (Ferraro, 2014). In 2017, US Steel released excessive amounts of hexavalent chromium from its wastewater treatment plants on different occasions and did not appropriately follow guidelines for the reporting and handling of such incidences (Massoels, 2017). In addition to the industrial presence, the Illinois–Indiana–Michigan area near Lake Michigan is a major hub for underground pipelines that carry gas, oil, and tar sands. Enbridge line 6B is a 293-mile pipeline, which carries millions of gallons of heavy crude oil and medium crude Canadian oil daily and a part of the pipeline runs from Griffith, Indiana (Lake County) to Marysville, Michigan. In 2010, a massive oil spill from this pipeline occurred in Kalamazoo, Michigan in the Kalamazoo River. More than a million gallons of tar sands leaked from a 6 ft rupture in the pipeline, in part, due to a delayed response by the company (Lu and Potter, 2012). The oil spread 40 miles downstream from the leak and contaminated approximately 4400 acres of land along the river (Michigan, 2010). In 2014, the British Petroleum (BP) refinery in Whiting, Indiana spilled approximately 1600 gal of oil into Lake Michigan. The mishap occurred at the refinery's largest and updated crude oil distillation unit, which was a main part of the massive $4 billion upgrade to refine tar sands from Canada.

The above incidences are examples of massive pollution releases from Northwest Indiana industries, which have invoked acute and long-term damage to the watershed and its ecosystems. It is unclear the extent to which these types of spills affect water quality, aquatic organisms, and the surrounding land since comprehensive, long-term data are difficult to locate or are simply nonexistent. While state and federal agencies monitor at the area of the industrial pollution spill, the efforts may be short-term and/or spatially incomplete, since these efforts require extensive time and money. In addition to major industrial pollution problems, many smaller industries, businesses, urban and suburban practices, wastewater treatment plants, septic systems, and others can introduce pollution to surface and groundwater in Northwest Indiana's watersheds. Many nonpoint sources of pollution, which are difficult to regulate (e.g., septic systems, degraded and leaking water and sewage lines), and emerging pollutants (e.g., microplastics), which are not yet regulated, can also degrade surface water quality (Archer et al., 2017; Pittman and Armitage, 2016). Even though the US EPAs Clean Water Act functions to protect surface water, the growing list of chemicals and contaminants often overwhelms the protectionary systems. The routine occurrence of smaller spills into waterways further substantiates the problem. Also, political drivers at the federal and state levels currently favor economic prosperity over environmental protections. The reduction in regulations that protect surface water demonstrate the unfortunate and dangerous vulnerability of freshwater protections to politics and special interests.

While laws are in place to limit the amount of pollution discharged to surface waters, the

1 INTRODUCTION

amount of monitoring and oversight done to ensure safe and healthy surface water is often far from sufficient to fully address problems within the watershed. Moreover, sustainable water-quality monitoring programs are costly, yet required to ensure regulations are met and fresh, surface water is protected (Peller et al., 2017). This, among many other factors, has led to an insufficient number of operative water-quality monitoring programs (Falco, 2017; Lindenmayer and Likens, 2009). Furthermore, the implementation of citizen or even academic-based monitoring programs has not been embraced by the regulatory community. This chapter describes an alternative, cost-effective option for routine water-quality monitoring, which was carried out in a Northwest Indiana Lake Michigan watershed. The 2-year EPA Environmental Education (EPA EE) project was undertaken to address the absence of consistent monitoring of the local watershed. The need to protect precious freshwater resources, especially the Great Lakes, should translate into the requirement for surface water monitoring of tributaries in a scientifically sound manner and at proper locations. The reported educationally based project set out to increase awareness and understanding of threats to the lake. Ultimately, society must value clean surface water/Great Lakes water enough to insist upon support for effective, long-term programs to ensure healthy water quality.

As part of an effort to assess pollution challenges to the watershed, university scientists and students explored emerging contaminants. An area of focus for the project turned toward the plastic waste problem, especially microplastics as they relate to water quality. Microplastics, defined as plastic materials less than 5 mm in size, are present in surface waters and the environment around the globe (Geyer et al., 2017). Even though plastic waste has been contaminating surface waters for decades, the assessment of these particulate contaminants has not become part of traditional water-quality monitoring.

The manufacture of synthetic polymers (plastics) dates back to the 1930s and 1940s. The widespread use of numerous plastic items, the dominance of plastic packaging and the low cost have contributed to the continual rise in the variety and quantity of materials made from synthetic polymers. By 1980, approximately 245 million tons of plastic were produced globally, in comparison to the 2016 production of 335 million tons (Geyer et al., 2017). The durability and persistence of these materials means they will remain in the environment for hundreds or thousands of years. Equally troubling is the outlook for production of plastics is anticipated to continue growing, resulting in greater amounts of plastic pollution. The current state of understanding of the ubiquitous macro and microplastics is critically low. Funding from the EPA EE project enabled the initiation of investigations into plastic pollution in the Northwest Indiana Lake Michigan watershed. Information about the science of microplastics in the watershed (emerging contaminants) became part of the educational program. This led to greater student awareness of the existing and growing problem and all the students began to identify their role in addressing this critical watershed issue.

Project goals included addressing important questions such as: (1) when governmental institutions are unable or unwilling to assume the responsibilities to protect water quality, what can be done to protect precious freshwaters? (2) Can educational programs be formulated in a strategic manner, where science teachers incorporate water-quality monitoring into their curriculum and focus on pollutants most likely to impair water quality or emerging pollutants? (3) Can these programs be supervised by scientists and also educate students and communities? We set out to answer these questions with funding from the EPA Environmental Education program: *Building bridges for environmental stewardship: schools, universities, and community collectively embracing the health of a local watershed* (Peller and Eberhardt, 2015).

EVALUATING WATER QUALITY TO PREVENT FUTURE DISASTERS

The city of Portage, Indiana, selected for the study, is on the border of Lake Michigan. Similar to many populated coastal areas, the water, beach, and ecosystem are subjected to numerous pollutants from major industries along the lake, intensely traveled highways, and agriculture and urban waste. While legacy pollution has adversely affected waterways in Northwest Indiana, restoration projects (i.e., Marquette Plan's Portage Lakefront (PL) and Riverwalk) have improved the health of the local ecosystem and have uplifted public perceptions of the natural areas near and along the lakeshore. This increased access and use does inevitably create new challenges for maintaining watershed resources.

Information and data from the project is presented in this chapter with thoughts and suggestions on how watershed monitoring using educational partnerships can be effective and sustainable in communities deficient in water-quality monitoring programs. Overall, many of the objectives of the project were met, and selected areas of the local watershed were well monitored over the 2-year period. As with most funded, short-term environmental science projects, the real challenge is finding a way to carry on the important monitoring, education, and stewardship for the watershed when the project funding is gone.

2 EXPERIMENTAL

Three university science professors worked with seven middle school science teachers, nonprofits, governmental, and community organizations to create and implement a 2-year educational watershed monitoring program in Portage, Indiana, a Lake Michigan coastal city. Approximately 600 students, from the middle schools and universities, participated in the comprehensive watershed monitoring program. Part of the project involved research by undergraduate science students. Another component consisted of university science education students working with middle school students as part of their university coursework. Middle school students were trained to collect traditional water-quality data in the field, analyze the data, and also think beyond the traditional parameters used to assess water quality. The student work was highlighted at the completion of each school year in a student posters session and program celebration. The nonprofit groups, Hoosier Environmental Council (HEC) and Save the Dunes, as well as the local USGS (LMERS), National Parks Service and Portage Parks Department contributed to the educational program.

2.1 Study Sites

The map and photograph shown in Fig. 1 display the Portage Lakefront (PL) Park along the Lake Michigan shoreline adjacent to Burns Waterway (BW), a highly studied waterway and the drainage basin for the Little Calumet River (Thupaki et al., 2013a). Three counties that include five wastewater treatment plants, smaller industries, and many nonpoint sources of pollution drain into this river (Olyphant et al., 2003). The water from the heavily loaded BW flows out to Lake Michigan, and regularly returns to the PL beach area due to the lake water currents (Thupaki et al., 2013b). In addition to the important municipal and industrial discharges into the Little Calumet, the watershed has numerous creeks and streams that flow into the river. Therefore, it is subject to a wide array of nonpoint source pollution, especially since the city of Portage has experienced significant urban expansion over the past decade. Most of these Lake Michigan tributaries are considered impaired for *Escherichia coli*, and also mercury and PCBs in fish, in addition to other pollutants, as denoted on the Indiana Department of Environmental Management website of Watersheds and Nonpoint Source Pollution.

Based on these aspects of the watershed and Lake Michigan, two creeks were chosen as

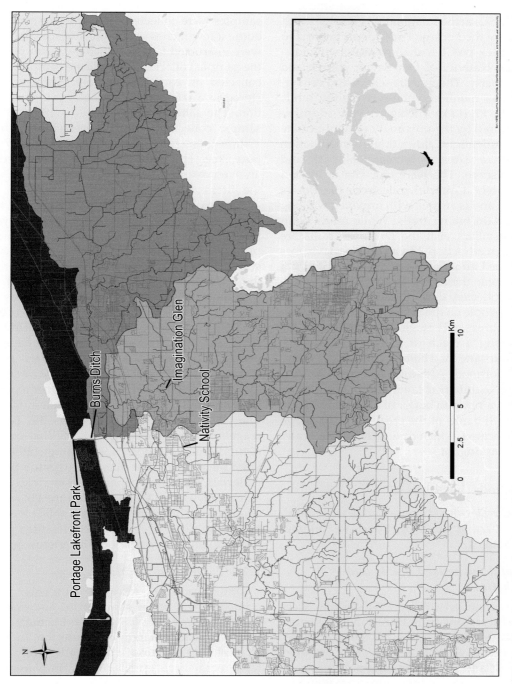

FIG. 1 Map of sampling locations with respect to Lake Michigan. Portage Lakefront beach, △; Burns Waterway, ◇; Watershed monitoring sites, ○. Google earth image showing Burns Waterway and Portage Lakefront beach.

sampling sites for the middle school students, Salt Creek (SC) at the Imagination Glen Park site, managed by the Portage Parks department and Willow Creek (WC), a shallow waterway adjacent to the Nativity of Our Savior church and school property. The middle school and university partners monitored these two sites, while BW and the PL (Lake Michigan) were additionally monitored by the university undergraduate research students. SC flows though Imagination Glen Park, a 276-acre recreational park and nature preserve. The sampling site is located approximately a quarter mile from the closest parking lot and is an area with unpaved walking and mountain biking trails. While many community members visit the park, the trails are not the main attraction. Most of the park visitors frequent the ballparks and playground and not the trails near the creek. During the times the students collected data, they observed only a few other people on the trails. WC flows through residential areas of Portage. The sampling site is located along a reach of the creek that flows under a four-lane road before entering a residential neighborhood. Although the creek is adjacent to the maintained lawn and parking lot of the church, a riparian buffer of native vegetation borders both banks of the creek. One discharge pipe was located, which drains the excess water from the parking lot.

2.2 Methods

Several water-quality parameters were monitored throughout the project and are summarized in Table 1.

2.2.1 Total Suspended Solids and Microplastics

Total suspended solids (TSS) samples were measured according to an EPA standard operating procedure. In brief, 500 mL sample bottles were filled with water from each creek or waterway. For SC and BW, a sampling pole was used to collect the water samples approximately 6 in.

below the surface. For WC, a shallow creek, the samples were collected directly in the sampling bottle a few inches below the surface. Samplers were instructed to not disturb the water prior to the sample collections. Duplicate samples were collected on each testing date. Samples were stored in a refrigerator or cooler prior to filtration. The samples were filtered within 48 h of sample collection using a preweighed, dried 47-mm glass filter fiber filter paper of 0.7-μm pore size. The collected solids were dried in an oven overnight at approximately 103°C and the filters were reweighed.

Approximately 6 months into the project, university researchers began investigating ways to further analyze the TSS filter contents for microplastics. Nylon filter paper (0.45 μm) was substituted for the glass fiber filter paper for the collection of TSS since the latter did not remain intact during the chemical processing procedure. After the TSS values were recorded, the TSS filters were viewed under the microscope and then further processed to assess the presence of microplastics. Each filter paper was put into a 40-mL centrifuge tube, followed by $FeCl_2$ (aq.), dilute HCl, and hydrogen peroxide, and gently heated for approximately 60 min. The chemical mixture, known as the Fenton reagent, creates hydroxyl radicals that oxidize natural organic materials, including cotton and wool microfibers, but does not degrade plastic materials. Each filter paper was then rinsed with deionized water (Milli-Q Ultrapure DI water), and the contents of the processed solids were filtered onto either 47- or 25-mm nylon filter paper (0.45 μm) and viewed under the microscope.

2.2.2 E. coli and Turbidity

Undergraduate research students were trained to test water samples for *E. coli* levels and turbidity at the USGS Lake Michigan Ecological Research Station in Chesterton, Indiana. In brief, 200-mL autoclaved bottles were used to collect water samples. *E. coli* measurements were performed on water samples (generally

2 EXPERIMENTAL

TABLE 1 Water-Quality Parameters Measured Throughout the Project

Water-quality parameter	Method	Monitors
Total suspended solids (TSS)	EPA Method 160.2	University and middle school students
Turbidity	Turbidimeter and turbidity tube	University and middle school students
E. coli	Colilert-18 IDEXX	University
Water temperature	YSI Multiparmeter Sonde	University
Dissolved oxygen	YSI Multiparmeter Sonde	University
Conductivity	YSI Multiparmeter Sonde	University
Anions (Cl^-, NO_3^-, SO_4^{2-})	Ion Chromatography	University
pH	YSI Multiparmeter Sonde	University
Air temperature	Thermometer	University and middle school students
Wind direction and speed	Anemometer	University and middle school students
Water flow rate	Flotation object, timer	University and middle school students
Weather conditions (additional)	Observations	University and middle school students
Invasive species	Observations	University and middle school students
Creek bank conditions	Observations	University and middle school students
Macroinvertebrates—PTI rating	Collections and identification of macroinvertebrates	University and middle school students
Pollution assessment	Collection and description	University and middle school students
Microfibers	TSS filters	University

25–100 mL) using the IDEXX Colilert-18 and Quanti-Tray 2000 method (IDEXX Laboratories, Westbrook, Maine), a defined substrate technology, with results provided as most probable number (MPN)/100 mL. The university students used a Hach turbidimeter in the lab to measure turbidity from collected water samples. The samples were homogenized by inverting the samples tubes several times prior to the measurement. In the field, turbidity measurements were performed by the middle school students using a turbidity tube. Again, water was retrieved either directly from the shallow WC or with the aid of a sampling pole in SC (Imagination Glen). Students were instructed to avoid stirring up the sediment during sampling. For the on-site turbidity measurement,

four or five students determined the depth level in centimeters of water when the tube base marking was visible.

2.2.3 Water Temperature, pH, Conductivity, Dissolved Oxygen, and Anions

A sonde (YSI Inc./Xylem Inc.) was used in the field to measure pH, water temperature, conductivity, and dissolved oxygen. These measurements were performed by the university research students. The instrument was calibrated before each outing as described in the user's manual. The instrument was not fully reliable, and the dataset for these parameters was not completed.

Ion chromatography was used to measure the concentrations of chloride, sulfate, nitrate, and nitrite in the filtered water samples. Water samples were collected in 100-mL polypropylene bottles and filtered using either a 0.45- or 0.22-μm filter paper. The filtered water samples were then analyzed using a Waters LC system equipped with an IC-Pak anion column and conductivity detector. The mobile phase consisted of diluted borate gluconate mixed with *n*-butanol and acetonitrile, as specified by the Waters manual. Each 10-μL water sample was run for 22 min using an isocratic flow of 0.9 mL/min. Water samples were compared to standards to determine the concentrations of chloride, sulfate, and nitrate/nitrite.

2.2.4 Macroinvertebrates and Additional Field Observations

Aquatic macroinvertebrates were collected in an ad hoc manner using standard D-nets. Middle school students took turns dipping nets along the banks, riffles, and muddy river bottom and through woody debris in the river. Organisms were placed in shallow white trays, identified in the field and returned to the river after data were recorded. A protocol developed by Hoosier Riverwatch (http://www.hoosierriverwatch.com/) for aquatic macroinvertebrates was used to identify each living organism to higher-order taxa and group these taxa by four pollution tolerance categories. Students practiced using a simple identification key provided by Hoosier Riverwatch in their classrooms before their river monitoring trips. During the field monitoring, students also had help identifying organisms from a trained student assistant or teacher. If any fish were inadvertently captured during monitoring, they were immediately returned to the river. Collection differences were minimized by having the same number of students collecting with D-nets for approximately the same amount of time in the same location of the river for each monitoring trip.

The students also tallied observations on creek bank conditions, additional weather data, pollution, and invasive species.

3 RESULTS AND DISCUSSION

3.1 Field and Lab Water-Quality Data

Several water-quality parameters were monitored throughout the project and are summarized in Table 1. TSS were measured throughout the project (excluding January–March, due to weather conditions) from the student field sites, SC and WC, and the larger waterway, BW. The water from SC was typically higher in suspended solids than the other sites, and the highest values were measured in the late spring and early summer and associated with periods of heightened precipitation. The same trends were found with turbidity measurements, another measure of water clarity. In general, higher levels of suspended solids or turbidity reduce the depth of sunlight penetration through water (Grobbelaar, 2009). Particles in the local stream waters originate from a number of sources, and the short-term changes in suspended particulates are mostly followed by returns to average levels. Overall, short-term trends are less meaningful than long-term monitoring of these parameters, and the long-term changes in stream embeddedness are poorly understood. Significant, persistent changes in TSS or turbidity values signify major impacts on the surface waters, which require attention since increased turbidity is a form of pollution (Grobbelaar, 2009).

In Fig. 3, the turbidity data from measurements performed by the university undergraduate research students and the middle school students at the SC location are graphed together. As expected, lower turbidity measurements mostly correspond with lower masses of TSS (Fig. 2). The middle school student turbidity data, measured using a turbidity tube, is noted by circles in Fig. 3. When the clarity of the water was good, the students were able to visualize the marking on the bottom of the tube at greater depths of water, or higher height measurements. The expected inverse relationship to the turbidimeter measurements is noted, since higher

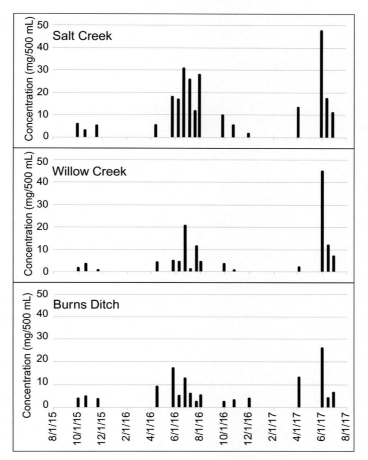

FIG. 2 Total suspended solids determined from water collected at Burns Waterway (BW), Salt Creek (SC), and Willow Creek (WC), in mg/500 mL.

values from the turbidity tube correlate to lower turbidity values. Student data show the same trends as the data taken by the more experienced water-quality monitors and more technical laboratory equipment, demonstrating that students can be trained to collect water clarity data using a simple turbidity tube. This is one set of data that can be routinely collected long term by middle school science teachers and students. Water-clarity data collected using citizen science has been demonstrated successfully in other volunteer programs (Lottig et al., 2014).

Additional parameters that can be measured by student water-quality monitors (citizen scientists) include water and air temperatures, descriptive water conditions, water depth, and water flow. As an example, the water temperature data collected at SC over the 2-year period of the project by the middle school students is shown in Fig. 4. Water temperature is a critical water-quality parameter in monitoring projects for assessing the changes in temperature with the changing climate and other local watershed alterations, such as new industries or heavier

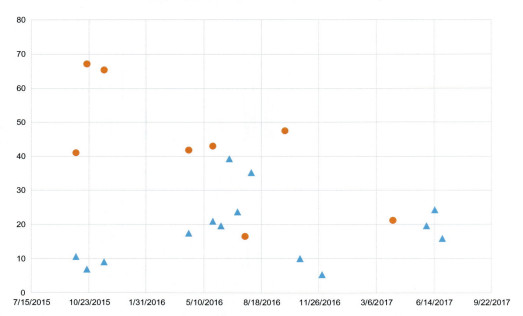

FIG. 3 Turbidity measurements from a turbidimeter (△) in Nephelometric Turbidity Units (NTUs) and by middle school students using a turbidity tube (●) and recorded as water height in cm from Salt Creek.

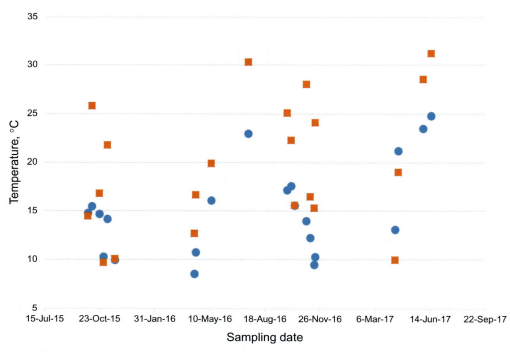

FIG. 4 Air (■) and water (●) temperatures (°C) measured at the Salt Creek location from October 1, 2015 through June 15, 2017.

loaded wastewater treatment plants. A recent document compiled by members of the scientific community in Northwest Indiana highlights concerns of climate change on the local ecosystems (Hook et al., 2018). Temperature data collected systematically over the course of many years is important in calculating ramifications of climate change and students can contribute to this necessary database.

Students collected aquatic macroinvertebrate data as part of the water-quality monitoring protocol. Despite advance training in identification and assistance from an instructor in the field, the data reported by students often could not be reconciled with known site characteristics, most likely due to the low number of samplings. A useful way of detecting water-quality changes over time through macroinvertebrate identification and classification could involve training middle school science teachers who conduct regular, seasonal monitoring. When teachers are trained and experienced, they are able to collect and organize the data even with changing groups of students each year. For example, if the monitoring site regularly yielded many sensitive taxa like Ephemeroptera (mayflies) or Plecoptera (stoneflies) and these were not found during a season of monitoring, this could be a sign of water-quality problems that required further attention. In addition, many regions are under constant threat of new invading aquatic nuisance species. Systematic monitoring of the same sites would be an excellent way to detect an invasion by a new species as well as providing before- and after-invasion monitoring data to examine the effects of these invaders on the ecosystem.

Students and other trained citizen scientists can be instrumental in the collection of water samples, which can be stored and later analyzed by scientists or trained professionals. During the 2-year project, the collected water samples were analyzed for chloride, sulfate, and nitrate concentrations using ion chromatography. The salt level or the salinity of a freshwater body is a very useful water-quality indicator and can be determined by measuring chloride ion levels. Chloride is a persistent ion since it is not taken up by any natural mechanism, and it accumulates in end-point waters, such as the Great Lakes. Due to the addition of chloride to tributaries, chloride ion levels have been slowly rising in the Great Lakes over the past several decades (Chapra et al., 2009). These rising levels are concerning since this type of contamination is permanent and certain organisms are sensitive to substantial changes in chloride concentration. Rising levels of chloride can be used as a warning indicator of the overall health of a freshwater body; therefore, chloride should be monitored regularly. Only recently, a comprehensive assessment of chloride ion data in freshwater lakes was compiled (Dugan et al., 2017). Significant sources of chloride ion include road salt, fertilizers, wastewater, and industrial effluents (Lax et al., 2017). Therefore, variations in chloride concentrations are closely linked to human activities and behavior. To that point, one factor contributing to rising chloride ion concentrations at the watershed level is the increase in impervious land cover. Given the totality of chloride inputs, the study predicts that many freshwater lakes will reach a dangerous level for aquatic life, deemed as 230 mg/L by the US EPA, in the next 50 years (Dugan et al., 2017).

Chloride ion concentrations were measured regularly at the two creek locations, BW and, in the summer, at the Lake Michigan shoreline location (PL). Fig. 5 shows the chloride ion data collected during the project. The highest measured chloride ion concentrations were consistently found in the WC water samples, including the highest measured value of 143 mg/L (or ppm). This site also exhibited the highest range in chloride concentrations, possibly reflecting the sensitivity of small creeks, and there was no apparent seasonality in the chloride data. Chloride ion data from SC revealed consistently higher values in summer months (June, July) during the periods of measurement. BW, which flows directly into Lake Michigan, exhibited the lowest variation

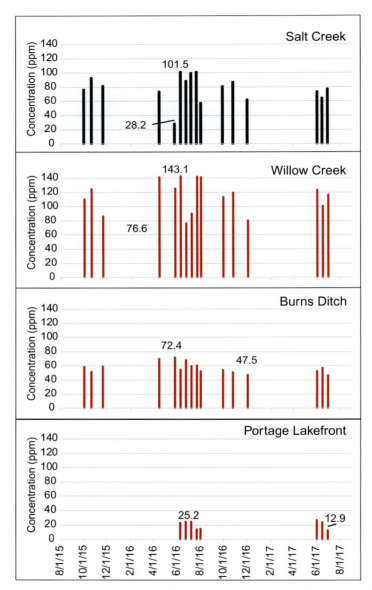

FIG. 5 Chloride ion concentrations in ppm, determined from ion chromatography, for the individual sampling sites.

in chloride concentrations. Lake Michigan water just west of this outflow area was analyzed at PL for anions only during the summer months; the chloride ion concentrations were lowest in the lake and exhibited the smallest range of values of the four sites, 13–27 ppm. The open waters of Lake Michigan in 2006 had a measured average value of 11.7 ppm. This level is predicted to continue rising with consistent chloride ion loading to the lakes (Chapra et al., 2009). Other recent studies conclude that chloride ion concentrations are rising faster than expected in river tributaries and exceeding the EPA's critical level of 230 mg/L. Inputs are greatest in the winter and spring months, likely from road salts (Corsi et al., 2015; Peller et al., 2013). Given these important concerns of chloride ion contamination in freshwater, it is critical to monitor this water-quality parameter in Great Lakes tributaries throughout all seasons.

The concentration of nitrate was measured and recorded throughout the project at the four sampling sites (Fig. 6). Samples from the shallow WC were consistently below the level of nitrate detection using the described ion chromatography method. This result is surprising in

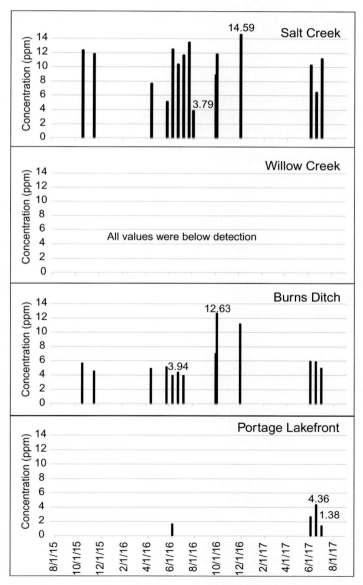

FIG. 6 Nitrate ion concentrations in ppm, determined from ion chromatography, for the individual sampling sites.

any urban area but may reflect the presence of a natural riparian buffer along the creek. The SC water samples contained the highest nitrate levels, ranging from 4 to 15 ppm. Burns Ditch samples exhibited the greatest range of values from below background to 13 ppm, measured October 2016. Though limited sampling was performed at PL Park, all measured concentrations were less than 5 ppm. Most clean, freshwater contains 1–2 mg/L (ppm) of nitrate, the EPA's limit for drinking water, and a level of 10 ppm is considered high from a public health standpoint. Sources of nitrate include wastewater effluent, runoff from animal waste, and fertilizers. The main consequence of too much dissolved nitrogen (nitrate, nitrite, ammonia) is eutrophication of the body of water, which stimulates the growth of certain organisms such as cyanobacteria and diatoms (Camargo and Alonso, 2006).

Sulfate concentrations were measured for the water samples collected throughout the project and are shown in Fig. 7. The WC and SC tributaries had similar ranges of sulfate throughout the sampling period of this study. The sulfate concentrations at Burns Ditch were notably consistent, measured around 50 ppm. Lake Michigan water samples from PL had the lowest sulfate concentrations and low variability in values. Sources of sulfate are both natural (sulfate minerals) and anthropogenic. Contributing sources of sulfate include fertilizer runoff, wastewater treatment plants, and many industrial processes. In regions where fossil fuels are combusted and metals are processed, such as Northwest Indiana, atmospheric sulfur dioxide may contribute to dissolved sulfate levels. At levels above 250 ppm in drinking water, sulfate can impart a bitter taste. When sulfate levels in drinking water are above 750 ppm, it can induce a laxative effect. When sulfate is used in the treatment of water, the treated water is often higher than the source water. The monitoring of sulfate over the long term in the SC watershed can be a means to check if additional sources of sulfate are added to the normal sulfate load in the creek.

E. coli is a fecal indicator bacteria used to determine recreational freshwater safety (Nevers et al., 2013). At levels above 235 MPN (most probable number)/100 mL water, the likelihood of pathogenic microorganisms that cause disease is high enough to be considered a public health concern. The EPA Recreational Water-Quality Criteria states that the public must be alerted when *E. coli* levels are above this value, as a warning that the water may not be safe for bathing or swimming. The data collected during the 2-year project are summarized in Table 2. *E. coli* counts in the creeks were mostly above the threshold level. Like many Lake Michigan tributaries in the Indiana Lake Michigan watershed, these surface waters are not considered safe for recreational activities. The *E. coli* counts for BW, which is magnitudes larger in volume than SC and WC, were mostly below the 235 MPN on the dates of the water sampling. While *E. coli* is a well-scrutinized fecal indicator, there are natural sources of *E. coli*, namely, animal waste, which can account for high levels in the shallow creek waters (Byappanahalli et al., 2006).

3.2 Implementing Emerging Contaminants of the Watershed in Watershed Monitoring: Plastics

Increases in disposable, nonbiodegradable materials have led to huge amounts of plastic pollution scattered in natural environments, especially bodies of water. The mostly lightweight waste often fragments and moves until it reaches surface water environments or it originates in wastewater discharge. The presence of massive amounts of microplastics in bodies of water worldwide should be a strong argument for the incorporation of these synthetic materials in water-quality monitoring protocol (do Sul and Costa, 2014; Eerkes-Medrano et al., 2015). The increasing number of published studies on microplastics in the environment conveys the rising interest and concern. While most studies on microplastic pollution have focused on the

3 RESULTS AND DISCUSSION

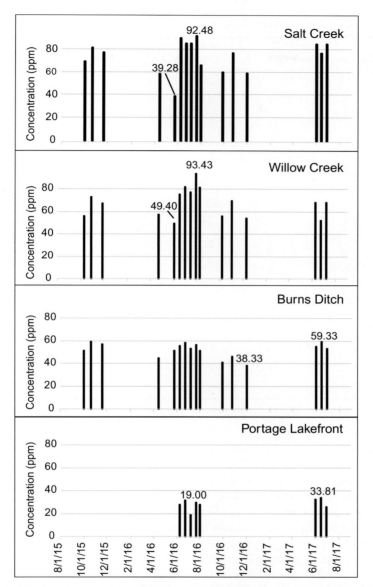

FIG. 7   Sulfate ion concentrations in ppm, determined from ion chromatography, for the individual sampling sites.

marine environment, recent scientific reports and reviews conclude that more research is necessary, especially in freshwater ecosystems (Eerkes-Medrano et al., 2015; GSAMP, 2015). Research has established that microplastics are ubiquitous in surface water samples (Imhof et al., 2016; Ivleva et al., 2017; Vandermeersch et al., 2015; Ziccardi et al., 2016). Rivers contribute an estimated 1.15–2.41 million tons of plastic waste into the ocean each year, and most of it enters from late spring through the fall (Lebreton et al., 2017). Plastic materials can release

TABLE 2 *E. coli* Counts (MPN/100 mL) From Water Samples Collected at the Study Sites

Date (↓) and location (→)	Salt Creek (SC)	Willow Creek (WC)	Burns Waterway (BW)	Portage Lakefront
10/1/2015	248.1	248.1	57.1	
10/20/2015	232	1095	35	
11/19/2015	2092	213	202	
4/14/2016	64	186	182	
5/25/2016	3266	373	345	
6/9/2016	387	83	50	2
6/23/2016	2190	5475	276	
7/7/2016	1844	1850	293	
8/1/2016	142	213	133	1
9/28/2016	3594	1743	57	
10/24/2016	188	977	81	
12/1/2016	149	263	199	
4/4/2017	597	93	2950	
6/1/2017	245	137	34	10
6/15/2017	876	1273	465	61
6/29/2017	404	217	164	14

Values above the RWQ standard of 235 indicate Lake Michigan beach advisories and are highlighted in gray.

chemicals and can also adsorb and possibly concentrate hydrophobic water contaminants; the overall ecological health consequences of these synthetic materials are simply not well known. As microplastic pollution potentially negatively impacts the freshwater aquatic biota and can be a vector for transfer of persistent pollutants into the aquatic food chain, it is imperative that scientific investigations are conducted on the presence and fate of these synthetic materials. As part of the EPA EE project, researchers and students added the study of the plastic pollution problem on the local level to the collected data during the second year of the project.

The middle school students collected and tallied plastic pollution at the study sites and categorized it using recycle codes. For the materials without the numerical codes, the students used water and a calcium chloride solution to estimate the density of the plastic waste. Most of the plastic garbage was classified in the categories of food wrappings and beverage containers. The materials that were recyclable were taken back to the school and added to appropriate recycling containers and the remaining garbage was thrown away. For this part of the watershed project, a plastic waste data set was initiated.

The undergraduate research students began investigations on microplastics in the watershed. Over the course of several months, sampling and laboratory protocol were established after testing a few different processing methods from the literature. Many blank samples were added to

3 RESULTS AND DISCUSSION

the sediment and water sampling and processing protocol, in addition to cotton laboratory coats and a laminar flow hood to reduce potential external contamination. Stereomicroscopic analyses were mainly used to identify and count microplastics, but Raman spectroscopy verified the synthetic nature of select samples. From the analyzed samples, it was clear that the most easily identified and most prominent microplastics in the local environment were synthetic microfibers, those made of materials such as polyester and nylon. By the end of the project, the identification and quantification of synthetic microfibers was deemed most scientifically accurate means to assess the microplastic pollution. The SC study site location is below two wastewater treatment plants, a main source of microfiber pollution; therefore, it was not surprising to consistently detect the synthetic fibers in the samples from the water and creek sediment. Since validation of the experimental methodology took most of the project time, the data for these studies was not complete to report and is ongoing. This part of the watershed monitoring highlighted the fact that most of the collected pollution was made of plastic, and the preliminary investigations of microplastic pollution support efforts to add plastic data to water-quality monitoring protocol. For water samples, the can be done using the TSS filters.

3.3 Education, Awareness, Stewardship

At the start of the project, the middle school science teachers attended training on methods for surface water-quality analysis. The required equipment and supplies were purchased through the funded program and continue to be used for watershed monitoring. The training took place in the classroom, at the local creek, and along the shore of Lake Michigan and was a beneficial means for ongoing teacher education. Throughout the project, the university professors, other local experts on surface water, and university

science education students collaborated to provide meaningful instruction and support for the teachers.

As part of the comprehensive program, students, parents, and staff took part in the Alliance for the Great Lakes annual beach cleanup. In this activity, participants collected and tallied the types of waste collected. This was especially enlightening for students and parents to see the variety of waste, especially the amount of plastics waste in the water and along the Lake Michigan shoreline. The data collected by the students was submitted to the larger database maintained by the Alliance for the Great Lakes.

Over the course of two school years, the students made regular visits to two sites in the local watershed. Neither site is monitored by the state; it is important to note that regional monitoring by the state occurs every 9 years. While the students were aware of these locations from their visits to the park for sporting events and social outings (SC) or simply the creek's presence on the school grounds (WC), they were introduced to new ways of observing the areas from the guidance of the university professors and students. The middle school teachers and students became watershed monitors and collected a variety of useful water-quality data. They were introduced to scientific methods for determining water quality; macroinvertebrate, and sediment sampling were carried out on a more limited scale to prevent damage to the ecosystem.

At the end of each school year, groups of students analyzed different data sets and researched their topics. They organized the data and created posters that they presented in their classes and at each year's end poster session. The students shared their research with each other and with the public. The funded project provided unique experiences for both middle school staff and students. The staff benefited from the expertise of those who specialize in this research; the younger students were able to carry out field science first-hand, and learn about the research performed by university students. Since the middle school students analyzed authentic data, they were able to see that their habits and actions impact their local watershed. They gained a better appreciation of water's vitality and value; moreover, the students discovered what effect their daily decisions had on water both locally and globally. Most of the eighth-grade students shared this learning experience with their parents.

For watersheds that are not regularly monitored or for areas where surface water is not monitored, students and teachers can function as citizen scientists when equipment and supplies are available and when experts train and supervise the collection of data. There are many benefits for the science teachers and students from an educational perspective. The scientifically based monitoring is a means to assess the health of surface water and the greater watershed. Since the students also collected garbage, this type of program can both utilize and teach younger students to take ownership of their local, natural environment.

4 CONCLUSIONS

In states where scientifically rigorous and consistent water-quality monitoring are not in place, scientists at universities can work with community partners to design and implement meaningful monitoring through appropriate data collection and analysis. Moreover, science-based monitoring within local communities meets the educational goals for school systems and community partners (e.g., park departments, state parks, nonprofits). All science programs require equipment, supplies, and professionals, which translate into a need for funding; this is the major obstacle for regular, robust water-quality monitoring. Through the EPA Environmental Education program, a grant-based project with a typically short timeline (few years) for environmental monitoring,

we have initiated a long-term partnership between university science professors and middle school science teachers that is expected to generate scientifically rigorous data for the long term.

In addition to the collection of basic water-quality parameters, watershed monitoring studies should also strive to identify emerging issues. It is highly beneficial for students to experience important lessons related to the nature of science and scientific responses to new information. Water-quality monitoring is an ideal area of science that allows educators to move away from the standard structure of the scientific model and integrate interactive teaching strategies that emphasize observation, experimentation, and reasoning. The exploration of the emerging issue of micro- and macro-plastics in freshwater stream systems evolved as part of this EPA-funded study. The project stimulated the design and implementation of experiments that examine the behavioral response of organisms to plastics. This work continues in both the middle school and university level biology curricula.

From the reported study, it became clear that there is a need to better integrate physical data related to the streams and weather/climate conditions with the chemical and biological data. Northwest Indiana is the only area of the state that is part of the Great Lakes (Lake Michigan) watershed. Differences in geology and sediment properties can contribute to the variations in chemical, biological, and physical parameters at individual sites. Also, differences in the physical properties of the watershed have the potential to impact downstream sites, in ways that remain poorly understood. In many areas of the country, significant shifts in land management and use have taken place over the past decades. These changes affect the watershed and require agencies (such as the Indiana Department of Environmental Management) to provide meaningful data to demonstrate the efficacy of restoration and naturalization projects.

This would emphasize the need to better assess the balance between public access and environmental impacts.

In order to protect freshwater, scientifically scrutinized and long-term water-quality and watershed monitoring must take place regularly. The state of Indiana does not currently have the resources to conduct monitoring on a regular basis and this is potentially problematic for freshwater health. Educational and community partnerships, such as the one described in this chapter, can be developed and supported by municipalities or at the state level as a means to ensure the protection of critical freshwater resources for a considerably lower cost (Frankenberger and Esman, 2012). The input of local or state resources is especially important to incentivize projects that are not professionally beneficial to university professors (Peller et al., 2017). These types of projects must be sustained in order to collect and compile important data for appropriate decision-making. These types of partnerships also provide education, awareness, and opportunity to explore new challenges, whether local or widespread.

Acknowledgments

The project was made possible through the United States Environmental Protection Agency Environmental Education grant #00E01498, followed by a National Science Foundation EAGER grant, #1744004. Ion analyses were performed using a liquid chromatography system purchase through funding from the National Science Foundation MRI grant #1531360.
The authors gratefully acknowledge the many others: teachers, school principals, local nonprofit, and government agencies in Portage and Northwest Indiana that contributed to the project.

References

Archer, E., Petrie, B., Kasprzyk-Hordern, B., Wolfaardt, G.M., 2017. The fate of pharmaceuticals and personal care products (PPCPs), endocrine disrupting contaminants (EDCs), metabolites and illicit drugs in a WWTW and environmental waters. Chemosphere 174, 437–446.

Arnell, N.W., Lloyd-Hughes, B., 2014. The global-scale impacts of climate change on water resources and flooding under new climate and socio-economic scenarios. Clim. Chang. 122 (1–2), 127–140.

Byappanahalli, M.N., Whitman, R.L., Shively, D.A., Sadowsky, M.J., Ishii, S., 2006. Population structure, persistence, and seasonality of autochthonous *Escherichia coli* in temperate, coastal forest soil from a great lakes watershed. Environ. Microbiol. 8 (3), 504–513.

Camargo, J.A., Alonso, A., 2006. Ecological and toxicological effects of inorganic nitrogen pollution in aquatic ecosystems: a global assessment. Environ. Int. 32 (6), 831–849.

Chapra, S.C., Dove, A., Rockwell, D.C., 2009. Great Lakes chloride trends: long-term mass balance and loading analysis. J. Great Lakes Res. 35 (2), 272–284.

Corsi, S.R., De Cicco, L.A., Lutz, M.A., Hirsch, R.M., 2015. River chloride trends in snow-affected urban watersheds: increasing concentrations outpace urban growth rate and are common among all seasons. Sci. Total Environ. 508, 488–497.

do Sul, J.A.I., Costa, M.F., 2014. The present and future of microplastic pollution in the marine environment. Environ. Pollut. 185, 352–364.

Dugan, H.A., Summers, J.C., Skaff, N.K., Krivak-Tetley, F.E., Doubek, J.P., Burke, S.M., Bartlett, S.L., Arvola, L., Jarjanazi, H., Korponai, J., Kleeberg, A., Monet, G., Monteith, D., Moore, K., Rogora, M., Hanson, P.C., Weathers, K.C., 2017. Data descriptor: long-term chloride concentrations in North American and European freshwater lakes. Sci. Data 4, 11.

Eerkes-Medrano, D., Thompson, R.C., Aldridge, D.C., 2015. Microplastics in freshwater systems: a review of the emerging threats, identification of knowledge gaps and prioritisation of research needs. Water Res. 75, 63–82.

Falco, E., 2017. Protection of coastal areas in Italy: where do national landscape and urban planning legislation fail? Land Use Policy 66, 80–89.

Ferraro, K., 2014. Assessment of Environmental Justice Needs in Northern Lake County Communities. Hoosier Environmental Council, Indianapolis.

Frankenberger, J., Esman, L., 2012. Monitoring Water in Indiana: Choices for Nonpoint Source and Other Watershed Projects. Purdue University and Indiana Department of Environmental Management, pp. 1–134.

Geyer, R., Jambeck, J.R., Law, K.L., 2017. Production, use, and fate of all plastics ever made. Sci. Adv. 3 (7), 5.

Gotz, C.W., Stamm, C., Fenner, K., Singer, H., Scharer, M., Hollender, J., 2010. Targeting aquatic microcontaminants for monitoring: exposure categorization and application to the swiss situation. Environ. Sci. Pollut. Res. 17 (2), 341–354.

Grobbelaar, J.U., 2009. Turbidity. In: Elias, S.A. (Ed.), Encyclopedia of Inland Waters. Academic Press, pp. 699–704.

GSAMP, 2015. Sources, Fate and Effects of Microplastics in the Marine Environment: A Global Assessment. IMO/FAO/UNESCO-IOC/UNIDO/WMO/IAEA/UN/UNEP/UNDP Joint Group of Experts on the Scientific Aspects of Marine Environmental Protection.

Hook, T., Foley, C., Collingworth, P., Dorworth, L., Fisher, B., Hoverman, J., LaRue, E., Pryon, M., Tank, J., Widhalm, M., Duke, J., 2018. Aquatic ecosystems in a shifting Indiana climate: a report from the Indiana climate change impacts assessment. In: Aquatic Ecosystems Report. Purdue University.

Imhof, H.K., Laforsch, C., Wiesheu, A.C., Schmid, J., Anger, P.M., Niessner, R., Ivleva, N.P., 2016. Pigments and plastic in limnetic ecosystems: a qualitative and quantitative study on microparticles of different size classes. Water Res. 98, 64–74.

Ivleva, N.P., Wiesheu, A.C., Niessner, R., 2017. Microplastic in aquatic ecosystems. Angew. Chem. Int. Ed. 56 (7), 1720–1739.

Lax, S.M., Peterson, E.W., Van der Hoven, S.J., 2017. Stream chloride concentrations as a function of land use: a comparison of an agricultural watershed to an urban agricultural watershed. Environ. Earth Sci. 76 (20), 12.

Lebreton, L.C.M., Van der Zwet, J., Damsteeg, J.W., Slat, B., Andrady, A., Reisser, J., 2017. River plastic emissions to the world's oceans. Nat. Commun. 8, 10.

Lindenmayer, D., Likens, G., 2009. Adaptive monitoring: a new paradigm for long-term research and monitoring. Trends Ecol. Evol. 24 (9), 482–486.

Lottig, N.R., Wagner, T., Henry, E.N., Cheruvelil, K.S., Webster, K.E., Downing, J.A., Stow, C.A., 2014. Long-term citizen-collected data reveal geographical patterns and temporal trends in lake water clarity. PLoS One 9 (4), 8.

Lu, V., Potter, M., 2012. Enbridge Handled Oil Spill Like 'Keystone Cops': Safety Board. thestar.com.

Mansour, F., Al-Hindi, M., Saad, W., Salam, D., 2016. Environmental risk analysis and prioritization of pharmaceuticals in a developing world context. Sci. Total Environ. 557, 31–43.

Massoels, R., 2017. Inspection Summary/Noncompliance Letter. IDEM. Indianapolis, IN.

Michigan, S.o., 2010. Exhibit A: Spill Area Map.

Nevers, M.B., Byappanahalli, M.N., Whitman, R.L., 2013. Choices in recreational water quality monitoring: new opportunities and health risk trade-offs. Environ. Sci. Technol. 47, 3073–3081.

Olyphant, G.A., Thomas, J., Whitman, R.L., Harper, D., 2003. Characterization and statistical modeling of bacterial (*Escherichia coli*) outflows from watersheds that discharge into southern Lake Michigan. Environ. Monit. Assess. 81 (1–3), 289–300.

Peller, J.R., Eberhardt, L., 2015. Building Bridges for Environmental Stewardship: Schools, Universities and Community Collectively Embracing the Health of a Local Watershed. EPA, United States.

Peller, J.R., Argyilan, E.P., Cox, J.J., Grabos, N., 2013. Analytical measurements to improve nonpoint pollution assessments in Indiana's Lake Michigan Watershed. In: Ahuja, S. (Ed.), Monitoring Water Quality: Pollution Assessment, Analysis, and Remediation. Elsevier, pp. 171–186.

Peller, J.R., Whitman, R.L., Schoer, J.K., 2017. Can incongruent studies effectively characterize long-term water quality? In: Ajuha, S. (Ed.), Chemistry and Water: The Science Behind Sustaining the World's Most Crucial Resource. Elsevier, pp. 393–420.

Pittman, J., Armitage, D., 2016. Governance across the land-sea interface: a systematic review. Environ. Sci. Pol. 64, 9–17.

Schoon, K.J., 2003. Calumet Beginnings. Indiana University Press, Bloomington, IN.

Thupaki, P., Phanikumar, M.S., Whitman, R.L., 2013a. Solute dispersion in the coastal boundary layer of southern Lake Michigan. J. Geophys. Res. Oceans 118 (3), 1606–1617.

Thupaki, P., Phanikumar, M.S., Schwab, D.J., Nevers, M.B., Whitman, R.L., 2013b. Evaluating the role of sediment-bacteria interactions on *Escherichia coli* concentrations at beaches in southern Lake Michigan. J. Geophys. Res. Oceans 118 (12), 7049–7065.

Vandermeersch, G., Van Cauwenberghe, L., Janssen, C.R., Marques, A., Granby, K., Fait, G., Kotterman, M.J.J., Diogene, J., Bekaert, K., Robbens, J., Devriese, L., 2015. A critical view on microplastic quantification in aquatic organisms. Environ. Res. 143, 46–55.

Ziccardi, L.M., Edgington, A., Hentz, K., Kulacki, K.J., Driscoll, S.K., 2016. Microplastics as vectors for bioaccumulation of hydrophobic organic chemicals in the marine environment: a state-of-the-science review. Environ. Toxicol. Chem. 35 (7), 1667–1676.

CHAPTER

8

Evaluation of the Toxicity of the *Deepwater Horizon* Oil and Associated Dispersant on Early Life Stages of the Eastern Oyster, *Crassostrea virginica*

Julien Vignier[a,†], Aswani Volety[a,b],*, Philippe Soudant[c], Fu-lin Chu[d], Ai Ning Loh[a,b], Myrina Boulais[b,c], René Robert[e], Jeffrey Morris[f], Claire Lay[f], Michelle Krasnec[f]

[a]Department of Marine and Ecological Sciences, College of Arts and Sciences, Florida Gulf Coast University, Fort Myers, FL, United States [b]University of North Carolina Wilmington, College of Arts and Sciences & Center for Marine Science, Wilmington, NC, United States [c]Laboratoire des Sciences de l'Environnement Marin (UMR 6539-LEMAR), IUEM-UBO, Technopole Brest Iroise, Plouzané, France [d]Virginia Institute of Marine Science (VIMS), College of William and Mary, Department of Aquatic Health Sciences, Gloucester Point, VA, United States [e]Ifremer, Unité Littoral, Centre Bretagne - ZI de la Pointe du Diable - CS 10070, Plouzané, France [f]Abt Associates, Boulder, CO, United States
*Corresponding author: E-mail: voletya@uncw.edu

[†] Current address: Cawthron Institute, Nelson, New Zealand.

1 THE *DEEPWATER HORIZON* OIL SPILL INCIDENT

1.1 Context

On April 20, 2010, the explosion of the *Deepwater Horizon* (DWH) oil-drilling platform, located 50 miles (77 km) off the coast of Louisiana, killed 11 workers and led to the largest oil-related environmental disaster in US history (Carriger and Barron, 2011; National Commission on the BP deep ocean horizon oil spill and offshore drilling, 2011). An oil leak, resulting from a failed blowout preventer, was discovered 2 days later at a depth of approximately 5000 ft. (1500 m) at the *Macondo-1* well (28°73′67″N, 88°38′69″W). Until the well was capped and the flow finally stopped, 87 days after the DWH oil rig sank, an estimated 3.19 million barrels (about 507 million L) of light Louisiana crude oil were released into the Gulf of Mexico (GoM) (OSAT-1, 2010; US District Court, 2014, 2015).

As a spill mitigation response, an estimated 8 million L of chemical dispersants were directly injected at the wellhead into the oil and gas plume at 1500 m depth, and at the surface to disperse oil slicks (Kujawinski et al., 2011; US Coast Guard, 2011). Although subsurface application enabled the retention of a considerable portion of oil in the water column and the creation of a "plume," oil also moved to the upper surface waters to form a slick (Camilli et al., 2010; OSAT-1, 2010). The cumulative surface coverage of that oil slick was estimated at $\approx 112{,}000\,km^2$ (Environmental Response Management Application, 2015; Fig. 1). Despite the 4.5 million L of dispersant that were applied to the floating slick (US Coast Guard, 2011) with the intent to protect the shorelines from oil contamination, oil reached the coastal areas where it polluted beaches, bays, estuaries, and marshes from eastern Texas to the Florida Panhandle. The coastal extent of the spill was estimated to be at least 2113 km of shoreline visibly oiled (Michel et al., 2013) (Fig. 1).

Prior to the DWH oil spill incident, no deepwater application of dispersant was ever conducted and behavior of such quantity of dispersant at 1500 m depth, its environmental fate and its potential toxicity on pelagic and benthic organisms were poorly understood.

1.2 The Natural Resource Damage Assessment (NRDA)

The DWH Trustees—the US Department of Commerce; the US Department of the Interior; the US Environmental Protection Agency; the US Department of Agriculture; and designated agencies representing each of the five Gulf states (Alabama, Florida, Louisiana, Mississippi, and Texas)—undertook a natural resource damage assessment, or NRDA, to evaluate the nature and extent of adverse effects of the DWH incident on natural resources and their services. Several large-scale field efforts, including over 90 offshore cruises, 30,000 samples of water, sediment, tissues, and oil from the deep ocean, offshore and coastal areas, were conducted to assess shoreline and wildlife oiling, coastal waters, and sediments (Deepwater Horizon NRDA Trustees, 2016). Over 7000 km of shoreline have been surveyed for oil impacts (Fig. 1). Other activities included complex modeling of oil distribution and fate and toxicity testing of multiple species. Ultimately, the NRDA aimed to assess injuries, damages, and restoration options for the DWH oil spill.

As a result of this extensive, multi-year NRDA, the Trustees concluded that the DWH oil spill and related oil spill response actions caused a wide array of injuries to natural resources and the services they provide throughout a large area of the northern GoM. These conclusions were based on the scientific findings of the studies performed by the Trustees as part of the NRDA and on data collected during the oil spill response, together with supplemental findings published by the scientific community (Deepwater Horizon NRDA Trustees, 2016).

2 CHEMICAL ASPECTS

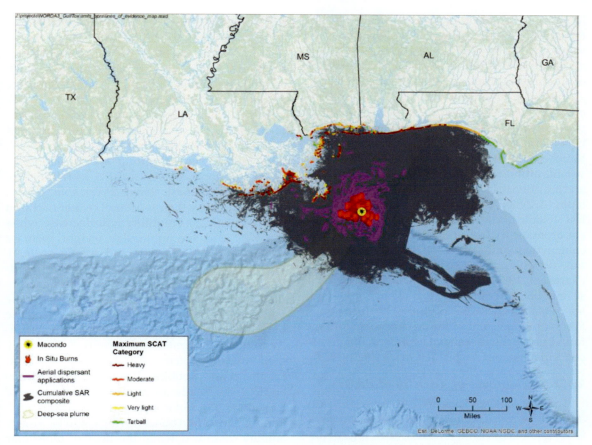

FIG. 1 Cumulative DWH oil footprint in the northern Gulf of Mexico and shoreline impact of the spill. Most surface slicks traveled toward shore, impacting at least 2100 km of coastline; and some slicks followed currents to the southeast. A deep-sea plume migrated more than 400 km (250 miles) southwest of the well. In response to the surface slicks, more than 400 flights applied chemical dispersant, and more than 400 fires were set to burn off surface oil. *Source:* Abt Associates. *From DWH NRDA Trustees, Chapter 4, 2016. http://www.gulfspillrestoration.noaa.gov/restoration-planning/gulf-plan).*

The main goal of this study was to evaluate, in the laboratory, the toxicity of crude oil released during the DWH oil spill and dispersant (Corexit 9500A) on early life stages of the Eastern oyster, *Crassostrea virginica*. Using acute exposure, lethal and sublethal responses to oil and/or dispersant of different early life stages were determined to assess the injury of the DWH oil spill on natural population of Eastern oyster.

2 CHEMICAL ASPECTS

2.1 The Deepwater Horizon Crude Oil

2.1.1 Composition and PAHs

Crude oils are complex mixtures containing thousands of organic (carbon-containing) compounds, and most of which contain only carbon and hydrogen (i.e., hydrocarbons). Hydrocarbons in crude oil range from light, volatile chemicals like those in gasoline to heavy, recalcitrant chemicals like those found in tar or asphalt. Light Louisiana sweet crude oils such as the DWH oil, is a low sulfur crude oil (Wang et al., 2003). The more toxic compounds in crude oil are aromatic chemicals— a subset of organic compounds that share a common chemical structure, namely, at least one benzene ring. These include mono-aromatic volatile organic compounds (VOCs) such as benzene, toluene, ethylbenzene, and xylenes (BTEX). These volatile aromatic hydrocarbons readily evaporate and are often responsible for the odors from petroleum. Another group of aromatic compounds is less volatile; these compounds are called polycyclic aromatic hydrocarbons (PAHs) because they contain two or more benzene rings. PAHs are neutral, nonpolar, stable, and lipophilic molecules (Varanasi, 1989). The larger PAHs, which have several benzene rings, are less water-soluble and less volatile. Because of these characteristics, PAHs are persistent in the environment and are found primarily in soil and sediment, but also in particulate matter suspended in air (Neff, 1985).

2.1.2 Mechanisms of Toxicity of PAHs

Toxicity of exposure to PAHs is generally defined as mechanisms which are responsible for negative effects on a variety of biological disruption such as growth, reproduction, and survival (Capuzzo et al., 1988). PAHs do not have one type of toxic action on aquatic organisms, but several, which depend on the compound, the type of exposure (acute or chronic), the organism tested, and the environmental conditions involved. The main mode of PAH toxicity in short-term acute toxicity exposure is the nonpolar narcosis (or baseline toxicity). It results from the accumulation of PAHs in biological membranes of an organism-causing disturbance of the membrane structure (Van Brummelen et al., 1998). Other modes of toxicity of PAHs include phototoxicity, biotransformation and subsequent adduct formation and disruption of hormone regulation (Landrum et al., 1987; Payne et al., 1987; Varanasi and Malins, 1977).

2.1.3 Bioavailability and Uptake of PAHs

The fate of PAHs in aquatic ecosystems is mainly determined by (i) their molecular structure and solubility and (ii) abiotic processes such as biodegradation, sedimentation, evaporation, or photochemical oxidation (see Section 2.1.3), which will determine the actual concentrations of PAHs to which organisms will be exposed (Lee et al., 1978; Van Brummelen et al., 1998). Although PAHs as a group are considered to be hydrophobic, they possess a wide range of solubility (Varanasi, 1989). PAH solubility is characterized by their n-octanol/water partition coefficient (K_{ow}): solubility will decrease while the K_{ow} and the molecular weight increase (Djomo et al., 1996; Porte and Albaiges, 1994). The partitioning behavior of PAHs between sediments, overlying water (porewater), soil, particulate or colloidal matter, and dissolved organic matter is one of the major factors influencing the bioavailability of PAHs and their binding affinity to these substrates is determined mainly by their hydrophobicity (Di Toro et al., 1991; Landrum and Robbins, 1990; Lee, 1977; Means et al., 1980; Readman et al., 1982). Because of this, lower molecular weight (LMW) PAHs, containing one or two benzene rings, will be preferentially dissolved while the heavier molecular weight (HMW) PAHs, containing three or more benzene rings, will be adsorbed onto or associated with particles (Baumard et al., 1999).

Uptake of PAHs by organisms is thus governed by its bioavailability, and it may take place from aqueous systems via passive diffusion through the gills or the skin, and from dietary sources via the gastrointestinal tract (Barron, 1995; Croxton et al., 2012; Power and Chapman, 1992). The majority of PAH body burden measurements have been made on bivalves. Given their biological characteristics (sessile, filter-feeder, benthic), bivalves can rapidly accumulate PAHs; but, they have little capacity for PAH metabolism (Meador et al., 1995; Neff et al., 1976; Stegeman and Teal, 1973; Viarengo et al., 1981) compared to fish, which have low

2 CHEMICAL ASPECTS

body burden of PAHs due to their ability to metabolize PAHs (Neff, 1979). Consequently, contamination by high-molecular weight PAHs, including those with carcinogenic potential, are of great concern to human consumers, because they are usually observed in shellfish, particularly in bivalve mollusks. For this reason, PAHs are the best indicator of the potential toxicity of spilled crude oil to water column and benthic organisms such as oysters (Anderson, 1977; Neff, 1979; Neff and Stubblefield, 1995).

2.1.4 Weathering Processes

When "fresh" oil is released into the environment, it undergoes continuous compositional changes through natural processes, collectively referred to as weathering. Weathering processes include evaporation, dissolution, emulsification, sedimentation, and photooxidation (Daling et al., 2014; Landrum et al., 1987; Leahy and Colwell, 1990; Wang and Stout, 2007). Weathered oil is considered to have less potential for causing adverse toxic effects. The surface-collected DWH oil used in our research (Slick A) has all the characteristics of a naturally weathered crude oil, with the most volatile organic components (e.g., BTEX) mostly reduced or lost by evaporation and dissolution (Daling et al., 2014), while a fraction of PAH alkylated homologues (alkylated PAHs) is produced (Wang and Fingas, 2003).

Another weathering process which can change significantly the chemical composition of crude oil is biodegradation by hydrocarbon-eating bacteria (*Acinetobacter* and *Marinobacter*) (Atlas, 1995). Light crude oils such as DWH oil due to their higher proportion of simpler LMW hydrocarbons are more readily biodegraded than heavy crude oil. Hydrocarbon biodegradation takes place at the hydrocarbon–water interface. Thus, the surface area to volume ratio of the oil can significantly affect the biodegradation rate (National Research Council (U.S), 2005).

2.2 The Chemical Dispersant Corexit 9500A®

2.2.1 Composition

Chemical dispersants such as Corexit 9500A® are complex mixtures, containing hazardous substances including petroleum distillates (solvent), propylene glycol (stabilizer), organic sulfonic acid salt or Dioctyl Sodium Sulfosuccinate or DOSS (surfactant), sorbitan, and other ingredients (Nalco Energy Services, 2012). The primary anionic surfactant, DOSS, was selected as the most appropriate indicator compound for Corexit 9500A® contamination in water, sediment, and tissues, due to its bioactivity, low volatility, and its potential to persist in the environment (OSAT-1, 2010).

2.2.2 Mode of Action

Surfactant components (DOSS) are amphiphile molecules possessing both hydrophobic and hydrophilic groups that act to decrease tension between the water and oil interface and to stimulate the development of small oil-surfactants micelles <100 μm (National Research Council (U. S), 2005; Singer et al., 1996), therefore facilitating the downward mixing of oil into the water (Li and Garrett, 1998). By using dispersants, oil slicks can be dispersed to concentrations below toxicity thresholds limits and become more accessible to hydrocarbon-degrading bacteria, thus enhancing the rates of biodegradation (Lessard and DeMarco, 2000; Page et al., 2000; Venosa and Zhu, 2003).

2.3 Water Accommodated Fraction as an Exposure Solution

To assess the potential aquatic toxicity of the DWH crude oil, it is necessary to understand the types and concentrations of chemicals likely to dissolve into the water, because organisms are likely to be exposed to these compounds. As mentioned previously, PAHs are characterized

EVALUATING WATER QUALITY TO PREVENT FUTURE DISASTERS

by low to very low solubility. Therefore, special attention must be paid to the way the exposure solutions are prepared. In the natural environment, prolonged turbulent mixing of crude oil by wind, currents, and waves can result in the release of water accommodated fraction (WAF) (Barron et al., 1999; Rossi et al., 1976). The "WAF approach" was chosen to provide the best estimate of what may solubilize from the oil into the overlying water (Barron et al., 1999; Singer et al., 1991, 2000). To minimize the variability between research groups and make results more directly comparable, some researchers attempted to standardize testing conditions and created the CROSERF (*Chemical Response to Oil Spills: Ecological Effects Research Forum*). Utilizing the methods from Singer et al. (1991), CROSERF also sought to standardize a suite of tests on marine organisms (sensitive early life stages) and the preparation of both the WAF of crude oil and the chemically enhanced water accommodated fraction (CEWAF via Corexit 9500A) (Singer et al., 2000). In this context, the method of preparation of WAF adopted in the current study followed procedure detailed in the *Quality Assurance Project Plan: Deepwater Horizon Laboratory Toxicity Testing* (Version 4; February 4, 2014; Appendix A, pp. A1– A9), NOAA.

3 THE EASTERN OYSTER CRASSOSTREA VIRGINICA

3.1 Oyster as a Model Organism

The Eastern oyster, *Crassostrea virginica*, is ubiquitous, and an ecologically and economically significant bivalve mollusk. It is one of the most commercially important shellfish species propagating along the east coasts of the United States, from Maine to the GoM (Galtsoff, 1964) and an ecologically vital species for the GoM region. Oyster reefs, which are the result of successive settlement of larvae onto existing reef structure, provide food, shelter,

and habitat for many fish and shellfish species, improve water quality, stabilize bottom areas, and influence water circulation patterns within estuaries (Newell, 2004; Volety et al., 2014; Wells, 1961). In addition to its ecological significance, it is also an economically important species. In 2012, total landings of *C. virginica* represented a value of $104 million in the United States from which $74 million originated in coastal regions of the northern GoM (National Marine Fisheries Service, 2012). In the northern part of the GoM, oyster spawning season typically occurs from mid-spring through late fall when water temperature is above 25°C (Ingle, 1951), with two peaks of settlement in early and late summer (Supan, 1983). From April 20th until the final capping of the leak on July 15th, DWH crude oil spilled from the Macondo well (US District Court, 2014), a period coinciding with the natural spawning and recruitment season of eastern oysters in the northern GoM.

Oysters are an excellent sentinel organism. The 1970s saw the beginning of a widespread research effort focusing on the effects of crude oil on bivalve species, and particularly on mussels and oysters. Oil components such as PAHs are highly resistant to degradation in the environment, and may accumulate in animal tissues and interfere with normal metabolic processes that affect growth, development, and reproduction (Capuzzo et al., 1988). Due to their biological characteristics (sedentary and filter-feeder), oysters are effective in taking up hydrocarbons via filtration and ingestion of particulate matter. Therefore, they may accumulate these hydrocarbons in high concentrations, and thus be susceptible to the negative effects of these pollutants. Results from environmental research in the United States suggested that mussels and oysters may be valuable as sentinel organisms for indicating levels of pollutants in estuarine and coastal marine waters on a worldwide scale. Subsequently, the US Mussel Watch program was created (Farrington et al., 1983; Goldberg et al., 1978). Meanwhile, off the coast of Brittany, the *Amoco Cadiz* ran aground in 1978 and caused the largest oil spill ever registered in the world (Berthou et al., 1987). On account of the extensive damage caused by this disaster, many research and monitoring programs were immediately carried out to determine the long-term effects of oil exposure on oysters (Balouet et al., 1986; Berthou et al., 1987; Neff and Haensly, 1982). Berthou et al. (1987) found that 7 years after the spill, oysters from the field were still contaminated with oil. They also described a rapid accumulation of hydrocarbons followed by an initial rapid loss of alkanes and low molecular weight hydrocarbons, but HMWs hydrocarbons were persistent in the tissues. Their observations followed other studies, which related to the two-phase depuration with the mobilization of lipid reserves (Neff, 1979). Biological effects of oil can be manifested at biochemical, cellular, and organismal levels before disturbances at the population level develop (Capuzzo et al., 1988). Adaptive responses such as resistance or metabolization of toxicants can exist in nature (Capuzzo, 1996) but are highly taxa-related. Consequently, all responses are not disruptive and do not necessarily result in deleterious impact at the next level of organization.

3.2 Use of Oyster Larvae in Ecotoxicology

3.2.1 The Bivalve Embryo-Toxicity Assay

Early life stages are more sensitive to contaminants than adults and represent a critical period in the life cycle of a marine organism such as oyster (Chapman and Long, 1983; Connor, 1972; His and Beiras, 1999; His and Robert, 1985; Thorson, 1950). In order to rapidly assess water quality, Woelke (1972) proposed for the first time a standard method for conducting 48-h bioassays using *Crassostrea gigas* embryos. Consequently, early stages of oyster (embryos or larvae), due to their small size, high number produced in a spawning event and the extensive information on their culture in the lab, were selected as a reliable model

organism in ecotoxicological studies to assess petroleum hydrocarbon pollution (Chapman and Long, 1983; Chapman and Morgan, 1983; Geffard et al., 2001, 2002a,b, 2003; His et al., 1996; His and Beiras, 1999; Loosanoff and Davis, 1963; Renzoni, 1975; Thain, 1992).

The standard bivalve embryo toxicity assay consists of exposing bivalve embryos after fertilization to contaminants or effluents, typically for 48h (Chapman and Long, 1983; Woelke, 1972). The usual endpoint or response measured after 48h of exposure is the percentage of abnormal (including deformed or absence of shells, or developmental failure) and dead larvae (His et al., 1997). It is generally expressed as a median effective concentration resulting in 50% abnormal larvae (EC_{50}) or median lethal concentration (LC_{50}).

3.2.2 Oil/PAHs Toxicity on Oyster Larvae

Although petroleum is one of the most studied pollutants of the aquatic environment and oyster species have been studied extensively in ecotoxicology, literature on the toxicity of crude oil and PAHs on early life stages of *C. virginica* is relatively scarce. Several authors have reported that acute exposure of early life stages of oysters to oil and/or PAH could result in reduced fertilization success, abnormal embryonic development, growth, and ultimately increased mortality (Finch et al., 2016; Geffard et al., 2001, 2002a,b, 2003; Jeong and Cho, 2005; Langdon et al., 2016; Laramore et al., 2014; Lyons et al., 2002; Nogueira et al., 2017; Renzoni, 1975; Stefansson et al., 2016; Wessel et al., 2007). Early work from Renzoni (1975) found that after exposing *C. virginica* sperm and eggs to different type of crude oils for 1h before fertilization, fertilization was depressed in a dose-dependent way, with rates falling to 60%–70% at 1000ppm. Jeong and Cho (2005) demonstrated that concentration of 10 individual PAHs from 50 to 200 μg L^{-1} could affect fertilization capabilities of *C. gigas* sperm and impact fertilization success and larval development. Another study by Wessel et al. (2007) indicated that oyster embryos, exposed to a mutagenic and teratogenic PAH, benzo(a)pyrene, showed abnormal development due to DNA strand breakage. More recently, several authors demonstrated that DWH oil (WAF) and dispersed oil (CEWAF) could affect negatively the early development of *C. virginica* by decreasing fertilization success, the frequency of normal larval development, shell growth, and even settlement success (Langdon et al., 2016; Laramore et al., 2014; Stefansson et al., 2016). Nogueira et al. (2017) evaluated the effects of phenanthrene (PHE) on the embryogenesis and larval development of *C. gigas*. The authors found that PHE (0.02 and 2.0 $\mu g L^{-1}$) impaired embryo-larval development, increased mantle and shell abnormalities in D-larvae, and caused changes on gene transcription in *C. gigas* larvae (Nogueira et al., 2017).

Various studies have examined the effects of experimental exposure of oyster larvae to PAH-contaminated sediment (Chapman and Morgan, 1983; Geffard et al., 2001, 2002a,b, 2003; His et al., 1997; Phelps and Warner, 1990; Thain, 1992). For example, Geffard et al. (2001) demonstrated that PAH released by contaminated sediment could have a deleterious effect on oyster embryos, and that embryotoxicity assays were more sensitive to PAHs than spermiotoxicity assays, resulting in less abnormal *C. gigas* larvae. They also reported that larval growth assay was more sensitive than the embryotoxicity assay (Geffard et al., 2002b). Besides, they found that some fraction of oil could be adsorbed on particle and become bioavailable to oyster larvae (Geffard et al., 2002b). Although oyster metamorphosis bioassay has been proved to be a rapid, sensitive, and reliable method in toxicological assessment (Phelps and Warner, 1990), few studies have used it as a biological response to study oil/PAH toxicity on oyster larvae. His et al. (1997) examined the toxicity of sediment-associated PAH on pediveliger stages of *C. gigas* and revealed a drastic reduction of metamorphosis, a critical process in the early life of oyster.

3.2.3 *Dispersant Toxicity on Oyster Larvae*

One of the first studies on dispersant toxicity was probably that of Hidu (1965) who investigated the effects of synthetic surfactants on the larvae of hard clams (*Mercenaria mercenaria*) and eastern oysters using embryotoxicity and larval growth assays. He found that larval growth was the most sensitive endpoint compared to embryonic development, and that concentration of surfactant of 10 ppm could be deleterious. Little information, however, exists on the effects of Corexit dispersants on oyster larvae. Most of the studies have been conducted on toxicological model of crustacean species such as mysid shrimps, amphipods, artemia, rotifers, or fish species such as inland silversides or trout (review from George-Ares and Clark, 2000). Fucik (1994) reported significant toxicity of Corexit 9527, another type of chemical dispersant, on eastern oyster embryos with $LC50_{96h}$ value of 4.9 $mg L^{-1}$.

4 MATERIAL AND METHODS

4.1 Preparation of Exposure Solutions

Surface oil, also referred to as "Slick A" was collected on the July 29, 2010, from the hold of barge number CTC02404, which received surface slick oil from various skimmer vessels near the Macondo well and was obtained under chain of custody (sample CTC02404-02). Slick A incurred significant natural weathering at sea before collection.

The dispersant Corexit 9500A® (Nalco Environmental Solutions LLC, Sugar Land, TX, USA) was provided by the DWH Trustees. For all exposure solutions, contaminants were added to UV-sterilized and 0.1 μm-filtered seawater (FSW), maintained at a salinity of 20–25 PSU.

4.1.1 High-energy Water Accommodated Fractions

The oil-only exposure solutions or high-energy water accommodated fractions (HEWAFs) were prepared at 25°C under fluorescent lights to avoid photo-reactivity (Landrum et al., 1987). To artificially recreate the action of waves and currents, 2 L of FSW and 4 g of slick oil (with a gastight syringe) were added to a stainless-steel blender pitcher (Waring™CB15, Waring Commercial, Torrington, CT). After 30 s at the lowest blending speed, the solution was transferred to a 2-L aspirator bottle and left to settle for at least 1 h to separate the residual floating oil (Incardona et al., 2013). The bottom layer of the mixture (or accommodated fraction) was then carefully drained from the aspirator bottle and FSW was added to this stock to prepare dilutions for exposure treatments. Preparations were not filtered, so dilutions contained particulate oil in addition to dissolved PAHs.

4.1.2 CEWAF

The oil/dispersant mixtures or chemically enhanced water accommodated fractions (CEWAFs) were also prepared at 25°C under fluorescent lights. Four grams of slick oil and 400 mg of dispersant (10:1 w:w) were added to an aspirator bottle filled with 2 L of FSW. The oil/dispersant mixtures were added with a gastight syringe, and stirred at a vortex adjusted to 25% using a stirring rod and a magnetic stirrer for 18 h. To allow for the separation of the solution from the residual floating oil, the oil and dispersant mixture was left to stand for 3 h prior to use, and the stock solution (or accommodated fraction) was carefully drained.

4.1.3 Corexit Only

Dispersant Corexit exposure solutions were prepared as described for CEWAF above, except that no oil was added and the mixture was not settled. The dispersant stock was collected by draining the aspirator bottle and, to obtain different exposure concentrations, the stock solution was diluted with FSW.

4.2 Analytical Chemistry

Chemical analyses of hydrocarbon constituents of the different HEWAF, CEWAF, dispersant concentrations, and the FSW control were performed by ALS Environment (Kelso, WA, USA). The 250-mL unfiltered water samples were collected for every test and stored in amber-bottles at 4°C until shipment to the analytical laboratory by expedited courier. Samples were then extracted and processed for GC–MS. PAHs including alkyl homologues were determined by gas chromatography with low-resolution mass spectrometry using selected ion monitoring (GC/MS-SIM) and a sum of 50 different PAHs (tPAH50) was quantified. The analytical procedure was based on EPA Method 8270D with the GC and MS operating conditions optimized for separation and sensitivity of the targeted analytes. Additional information regarding the PAH analytes and the tPAH50 sum can be found in Forth et al. (2017); whereas, more details regarding the methods used (e.g., standards used, QC criteria for surrogate recovery, internal standards, and spiked blanks) can be found in the analytical QAPP provided by the analytical laboratory and applied to all samples analyzed for the Deepwater Horizon Natural Resource Damage Assessment (DWH NRDA): https://pub-dwhdatadiver.orr.noaa.gov/dwh-ar-documents/945/DWH-AR0101767.pdf

4.3 Experimental Oysters

4.3.1 Brood Stock Collection and Conditioning

Adult specimens of *C. virginica* (mean weight of 75 g ± 20) were collected from natural populations, unexposed to the spill, in Estero Bay, Florida (Lat. 26°19′50″N, Long. 81°50′15″W). Oyster brood stock were maintained in the hatchery at 22°C ± 1 following Loosanoff and Davis (1963), in a flow-through system supplied with coarsely filtered (30-μm sand filter) seawater at ambient salinity (20–30 PSU) and fed a mixture of laboratory-cultured fresh microalgae (*Tetraselmis chui*, *Chaetoceros* sp., and *Tisochrysis lutea*) for conditioning (Utting and Millican, 1997).

4.3.2 Spawning, Gamete Collection, and Larval Rearing

Sexually mature oysters were induced to spawn by thermal stimulation. When thermal shock was unsuccessful, gonad stripping was performed. After verification under a microscope, gametes from at least 2–3 adults were washed, counted, and pooled for fertilization. Fertilized embryos were then stocked in the experimental hatchery in 50-L rearing tanks, at a density of $40 \, \text{mL}^{-1}$, until hatching the next day. Newly hatched D-larvae were counted and stocked at a density of $10 \, \text{mL}^{-1}$ in the larval rearing tanks, filled up with UV-sterilized and 0.1 μm FSW. Larval rearing was carried out for about 15 days at 28°C, with a water renewal every other day, until larvae reached the pediveliger stage. Settlement of competent larvae was conducted by providing microclutch on a suspended sieve. Through the whole larval cycle, oysters were fed a mixture of cultured microalgae (*T. lutea*, *Chaetoceros muelleri*, and *T. chui*) according to Helm and Bourne (2004).

4.4 Acute Exposure Procedure

Most exposures were conducted in 400-mL sterile glass beakers, filled with 200 mL of increasing concentrations of HEWAF, CEWAF, or dispersant only. The range of doses tested and exposure duration were chosen based on results observed during preliminary range finding tests. Ranges of concentration tested (nominal and tPAH50 content) for the different life stages are shown in Table 1. Four replicates were carried out per concentration and controls consisted of exposure to FSW only. Exposure beakers were maintained in darkness at $26 \pm 1°C$ and at a salinity of 22 ± 2 PSU for 96 h (4 days), with no renewal of exposure solutions (exception being the spat assay, see Section 4.4.5). Freshly cultured phytoplankton (*T. lutea* and *C. muelleri*) was added to each exposure beaker at day 1 (5×10^4 cells mL^{-1})

TABLE 1 Range of Nominal Concentrations ($\text{mg} \, \text{L}^{-1}$) of Test Solutions Used for Exposures of Gamete, Embryos, 1-Day-Old Veliger Larvae, 10-Day-Old Umbo Larvae, 14-Day-Old Pediveliger Larvae, and 2-Month-Old Spat, and Corresponding Measured Concentrations of a Sum of 50 PAHs (tPAH50) Expressed in $\mu\text{g} \, \text{L}^{-1}$ for Oil (HEWAF) and Dispersed Oil (CEWAF)

Stage exposed	Gamete/embryo	Veliger/umbo	Pediveliger	Spat
Oil preparation	Nominal ($\text{mg} \, \text{L}^{-1}$) $\rightarrow$ tPAH50 ($\mu\text{g} \, \text{L}^{-1}$)			
HEWAF	$0 \rightarrow 0.01$	$0 \rightarrow 0.5$	$0 \rightarrow 0.1$	$0 \rightarrow 0.03$
	$62.5 \rightarrow 108$	$62.5 \rightarrow 95$	$31.25 \rightarrow 48$	$10 \rightarrow 7$
	$125 \rightarrow 198$	$125 \rightarrow 202$	$62.5 \rightarrow 113$	$50 \rightarrow 66$
	$250 \rightarrow 417$	$250 \rightarrow 390$	$125 \rightarrow 191$	$100 \rightarrow 147$
	$500 \rightarrow 839$	$500 \rightarrow 762$	$250 \rightarrow 399$	$500 \rightarrow 904$
	$1000 \rightarrow 1635$	$1000 \rightarrow 1605$	$500 \rightarrow 719$	$1000 \rightarrow 3450$
		$2000 \rightarrow 2985$		
CEWAF	$0 \rightarrow 0.01$	$0 \rightarrow 0.4$	$0 \rightarrow 0.8$	—
	$62.5 \rightarrow 1.3$	$62.5 \rightarrow 14$	$31.25 \rightarrow 10$	
	$125 \rightarrow 3.3$	$125 \rightarrow 25$	$62.5 \rightarrow 19$	
	$250 \rightarrow 6.4$	$250 \rightarrow 45$	$125 \rightarrow 44$	
	$500 \rightarrow 14.2$	$500 \rightarrow 91$	$250 \rightarrow 81$	
	$1000 \rightarrow 26.2$	$1000 \rightarrow 179$	$500 \rightarrow 177$	

4 MATERIAL AND METHODS

and day 3 (1×10^5 cells mL^{-1}) for the gamete/embryo assays, and at day 0 and day 2 (1×10^5 cells mL^{-1}) for the larval assays.

4.4.1 Fertilization Assay

Briefly, sperm and unfertilized oocytes were exposed separately to various concentrations of toxicants for 30 min. Each 10-mL replicate of sperm (dense solution of 1.5×10^7 to 2.5×10^7 cells mL^{-1}) was incubated in the test solution (40 mL HEWAF, CEWAF, or dispersant solution in 50-mL beakers). Oocytes (4000 per beaker) were also incubated in the test solution (200 mL HEWAF, CEWAF, or dispersant in 400-mL beakers). After the 30-min incubation, oocytes from each exposure replicate were fertilized with 10 mL of sperm from corresponding sperm-exposure replicates. Each test chamber was subsampled 1-h postfertilization and samples were preserved in 10% buffered formalin for later examination. To determine fertilization success, at least 100 embryos per treatment were examined for cell cleavage.

Concomitantly, cellular functions of both spermatozoa and oocytes were assessed using flow-cytometry assays. Procedures are detailed in Volety et al. (2016) and Vignier et al. (2017).

4.4.2 Embryo Assay

While the gametes were incubated in exposure solutions, the remaining unexposed oocyte solution was fertilized with 50 mL of the remaining unexposed sperm. One hour after fertilization, when the two- to four-cell stage was reached, embryos were counted and transferred into the exposure solutions at a density of $\approx$20 mL^{-1} (4000 embryos per beaker).

After 24 h of fertilization, a 10 mL subsample was collected from each exposure beaker and preserved with 10% buffered formalin for assessment of abnormalities. A minimum of 50 randomly selected individuals from each replicate were examined under a microscope to determine the percentage of normality, and abnormality of exposed embryos, categories

including (Fig. 6): (1) segmented eggs, normal embryos, or malformed embryos that did not reach the D-larval stage; and (2) D-larvae with either a convex hinge, indented shell margins, incomplete shells, a protruded velum, or an extrusion of mantle (His et al., 1997). After 96 h of exposure, the final survival was assessed by filtering the content of each beaker through a 40-μm nylon mesh to collect the larvae, and the number of dead larvae, that is, translucent and/or open shells, was also estimated under magnification. Final survival was calculated by comparing the initial number of stocked embryos and the final number of survivors at 96-h.

4.4.3 Larval Assay

One-day-old veliger larvae were collected from the hatchery, counted, and distributed at a density of 15 larvae mL^{-1} into the exposure solutions ($\approx$3000 larvae per beaker). Gentle aeration ($\approx$1 bubble s^{-1}) was provided for each beaker to maintain dissolved oxygen (DO) levels above 4 mg L^{-1}. Fresh cultured microalgae (*T. lutea* and *C. muelleri*) were added to each beaker at days 0 and 2 (1×10^5 cells mL^{-1}).

Ten-day-old umbo larvae were collected from hatchery (same cohort of larvae used for veliger assay), counted and then loaded at a density of 10 larvae mL^{-1} into the exposure solutions ($\approx$2000 larvae per beaker). Acute exposure was performed using the same protocol described above, and employing the same nominal exposure concentrations (Table 1).

Larvae from both assays were collected at test initiation and 48 h later by taking a 10-mL subsample from each exposure beaker and preserved with 10% buffered formalin for measurements of shell length. After 96 h of exposure, larvae from each beaker were concentrated by filtering larvae through a 35 μm mesh, preserved with 10% buffered formalin for later examinations of survival and shell measurements and to obtain a final volume of 30 mL. Final survival (comparing the initial number of stocked larvae and the final number of survivors) was assessed by taking five 300 µL subsamples from each of the four replicates of the 30 mL concentrate after homogenization, and examined under a microscope to evaluate live and dead larvae (translucent shell and/or opened valves). At each sampling time (0, 48, and 96 h), shell lengths of 25–50 randomly selected live larvae from each replicate were measured using an inverted microscope (Olympus IX73) equipped with a camera Olympus DP73, and the CellSens Software.

4.4.4 Settlement Assay

Fourteen-day-old pediveliger larvae were distributed at ≈1000 individuals into 600-mL beakers filled with 450 mL of the different exposure concentrations of HEWAF and CEWAF. Exposure consisted of five to six nominal concentrations and a FSW control (Table 1), with four replicates per concentration.

Pediveligers were exposed, for 72 h, in a static system at 23°C ± 2 and 23 PSU ± 1 with no renewal of contaminant. Two settlement plates consisting of HardieBacker cement board tiles (120 × 58 mm), previously soaked in seawater for a minimum of 2 weeks, were set up vertically in the water column of each container. Gentle aeration (≈1 bubble s^{-1}) was supplied to each beaker for 30 min every 2 h using a timer-controlled air pump, to maintain DO levels >4 mg L^{-1}. Fresh cultured microalgae (*T. lutea* and *C. muelleri*) were added to each exposure beaker at days 0 and 2 (1×10^5 cells mL^{-1}).

After 72 h of exposure, developmental success of pediveliger was determined by their progression to settlement as well as mortality. Settlement plates and container walls were examined under a dissecting microscope, and newly settled oysters counted. Larval survival was estimated by taking three 1 mL subsamples from the 30 mL concentrate of larvae and using the same protocol previously described (Vignier et al., 2016). Newly settled larvae were identified by their larger size (>400 µm) and their attachment to the substrate, and by the transition from rounded pediveliger to a flat shape with the new dissoconch. Settlement success was calculated by comparing total settled larvae on tiles vs the total number of pediveligers unsettled, and a median effective concentration (EC50) inhibiting settlement was determined for HEWAF and CEWAF (Table 2).

4.4.5 Spat Assay

Two-month-old single-seed oyster spat were haphazardly distributed at a density of 15 per beaker, in triplicate (five replicates for the control). Spat were exposed for 10 days to increasing concentrations of HEWAF (Table 1). Exposure was performed in 600-mL glass beakers filled up with 500 mL of solution, under static-renewal conditions, with the exposure media renewed every other day. Beakers were cleaned using deionized water and Kim-Wipes®, prior to refilling with new exposure media. Gentle central aeration by means of a glass pipette was supplied to each beaker at 100 mL min^{-1}. Spat from each beaker were fed daily with 2 mL of Instant Algae®. Every other day, spat were collected on a 2-mm stainless steel sieve, and examined using a dissecting microscope for dead spat, measured by their failure to close their valves in seawater.

At the end of the 10-day exposure to HEWAF, 15 live spat from each concentration were randomly selected for histological examination. Histological preparation followed a procedure extensively detailed in Vignier et al. (2018). Using an Olympus IX73 inverted microscope equipped with an Olympus DP73 camera and the CellSens image analysis software, digestive tubules (DGTs) were measured at random from histological sections of six surviving animals haphazardly selected from each treatment. Morphometric analysis followed a protocol adapted and modified from Cajaraville et al. (1989) and Winstead (1995). Histological cross sections of the digestive gland were divided into four fields of observation per animal. Eight randomly selected tubules from each field (32 total tubules per spat) were measured. Two sets of measurements, internal (luminal) surface, and total

TABLE 2 Median Effective Concentrations (EC50), 20% Inhibition Concentrations (IC20) and Median Lethal Concentrations (LC50) Determined in Gamete (1hPF), Embryo (1 hPF to day 4), Veliger (day 1–5), Umbo (day 10–14), and Pediveliger Larvae (day 15–18) Following Exposures to HEWAF, CEWAF, and the Corexit Dispersant (1:10 Oil Ratio)

Exposure solution	Gamete EC50$_{1h}$ (fertilization)	Embryo		Veliger		Umbo		Pediveliger	
		EC50$_{24h}$ (abnormal)	LC50$_{96h}$	IC20$_{96h}$ (growth)	LC50$_{96h}$	IC20$_{96h}$ (growth)	LC50$_{96h}$	EC50$_{72h}$ (settlement)	LC50$_{72h}$
HEWAF (μg tPAH50L^{-1})	**2250**[a] (1770–6460)	**342** (242–504)	**220** (216–224)	**106** (75–137)	**715** (NC)	**61** (41–80)	**2814** (2738–2875)	**1.7** (NC)	**1530** (1370–1760)
CEWAF (μg tPAH50L^{-1})	**29.9**[a] (27.2–35.6)	**15.6** (14.9–16.5)	**17.7** (17.5–17.9)	**1.1** (NC)	**41.8** (41.2–42.4)	**8.6** (3.5–14.5)	**72** (71–73)	**35** (31–39)	**88** (85–91)
Corexit (mg L^{-1})	**11.5**[a] (10.2–12.7)	**5.67** (NC)	**2.7** (2.6–2.8)	**3.5** (0.7–4.9)	**22.9** (22.5–23.3)	**10.7** (7.6–14)	**58** (57–59)	—	—

[a] *The effect concentration exceeded the range tested (i.e., higher than 100% lethal concentration).*

NC: Could not be calculated due to a lack of intermediate responses; 1hPF: 1-hour post-fertilization?

EC50, IC20, and LC50 values are expressed as measured concentrations of a sum of 50 PAHs (tPAH50) in μg L^{-1} for oil alone (HEWAF) and dispersed oil (CEWAF), and as nominal concentration of Corexit (in mg L^{-1}) for dispersant alone ±95% Confidence interval (CI). EC50$_{1h}$ (fertilization) corresponds to the concentration causing 50% inhibition of fertilization success for gametes exposed for 1 h to toxicants; EC50$_{24h}$ (abnormality) corresponds to the concentration causing 50% of abnormal development in embryos exposed for 24 h to toxicants from 1hPF; IC20$_{96h}$ (growth) corresponds to the concentration of toxicants reducing shell lengths by 20% in larvae exposed for 96 h; EC50$_{72h}$ (settlement) corresponds to the concentration of toxicants causing 50% inhibition of settlement success in pediveligers exposed for 72 h to toxicants.

tubule surface for each tubule were determined using the CellSens image analysis software, and tubule luminal ratio (luminal surface/total surface) was calculated.

5 KEY FINDINGS

Oil caused a range of adverse effects to early life stages of oysters. Short-term lethality or "acute lethality" was the most severe of those effects. Lethal responses were used by the trustees for purposes of quantifying water column resource injuries as part of the NRDA. However, sublethal toxicity occurs at lower oil concentrations than mortality and the sublethal responses observed in the laboratory have the implication that reduced survival and reproduction can occur in the natural environment. Here, both lethal and sublethal responses of Eastern oyster to oil and/or dispersant are presented to, better estimate the full scope of injuries.

5.1 Lethal Effects Depend on Exposure Duration, Concentrations, and Life Stage Tested

Laboratory-based acute exposures to WAFs of oil and/or dispersant generally led to dose-dependent mortalities. Lethal effects concentrations for dispersant occurred in the parts-per-million (ppm, or mg L^{-1}) range of Corexit in water [i.e., 2.7–58], whereas effects concentrations for TPAH50 were in the parts-per-billion (ppb, or µg L^{-1}) range [i.e., 17.7–2814] (Table 2).

Our study also showed that the toxicity of CEWAFs (containing dispersant) was greater (i.e., lower LC values) than in HEWAFs (without dispersant) and that CEWAF and dispersant alone resulted in the highest impact on survival of oysters, regardless of life stages tested (Table 2). At equivalent nominal concentration tested, oyster survival followed similar trends

after exposure to CEWAF or dispersant only, suggesting that most of the toxic effect induced by the dispersed oil can be attributed to the dispersant fraction of the CEWAF (i.e., 1:10) (Fig. 2) (Vignier et al., 2015, 2016). Similar results were reported with embryo and larval stages of Eastern oyster exposed to chemically dispersed oil and dispersant (Fucik, 1994; Langdon et al., 2016; Laramore et al., 2014). The main ingredient of the Corexit, the surfactant DOSS, and its surface-active effects on bio-membrane is likely responsible for the toxic effects observed following 72–96h CEWAF and dispersant exposures (Singer et al., 1991). Given the sensitivity of oyster to the dispersant Corexit 9500A, caution should be used when deciding to use chemical dispersion as a remedial action for an oil spill, especially in a coastal marine environment.

Lastly, mortalities of oysters following incubations with oil and dispersant were stage-dependent, that is, sensitivity to oil/PAH and dispersant was greater for early developmental stages (Table 2). This stage-related sensitivity can be summarized as follows, from most sensitive to less sensitive:

Gametes > Embryos > Veliger > Umbo > Pediveliger > Spat

Many authors have reported that early life stages of aquatic organisms like oyster are more sensitive to xenobiotics than adults and represent a sensitive period in their life cycle (Chapman and Long, 1983; Connor, 1972; Thorson, 1950).

5.2 Sublethal Effects

5.2.1 Oil and Dispersant Affected Fertilization Success

Fertilization is a sensitive process in the early development of oysters. Acute exposure of gametes throughout the fertilization process to oil/HEWAF, dispersed oil/CEWAF, or

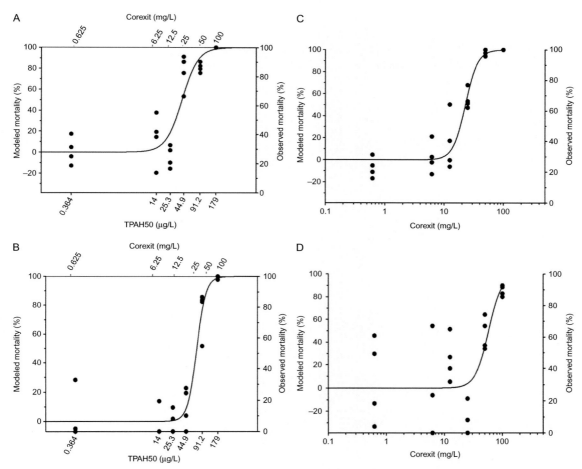

FIG. 2 Dose–response curves for CEWAF (A and B) and Corexit alone (C and D) exposures of 1-day-old veliger larvae (A and C) and 10-day old umbo larvae (B and D). Observed mortalities (in %) were reported after 96-h of exposure, for four replicates per treatment, and calculated from initial stocking numbers and final number of survivors. Model for CEWAF mortalities (A and B) was fitted to measured TPAH50 (sum of 50 PAHs) exposure concentrations ($\mu g\,L^{-1}$), and the corresponding nominal concentration of dispersant ($mg\,L^{-1}$) is shown. Oil to dispersant ratio in the CEWAF was 10:1, meaning nominal doses of CEWAF ranged from 62.5/6.25 to 1000/100 $mg\,L^{-1}$. Modeled mortalities for Corexit only (C and D) were fitted to TPAH50 exposure concentrations ($\mu g\,L^{-1}$) and nominal Corexit ($mg\,L^{-1}$), respectively. *From Vignier, J., Soudant, P., Chu, F. L.E, Morris, J.M., Carney, Lay, C., Krasnec, M.O., Robert, R., Volety, A.K., 2016. Lethal and sublethal effects of* Deepwater Horizon *slick oil and dispersant on oyster (*Crassostrea virginica*) larvae. Mar. Environ. Res. 120, 20–31.*

dispersant/Corexit resulted in a significant decline in fertilized embryos at concentrations of 1635 μg tPAH50 L^{-1}, 14 μg tPAH50 L^{-1}, and 5 mg Corexit L^{-1}, respectively (Fig. 3). In addition, median effective levels ($EC50_{1h}$) of HEWAF, CEWAF, and dispersant on fertilization were 2250 tPAH50 L^{-1}, 29.9 tPAH50 L^{-1}, and 11.5 mg Corexit L^{-1}, respectively (Table 2). As shown previously with lethality, the toxicity of CEWAF on fertilization was greater than in HEWAF (Table 2). Moreover, fertilization success was inhibited by CEWAF and Corexit in a similar

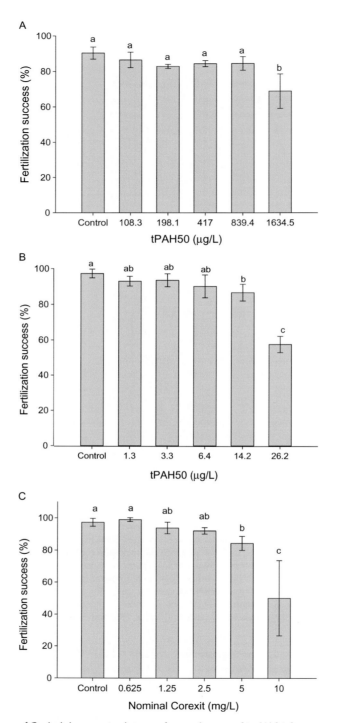

FIG. 3 Fertilization success of *C. virginica* gametes (eggs and sperm) exposed to (A) high energy water accommodated fraction (HEWAF), (B) chemically enhanced water accommodated fraction (CEWAF), and (C) dispersant (Corexit 9500A®). Data are presented as mean percentage ± standard deviation (SD). *Different letters* denote a significant difference at $P \leq 0.05$ (ANOVA, Tukey HSD post hoc test). *From Vignier, J., Donaghy, L., Soudant, P., Chu, F.L.E, Morris, J.M., Carney, M.W., Lay, C., Krasnec, M., Robert, R., Volety, A.K., 2015. Impacts of Deepwater Horizon oil and associated dispersant on early development of the Eastern oyster,* Crassostrea virginica. Mar. Pollut. Bull. *100, 426–437. https:/doi.org/10.1016/j.marpolbul.2015.08.011.*

manner and at equivalent nominal concentrations, indicating again that the chemical properties of the dispersant are likely responsible for the majority of the fertilization inhibition observed with the CEWAF exposure (Fig. 3, Table 2). Similarly, Laramore et al. (2014) found that acute exposure of Eastern oyster gamete to CEWAF of DWH oil (+Corexit 9500A) decreased the frequency of fertilized embryos.

Flow-cytometry analyses revealed that certain chemical properties of Corexit, in addition to the dissolved PAHs contained in CEWAF, induced the most severe effects on cellular mechanisms potentially involved in fertilization processes (Vignier et al., 2017; Volety et al., 2016). Following exposure to CEWAF and Corexit alone, a reduction of sperm viability, an alteration of acrosomal integrity (Fig. 4), and an inhibition of intracellular reactive oxygen species (ROS) production in spermatozoa were observed (Fig. 5). These results suggest that impaired fertilization following exposure to oil and/or dispersant may

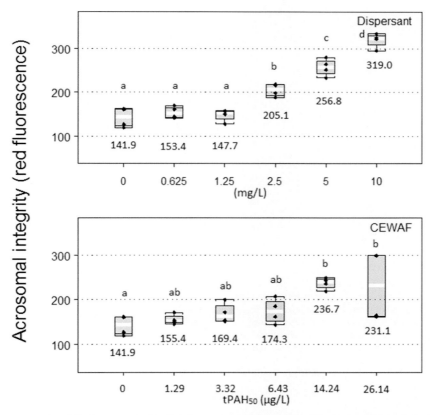

FIG. 4 Acrosome-associated fluorescence in *C. virginica* sperm cells exposed to dispersant (*top*) and chemically enhanced water accommodated fraction (CEWAF) (*bottom*). The *horizontal lines* of each box indicate the minimum value, estimated 25th quantile, estimated median (*white* band), estimated 75th quantile, and the maximum value from bottom to top, respectively, within each treatment group. Individual values are displayed as *black symbols*. Means are reported on the figure and *different letters* indicate significantly different results between conditions ($n=4$, $P<0.01$). From Vignier, J., Soudant, P., Chu, F.L.E, Morris, J.M., Carney, Lay, C., Krasnec, M.O., Robert, R., Volety, A.K., 2016. Lethal and sublethal effects of Deepwater Horizon *slick oil and dispersant on oyster* (Crassostrea virginica) *larvae*. Mar. Environ. Res. 120, 20–31.

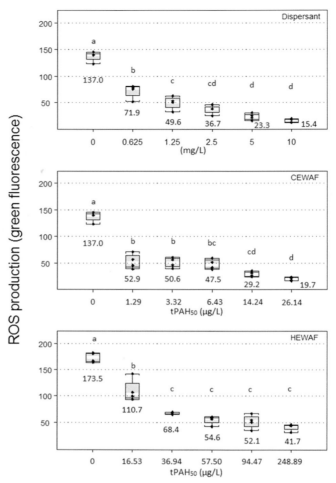

FIG. 5 Oxidative activity in *C. virginica* spermatozoa exposed to dispersant (*top*), and chemically enhanced water accommodated fraction (CEWAF) (*bottom*). The *horizontal lines* of each box indicate the minimum value, estimated 25th quantile, estimated median (*white* band), estimated 75th quantile, and the maximum value from bottom to top, respectively, within each treatment group. Individual values are displayed as *black symbols*. Means are reported on the figure and *different letters* indicate significantly different results between conditions ($n=4, P<0.001$). From Vignier, J., Soudant, P., Chu, F.L.E, Morris, J.M., Carney, Lay, C., Krasnec, M.O., Robert, R., Volety, A.K., 2016. Lethal and sublethal effects of Deepwater Horizon *slick oil and dispersant on oyster* (Crassostrea virginica) *larvae.* Mar. Environ. Res. 120, 20–31.

partially due to the alterations of cellular functions of spermatozoa (Volety et al., 2016). In a separate experiment, ROS production in exposed oocytes tended to increase with increasing CEWAF concentrations suggesting that exposure to dispersed oil caused an oxidative stress (Vignier et al., 2017). Oxidative stress has been shown to cause toxic effects on human oocytes and has been linked to reduction of fertilization success (Tamura et al., 2008). Numerous authors have shown that environmental stresses, such as pollutants, biotoxins, pathogens, heat shock, UV exposure, or hypoxia, induced stimulation of ROS production in bivalve tissues, including oocytes (Hégaret et al., 2003; Le Goïc et al., 2014; Roesijadi et al., 1997; Rolton et al., 2015). Oxidative stress and membrane impairments in bivalve tissues were described following exposure to petroleum hydrocarbons (Downs et al., 2002; Livingstone et al., 1989). Contrastingly, HEWAF did not affect sperm viability per se, but aggregation of the particulate fraction, the oil droplets, with sperm resulted in a reduction of "free" sperm cells available to fertilize eggs (Vignier et al., 2017).

5.2.2 Oil and Dispersant Affected Embryogenesis

Continuous acute 24-h exposure of oyster embryos throughout embryogenesis to HEWAF, CEWAF, and Corexit resulted in high rates of morphological abnormalities (Fig. 6). This finding is in agreement with studies which reported increased instances of abnormal early development in oysters previously exposed to crude oil/PAHs as embryos (Geffard et al., 2001; Laramore et al., 2014; Le Gore, 1974; Lyons et al., 2002; Nogueira et al., 2017; Stefansson et al., 2016; Wessel et al., 2007). Effective concentrations of HEWAF, CEWAF, and dispersant causing 50% of abnormal larvae ranged from 342 tPAH50 L^{-1}, 15.6 tPAH50 L^{-1}, and 5.7 mg Corexit L^{-1}, respectively, after 24-h exposure (Table 2). Our results suggest that oil associated PAH and dispersant may interfere with the mechanisms involved in shell secretion, which begins early in embryogenesis (Eyster and Morse, 1984). More specifically, they may disrupt and inhibit certain enzymatic activities, such as carbonic anhydrase, involved in shell formation and calcium transport pathways (Galtsoff, 1969; Hinkle et al., 1987). Dissolved PAHs and/or toxic compounds contained in Corexit could also interfere with embryos membrane fluidity (Eyster and Morse, 1984; Singer et al., 1996; Van Brummelen et al., 1998), consequently damage DNA, thus resulting in abnormal development of embryos

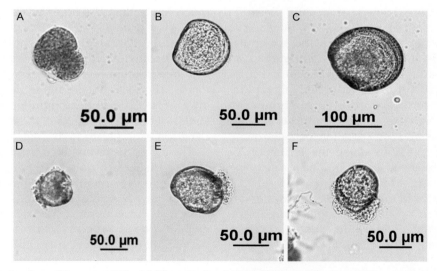

FIG. 6 Normal embryo (A), normal 24-h-old (B), and normal 96-h-old larvae (C); abnormal larvae (D, E, and F) observed after 24h exposure to HEWAF. *From Vignier, J., Donaghy, L., Soudant, P., Chu, F.L.E, Morris, J.M., Carney, M.W., Lay, C., Krasnec, M., Robert, R., Volety, A.K., 2015. Impacts of Deepwater Horizon oil and associated dispersant on early development of the Eastern oyster,* Crassostrea virginica. *Mar. Pollut. Bull. 100, 426–437. https://doi.org/10.1016/j.marpolbul.2015.08.011.*

(Wessel et al., 2007). Recent work from Nogueira et al. (2017) reported that PHE exposure of *C. gigas* embryos affected shell formation and caused abnormal development; these impairments were likely attributable to a toxic effect of PHE on the transcriptomic expression of genes involved in shell calcification.

5.2.3 Oil and Dispersant Affected Larval Growth

A consistent decline of shell lengths was observed in larvae exposed to oil and/or dispersant (Fig. 7). Of all the endpoints investigated, larval shell length was one of the most sensitive, with $IC20_{96h}$ of CEWAF exposure ranging as low as 1.1 and 8.6 ppb of tPAH50 for veliger and umbo, respectively (Table 2). As larval growth is strongly correlated and dependent on feeding ability (Strathmann, 1987), our results suggest that oil/dispersant could reduce the fitness of affected larvae by reducing their filtration rates. Narcotic effects from certain dissolved volatile PAHs (e.g., Naphtalene) are sublethal responses commonly observed in marine plankton species after oil exposure (Almeda et al., 2013; Berdugo et al., 1977). Nonpolar narcosis is a nonspecific mode of toxicity resulting from the accumulation of PAHs in biological membranes of an organism (Barron et al., 2004). Disturbance of membrane structure can occur relatively rapidly during short-term exposure (Van Brummelen et al., 1998). Narcosis usually manifests as sluggish behavior and/or cessation of swimming, which could possibly cause the reduction of feeding efficiency observed in the exposed oyster larvae in our studies.

Filter-feeding plankton species like oyster larvae could ingest oil droplets, which are in the same size range as their food spectrum (Almeda et al., 2014; Hansen et al., 2012; Lee et al., 1978). In fact, oil droplets were observed within the gut of exposed umbo larvae, confirming that direct ingestion of particulate oil by oyster larvae is another potential route of exposure.

In view of these results, one can conclude that reduction of feeding efficiency and associated growth impairments were caused by both dissolved PAHs (nonpolar narcosis) and the particulate form of oil, which could physically coat the velum, aggregate with algae, and directly be ingested by the larvae.

5.2.4 Oil and Dispersant Affected Settlement Success

Metamorphosis and settlement success of pediveliger larvae were significantly impaired by HEWAF and CEWAF (Fig. 8). Effective concentrations of HEWAF and CEWAF inhibiting settlement success ($EC50_{72h}$) were estimated at 1.7 and 35 µg tPAH50 L^{-1}, respectively (Table 2).

This result implies that relatively low concentrations of PAHs, at levels realistically found in the environment at the time of the DWH oil spill, could have detrimental consequences on metamorphosis/settlement of *C. virginica* larvae. Our study also demonstrated that HEWAF solutions reduced larval settlement at concentrations much lower than the doses of HEWAF inhibiting larval growth (1.7 vs 106 µg tPAH50 L^{-1}), suggesting that larval settlement inhibition is a very sensitive endpoint, which should be included in toxicological assessment of contaminants. The sensitivity of the settlement assay to oil/PAHs had also been reported by His et al. (1997) with *C. gigas* and more recently by Langdon et al. (2016) with *C. virginica*.

The mechanisms involved in the toxicity of oil and/or dispersant on settlement are not clear. As previously mentioned, dissolved PAHs and/or surface-active compounds of the dispersant may have exerted a toxic effect on the metamorphosis process. More realistically, the particulate oil droplets, especially abundant in the HEWAF, could have coated the settlement substrates, subsequently impeded settlement of competent larvae, as described by several authors (Banks and Brown, 2002; Smith and

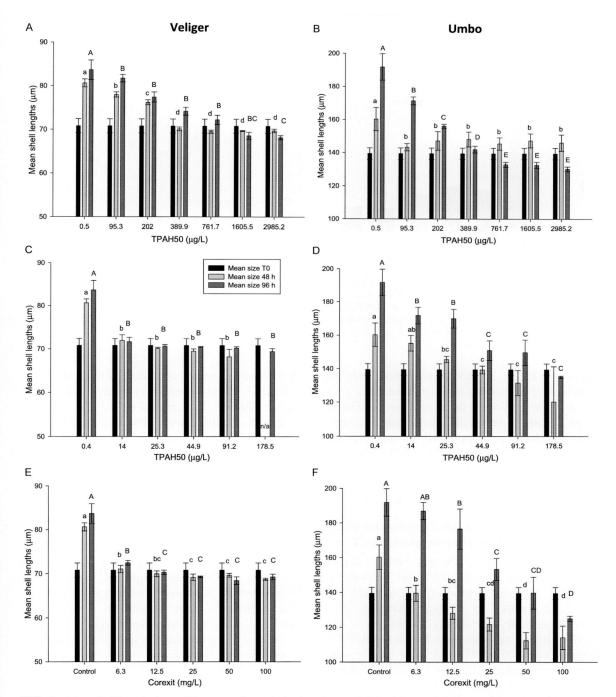

FIG. 7 Mean shell lengths (μm ±SD; $n=4$ replicates) of veliger larvae (1-day-old) and umbo larvae (10-day-old) after 96 h of exposure to HEWAF (A and B), CEWAF (C and D) and dispersant Corexit (E and F), expressed as measured TPAH50 concentrations (μg L^{-1}) for oil solutions, or nominal Corexit (mg L^{-1}) for dispersant solutions. TPAH50 values for umbo tests are nominal. n/a: no live larvae were observed, that is, 100% mortality. *Different letters* denote statistical difference at $\alpha=0.05$ (ANOVA). Tukey post hoc tests were performed for exposure B, E, and F; Dunnett's post hoc tests were performed for exposure A, C, and D. *From Vignier, J., Soudant, P., Chu, F.L.E, Morris, J.M., Carney, Lay, C., Krasnec, M.O., Robert, R., Volety, A.K., 2016. Lethal and sublethal effects of* Deepwater Horizon *slick oil and dispersant on oyster (*Crassostrea virginica*) larvae. Mar. Environ. Res. 120, 20–31.*

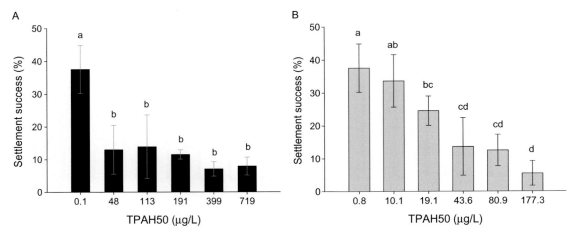

FIG. 8 Mean percent settlement success (±SD) of pediveliger larvae after 72h of acute exposure to HEWAF (A), and CEWAF (B), expressed as measured TPAH50 concentrations (µg L^{-1}). *Different letters* denote significant differences between treatments (ANOVA, Tukey: $P<0.05$). *From Vignier, J., Soudant, P., Chu, F.L.E, Morris, J.M., Carney, Lay, C., Krasnec, M.O., Robert, R., Volety, A.K., 2016. Lethal and sublethal effects of* Deepwater Horizon *slick oil and dispersant on oyster (*Crassostrea virginica*) larvae. Mar. Environ. Res. 120, 20–31.*

Hackney, 1989). It should be noted that any impairment of feeding and/or substantial retardation of growth could cause an absence of settlement. During its larval life, oysters need to ingest sufficient food to supply all of the nutrients required for metamorphosis (Hickman and Gruffydd, 1971). Any nutritional deficiency could cause prolongation of the pelagic life (Gallager and Mann, 1986; Holland and Spencer, 1973).

5.2.5 Oil Disrupted Digestive Processes and Induce an Inflammatory Response in Tissue

Oyster spat (~10–15mm) were acutely exposed for 10 days to DWH oil prepared as HEWAF. Although a broad range of PAHs was tested (i.e., from 7 to 3450µg tPAH50 L^{-1}), no dose-dependent mortality was observed, suggesting that oyster spat are tolerant to oil/PAHs (Vignier et al., 2018). Thus, additional endpoints were investigated. Results from histological analyses of oil-exposed tissues revealed that bothdissolved and particulate fraction of oil induced sublethal effects, occurred from concentrations of 7–3450µg tPAH50 L^{-1} (Figs. 9 and 10).

Specifically, a morphometric study of the DGT indicated a dose-dependent response to oil exposure on lumen dilation and epithelium thinning of the DGT, with lowest observed effective concentration (LOEC) reaching 66µg tPAH50 L^{-1} (Fig. 9). This finding suggests that structural changes occurred in the digestive gland of exposed oysters likely attributed to an oil-related stress.

In addition, histological observations showed that tissues (gills, palp, connective tissue, and digestive gland) in contact with HEWAF were adversely impacted at doses as low as ≥ 7 µg tPAH50 L^{-1} (Fig. 10). Affected spat tissues exhibited pathological symptoms, typical of an inflammatory response, hemocyte diapedesis and infiltration, syncytia, ulcer, lesions, and necrosis of the digestive diverticula or epithelium sloughing (Vignier et al., 2018). Exposure of the digestive system via ingested oil droplets and/or adsorbed dissolved PAHs resulted in a severe degeneration of the digestive diverticula (atrophy of digestive tubules) and an inflammatory response mostly located around the alimentary canal and the connective tissue.

5 KEY FINDINGS

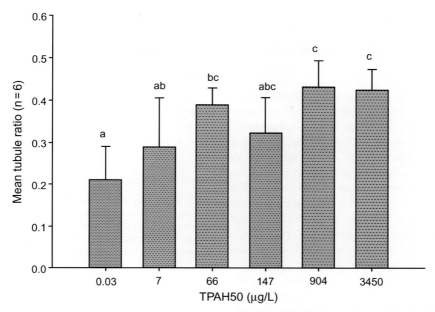

FIG. 9 Mean tubule luminal ratio (±SD) from six oysters exposed for 10 days to increasing concentrations of HEWAF, expressed in μg tPAH50 L^{-1}. *Different letters* denote a statistical difference (ANOVA: $P=0.05$). *From Vignier, J., Rolton, A., Soudant, P., Fu-lin, E.C., Robert, R., Volety, A.K., 2018. Evaluation of toxicity of Deepwater Horizon slick oil on spat of the oyster Crassostrea virginica. Environ. Sci. Pollut. Res. 25 (2), 1176–1190.*

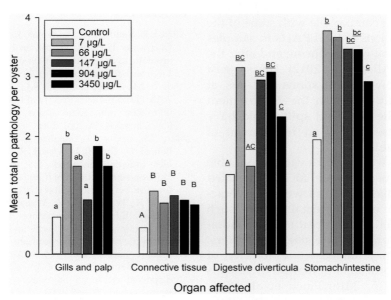

FIG. 10 (A) Organ distribution of total pathological conditions observed per oyster ($n \geq 9$) following 10 days of exposure to increasing concentrations of HEWAF, expressed in μg tPAH50 L^{-1}. *Different letters* within each organ/pathological condition denote statistical difference at $P \leq 0.1$ (Mann–Whitney U-test). Lettering differed between categories because separate analyses were performed for each organ/condition. *From Vignier, J., Rolton, A., Soudant, P., Fu-lin, E.C., Robert, R., Volety, A.K., 2018. Evaluation of toxicity of Deepwater Horizon slick oil on spat of the oyster Crassostrea virginica. Environ. Sci. Pollut. Res. 25 (2), 1176–1190.*

6 CONCLUSIONS

The present study revealed that sublethal concentrations of surface-collected DWH oil and associated dispersant (CEWAF and Corexit only) could affect reproduction, early development and recruitment of the ecologically and commercially important *C. virginica*. Thus, application of the dispersant Corexit 9500A, at the surface or underwater, in response to an oil spill should be considered cautiously and all environmental trade-offs should be evaluated, especially when dispersants are applied during periods when organisms such as oysters are going through reproductive phase. However, the cumulative surface area of the ocean over which dispersant was applied was only 0.06% of the cumulative area of surface oiling (Deepwater Horizon NRDA Trustees (2016)). Consequently, any potential contribution from dispersant to total toxic effects would have been minimal relative to the injury caused by oil in areas beyond where oysters live. More importantly, effective concentrations for larval growth and settlement endpoints determined in the laboratory ($\sim$1–2 µg tPAH50L^{-1}) were comparable with some of the lowest concentrations of total PAHs in samples collected in coastal waters during and after the DWH incident (Allan et al., 2012). This suggest that relatively low concentrations of PAHs, at levels realistically found in the environment at the time of the DWH oil spill, could have detrimental consequences on recruitment of *C. virginica* larvae.

Furthermore, we did not evaluate possible chronic effects of exposure to oil-associated PAHs and/or dispersant, as well as interactions with other environmental factors, such as other xenobiotics, temperature, UV radiation, or salinity on early life stages of oysters. These factors may increase the toxicity of petroleum hydrocarbons to marine organisms (Alloy et al., 2015, 2016; Finch et al., 2016; Lyons et al., 2002; Pelletier et al., 1997; Ramachandran et al., 2006). In addition, various concentrations of Corexit can directly, or in concert with PAHs, can impact responses of oyster life stages. Finally, these studies have not looked at trans-generational impacts of PAHs (e.g., exposure of adults and impacts of gametogenesis and survival of F-2 generation). These areas can form the basis for future studies examining the impacts of PAHs and dispersants on shellfish.

Acknowledgments

This work was supported by funds provided as part of the natural resource damage assessment for the *Deepwater Horizon* oil spill. Data presented here are a subset of a larger toxicological database that is being generated as part of the *Deepwater Horizon* Natural Resource Damage Assessment. The authors thank Abt Associates, Michael Carney, and Ronald Hall, for their support with technical assistance. We also would like to thank the graduate students and staff at the Vester Field Station/FGCU, especially Brooke Denkert, Anne Rolton, Jeffrey Devine, Kelsey McEachern, Emily Nickols, Molly Rybovich, John Roberts, Vaiola Osne, Audrey Barbe, and Gaelle Richard for their technical assistance.

References

Allan, S.E., Smith, B.W., Anderson, K.A., 2012. Impact of the Deepwater Horizon oil spill on bioavailable polycyclic aromatic hydrocarbons in Gulf of Mexico coastal waters. Environ. Sci. Technol. 46 (4), 2033–2039.

Alloy, M.M., Boube, I., Griffitt, R.J., Oris, J.T., Roberts, A.P., 2015. Photo-induced toxicity of Deepwater Horizon slick oil to blue crab (*Callinectes sapidus*) larvae. Environ. Toxicol. Chem. 34 (9), 2061–2066.

Alloy, M., Baxter, D., Stieglitz, J., Mager, E., Hoenig, R., Benetti, D., Grosell, M., Oris, J., Roberts, A., 2016. Ultraviolet radiation enhances the toxicity of Deepwater Horizon oil to mahi-mahi (*Coryphaena hippurus*) embryos. Environ. Sci. Technol. 50 (4), 2011–2017.

Almeda, R., Wambaugh, Z., Wang, Z., Hyatt, C., Liu, Z., Buskey, E.J., 2013. Interactions between zooplankton and crude oil: toxic effects and bioaccumulation of polycyclic aromatic hydrocarbons. PLoS One 8(6).

Almeda, R., Baca, S., Hyatt, C., Buskey, E.J., 2014. Ingestion and sublethal effects of physically and chemically dispersed crude oil on marine planktonic copepods. Ecotoxicology 23, 988–1003.

Anderson, J.W., 1977. Responses to sublethal levels of petroleum hydrocarbons: are they sensitive indicators and do

REFERENCES

they correlate with tissue concentration. In: Wolfe, D.A. (Ed.), Fate and Effects of Petroleum Hydrocarbons in Marine Organisms and Ecosystems. Pergamon Press, pp. 95–114.

Atlas, R.M., 1995. Bioremediation of petroleum pollutants. Int. Biodeter. Biodegr. 35 (1), 317–327.

Balouet, G., Poder, M., Cahour, A., Auffret, M., 1986. Proliferative hemocytic condition in European flat oysters (*Ostrea edulis*) from Breton coasts: a 6-year survey. J. Invertebr. Pathol. 48 (2), 208–215.

Banks, P.D., Brown, K.M., 2002. Hydrocarbon effects on fouling assemblages: the importance of taxonomic differences, seasonal, and tidal variation. Mar. Environ. Res. 53 (3), 311–326.

Barron, M.G., 1995. Bioaccumulation and bioconcentration in aquatic organisms. In: Hoffman, D.J., Rattner, B.A., Burton Jr., G.A., Cairns Jr., J. (Eds.), Handbook of Ecotoxicology, p. 625.

Barron, M.G., Podrabsky, T., Ogle, S., Ricker, R.W., 1999. Are aromatic hydrocarbons the primary determinant of petroleum toxicity to aquatic organisms? Aquat. Toxicol. 46 (3), 253–268.

Barron, M.G., Carls, M.G., Heintz, R., Rice, S.D., 2004. Evaluation of fish early life-stage toxicity models of chronic embryonic exposures to complex polycyclic aromatic hydrocarbon mixtures. Toxicol. Sci. 78 (1), 60–67.

Baumard, P., Budzinski, H., Garrigues, P., Narbonne, J.F., Burgeot, T., Michel, X., Bellocq, J., 1999. Polycyclic aromatic hydrocarbon (PAH) burden of mussels (*Mytilus sp.*) in different marine environments in relation with sediment PAH contamination, and bioavailability. Mar. Environ. Res. 47 (5), 415–439.

Berdugo, V., Harris, R.P., O'Hara, S.C., 1977. The effect of petroleum hydrocarbons on reproduction of an estuarine planktonic copepod in laboratory cultures. Mar. Pollut. Bull. 8 (6), 138–143.

Berthou, F., Balouet, G., Bodennec, G., Marchand, M., 1987. The occurrence of hydrocarbons and histopathological abnormalities in oysters for seven years following the wreck of the *Amoco Cadiz* in Brittany (France). Mar. Environ. Res. 23 (2), 103–133.

Cajaraville, M., Díez, G., Larrea, P., Marigómez, J.A., Angulo, E., 1989. Planimetric parameters of the digestive tubules of *Mytilus edulis*: a sensitive tool for monitoring petroleum hydrocarbon toxicity. In: Seminario Internacional do mexillon. Toxa, O Grove (Spain). 6-10 Nov 1989. Publicacions do Seminario de Estudos Galegos, Xunta de Galicia, p. 146.

Camilli, R., Reddy, C.M., Yoerger, D.R., Van Mooy, B.A., Jakuba, M.V., Kinsey, J.C., McIntyre, C.P., Sylva, S.P., Maloney, J.V., 2010. Tracking hydrocarbon plume transport and biodegradation at Deepwater Horizon. Science 330 (6001), 201–204.

Capuzzo, J.M., 1996. The bioaccumulation and biological effects of lipophilic organic contaminants. In: The Eastern Oyster *Crassostrea virginica*. MD Sea Grant Publication, pp. 539–557.

Capuzzo, J.M., Moore, M.N., Widdows, J., 1988. Effects of toxic chemicals in the marine environment: predictions of impacts from laboratory studies. Aquat. Toxicol. 11 (3), 303–311.

Carriger, J.F., Barron, M.G., 2011. Minimizing risks from spilled oil to ecosystem services using influence diagrams: the Deepwater Horizon spill response. Environ. Sci. Technol. 45 (18), 7631–7639.

Chapman, P.M., Long, E.R., 1983. The use of bioassays as part of a comprehensive approach to marine pollution assessment. Mar. Pollut. Bull. 14 (3), 81–84.

Chapman, P.M., Morgan, J.D., 1983. Sediment bioassays with oyster larvae. Bull. Environ. Contam. Toxicol. 31 (4), 438–444.

Connor, P.M., 1972. Acute toxicity of heavy metals to some marine larvae. Mar. Pollut. Bull. 3 (12), 190–192.

Croxton, A.N., Wikfors, G.H., Schulterbrandt-Gragg, R.D., 2012. Immunomodulation in eastern oysters, *Crassostrea virginica*, exposed to a PAH-contaminated, microphytobenthic diatom. Aquat. Toxicol. 118, 27–36.

Daling, P.S., Leirvik, F., Almås, I.K., Brandvik, P.J., Hansen, B.H., Lewis, A., Reed, M., 2014. Surface weathering and dispersibility of MC252 crude oil. Mar. Pollut. Bull. 87 (1), 300–310.

Deepwater Horizon Natural Resource Damage Assessment Trustees, 2016. Injury to natural resources. In: Deepwater Horizon oil spill: Final Programmatic Damage Assessment and Restoration Plan and Final Programmatic Environmental Impact Statement. NOAA, Chapter 4. Retrieved from, http://www.gulfspillrestoration.noaa.gov/restoration-planning/gulf-plan.

Di Toro, D.M., Zarba, C.S., Hansen, D.J., Berry, W.J., Swartz, R.C., Cowan, C.E., Pavlou, S.P., Allen, H.E., Thomas, N.A., Paquin, P.R., 1991. Technical basis for establishing sediment quality criteria for nonionic organic chemicals using equilibrium partitioning. Environ. Toxicol. Chem. 10 (12), 1541–1583.

Djomo, J.E., Garrigues, P., Narbonne, J.F., 1996. Uptake and depuration of polycyclic aromatic hydrocarbons from sediment by the zebrafish (Brachydanio rerio). Environ. Toxicol. Chem. 15 (7), 1177–1181.

Downs, C.A., Shigenaka, G., Fauth, J.E., Robinson, C.E., Huang, A., 2002. Cellular physiological assessment of bivalves after chronic exposure to spilled Exxon Valdez crude oil using a novel molecular diagnostic biotechnology. Environ. Sci. Technol. 36 (13), 2987–2993.

Environmental Response Management Application, 2015. Deepwater Gulf Response. National Oceanic and Atmospheric Administration. Available: http://response.restoration.noaa.gov/maps-and-spatial-data/environmental-response-management-application-erma/erma-gulf-response.html. (accessed 7/30/2015).

Eyster, L.S., Morse, M.P., 1984. Early shell formation during molluscan embryogenesis, with new studies on the surf clam, *Spisula solidissima*. Am. Zool. 24 (4), 871–882.

Farrington, J.W., Goldberg, E.D., Risebrough, R.W., Martin, J.H., Bowen, V.T., 1983. US" mussel watch" 1976-1978: an overview of the trace-metal, DDE, PCB, hydrocarbon and artificial radionuclide data. Environ. Sci. Technol. 17 (8), 490–496.

Finch, B.E., Stefansson, E.S., Langdon, C.J., Pargee, S.M., Blunt, S.M., Gage, S.J., Stubblefield, W.A., 2016. Photo-enhanced toxicity of two weathered Macondo crude oils to early life stages of the eastern oyster (*Crassostrea virginica*). Mar. Pollut. Bull. 113 (1–2), 316–323.

Forth, H.P., Mitchelmore, C.L., Morris, J.M., Lay, C.R., Lipton, J., 2017. Characterization of dissolved and particulate phases of water accommodated fractions used to conduct aquatic toxicity testing in support of the *Deepwater Horizon* natural resource damage assessment. Environ. Toxicol. Chem. 36, 1460–1472. https://doi.org/10.1002/etc.3803.

Fucik, K., 1994. Dispersed Oil Toxicity Tests With Species Indigenous to the Gulf of Mexico. OCS reports, U. S. Minerals Management Service, p. 94.

Gallager, S.M., Mann, R., 1986. Growth and survival of larvae of *Mercenaria mercenaria* (L.) and *Crassostrea virginica* (Gmelin) relative to broodstock conditioning and lipid content of eggs. Aquaculture 56 (2), 105–121.

Galtsoff, P.S., 1964. The American oyster, *Crassostrea virginica* (Gmelin). Fish. Bull. 64, 1–480.

Galtsoff, P.S., 1969. Anomalies and malformation in the shells of *Crassostrea virginica*. Natl. Cancer Inst. Monogr. 31, 575–580.

Geffard, O., Budzinski, H., Augagneur, S., Seaman, M.N., His, E., 2001. Assessment of sediment contamination by spermiotoxicity and embryotoxicity bioassays with sea urchins (*Paracentrotus lividus*) and oysters (*Crassostrea gigas*). Environ. Toxicol. Chem. 20 (7), 1605–1611.

Geffard, O., Budzinski, H., His, E., Seaman, M.N., Garrigues, P., 2002a. Relationships between contaminant levels in marine sediments and their biological effects on embryos of oysters, *Crassostrea gigas*. Environ. Toxicol. Chem. 21 (11), 2310–2318.

Geffard, O., Budzinski, H., His, E., 2002b. The effects of elutriates from PAH and heavy metal polluted sediments on *Crassostrea gigas* (Thunberg) embryogenesis, larval growth and bio-accumulation by the larvae of pollutants from sedimentary origin. Ecotoxicology 11 (6), 403–416.

Geffard, O., Geffard, A., His, E., Budzinski, H., 2003. Assessment of the bioavailability and toxicity of sediment-associated polycyclic aromatic hydrocarbons and heavy metals applied to *Crassostrea gigas* embryos and larvae. Mar. Pollut. Bull. 46 (4), 481–490.

George-Ares, A., Clark, J.R., 2000. Aquatic toxicity of two Corexit® dispersants. Chemosphere 40 (8), 897–906.

Goldberg, E.D., Bowen, V.T., Farrington, J.W., Harvey, G., Martin, J.H., Parker, P.L., Risebrough, R.W., Robertson, W., Schneider, E., Gamble, E., 1978. The mussel watch. Environ. Conserv. 5 (02), 101–125.

Hansen, B.H., Altin, D., Olsen, A.J., Nordtug, T., 2012. Acute toxicity of naturally and chemically dispersed oil on the filter-feeding copepod *Calanus finmarchicus*. Ecotoxicol. Environ. Saf. 86, 38–46.

Hégaret, H., Wikfors, G.H., Soudant, P., 2003. Flow cytometric analysis of haemocytes from eastern oysters, *Crassostrea virginica*, subjected to a sudden temperature elevation: II. Haemocyte functions: aggregation, viability, phagocytosis, and respiratory burst. J. Exp. Mar. Biol. Ecol. 293 (2), 249–265.

Helm, M.M., Bourne, N., 2004. In: Lovatelli, A. (Ed.), Hatchery Culture of Bivalves: A Practical Manual. vol. 471.Rome: Food and agriculture organization of the United Nations.

Hickman, R.W., Gruffydd, L.D., 1971. The histology of the larvae of *Ostrea edulis* during metamorphosis. In: Crisp, D.J. (Ed.), Proceedings of the 4th Eur. Mar. Biol. Symp. Cambridge University Press, Cambridge, pp. 281–284.

Hidu, H., 1965. Effects of synthetic surfactants on the larvae of clams (*M. mercenaria*) and oysters (*C. virginica*). J. Water Pollut. Control Fed. 37, 262–270.

Hinkle, P.M., Kinsella, P.A., Osterhoudt, K.C., 1987. Cadmium uptake and toxicity via voltage-sensitive calcium channels. J. Biol. Chem. 262 (34), 16333–16337.

His, E., Beiras, R., 1999. The assessment of marine pollution-bioassays with bivalve embryos and larvae. Adv. Mar. Biol. 37, 1.

His, E., Robert, R., 1985. Utilisation des élevages larvaires de Crassostrea gigas en écotoxicologie marine. In: Actes du Congrès, Wimereux, 4–8 Novembre 1985, Haliotis 15, pp. 301–308. https://archimer.ifremer.fr/doc/00000/6184/.

His, E., Seaman, M.N.L., Beiras, R., 1996. A simplification of the bivalve embryogenesis and larval development bioassay method for water quality assessment. Water Res. 31 (2), 351–355. https://doi.org/10.1016/S0043-1354(96)00244-8.

His, É., Budzinski, H., Geffard, O., Beiras, R., 1997. Action d'un sédiment pollué par les hydrocarbures sur la métamorphose de l'huître japonaise, *Crassostrea gigas* (Thunberg). C. R. Acad. Sci. III 320 (10), 797–803.

Holland, D.L., Spencer, B.E., 1973. Biochemical changes in fed and starved oysters, *Ostrea edulis* L. during larval development, metamorphosis and early spat growth. J. Mar. Biol. Assoc. U. K. 53 (02), 287–298.

Incardona, J.P., Swarts, T.L., Edmunds, R.C., Linbo, T.L., Aquilina-Beck, A., Sloan, C.A., Gardner, L.D., Block, B.A., Scholz, N.L., 2013. *Exxon Valdez* to *Deepwater Horizon*: comparable toxicity of both crude oils to fish early life stages. Aquat. Toxicol. 142, 303–316.

REFERENCES

Ingle, R.M., 1951. Spawning and setting of oysters in relation to seasonal environmental changes. Bull. Mar. Sci. Gulf Caribb. 1, 111–135.

Jeong, W.G., Cho, S.M., 2005. The effects of polycyclic aromatic hydrocarbon exposure on the fertilization and larval development of the Pacific oyster, *Crassostrea gigas*. J. Shellfish. Res. 24 (1), 209.

Kujawinski, E.B., Kido Soule, M.C., Valentine, D.L., Boysen, A.K., Longnecker, K., Redmond, M.C., 2011. Fate of dispersants associated with the Deepwater Horizon oil spill. Environ. Sci. Technol. 45 (4), 1298–1306.

Landrum, P.F., Robbins, J.A., 1990. Bioavailability of sediment-associated contaminants to benthic invertebrates. In: Sediments: Chemistry and Toxicity of In-Place Pollutants. CRC Press, Inc., Boca Raton, Florida, USA, pp. 237–263.

Landrum, P.F., Giesy, J.P., Oris, J.T., Allred, P.M., 1987. Photo-induced toxicity of polycyclic aromatic hydrocarbons to aquatic organisms. In: Oil in Freshwater: Chemistry, Biology, Countermeasure Technology. Pergamon Press, Elmsford, NY, pp. 304–318.

Langdon, C.J., Stefansson, E.S., Pargee, S.M., Blunt, S.M., Gage, S.J., Stubblefield, W.A., 2016. Chronic effects of non-weathered and weathered crude oil and dispersant associated with the Deepwater Horizon incident on development of larvae of the eastern oyster, *Crassostrea virginica*. Environ. Toxicol. Chem. 35 (8), 2029–2040.

Laramore, S., Krebs, W., Garr, A., 2014. Effects of Macondo canyon 252 oil (naturally and chemically dispersed) on larval *Crassostrea virginica* (Gmelin, 1791). J. Shellfish. Res. 33 (3), 709–718.

Le Goïc, N., Hégaret, H., Boulais, M., Béguel, J.P., Lambert, C., Fabioux, C., Soudant, P., 2014. Flow cytometric assessment of morphology, viability, and production of reactive oxygen species of *Crassostrea gigas* oocytes. Application to toxic dinoflagellate (*Alexandrium minutum*) exposure. Cytometry A 85 (12), 1049–1056.

Le Gore, R.S., 1974. The Effect of Alaskan Crude Oil and Selected Hydrocarbon Compounds on Embryonic Development of the Pacific Oyster, *Crassostrea Gigas*. Ph. D. Thesis.

Leahy, J.G., Colwell, R.R., 1990. Microbial degradation of hydrocarbons in the environment. Microbiol. Rev. 54 (3), 305–315.

Lee, R.F., 1977. Fate of petroleum components in estuarine waters of the southeastern United States. In: International Oil Spill Conference. vol. 1977. American Petroleum Institute, pp. 611–616. no. 1. March.

Lee, R.F., Gardner, W.S., Anderson, J.W., Blaylock, J.W., Barwell-Clarke, J., 1978. Fate of polycyclic aromatic hydrocarbons in controlled ecosystem enclosures. Environ. Sci. Technol. 12 (7), 832–838.

Lessard, R.R., DeMarco, G., 2000. The significance of oil spill dispersants. Spill Sci. Technol. Bull. 6 (1), 59–68.

Li, M., Garrett, C., 1998. The relationship between oil droplet size and upper ocean turbulence. Mar. Pollut. Bull. 36 (12), 961–970.

Livingstone, D.R., Garcia Martinez, P., Winston, G.W., 1989. Menadione-stimulated oxyradical formation in digestive gland musomes of the common mussel, *Mytilus edulis* L. Aquat. Toxicol. 15 (3), 213–236.

Loosanoff, V.L., Davis, H.C., 1963. Rearing of bivalve mollusks. Adv. Mar. Biol. 1, 1–136.

Lyons, B.P., Pascoe, C.K., McFadzen, I.R.B., 2002. Phototoxicity of pyrene and benzo [*a*] pyrene to embryo-larval stages of the pacific oyster *Crassostrea gigas*. Mar. Environ. Res. 54 (3), 627–631.

Meador, J.P., Stein, J.E., Reichert, W.L., Varanasi, U., 1995. Bioaccumulation of polycyclic aromatic hydrocarbons by marine organisms. In: Reviews of Environmental Contamination and Toxicology. Springer, New York, pp. 79–165.

Means, J.C., Wood, S.G., Hassett, J.J., Banwart, W.L., 1980. Sorption of polynuclear aromatic hydrocarbons by sediments and soils. Environ. Sci. Technol. 14 (12), 1524–1528.

Michel, J., Owens, E.H., Zengel, S., Graham, A., Nixon, Z., Allard, T., Holton, W., Reimer, P.D., Lamarche, A., White, M., Rutherford, N., Childs, C., Mauseth, G., Challenger, G., Taylor, E., 2013. Extent and degree of shoreline oiling: *Deepwater Horizon* oil spill, Gulf of Mexico, USA. PLoS ONE 8 (6), e65087. https://doi.org/10.1371/journal.pone.0065087.

Nalco Energy Services, 2012. Material Safety Data Sheet for Corexit 9500A®. Available at http:www.nalcoesllc.com nes documents MSDS NES-LLC-COREXIT-EC9500A-March_2012.pdf (last consulted on August 10th 2017).

National Commission on the BP deep ocean horizon oil spill and offshore drilling, 2011. Deep Water: The Gulf Oil Disaster and the Future of Offshore Drilling. Report to the president. www.oilspillcommission.gov. Accessed June 2014.

National Marine Fisheries Service, 2012. Annual Commercial Landing Statistics, Fisheries Statistics. NOAA, NMFS. Available at http://www.st.nmfs.noaa.gov/st1/commercial/landings/annual_landings.html. (last consulted in July 2016).

National Research Council (U.S), 2005. Committee on Understanding Oil Spill Dispersants: Efficacy and Effects. Oil Spill Dispersants: Efficacy and Effects. The National Academies Press, Washington, DC.

Neff, J.M., 1979. Polycyclic Aromatic Hydrocarbons in the Aquatic Environment: Sources, Fates and Biological Effects. Applied Sciences Publishers LTD, London.

Neff, J.M., 1985. Polycyclic aromatic hydrocarbons. In: Fundamentals of Aquatic Toxicology: Methods and Applications. Hemisphere Publishing Corporation, Washington DC, pp. 416–454. 1985. 2 fig, 7 tab, 140 ref.

Neff, J.M., Haensly, W.E., 1982. Long-term impact of the Amoco Cadiz crude oil spill on oysters *Crassostrea gigas* and plaice Pleuronectes platessa from Aber Benoit and

EVALUATING WATER QUALITY TO PREVENT FUTURE DISASTERS

Aber Wrac'h, Brittany, France. In: Ecological Study of the AMOCO Cadiz Oil Spill, vol. 2, NOAA/CNEXO Joint Scientific Commission, Boulder, pp. 59–328.

Neff, J.M., Stubblefield, W.A., 1995. Chemical and toxicological evaluation of water quality following the Exxon Valdez oil spill. ASTM Spec. Tech. Publ. 1219, 141–177.

Neff, J.M., Cox, B.A., Dixit, D., Anderson, J.W., 1976. Accumulation and release of petroleum-derived aromatic hydrocarbons by four species of marine animals. Mar. Biol. 38 (3), 279–289.

Newell, R.I., 2004. Ecosystem influences of natural and cultivated populations of suspension-feeding bivalve molluscs: a review. J. Shellfish. Res. 23 (1), 51–62.

Nogueira, D.J., Mattos, J.J., Dybas, P.R., Flores-Nunes, F., Sasaki, S.T., Taniguchi, S., Schmidt, E.C., Bouzon, Z.L., Bicego, M.C., Melo, C.M., Toledo-Silva, G., 2017. Effects of phenanthrene on early development of the Pacific oyster Crassostrea gigas (Thunberg, 1789). Aquat. Toxicol. 191, 50–61.

OSAT-1, 2010. Summary Report for Sub-Sea and Sub-Surface Oil and Dispersant Detection: Sampling and Monitoring. Prepared for Paul F. Zunkunft, Federal On-Scene Coordinator, Deepwater Horizon MC252. December 17.

Page, C.A., Bonner, J.S., Sumner, P.L., McDonald, T.J., Autenrieth, R.L., Fuller, C.B., 2000. Behavior of a chemically-dispersed oil and a whole oil on a near-shore environment. Water Res. 34 (9), 2507–2516.

Payne, J.F., Fancey, L.L., Rahimtula, A.D., Porter, E.L., 1987. Review and perspective on the use of mixed-function oxygenase enzymes in biological monitoring. Comp. Biochem. Physiol. C Comp. Pharmacol. 86 (2), 233–245.

Pelletier, M.C., Burgess, R.M., Ho, K.T., Kuhn, A., McKinney, R.A., Ryba, S.A., 1997. Phototoxicity of individual polycyclic aromatic hydrocarbons and petroleum to marine invertebrate larvae and juveniles. Environ. Toxicol. Chem. 16 (10), 2190–2199.

Phelps, H.L., Warner, K.A., 1990. Estuarine sediment bioassay with oyster pediveliger larvae (Crassostrea gigas). Bull. Environ. Contam. Toxicol. 44 (2), 197–204.

Porte, C., Albaiges, J., 1994. Bioaccumulation patterns of hydrocarbons and polychlorinated biphenyls in bivalves, crustaceans, and fishes. Arch. Environ. Contam. Toxicol. 26 (3), 273–281.

Power, E.A., Chapman, P.M., 1992. Assessing sediment quality. In: Sediment Toxicity Assessment. Lewis Publishers, Boca Raton, Florida, pp. 1–18. 1992. 3 fig, 1 tab, 42 ref.

Ramachandran, S.D., Sweezey, M.J., Hodson, P.V., Boudreau, M., Courtenay, S.C., Lee, K., King, T., Dixon, J.A., 2006. Influence of salinity and fish species on PAH uptake from dispersed crude oil. Mar. Pollut. Bull. 52 (10), 1182–1189.

Readman, J.W., Mantoura, R.F.C., Rhead, M.M., Brown, L., 1982. Aquatic distribution and heterotrophic degradation of polycyclic aromatic hydrocarbons (PAH) in the Tamar Estuary. Estuar. Coast. Shelf Sci. 14 (4), 369–389.

Renzoni, A., 1975. Toxicity of three oils to bivalve gametes and larvae. Mar. Pollut. Bull. 6 (8), 125.

Roesijadi, G., Brubacher, L.L., Unger, M.E., Anderson, R.S., 1997. Metallothionein mRNA induction and generation of reactive oxygen species in molluscan hemocytes exposed to cadmium in vitro. Comp. Biochem. Physiol. C Pharmacol. Toxicol. Endocrinol. 118 (2), 171–176.

Rolton, A., Soudant, P., Vignier, J., Pierce, R., Henry, M., Shumway, S.E., Bricelj, M.V., Volety, A.K., 2015. Susceptibility of gametes and embryos of the eastern oyster, Crassostrea virginica, to Karenia brevis and its toxins. Toxicon 99, 6–15.

Rossi, S.S., Anderson, J.W., Ward, G.S., 1976. Toxicity of water-soluble fractions of four test oils for the polychaetous annelids, Neanthes arenaceodentata and Capitella capitata. Environ. Pollut. 10 (1), 9–18.

Singer, M.M., Smalheer, D.L., Tjeerdema, R.S., Martin, M., 1991. Effects of spiked exposure to an oil dispersant on the early life stages of four marine species. Environ. Toxicol. Chem. 10 (10), 1367–1374.

Singer, M.M., George, S., Jacobson, S., Lee, I., Weetman, L.L., Tjeerdema, R.S., Sowby, M.L., 1996. Comparison of acute aquatic effects of the oil dispersant Corexit 9500 with those of other Corexit series dispersants. Ecotoxicol. Environ. Saf. 35 (2), 183–189.

Singer, M.M., Aurand, D., Bragin, G.E., Clark, J.R., Coelho, G.M., Sowby, M.L., Tjeerdema, R.S., 2000. Standardization of the preparation and quantitation of water-accommodated fractions of petroleum for toxicity testing. Mar. Pollut. Bull. 40 (11), 1007–1016.

Smith, C.M., Hackney, C.T., 1989. The effects of hydrocarbons on the setting of the American oyster, Crassostrea virginica, in intertidal habitats in southeastern North Carolina. Estuaries 12 (1), 42–48.

Stefansson, E.S., Langdon, C.J., Pargee, S.M., Blunt, S.M., Gage, S.J., Stubblefield, W.A., 2016. Acute effects of non-weathered and weathered crude oil and dispersant associated with the Deepwater Horizon incident on the development of marine bivalve and echinoderm larvae. Environ. Toxicol. Chem. 35 (8), 2016–2028.

Stegeman, J.J., Teal, J.M., 1973. Accumulation, release and retention of petroleum hydrocarbons by the oyster Crassostrea virginica. Mar. Biol. 22 (1), 37–44.

Strathmann, R.R., 1987. Larval feeding. In: Reproduction of Marine Invertebrates. vol. 9, pp. 465–550.

Supan, J., 1983. Evaluation of a leased oyster bottom in Mississippi sound. Gulf Res. Rep. 7 (3), 261–266.

Tamura, H., Takasaki, A., Miwa, I., Taniguchi, K., Maekawa, R., Asada, H., Sugino, N., 2008. Oxidative stress impairs oocyte quality and melatonin protects oocytes from free radical damage and improves fertilization rate. J. Pineal Res. 44 (3), 280–287.

REFERENCES

Thain, J.E., 1992. Use of the oyster *Crassostrea gigas* embryo bioassay on water and sediment elutriate samples from the German bight. Mar. Ecol. Prog. Ser. 91 (1), 211–213. Oldendorf.

Thorson, G., 1950. Reproductive and larval ecology of marine bottom invertebrates. Biol. Rev. 25 (1), 1–45.

U.S. Coast Guard, 2011. On Scene Coordinator Report: Deepwater Horizon Oil Spill. Submitted to the National Response Team. September, U.S. Department of Homeland Security, U.S. Coast Guard, Washington, DC. Available: http://www.uscg.mil/foia/docs/dwh/fosc_dwh_report.pdf. Accessed July 7, 2015.

U.S. District Court, 2014. In Re: Oil Spill by the Oil Rig "Deepwater Horizon" in the Gulf of Mexico, on April 20, 2010, No. MDL 2179, Section 7 (Revised September 9, 2014) ("Findings of Fact and Conclusions of Law: Phase One Trial"), Figure 1. United States District Court for the Eastern District of Louisiana.

U.S. District Court, 2015. In Re: Oil Spill by the Oil Rig "Deepwater Horizon" in the Gulf of Mexico, on April 20, 2010, No. MDL 2179, 2015 WL 225421 (La. E.D. Jan. 15, 2015) ("Findings of Fact and Conclusions of Law: Phase Two Trial"). (United States District Court for the eastern district of Louisiana).

Utting, S.D., Millican, P.F., 1997. Techniques for the hatchery conditioning of bivalve broodstock and the subsequent effect on egg quality and larval viability. Aquaculture 155 (1), 45–54.

Van Brummelen, T.C., Van Hattum, B., Crommentuijn, T., Kalf, D.F., 1998. Bioavailability and ecotoxicity of PAHs. In: Neilson, A., Hutzinger, O. (Eds.), In: PAHs and Related Compounds. The Handbook of Environmental Chemistry 3, Springer-Verlag, pp. 203–263. part J. https://research.vu.nl/en/publications/81e117fd-8f40-422c-8dcd-8f3c6d23f8cc.

Varanasi, U., 1989. Metabolism of Polycyclic Aromatic Hydrocarbons in the Aquatic Environment. CRC Press.

Varanasi, U., Malins, D.C., 1977. Metabolism of petroleum hydrocarbons: accumulation and biotransformation in marine organisms. In: Biological Effects. Elsevier, pp. 175–270.

Venosa, A.D., Zhu, X., 2003. Biodegradation of crude oil contaminating marine shorelines and freshwater wetlands. Spill Sci. Technol. Bull. 8 (2), 163–178.

Viarengo, A., Zanicchi, G., Moore, M.N., Orunesu, M., 1981. Accumulation and detoxication of copper by the mussel, *Mytilus galloprovincialis* Lam: a study of the subcellular distribution in the digestive gland cells. Aquat. Toxicol. 1 (3), 147–157.

Vignier, J., Donaghy, L., Soudant, P., Chu, F.L.E., Morris, J.M., Carney, M.W., Lay, C., Krasnec, M., Robert, R., Volety, A.K., 2015. Impacts of Deepwater Horizon oil and associated dispersant on early development of the Eastern

oyster, *Crassostrea virginica*. Mar. Pollut. Bull. 100, 426–437. https://doi.org/10.1016/j.marpolbul.2015.08.011.

Vignier, J., Soudant, P., Chu, F.L.E., Morris, J.M., Carney, Lay, C., Krasnec, M.O., Robert, R., Volety, A.K., 2016. Lethal and sublethal effects of *Deepwater Horizon* slick oil and dispersant on oyster (*Crassostrea virginica*) larvae. Mar. Environ. Res. 120, 20–31.

Vignier, J., Volety, A.K., Rolton, A., Le Goïc, N., Chu, F.L., Robert, R., Soudant, P., 2017. Sensitivity of eastern oyster (*Crassostrea virginica*) spermatozoa and oocytes to dispersed oil: cellular responses and impacts on fertilization and embryogenesis. Environ. Pollut. 225, 270–282.

Vignier, J., Rolton, A., Soudant, P., Fu-lin, E.C., Robert, R., Volety, A.K., 2018. Evaluation of toxicity of Deepwater Horizon slick oil on spat of the oyster *Crassostrea virginica*. Environ. Sci. Pollut. Res. 25 (2), 1176–1190.

Volety, A.K., Haynes, L., Goodman, P., Gorman, P., 2014. Ecological condition and value of oyster reefs of the Southwest Florida shelf ecosystem. Ecol. Indic. 44, 108–119. https://doi.org/10.1016/j.ecolind.2014.03.012.

Volety, A., Boulais, M., Donaghy, L., Vignier, J., Loh, A.N., Soudant, P., 2016. Application of flow cytometry to assess Deepwater Horizon oil toxicity on the Eastern oyster *Crassostrea virginica* spermatozoa. J. Shellfish. Res. 35 (1), 91–99.

Wang, Z., Fingas, M.F., 2003. Development of oil hydrocarbon fingerprinting and identification techniques. Mar. Pollut. Bull. 47 (9–12), 423–452.

Wang, Z.D., Stout, S.A., 2007. Chemical fingerprinting of spilled or discharged petroleum—methods and factors affecting petroleum fingerprints in the environment. In: Oil Spill Environmental Forensics–Fingerprinting and Source Identification. Elsevier, London, pp. 1–53.

Wang, Z., Hollebone, B.P., Fingas, M., Fieldhouse, B., Sigouin, L., Landriault, M., Weaver, J.W., 2003. Characteristics of Spilled Oils, Fuels, and Petroleum Products: 1. Composition and Properties of Selected Oils. US EPA Report EPA/600-R/03, US EPA, 72. https://www.researchgate.net/profile/Merv_Fingas/publication/265189604_Characteristics_of_Spilled_Oils_Fuels_and_Petroleum_Products_1_Composition_and_Properties_of_Selected_Oils/links/5485a5a60cf2437065c9eb89.pdf.

Wells, H.W., 1961. The fauna of oyster beds, with special reference to the salinity factor. Ecol. Monogr. 31 (3), 239–266.

Wessel, N., Rousseau, S., Caisey, X., Quiniou, F., Akcha, F., 2007. Investigating the relationship between embryotoxic and genotoxic effects of benzo [*a*] pyrene, 17α-ethinylestradiol and endosulfan on *Crassostrea gigas* embryos. Aquat. Toxicol. 85 (2), 133–142.

Winstead, J.T., 1995. Digestive tubule atrophy in eastern oysters, *Crassostrea virginica* (Gmelin, 1791), exposed to salinity and starvation stress. J. Shellfish. Res. 14 (1), 105–112.

Woelke, C.E., 1972. Development of a receiving water quality bioassay criterion based on the 48-hour Pacific oyster (*Crassostrea gigas*) embryo. In: Washington Department of Fisheries-Technical Report 9. http://agris.fao.org/agris-search/search.do?recordID=US201300506855.

Further Reading

De Lamirande, E., Jiang, H., Zini, A., Kodama, H., Gagnon, C., 1997. Reactive oxygen species and sperm physiology. Rev. Reprod. 2 (1), 48–54.

CHAPTER

9

Analytical Methods for the Comprehensive Characterization of Produced Water

Tiffany Liden[a], Inês C. Santos[a,b], Zacariah L. Hildenbrand[b,c],*, Kevin A. Schug[a,b],*

[a]Department of Chemistry and Biochemistry, The University of Texas at Arlington, Arlington, TX, United States [b]Affiliate of Collaborative Laboratories for Environmental Analysis and Remediation, The University of Texas at Arlington, Arlington, TX, United States [c]Inform Environmental, LLC, Dallas, TX, United States
*Corresponding authors: E-mail: zac@informenv.com, kschug@uta.edu

1 INTRODUCTION

Currently, the American economy relies heavily on oil and natural gas resources to support approximately 81% of the nation's growing energy demands (US Department of Energy, 2015). Most recently, this has been accomplished through the expansion of unconventional oil and gas development (UD), which in itself has been one of the most controversial topics in the climate change era. Through this process, water is injected into the production well at a high rate and pressure in order to cause fissures in low porosity rock to allow oil and natural gas a path to migrate back to the surface. Once stimulation is completed, the water and the commodity are pumped back to the surface (Liden et al., 2017).

From 2000 to 2007, there was a 39% increase in unconventional natural gas production in the United States. The growth was a result of a combination of advanced methods in horizontal drilling and hydraulic fracturing (US Energy Information Administration, 2017). By 2015, two-thirds of the natural gas output in the United States came from hydraulically fractured production wells (Ground Water Protection Council and All Consulting, 2009; Perrin and Cook, 2016; Wang et al., 2014), with 92% of production growth being derived from unconventional resources in seven key oil and gas basins: Bakken (North Dakota and Montana), Niobrara (Colorado), Marcellus and Utica (Pennsylvania, Ohio, and West Virginia), Haynesville (Louisiana and East Texas), Eagle

Ford (South Texas), and the Permian Basin (West Texas and Southeast New Mexico).

The success of UD relies heavily on the accessibility to large volumes of water, frequently in the form of surface and groundwater. The amount of water used during hydraulic fracturing varies across the nation and is dependent on a number of factors, such as the intrinsic properties of the target formation, the type of well (conventional or unconventional), the depth of the well, and the form of water-stimulation used (Alessi et al., 2017; Gallegos et al., 2015; US Department of Energy, 2015). In 2011, 4.9 billion gallons of water was used to stimulate unconventional well production in Texas (Ground Water Protection Council and All Consulting, 2009; Shaffer et al., 2013). Overall this volume of water used is relatively low compared to other anthropogenic activities (i.e., agriculture and farming), accounting for less than 1% of the state's water usage. However, these volumes tend to be highly localized, which can lead to cones of groundwater depletion, especially in arid and semiarid regions.

An example of where this is occurring is the Permian Basin located in west Texas, which is one of the most prolific production basins in the world (Ewing et al., 2014). In 2012, this region accounted for two-thirds of Texas' and 14% of the nation's crude oil production, respectively (Permian Basin Information, 2017). Additionally, the expansion of horizontal wells leads to increased water usage. From 2005 to 2015, horizontal wells represented only a third of the unconventional wells, but they accounted for approximately two-thirds of the hydraulic fracturing water use (Scanlon et al., 2017). The national median water use for a horizontally drilled well is ~5 million gallons, which is approximately 5 times more than vertical wells in the Permian Basin. From 2013 to 2016, the water use per well went up by approximately 434% in the Permian Basin (Barclays and Columbia Water Center, 2017), with a median of 12 million gal/horizontal well stimulation

(Backstrom, 2018). This increase is attributed to drilling activity, particularly for horizontal unconventional wells, and completions where the average horizontal length was increased from 5700 to 6800 lateral ft. (Barclays and Columbia Water Center, 2017).

In addition to requiring large amount of water for hydrocarbon extraction, UD wells also produce large volumes of waste. This waste stream is comprised of two distinctive subcomponents, termed flowback water and produced water. Flowback water is extricated during the initial 2 weeks of production. Flowback water is primarily comprised of production chemicals. The longer the solution remains in the formation, the more the chemical composition will resemble the geology of the formation (Liden et al., 2017). At that point, the waste stream is referred to as produced water (Veil, 2015). Since produced water is the major component of the waste stream, the two components are often collectively referred to as produced water.

Produced water is typically disposed of into the subsurface via Class II wells, rather than treating the aqueous waste and releasing it back into the water system. Class II injections wells are the most popular choice for produced water management because of convenience and cost. This is best illustrated by the fact that 98% of the 21.2 billion bbl (58 million bbl/day) of produced water from oilfield activities generated in 2012 was disposed of via Class II Injection wells (Arthur et al., 2008; Clark and Veil, 2009; Igunnu and Chen, 2014). These wells are used to inject "nonhazardous" fluids associated with oil and gas production for disposal, hydrocarbon storage wells, or enhanced recovery (US Environmental Protection Agency, 2016a). Enhanced oil recovery accounted for 58% of overall disposal, which is beneficial to oil and gas production with respect to yield. However, this waste stream is primarily from conventional wells cycled back into conventional wells, as shown in Fig. 1 (Scanlon et al., 2017). The remaining 40% is disposed of through subsurface

1 INTRODUCTION

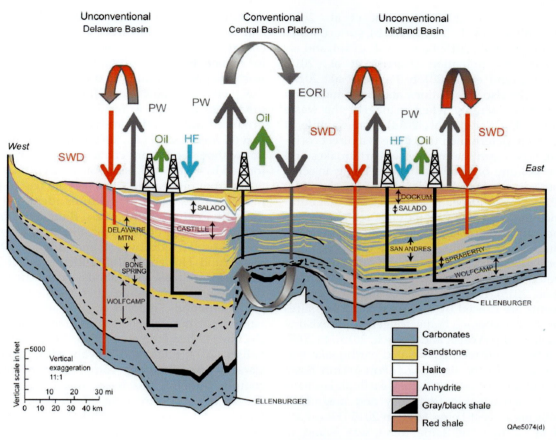

FIG. 1 East–west cross section along the southern margin of the Permian Basin. The dominant unconventional reservoirs include Wolfcamp and Bone Spring in the Delaware Basin and Wolfcamp and Spraberry in the Midland Basin. These reservoirs are much thicker in the Delaware Basin than in the Midland Basin. The Central Basin Platform includes predominantly conventional reservoirs. The *arrows* are used to schematize the inputs and production data. *Arrow sizes* are used to approximate relative volumes. For example, conventional wells generate ~13 times more produced water (PW) than oil; however, most of this water is recycled by injecting into EORI wells for water flooding. PW from conventional wells accounts for ~90% of total PW volume. Water injection for hydraulic fracturing (HF) is shown in *blue arrows*. HF water volumes for unconventional horizontal wells are ~2 times higher than oil production PW from unconventional wells is ~3 times oil production and is injected into saltwater disposal (SWD) wells in shallow horizons (e.g., San Andres in Midland Basin or Delaware Mtn. Grp. in Delaware Basin) or in deeper units (e.g., Ellenburger unit near the basement). The Dockum Aquifer is shown near the surface (Scanlon et al., 2017). Reproduced with permission from Scanlon, B.R., Reedy, R.C., Male, F., Walsh, M., 2017. Water issues related to transitioning from conventional to unconventional oil production in the Permian Basin. Environ. Sci. Technol. 51, 10903–10912. https://doi.org/10.1021/acs.est.7b02185; 2018, American Chemical Society https://pubs.acs.org/doi/abs/10.1021/acs.est.7b02185 Further permissions related to the material excerpted should be directed to the ACS.

injection via saltwater disposal wells (SWD) into nonoil-producing geological intervals located primarily in Texas, California, Oklahoma, and Kansas (Arthur et al., 2008; Clark and Veil, 2009).

Despite the economic benefits, the increase in UD across the American shale energy landscape has been not without concern and controversy. In addition to the potential for groundwater

(Darrah et al., 2014; Fontenot et al., 2013; Hildenbrand et al., 2015, 2016a,b), surface water (Drollette et al., 2015; Lauer et al., 2016), and air quality contamination (Harriss et al., 2015; Hildenbrand et al., 2016b; Payne et al., 2017), there are also trepidations about the induction of seismic activity when managing the waste from UD (Wang et al., 2014). For example, it has been determined that there is a direct correlation between the number of earthquakes and the volume of produced water being injected into disposal wells (Hornbach et al., 2015, 2016; van der Elst et al., 2016). Factors that influence the occurrence of induced seismic events include variations in geology (e.g., existence of faults, pore pressures, capacity of the geologic formation, and rock type) and the volume of water injected into the SWD well (Gallegos et al., 2015). It is believed that induced seismic events are caused by increased fluid pressures on already stressed faults, triggering preexisting faults to slip (Walsh and Zoback, 2015). In 2008, the first confirmed UD-related earthquake was measured in the Bend Arch-Fort Worth Basin in northern Texas. Since then, earthquakes have been increasing in frequency and magnitude with the largest being M4.0 in 2015 (Hornbach et al., 2016). The same trend was found in Oklahoma, where regulators called for a 40% reduction in the volume of saltwater being injected into seismic-active areas (Jula, 2016). The dramatic reduction in disposal volumes was in response to the 900 anthropologically induced earthquakes in 2015 that were of a magnitude of 3.0 or greater, as compared to the average of 1.2 seismic events annually before 2009. As such, the greatest challenge for SWDs is now keeping up with the disposal demands in high production regions, while monitoring and modulating injections rates. Moving forward, it is expected that the disposal option will be subject to growing constraints due to the concerns regarding earthquakes increasing the need for treatment options that allow for reuse of produced water (Parker et al., 2014).

With respect to environmental stewardship and greater sustainability within the shale energy basins, further innovation is required to reduce localized groundwater usage and waste disposal volumes. In order to accomplish this, efficient and cost-effective treatment modalities are vital to facilitating the reuse of produced water for the production well stimulation (Hildenbrand et al., 2018). However, this is not without significant challenge. A study performed by Khan et al., suggested that produced water from oil-bearing shale may be one of the most complex aqueous mixtures identified to date (Khan et al., 2016). These fluids contain a complex mixture of inorganic and organic compounds used as additives (Camarillo et al., 2016; Elsner and Hoelzer, 2016; Hildenbrand et al., 2015), as well as compounds extracted from the shale itself including salts, metals, oil and gas compounds, natural organic matter, and naturally occurring radioactive material (NORM) (Abualfaraj et al., 2014; Chapman and Palisch, 2014; Engle and Rowan, 2014). Understanding the complete composition of these complex fluid mixtures is essential to evaluating the efficacy of additives, the effectiveness of different water treatment technologies, as well as the impacts of exposure due to leaks and spills (Luek and Gonsior, 2017) (Fig. 2). Unfortunately, due to the limited information that is currently available, there are knowledge gaps related to exposure, hazard data, and analytical methods (Mueller, 2017). The lack of information and standardized methods has left unanswered questions (Ferrer and Thurman, 2015; Luek and Gonsior, 2017; Mueller, 2017; Rogers et al., 2015). For example, what are the most pertinent constituents of concern? How do we comprehensively test for these constituents? What chemical reaction by-products are formed under subsurface conditions and how can they vary over the lifetime of UD wells? How to identify and evaluate potential exposure?

Further to this point, the Energy Policy Acts of 2005 left reporting requirements for chemical

1 INTRODUCTION

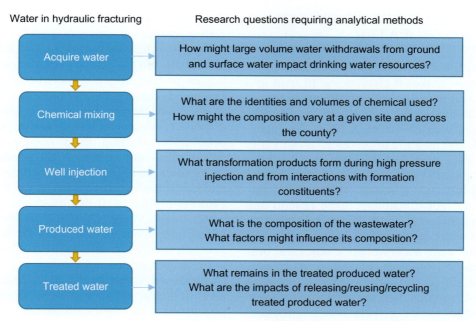

FIG. 2 Pertinent research questions held about the various stages if water procurement and waste management in UD.

additives used during production well stimulation up to the individual states (Konschnik and Dayalu, 2016). As of 2016, only 28 states had required companies to disclose the chemicals being used during UD, but the content of submission and methods of these disclosures still varies from state to state. Currently, Fracfocus.org contains the most comprehensive dataset on chemicals used in hydraulic fracturing; however, relatively little information is known about the additives used by production companies due to protected intellectual property and trade secrets. The protection of proprietary knowledge encourages innovation; however, it creates challenges in terms of determining the exposure safety, efficacy of treatment modalities, or the transformation products in produced water. In addition to the aspect of proprietary knowledge, concern can also arise from issues with nomenclature as many compounds generally have multiple names and/or incorrect CAS numbers during reporting (FracFocus, 2017). Currently, there are approximately 2500 different additives that could be found in wastewater from UD, and this does not account for chemicals from the formation or transformation by-products (Danforth, 2018). Of the known compounds, less than 20% have standard analytical methods or toxicity data.

This chapter addresses analytical methods required to comprehensively characterize produced water. The discussion will incorporate known contaminants found in produced water. These include chemical additives that are used during the stimulation process, such as alcohols used for surfactants, corrosion inhibitors, scale inhibitors, and biocides, as well as aliphatic and aromatic hydrocarbons (i.e., benzene, toluene, ethylbenzene, and xylenes (BTEX), and bacterial species that can reduce the permeability of production strata, residual oil and gas, and produce H_2S. Pertinent metals and minerals also need to be characterized. In addition, the chapter will touch on the outstanding concerns that currently remain unaddressed.

2 ANALYSIS OF PRODUCED WATER

The complex physicochemical nature of produced water makes it challenging to treat and analyze. Additionally, the composition can change over the lifetime of the well, which amplifies the challenges (Fakhru'l-Razi et al., 2009). The chemical profile and quantity of produced water varies for each shale play, well location, and depth, as well as the age of the reservoir and type of hydrocarbon produced (Ground Water Protection Council and All Consulting, 2009; Igunnu and Chen, 2014; Shih et al., 2016). When characterizing produced water, there are five primary categories to evaluate: basic water-quality metrics and bulk measurements, organics, biological, inorganics, and naturally occurring radioactive materials (NORM). Currently, there are not approved regulatory metrics for the management of produced water, and standardized analytical procedures specifically for characterizing produced water are not well established. Nonetheless, as interest continues to expand in this important field of research, so too have the applicable techniques.

2.1 Bulk Measurements and Basic Water-Quality Parameters

Bulk measurements do not allow for determination of specific chemicals or ions present; however, they are quick measurements that provide significant insight about the general state of the produced water matrix. These basic parameters can be used for process control with more detailed analyses being subsequently preformed if any notable fluctuations are observed. Typically, bulk measurements are more cost effective and time effective than advanced analytical techniques. The most informative bulk measurements with respect to the analysis of produced water include alkalinity, the relative abundance of dissolved and suspended solids, total organic carbon (TOC), and total nitrogen (TN).

Alkalinity is considered the buffering capacity of a solution or the ability to resist pH changes (Davis and Masten, 2014). In natural waters, it encompasses any weak bases such as hydrogen sulfate or hydrogen and dihydrogen phosphate, but the primary components are carbonate and hydrogen carbonate (bicarbonate). Produced waters generally have elevated levels of alkalinity largely due to the contribution of carbonate, which can induce scaling while also potentially affecting fluid stability and pH (Wasylishen and Fulton, 2012). Methods used to monitor alkalinity levels include titrations and colorimetric testing such as Environmental Protection Agency (EPA) methods 310.1 and 310.2, respectively (National Water Quality Monitoring Council, 2017; US Environmental Protection Agency, 2017a). EPA method 310.1 is preferable for the quantification of alkalinity in produced water, which uses titration with an endpoint of pH 4.5 and determination being performed electrometrically thereby eliminating interferences. On the contrary, colorimetric analysis using methyl orange, as called for by EPA method 310.2, can be problematic with produced water. Residual hydrocarbons, turbidity from solids, as well as discoloration from biological contaminants can leave produced water with a yellow hue that absorbs in the same range as methyl orange.

Solids, which also impact a solution's clarity and influences water quality, can be broken into groups based on the size of the particle. Large particles, which are typically greater than $1\mu M$, are classified as total suspended solids (TSS). These particles are representative of formation sands and clays, proppants, corrosion by-products, and precipitate crystals. Fine, solid particles may stabilize an emulsion if they are of correct size and abundance (Nalco Chemcial Comapny, 1988) leading to an increased turbidity or cloudiness of the solution. Constituents that contribute to turbidity are generally less than $1\mu M$ in size and can be present in the form of dissolved solids, colloids, and bacteria (Hildenbrand et al., 2018). Dissolved inorganic compounds or minerals are introduced into UD wastewater through contact with the formation or through the addition of clay stabilizers. The total dissolved solids (TDS) are primarily composed of charged particles, referred to as cations and anions. The salinity or concentration of inorganics can vary considerably across a given region. For example, produced water from the Barnett Shale has been found to have TDS levels ranging from 500 to 200,000 ppm (0.5%–20%), whereas produced water from the Marcellus Shale formation varies from 10,000 to 300,000 ppm (1%–30%) (Ground Water Protection Council and All Consulting, 2009). For the most accurate measurements of TSS and TDS, glass fiber, binder free filters with a nominal pore size of 1.5 μm are used to separate particles, and then, the weights of recovered materials are measured gravimetrically. Filter paper needs to be rinsed and dried before analysis, and a post wash of the filter paper is also required after separation is complete. Currently, the methods approved by the EPA Clean Water Act (CWA) for analysis of solids residue from domestic and industrial wastewater are Standard Method (SM) 2540 (A-F), ASTM International (ASTM) D5907-13, and USGS methods I-3750-85, I-1750-85, I-3765-85, and I-3753-85 (Fig. 3) (US Environmental Protection Agency, 2017a). While gravimetric analysis is preferred for solids, concentrations of TDS can be estimated by measuring the salinity or conductivity with probes, but this will not include dissolved constituents that do not have an electrical charge, such as neutral organics. As with TDS, TSS can be estimated from turbidity (Williamson and Crawford, 2011). Turbidity methods determine the amount of cloudiness in solution based on the transmission and scattering of light (The Editors of Encyclopedia Britannica, 2017). As a beam of light is passed through a turbid sample, the intensity is reduced by the scattering of light, which will depend on the concentrations and sizes of the particles in solution. Current analytical methods approved by the EPA's CWA to measure turbidity are EPA 180.1 Rev. 2.0, SM 2130 B-2011, ASTM D1889-00, and USGS I-3860-85, as shown in Fig. 3.

TOC and TN can also be classified as bulk measurements. TOC is an analytical technique that quantifies the concentration of organic molecules in solution. It provides a direct expression of TOC, in contrast to biological oxygen demand (BOD) or chemical oxygen demand (COD) methods, which are only proxy measurements for organic content, and TOC is a more convenient measurement to perform (ASTM International, 2018; Standard Methods Committee, 2011). To measure TOC, first the carbon is oxidized to form carbon dioxide, which is then measured (US Environmental Protection Agency, 2004). TOC can be classified into dissolved organic carbon (DOC) and suspended organic carbon, as well as purgeable and nonpurgeable organic carbon. The distinction between DOC and suspended organics is defined by the ability to separate the organic component from solution using a 0.45 μm filter. DOC will remain in solution while the suspended organic carbon will be removed from solution. Purgeable organic carbon, also referred to as volatile organic carbon, pertains to the fraction of TOC removed from an aqueous solution by gas stripping, whereas nonpurgeable organic carbon is the fraction of TOC, which remains in solution following gas stripping.

TN is a method that converts all nitrogen compounds, including nitrates, nitrites, ammonia, and organic nitrogen to NO, which is reacted with ozone to form excited NO_2 (ASTM International, 2016). As the NO_2 relaxes, it emits radiation that is measured photoelectrically. ASTM D8083 allows for the simultaneous analysis of TN for a sample if TOC is measured using ASTM D7573 (Fig. 3). While TOC and TN measurements do not provide the same level of specific information as other methods, they still provide valuable insight about water quality. They can be used to determine if mass balance has been achieved, based on singular or combined results using more specific and advanced analytical techniques.

Basic water-quality measurements, such as pH, conductivity, oxidation–reduction potential

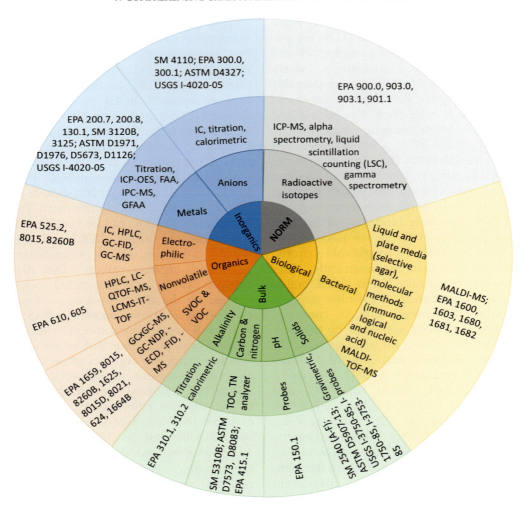

FIG. 3 Schematic of various analytical methods and tools that can be utilized to comprehensively characterize produced water (Ferrer and Thurman, 2015; Hildenbrand et al., 2018; Jo, 2001; Luek and Gonsior, 2017; Oetjen et al., 2017; US Environmental Protection Agency, 2017a).

(ORP), and temperature are also useful when beginning to characterize oilfield waste streams. For example, pH values can help evaluate the equilibrium state between bicarbonate scaling (elevated pH) compared to the potential of producing corrosive CO_2 (low pH). ORP values not only provide insight about the balance between metals and anions, but can also help judge the possible microbial content of the solution (Hildenbrand et al., 2018; Suslow, 2004). A higher reduction potential (more positive) is observed with species in solution that preferentially gain electrons. Conversely, species with a lower (more negative) reduction potential tend to lose electrons. From a microbial perspective, when cell membranes succumb to reduction by other species in solution, the integrity of the cell membrane is destroyed, leading to death. Collectively, these measurements are quick and can be performed in situ using a multimeric probe. Lastly, multimeric probes can also be used to collect multiple measurements simultaneously, including those that measure TDS, TSS, and turbidity.

2.2 Organic

Organic constituents found in produced water can be categorized into three groups based on solubility: insoluble, dispersed, and dissolved (Kajitvichyanukul et al., 2011). Insoluble organics or nonpolar hydrocarbons are the easiest to separate from produced water (Igunnu and Chen, 2014; National Energy Technology Laboratory, n.d.). These are the primary target for recovery of oil and gas operations. As the polarity of the molecules in solution increases, the organics are more soluble making them challenging to separate using conventional means such as gravity separation (Liden et al., 2017). These compounds represent dispersed and dissolved organics and are the most environmentally dangerous forms (Kajitvichyanukul et al., 2011). Diverse sample preparation and analytical techniques, including gas chromatography (GC) and liquid chromatography (LC), are required to identify and quantify individual organic compounds among the complex mixture of additives, oil and gas compounds, natural organic material, and transformation products (Ferrer and Thurman, 2015; Luek and Gonsior, 2017).

Higher TOC levels are indicative of increased organic content in water samples and will therefore, yield increased total GC and LC peak areas. However, mass balance between TOC and combined LC and GC techniques has been problematic for produced water. Several studies have worked to identify analytical methods, which are appropriate for quantifying organic compounds in produced water originating from diverse geological environments and industrial operations. However, as of yet, there lacks a comprehensive knowledge of all organic compounds and constituents within these fluids. For example, produced water from the Permian Basin was evaluated, and approximately 300 of the 1400 chemicals separated and detected were identified (Luek and Gonsior, 2017). Understanding these complex fluid mixtures is essential for understanding fluid treatment options and efficiencies for reuse in future UD, as well as exposure threats for the natural environment and humans.

To date, more than 200 oil and gas wastewaters have been analyzed, whereby GC has been the preferred analytical tool (Luek and Gonsior, 2017). According to one survey, 54% of studies reported to identify organics were based solely on the use of GC, while another 34% used LC only and 32% of the studies analyzed samples using both LC and GC for a more comprehensive characterization (Luek and Gonsior, 2017).

Analyses using GC generally target volatile and semivolatile organic compounds (VOC and SVOC) with the most frequently analyzed being BTEX, acetate, and acetone. The VOCs such as methanol and ethanol can be analyzed using headspace gas chromatography (HS-GC), thereby reducing the required sample preparation. Alcohols, which are used as solvents, surfactants, and corrosion inhibitors, are the dominant additives used in stimulation fluid (Elsner and Hoelzer, 2016; Konschnik and Dayalu, 2016; Luek and Gonsior, 2017). Overall, the four most frequently reported compounds used in production wells were methanol, isoparaffin, hydrochloric acid, and ethylene glycol (Konschnik and Dayalu, 2016). Alkylated phenols, commonly used as surfactants, have been analyzed using GCMS, following a DCM or ethyl acetate extraction (Orem et al., 2014; Thacker et al., 2015). Other surfactants routinely used in hydraulic fracturing, specifically ethoxylated compounds, have been analyzed using LC tandem-MS (MS/MS), following solid phase extraction (SPE) (Cluff et al., 2014; Thurman et al., 2014). SPE is based on passing the sample across a packed bed of sorbent, and it is useful for analyte extraction, concentration, and sample clean up. It is frequently used to extract semivolatile and nonvolatile analytes from a complex matrix. A recent analysis of produced water from the Cline Shale region revealed residual levels of surfactants, such as 2-butoxyethanol, which was analyzed in a targeted approach using GC–MS combined with an ethyl acetate extraction (Thacker et al., 2015). In the same study, N-alkyl-N,N-dimethylamine was identified using SPE and LC–MS/MS.

Hydrocarbons are the second most common additive reported by production companies which is frequently used as solvent (Konschnik and Dayalu, 2016). The determination of total petroleum hydrocarbons (TPH) can also be regarded as a bulk measurement (Carlton et al., 2017); however, it can provide more information than some of the other aforementioned bulk measurements. For example, TPH analyzed with EPA Method 1664B, as indicated in Fig. 3, can be used to determine the concentration of hydrocarbon species present in produced water; these will have originated either through their intentional addition to fluid pumped down-hole or from the formation, which the fluid contacted (Elsner and Hoelzer, 2016).

Polyethylene glycols (PEG) and polypropylene glycols (PPG) have been identified in produced water using LC–MS by some researchers (Heyob et al., 2017; Thurman et al., 2017). The nonionic composition of the aforementioned surfactants functions as weatherizers, emulsifiers, corrosion inhibitors, and wetting agents. Benzothiazole and other heterocyclic compounds have also been identified in wastewater from gas production using LC–MS (Orem et al., 2014).

One of the greatest challenges is a lack of standardized methods. With the continued expansion in UD, methods need to be developed for this type of wastewater (Oetjen et al., 2017). While standard methods for evaluating domestic and industrial wastewater have been used to analyze produced water, they have not been verified to perform as expected in produced water matrices, which can be up to four times saltier

than seawater. To date, the greatest opportunity for the elucidation of unidentified compounds in produced water is nontargeted analysis using LC (Luek and Gonsior, 2017).

2.3 Biological

So far, not much attention has been given to the microbial quality of produced water, particularly after treatment. Even so, recent studies have shown the levels of bacteria in produced water can vary with the shale and overall conditions used during hydraulic fracturing (Mouser et al., 2016). A diverse array of bacteria, including those within the taxa *γ-proteobacteria*, *α-proteobacteria*, *δ-proteobacteria*, *Clostridia*, *Synergistetes*, *Thermotogae*, *Spirochetes*, *Bacteroidetes*, and *Archae* have all been found in untreated flowback water samples (Murali Mohan et al., 2013b). Care must be taken to determine the types and levels of microorganisms present when evaluating the potential risk to human health and the reusability of this wastewater. Depending on the application, different standards and recommendation values for microbial levels are applicable (Table 1).

As shown in Table 1, currently, there are no standards with respect to acceptable levels of bacteria in the water used in hydraulic fracturing. However, it is known that the control of sulfate-reducing bacteria (SRB), iron-oxidizing bacteria (IOB), and acid-producing bacteria (APB) is of paramount importance. These microorganisms have previously been found in produced water (Cluff et al., 2014; Hildenbrand et al., 2018; Murali Mohan et al., 2013a; Struchtemeyer and Elshahed, 2012) and are known to form biofilms and produce acid that will lead to product "souring" and degradation. They can also cause corrosion of well casings, which can potentially lead to casing failure and unwanted connections between operational/waste fluids with the environment. Furthermore, some of the additives used in hydraulic fracturing provide a food source for

TABLE 1 Standards and maximum levels (colony-forming units (cfu)/100 mL) of microorganisms allowed in water according to its intended use (Office of Science and Technology et al., 2012; US Environmental Protection Agency, 2017b,c)

Bacteria (cfu/100 mL)	Desired water application				
	Drinking water	Agriculture	Recreation	Industry	Hydraulic fracturing
Total Coliforms	0	—	—	—	—
Fecal Coliforms	0	<10	—	—	—
E. coli	0	—	126	—	—
Enterococci	—	—	35	—	—
Heterotrophic plate count	<500	—	—	—	—
Legionella	0	—	—	—	—

bacteria that facilitates their proliferation (Daly et al., 2016). Therefore, the removal of microorganisms needs to be ensured since, if not well treated, the reuse of produced water can lead to their regrowth, and ultimately a negative impact on production well performance, or other intended uses of the fluid.

Different methods can be used to detect and quantify bacteria (National Research Council, 1999); however, the most common ones used in industry are plate methods where the sample is plated in different agar media and colonies are allowed to grow overnight. Identification is based on a colorimetric scheme and quantification of the number of colonies per sample volume used. Additionally, the colonies can be further isolated and characterized by microscopy, Gram staining, and biochemical tests. These methods are good for the identification of coliforms and heterotrophic bacteria. As it is known, only 1% of the microorganisms that occur in nature are able to grow under artificial conditions (Palková, 2004). Due to this reason, molecular methods have been developed to aid bacterial identification. The identification of unculturable and slow-growing microorganisms is possible, as their cultivation is not necessary and identification can be achieved in a timelier manner.

Molecular methods include immunological and nucleic acid-based techniques. Immunological methods rely on the use of antibodies, which specifically bind to bacterial antigens. An example is the enzyme-linked immunosorbent assay (ELISA), which uses a chromogenic reaction to detect the presence of a certain microorganism. In spite of its high-throughput capacity, this method does not provide good specificity since it requires the presence of large amounts of antigen and is unable to indicate the viability of the organisms (Peruski and Peruski, 2003).

Nucleic acid-based techniques, such as polymerase chain reaction (PCR), DNA/RNA sequencing, and others, can provide for high-throughput identification with a high degree of specificity. These methods use genetic sequences that are unique to each microorganism, usually conserved genes (housekeeping genes), for their identification. PCR amplifies the DNA or RNA and therefore allows for the detection of small amounts of genetic material. Nevertheless, this technique only works for microorganisms whose sequence is already known and is not ideal for the identification of unknown species. On the other hand, DNA/RNA sequencing provides for the identification of unknown microorganisms (Emerson et al., 2008). While viable and useful, these methods are expensive, and they do not provide any information on the metabolic state of the microorganism.

Recently, matrix-assisted laser desorption/ionization—time-of-flight—mass spectrometry (MALDI-TOF-MS) has been applied for the identification and differentiation of microorganisms based on their protein profile (Havlicek et al., 2013; Jo, 2001). This technique not only provides identification, it also delivers insight into the metabolic states of the detected cells. Beyond the initial capital cost, which is not trivial, subsequent analyses are cheaper and faster than nucleic acid techniques. Due to its detection limit and to ensure purity, the culturing of bacteria before analysis is required, and therefore, the appropriate culture media needs to be chosen. Preferably, broad-spectrum media should be used for the isolation of major bacterial groups. Of course, only culturable microorganisms can be analyzed by this technique, and this is certainly a drawback. However, the need for culturing is a common step for other conventional identification methods, as described before, and this step also ensures that only viable (i.e., live) organisms in the sample are identified.

Subsequent innovations should be emphasized for the sake of using MALDI-TOF-MS for the identification of unculturable bacteria and bacteria from environmental settings. Commercial databases used to support MALDI-TOF-MS identification of bacteria are clinically biased, given the desire to identify organisms associated with human infections. Despite this shortcoming, the applicability of this technique has also been demonstrated in the environmental field (Santos et al., 2016). Previous works show that MALDI-TOF-MS is a good option for bacterial identification and can be used as a tool to study environmental microbiomes (Martin et al., 2017; Santos et al., 2018). Identifications made by MALDI-TOF-MS were confirmed by 16S RNA biochemical sequencing methods. More recently, this method was used to characterize the microbiome of produced water and determine the efficiency of a wastewater treatment facility (Hildenbrand et al., 2018). A number of unique organophilic bacteria, such as *Pseudomonas*, *Paracoccus*, *Aeromonas*, and *Bacillus*, were identified. Importantly these bacteria exhibited resilience to traditional disinfection modalities.

2.4 Inorganic

The primary purpose of monitoring inorganic constituents in produced water is to preserve the life of equipment in the field and to maintain stable stimulation fluids. Conductivity provides a relative idea of the concentrations of dissolved solids in solution, but TDS measured using gravimetric analysis provides more accurate values of dissolved inorganic constituents. Sodium and chloride are the predominant ions in produced water, but, other cations and anions are of greater interest. For examples, barium, calcium, magnesium, and strontium are generally the largest contributors to problematic scale formation (Hu et al., 2016), which in itself can have devastating financial implications for any production well operator. Elevated levels of boron, iron, titanium, and zirconium can lead to over crosslinking, reduced gel viscosity, altered temperature stability, and inefficient proppant dispersion (Hildenbrand et al., 2018; Wasylishen and Fulton, 2012). Anions such as sulfate and carbonate can also be found in produced water, which readily interact with multivalent cations to form scale (Zielinski and Otton, 1999). Of these inorganic constituents, iron, magnesium, and calcium are the most pertinent given their affinity to interact with carbonate and sulfate minerals (Zielinski and Otton, 1999). Calcium is also a major contributor to hypersaline produced waters, in addition to boron, magnesium, iron, lithium, potassium, and manganese. Metals that have been detected in produced waters from the Permian Basin region include barium, beryllium, boron, calcium, cadmium, copper, iron, potassium, lithium, magnesium, manganese, and selenium (Khan et al., 2016).

Metals or cations can be measured individually or as group. Methods that focus on individual elements, such as EPA 130.1 and ASTM D1126, which is used to evaluate the hardness of water, traditionally prescribe the use of titrations or spectrophotometric methods (ASTM International, 2017a; US Environmental Protection Agency, 1971). Hardness, the potential to form scale, can be caused by any number of polyvalent cations; however, those other than calcium and magnesium are seldom present in more than trace amounts in produced water. Individual metal ions can also be evaluated using flame atomic absorption (FAA) or graphite furnace atomic absorption (GFAA) (Thermo-Fisher Scientific, 2017) using methods such as ASTM D1971 (ASTM International, 2018). The CWA approves various EPA methods for the analysis of metals (Fig. 3), such as Method 265.2 or 267.2, in domestic and industrial wastewater; in spite of that, there are only a handful of the metals approved using GFAA and FAA which do not include calcium, iron, or magnesium.

Inductively coupled plasma—optical emission spectroscopy (ICP-OES), also referred to as inductively coupled plasma—atomic emission spectroscopy (ICP-AES), allows scientists to confirm multiple elements at one time. These are prescribed for use in SMs 3120B, EPA 200.7, and ASTM D1976. ICP—mass spectrometry (ICP-MS) is also a technique that enables the simultaneous determination of multiple elements in a single analysis from a sample. ICP-MS is generally more sensitive than ICP-OES, and it can provide detection down to parts-per-trillion (ppt) concentration levels in a sample. Methods, such as EPA 200.8, SM 3125, ASTM 5673, and USGS I-4020-05 prescribe the use of ICP-MS (US Environmental Protection Agency, 2017a). Minimal sample preparation is required for these ICP-based elemental determination techniques, apart from filtration and/or dilution of the sample, and the addition of acids, prior to analysis (Ferrer and Thurman, 2015).

The high temperature of an ICP torch alleviates many interferences, which can hamper other techniques like GFAA and FAA.

For the analysis of pertinent anions in produced water (i.e., chloride, bromide, fluoride, nitrate, sulfate, and sulfide), ion chromatography is the preferred analytical technique (Ferrer and Thurman, 2015). Several SMs for the analysis of anions have been established for domestic and industrial waste, including SM 4110, EPA 300.0 and 300.1, as well as ASTM D4327 (ASTM International, 2017b; Standard Methods Committee, 2000; US Environmental Protection Agency, 2017a). Samples can potentially require drastic dilution due to the high concentrations of anions, particularly chloride; otherwise, there is an increased risk for inaccurate results. In one study, the PW samples from the Niobrara shale formation were diluted 30 times to bring the chloride concentration below 500 mg/L in order to accurately determine the anion concentrations (Coday et al., 2015). Therefore, when using IC to analyze PW with TDS levels up to 300,000 ppm and chloride accounting for 50% of the ions, samples should be diluted 300 times.

2.5 Naturally Occurring Radioactive Material (NORM)

Oilfield waters that are particularly rich in chloride enhance the solubility of other elements, including naturally occurring radioactive material (NORM) (Zielinski and Otton, 1999). The health concerns associated with radioactive material being brought up to the surface with produced water were first realized in the 1980s when unacceptable radiation levels were detected by scrap metal dealers (Zielinski and Otton, 1999). The most abundant NORM constituents in produced water are radium isotopes, ^{226}Ra and ^{228}Ra. However, shale and sandstone formations can also contain uranium, thallium, and thorium (Guerra et al., 2011; US Environmental Protection Agency, 2016b;

Zielinski and Otton, 1999). Radium accumulates on oilfield equipment through coprecipitation with scale deposits, which coat equipment (Akob et al., 2015; Guerra et al., 2011; Igunnu and Chen, 2014). The mineral barite ($BaSO_4$) is the most likely host of radium in the subsurface formation (Zielinski and Otton, 1999). Environments which host SRB, such as produced water, increase the solubility of barium sulfate, which can potentially liberate the radium, which was previously encapsulated (Akob et al., 2015). The half-lives of the two principal radioactive isotopes in produced water, ^{226}Ra and ^{228}Ra, are 1600 and 5.75 years, respectively. Approximately 10 half-lives are needed for the radioactivity to decay to reach negligible quantities (Rowan et al., 2011). When radium is brought to the surface, prolonged exposure becomes the primary concern due to the long half-lives of these species.

Rapid and robust methods to evaluate levels of NORM and thus the associated safety of personally handling that produced water are critical (Zhang et al., 2015). ^{226}Ra primarily emits alpha particles, which can be measured directly (Zhang et al., 2015). Conversely, ^{228}Ra levels are determined indirectly by measuring the beta-decay process of ^{228}Ac, which is a product of ^{228}Ra (Johnson, 1971; US Environmental Protection Agency, 2014). However, long counting times (24–48 h) are required for traditional methods, thereby limiting sample throughput (Rowan et al., 2011). As indicated in Fig. 3, there are currently three SMs for evaluating radioactive material in domestic and industrial wastewater (EPA Method 900.0, 903.0, and 903.1) (US Environmental Protection Agency, 2017a). Using the techniques described in these methods, alpha- and beta-particles can be identified using alpha spectrometry or liquid scintillation counting (LSC), respectively, to quantify emission rates. Gamma spectrometry is used to quantify the emission rates of γ rays; however, the EPA only approves method 901.1 for the analysis of drinking water, and not other water-based samples (US Environmental Protection Agency,

1980). One study successfully used ICP-MS for the rapid detection of ^{226}Ra. The result of which were compared to those obtained using gamma spectroscopy (Zhang et al., 2015). The study indicated that rapid ^{226}Ra analysis using ICP-MS is not only feasible, but is also incredibly robust with recoveries near 100% from produced water samples.

3 CONCLUSIONS

Unconventional oil and gas development is on the rise, as evidenced by the 1200 new wells drilled in the Permian Basin during 2017 (US Energy Information Administration, 2018). As this field expands, there is a need for increased environmental stewardship. With regard to handling the waste generated during unconventional drilling, a well-developed understanding of its biogeochemical composition is necessary. Little toxicity information is known about the 2500 additives used during well development and stimulation. The knowledge gaps have been perpetuated by the lack of standard analytical methods. Closing knowledge gaps can yield valuable insight into environmental and human exposure concerns, especially in the event of spills, leaks, and other contamination events.

Additionally, evidence indicates that the disposal of the wastewater into SWD wells has triggered seismicity in some areas (Gallegos et al., 2015; Hornbach et al., 2016). For example, from 2000 to 2011, there was an increase from approximately 21 seismic events to 134 events per year that measured greater than M3.0 (magnitude) on the Richter scale. The USGS concluded that the increase in the earthquakes is of anthropogenic origin and likely attributable to unconventional development (Wang et al., 2014). As such, recycling and reusing produced water is a relatively new operational paradigm shift, which is occurring. This will not only lead to reduced disposal volumes, it will also decrease the demand for groundwater usage.

The practice of treating produced water requires the effectiveness of prospective treatment technologies to be evaluated. However, currently there are no approved methods for the analysis of produced water, so scientists have primarily used methods approved for the analysis of domestic and industrial wastewater, or they have developed their own. As such, verifying the accuracy of these methods for produced water, especially where high salinity levels can potentially lead to analytical interferences, is of paramount importance. Moreover, the development of new, standardized, and robust methods with proper validation is necessary. Additionally, future research should encompass multiple analytical techniques, including GC, LC, MALDI, and ICP to provide a compressive characterization of produced water.

References

Abualfaraj, N., Gurian, P.L., Olson, M.S., 2014. Characterization of marcellus shale flowback water. Environ. Eng. Sci. 31, 514–524. https://doi.org/10.1089/ees.2014.0001.

Akob, D.M., Cozzarelli, I.M., Dunlap, D.S., Rowan, E.L., Lorah, M.M., 2015. Organic and inorganic composition and microbiology of produced waters from Pennsylvania shale gas wells. Appl. Geochem. 60, 116–125. https://doi.org/10.1016/j.apgeochem.2015.04.011.

Alessi, D.S., Zolfaghari, A., Kletke, S., Gehman, J., Allen, D.M., Goss, G.G., 2017. Comparative analysis of hydraulic fracturing wastewater practices in unconventional shale development: water sourcing, treatment, and disposal practices. Can. Water Res. J. 42, 105–121. https://doi.org/10.1080/07011784.2016.1238782.

Arthur, D.J., Bohm, B., Coughlin, B.J., Layne, M., Cornue, D., 2008. Evaluating the environmental implications of hydraulic fracturing in shale gas reservoirs. In: SPE Americas E&P Environmental and Safety Conference, Society of Petroleum Engineers (SPE), San Antonio, pp. 1–21. https://doi.org/10.2118/121038-MS.

ASTM International, 2016. ASTM D8083-16, Standard Test Method for Total Nitrogen, and Total Kjeldahl Nitrogen (TKN) by Calculation, In Water by High Temperature Catalytic Combustion and Chemiluminescence Detection. ASTM International, West Conshohocken, PA. www.astm.org.

ASTM International, 2017a. ASTM D1126-17. Standard Test Method for Hardness in Water. ASTM International, West Conshohocken, PA. www.astm.org.

ASTM International, 2017b. ASTM D4327-17. Standard Test Method for Anions in Water by Suppressed Ion Chromatography. ASTM International, West Conshohocken, PA. www.astm.org.

ASTM International, 2018. ASTM D7573-18ae1. Standard Test Method for Total Carbon and Organic Carbon in Water by High Temperature Catalytic Combustion and Infrared Detection. ASTM International, West Conshohocken, PA. www.astm.org.

Backstrom, J., 2018. Groundwater Regulations and Hydraulic Fracturing: Reporting Water Use in the Permian. Texas A&M, College Station, TX.

Barclays, Columbia Water Center, 2017. The Water Challenge: Preserving a Global Resource. Barclays Bank PLC

Camarillo, M.K., Domen, J.K., Stringfellow, W.T., 2016. Physical-chemical evaluation of hydraulic fracturing chemicals in the context of produced water treatment. J. Environ. Manag. 183, 164–174. https://doi.org/10.1016/j.jenvman.2016.08.065.

Carlton Jr., D.D., Hildenbrand, Z.L., Schug, K.A., 2017. Analytical approaches for high-resolution environmental investigation of unconventional oil and gas exploration. In: Hildenbrand, Z.L., Schug, K.A. (Eds.), Environmental Issues Concerning Hydraulic Fracturing. vol. 1.APMP. Academic Press, UK, pp. 193–226.

Chapman, M., Palisch, T., 2014. Fracture conductivity—design considerations and benefits in unconventional reservoirs. J. Pet. Sci. Eng. 124, 407–415.

Clark, C.E., Veil, J.A., 2009. Produced Water Volumes and Management Practices in the United States. Argonne National Laboratory.https://doi.org/10.2172/1007397.

Cluff, M.A., Hartsock, A., Macrae, J.D., Carter, K., Mouser, P.J., 2014. Temporal changes in microbial ecology and geochemistry in produced water from hydraulically fractured Marcellus shale gas wells. Environ. Sci. Technol. 48, 6508–6517. https://doi.org/10.1021/es501173p.

Coday, B.D., Almaraz, N., Cath, T.Y., 2015. Forward osmosis desalination of oil and gas wastewater: impacts of membrane selection and operating conditions on process performance. J. Membr. Sci. 488, 40–55. https://doi.org/10.1016/j.memsci.2015.03.059.

Daly, R.A., Borton, M.A., Wilkins, M.J., Hoyt, D.W., Kountz, D.J., Wolfe, R.A., Welch, S.A., Marcus, D.N., Trexler, R.V., MacRae, J.D., Krzycki, J.A., Cole, D.R., Mouser, P.J., Wrighton, K.C., 2016. Microbial metabolisms in a 2.5-km-deep ecosystem created by hydraulic fracturing in shales. Nat. Microbiol. 1, 16146.

Danforth, C., 2018. Could Wastewater From Oil and Gas Production Help Solve Our Water Crisis? Not Without Better Science. WWW Document, Environ. Def. Fund Energy Exch.

Darrah, T.H., Vengosh, A., Jackson, R.B., Warner, N.R., Poreda, R.J., 2014. Noble gases identify the mechanisms of fugitive gas contamination in drinking-water wells overlying the marcellus and barnett shales. Proc. Natl. Acad. Sci. 111, 14076–14081. https://doi.org/10.1073/pnas.1322107111.

Davis, M.L., Masten, S.J., 2014. Principles of Environmental Engineering and Science, third ed. McGraw-Hill, New York.

Drollette, B.D., Hoelzer, K., Warner, N.R., Darrah, T.H., Karatum, O., O'Connor, M.P., Nelson, R.K., Fernandez, L.A., Reddy, C.M., Vengosh, A., Jackson, R.B., Elsner, M., Plata, D.L., 2015. Elevated levels of diesel range organic compounds in groundwater near marcellus gas operations are derived from surface activities. Proc. Natl. Acad. Sci. U.S.A. 112, 13184–13189. https://doi.org/10.1073/pnas.1511474112.

Elsner, M., Hoelzer, K., 2016. Quantitative survey and structural classification of hydraulic fracturing chemicals reported in unconventional gas production. Environ. Sci. Technol. 50, 3290–3314. https://doi.org/10.1021/acs.est.5b02818.

Emerson, D., Agulto, L., Liu, H., Liu, L., 2008. Identifying and characterizing bacteria in an era of genomics and proteomics. Bioscience 58, 925–936.

Engle, M.A., Rowan, E.L., 2014. Geochemical evolution of produced waters from hydraulic fracturing of the marcellus shale, northern appalachian basin: a multivariate compositional data analysis approach. Int. J. Coal Geol. 126, 45–56. https://doi.org/10.1016/j.coal.2013.11.010.

Ewing, B.T., Waston, M.C., McInturff, T., McInturff, R., 2014. In: Uddameri, V., Morse, A., Tindle, K.J. (Eds.), Economic impact of the permian basin oil and gas industry. CRC Press, Boca Raton, pp. 1–80.

Fakhru'l-Razi, A., Pendashteh, A., Abdullah, L.C., Biak, D.R.A., Madaeni, S.S., Abidin, Z.Z., 2009. Review of technologies for oil and gas produced water treatment. J. Hazard. Mater. 170, 530–551. https://doi.org/10.1016/j.jhazmat.2009.05.044.

Ferrer, I., Thurman, M.E., 2015. Chemical constituents and analytical approaches for hydraulic fracturing waters. Trends Environ. Anal. Chem. 5, 18–25. https://doi.org/10.1016/j.teac.2015.01.003.

Fontenot, B.E., Hunt, L.R., Hildenbrand, Z.L., Carlton Jr., D.D., Oka, H., Walton, J.L., Hopkins, D., Osorio, A., Bjorndal, B., Hu, Q.H., Schug, K.A., 2013. An evaluation of water quality in private drinking water wells near natural gas extraction sits in the barnett shale formation. Environ. Sci. Technol. 47, 10032–10040. https://doi.org/10.1021/es4011724.

FracFocus, 2017. What Chemicals Are Used. WWW Document. https://fracfocus.org/chemical-use/what-chemicals-are-used.

Gallegos, T.J., Varela, B.A., Haines, S.S., Engle, M.A., 2015. Hydraulic fracturing water use variability in the United States and potential environmental implications. Water Resour. Res. 51, 5839–5845. https://doi.org/10.1002/2015WR017278.Received.

Ground Water Protection Council, All Consulting, 2009. Modern Shale Gas—A Primer. US Department of Energy

Guerra, K., Dahm, K., Dundorf, S., 2011. Oil and Gas Produced Water Management and Beneficial Use in the Western United States. US Department of the Interior Bureau of Reclamation.

Harriss, R., Alvarez, R.A., Lyon, D., Zavala-Araiza, D., Nelson, D., Hamburg, S.P., 2015. Using multi-scale measurements to improve methane emission estimates from oil and gas operations in the barnett shale region, Texas. Environ. Sci. Technol. 49, 7524–7526. https://doi.org/10.1021/acs.est.5b02305.

Havlicek, V., Lemr, K., Schug, K.A., 2013. Current trends in microbial diagnostics based on mass spectrometry. Anal. Chem. 85, 790–797. https://doi.org/10.1021/ac3031866.

Heyob, K.M., Blotevogel, J., Brooker, M., Evans, M.V., Lenhart, J.J., Wright, J., Lamendella, R., Borch, T., Mouser, P.J., 2017. Natural attenuation of nonionic surfactants used in hydraulic fracturing fluids: degradation rates, pathways, and mechanisms. Environ. Sci. Technol. 51, 13985–13994, AcsEst.7b01539. https://doi.org/10.1021/acs.est.7b01539.

Hildenbrand, Z.L., Carlton Jr., D.D., Fontenot, B.E., Meik, J.M., Walton, J.L., Taylor, J.T., Thacker, J.B., Korlie, S., Shelor, C.P., Henderson, D., Kadjo, A.F., Roelke, C.E., Hudak, P.F., Burton, T., Rifai, H.S., Schug, K.A., 2015. A comprehensive analysis of groundwater quality in the barnett shale region. Environ. Sci. Technol. 49, 8254–8262. https://doi.org/10.1021/acs.est.5b01526.

Hildenbrand, Z.L., Carlton Jr., D.D., Fontenot, B.E., Meik, J.M., Walton, J.L., Thacker, J.B., Korlie, S., Shelor, C.P., Kadjo, A.F., Clark, A., Usenko, S., Hamilton, J.S., Mach, P.M., Verbeck, G.F., Hudak, P., Schug, K.A., 2016a. Temporal variation in groundwater quality in the permian basin of texas, a region of increasing unconventional oil and gas development. Sci. Total Environ. 562, 906–913. https://doi.org/10.1016/j.scitotenv.2016.04.144.

Hildenbrand, Z.L., Mach, P.M., McBride, E.M., Dorreyatim, M.N., Taylor, J.T., Carlton Jr., D.D., Meik, J.M., Fontenot, B.E., Wright, K.C., Schug, K.A., Verbeck, G.F., 2016b. Point source attribution of ambient contamination events near unconventional oil and gas development. Sci. Total Environ. 573, 382–388. https://doi.org/10.1016/j.scitotenv.2016.08.118.

Hildenbrand, Z.L., Santos, I.C., Liden, T., Carlton Jr., D.D., Varona-Torres, E., Martin, M.S., Reyes, M.L., Mulla, S.R., Schug, K.A., 2018. Characterizing variable biogeochemical changes during the treatment of produced oilfield waste. Sci. Total Environ. 634, 1519–1529. https://doi.org/10.1016/j.scitotenv.2018.03.388.

Hornbach, M.J., DeShon, H.R., Ellsworth, W.L., Stump, B.W., Hayward, C., Frohlich, C., Oldham, H.R., Olson, J.E., Magnani, M.B., Brokaw, C., Luetgert, J.H., 2015. Causal factors for seismicity near Azle, Texas. Nat. Commun. 6, 1–11. https://doi.org/10.1038/ncomms7728.

Hornbach, M.J., Jones, M., Scales, M., DeShon, H.R., Magnani, M.B., Frohlich, C., Stump, B., Hayward, C., Layton, M., 2016. Ellenburger wastewater injection and seismicity in North Texas. Phys. Earth Planet. Inter. 261, 54–68. https://doi.org/10.1016/j.pepi.2016.06.012.

Hu, Y., Mackay, E., Ishkov, O., Strachan, A., 2016. Predicted and observed evolution of produced-brine compositions and implications for scale management. SPE Prod. Oper. 31, 270–279. https://doi.org/10.2118/169765-PA.

Igunnu, E.T., Chen, G.Z., 2014. Produced water treatment technologies. Int. J. Low Carbon Technol. 9, 157–177. https://doi.org/10.1093/ijlct/cts049.

Jo Jr., L., 2001. MALDI-TOF mass spectrometry of bacteria. Mass Spectrom. Rev. 20, 172–194. https://doi.org/10.1002/mas.10003.

Johnson, J., 1971. Determination of Radium-228 in Natural Waters. United States Geological Survey.

Jula, M., 2016. In Oklahoma, injecting less saltwater lowers quake rates. Sci. Magzine, American Association of Advancement of Science. 2016–2019.

Kajitvichyanukul, P., Hung, Y.-T., Wang, L., 2011. Membrane technologies for oil-water separation. In: Chen, J.P., Hung, T.-T., Shammas, N. (Eds.), Wang, L. Humana Press, Membrane and Desalination Technologies, pp. 661–690.

Khan, N.A., Engle, M.A., Dungan, B., Holguin, F.O., Xu, P., Carroll, K.C., 2016. Volatile-organic molecular characterization of shale-oil produced water from the Permian basin. Chemosphere 148, 126–136. https://doi.org/10.1016/j.chemosphere.2015.12.116.

Konschnik, K., Dayalu, A., 2016. Hydraulic fracturing chemicals reporting: analysis of available data and recommendations for policymakers. Energy Policy 88, 504–514. https://doi.org/10.1016/j.enpol.2015.11.002.

Lauer, N.E., Harkness, J.S., Vengosh, A., 2016. Brine spills associated with unconventional oil development in North Dakota. Environ. Sci. Technol. 50, 5389–5397. https://doi.org/10.1021/acs.est.5b06349.

Liden, T., Clark, B.G., Hildenbrand, Z.L., Schug, K.A., 2017. Unconventional oil and gas production: waste management and the water cycle. In: Schug, K.A., Hildenbrand, Z.L. (Eds.), Environmental Issues Concerning Hydraulic Fracturing. vol. 1.APMP. Academic Press, UK, pp. 17–45. https://doi.org/10.1016/bs.apmp.2017.08.012.

Luek, J.L., Gonsior, M., 2017. Organic compounds in hydraulic fracturing fluids and wastewaters: a review. Water Res. 123, 536–548. https://doi.org/10.1016/j.watres.2017.07.012.

Martin, M.S., Santos, I.C., Carlton Jr., D.D., Stigler-Granados, P., Hildenbrand, Z.L., Schug, K.A., 2017. Characterization of bacterial diversity in contaminated groundwater using matrix-assisted laser desorption/ionization time-of-flight mass spectrometry. Sci. Total Environ. 622–623, 1–10. https://doi.org/10.1016/j.scitotenv.2017.10.027.

Mouser, P.J., Borton, M., Darrah, T.H., Hartsock, A., Wrighton, K.C., 2016. Hydraulic fracturing offers view of microbial life in the deep terrestrial subsurface. FEMS Microb. Ecol. 92. fiw166-fiw166.

Mueller, D., 2017. Minding the Knowledge Gap. World Water Water Reuse Desalin, pp. 20–21.

Murali Mohan, A., Hartsock, A., Bibby, K.J., Hammack, R.W., Vidic, R.D., Gregory, K.B., 2013a. Microbial community changes in hydraulic fracturing fluids and produced water from shale gas extraction. Environ. Sci. Technol. 47, 13141–13150. https://doi.org/10.1021/es402928b.

Murali Mohan, A., Hartsock, A., Hammack, R.W., Vidic, R.D., Gregory, K.B., 2013b. Microbial communities in flowback water impoundments from hydraulic fracturing for recovery of shale gas. FEMS Microbiol. Ecol. 86, 567–580. https://doi.org/10.1111/1574-6941.12183.

Nalco Chemcial Comapny, 1988. Emulsion breaking. In: Kemmer, F.N. (Ed.), The NALCO Water Handbook. McGraw-Hill Book Co., pp. 11.1–11.18.

National Energy Technology Laboratory, n.d. Produced Water Management Technologies Descriptions [WWW Document], US Department of Energy. URL https://www.netl.doe.gov/research/coal/crosscutting/pwmis/tech-desc (accessed 1.10.17).

National Research Council, 1999. Identifying Future Drinking Water Contaminants. The National Academies Press, Washington, DC. https://doi.org/10.17226/9595.

National Water Quality Monitoring Council, 2017. Alkalinity by Titration. WWW Document Natl. Environ. Methos Index. URL. https://acwi.gov/monitoring/.

Oetjen, K., Giddings, C.G.S., McLaughlin, M., Nell, M., Blotevogel, J., Helbling, D.E., Mueller, D., Higgins, C.P., 2017. Emerging analytical methods for the characterization and quantification of organic contaminants in flowback and produced water. Trends Environ. Anal. Chem. 15, 12–23. https://doi.org/10.1016/j.teac.2017.07.002.

Office of Science and Technology, Health and Ecological, Criteria Division, U.S. Environmental Protection Agency, 2012. Recreational Water Quality Criteria.

Orem, W., Tatu, C., Varonka, M., Lerch, H., Bates, A., Engle, M.A., Crosby, L., McIntosh, J., 2014. Organic substances in produced and formation water from unconventional natural gas extraction in coal and shale. Int. J. Coal Geol. 126, 20–31. https://doi.org/10.1016/j.coal.2014.01.003.

Palková, Z., 2004. Multicellular microorganisms: laboratory versus nature. EMBO Rep. 5, 470–476. https://doi.org/10.1038/sj.embor.7400145.

Parker, K.M., Zeng, T., Harkness, J., Vengosh, A., Mitch, W.A., 2014. Enhanced formation of disinfection byproducts in shale gas wastewater-impacted drinking water supplies. Environ. Sci. Technol. 48, 11161–11169. https://doi.org/10.1021/es5028184.

Payne, B.F., Ackley, R., Wicker, A.P., Hildenbrand, Z.L., Carlton Jr., D.D., Schug, K.A., 2017. Characterization of methane plumes downwind of natural gas compressor stations in Pennsylvania and New York. Sci. Total Environ. 580, 1214–1221.

Permian Basin Information [WWW Document], 2017. Railroad Commission of Texas. URL http://www.rrc.state.tx.us/oil-gas/major-oil-gas-formations/permian-basin/ (accessed 1.1.17).

Perrin, J., Cook, T., 2016. Hydraulically Fractured Wells Provide Two-Thirds of U.S. Natural Gas Production. US Energy Information Administration.

Peruski, A.H., Peruski, L.F.J., 2003. Immunological methods for detection and identification of infectious diseases and biological warfare agents. Clin. Diagn. Lab. Immunol. 10, 506–513. https://doi.org/10.1128/CDLI.10.4.506-513.2003.

Rogers, J.D., Burke, T.L., Osborn, S.G., Ryan, J.N., 2015. A framework for identifying organic compounds of concern in hydraulic fracturing fluids based on their mobility and persistence in groundwater. Environ. Sci. Technol. Lett. 2, 158–164. https://doi.org/10.1021/acs.estlett.5b00090.

Rowan, E.L., Engle, M.A., Kirby, C.S., Kraemer, T.F., 2011. Radium Content of Oil- and Gas-Field Produced Waters in the Northern Appalachian Basin (USA): Summary and Discussion of Data. USGS Scientific Investigations Report.

Santos, I.C., Hildenbrand, Z.L., Schug, K.A., 2016. Applications of MALDI-TOF MS in environmental microbiology. Analyst 141, 2827–2837. https://doi.org/10.1039/c6an00131a.

Santos, I.C., Martin, M.S., Reyes, M.L., Carlton Jr., D.D., Stigler-Granados, P., Valerio, M.A., Whitworth, K.W., Hildenbrand, Z.L., Schug, K.A., 2018. Exploring the links between groundwater quality and bacterial communities near oil and gas extraction activities. Sci. Total Environ. 618, 165–173. https://doi.org/10.1016/j.scitotenv.2017.10.264.

Scanlon, B.R., Reedy, R.C., Male, F., Walsh, M., 2017. Water issues related to transitioning from conventional to unconventional oil production in the permian basin. Environ. Sci. Technol. 51, 10903–10912. https://doi.org/10.1021/acs.est.7b02185.

Shaffer, D., Chavez, L.H.A., Ben-Sasson, M., Castrillo, S.R., Yin Yip, N., Elimelech, M., 2013. Desalination and reuse of high-salinity shale gas produced water: drivers, technologies, and future directions. Environ. Sci. Technol. 47, 9569–9583. https://doi.org/10.1021/es401966e.

Shih, J., Swiedler, E., Krupnick, A., 2016. A Model for Shale Gas Wastewater Management. Resources for the Future, pp. 1–68.

Standard Methods Committee, 2000. Standard Method 4110 Determination of Anions by Ion Chromatography. American Water Works Association.

Standard Methods Committee, 2011. 5310 Total Organic Carbon. American Water Works Association.

Struchtemeyer, C.G., Elshahed, M.S., 2012. Bacterial communities associated with hydraulic fracturing fluids in thermogenic natural gas wells in north central Texas, USA. FEMS Microb. Ecol. 81, 13–25. https://doi.org/10.1111/j.1574-6941.2011.01196.x.

Suslow, T.V., 2004. Oxidation-Reduction Potential (ORP) for Water Disinfection Monitoring, Control, and Documentation. ANR Publ. 8149, pp. 1–5.

Thacker, J.B., Carlton Jr., D.D., Hildenbrand, Z.L., Kadjo, A., Schug, K.A., 2015. Chemical analysis of wastewater from unconventional drilling operations. Water 7, 1568–1579. https://doi.org/10.3390/w7041568.

The Editors of Encyclopedia Britannica, 2017. Nephelometry and Turbidimetry. Encycl. Br.

Thermo-Fisher Scientific, 2017. Metal contaminants. WWW Document. In: Environmental Contaminant Analysis. Thermo-Fisher Scientific. https://www.thermofisher.com/us/en/home/industrial/environmental/environmental-contaminant-analysis/metal-contaminants.html.

Thurman, M.E., Ferrer, I., Blotevogel, J., Borch, T., 2014. Analysis of hydraulic fracturing flowback and produced waters using accurate mass: identification of ethoxylated surfactants. Anal. Chem. 86, 9653–9661. https://doi.org/10.1021/ac502163k.

Thurman, M.E., Ferrer, I., Rosenblum, J., Linden, K., Ryan, J.N., 2017. Identification of polypropylene glycols and polyethylene glycol carboxylates in flowback and produced water from hydraulic fracturing. J. Hazard. Mater. 323, 11–17. https://doi.org/10.1016/j.jhazmat.2016.02.041.

U.S Environmental Protection Agency, 2014. Method 9320. US EPA.

U.S. Department of Energy, 2015. Advancing Systems and Technologies to Produce Cleaner Fuels Technology Assessments: Unconventional Oil and Gas, Quadrennial Technology Review 2015. United State Department of Energy.

U.S. Energy Information Administration, 2017. Drilling Productivity Report.

U.S. Energy Information Administration, 2018. Drilling Productivity Report.

U.S. Environmental Protection Agency, 1971. Method 130.1: Hardness, Total (mg/L as CaCO3) (Colorimetric, Automated EDTA) by Spectrophotometer. US EPA.

U.S. Environmental Protection Agency, 1980. Gamma Emitting Radionuclides in Drinking Water Method 901.1. US EPA.

U.S. Environmental Protection Agency, 2004. Method 415.3—Measurement of Total Organic Carbon, Dissolved

REFERENCES

Organic Carbon and Specific UV Absorbance at 254 nm in Source Water and Drinking Water. US EPA.

U.S. Environmental Protection Agency, 2016a. Class II Oil and Gas Related Injection Wells. [WWW Document]. URL. https://www.epa.gov/uic/class-ii-oil-and-gas-related-injection-wells (accessed 1.1.16).

U.S. Environmental Protection Agency, 2016b. Hydraulic fracturing for oil and gas: impacts from the hydraulic fracturing water cycle on drinking water resources in the United States (main report - EPA/600/R-16/236fa). In: EPA's Study of Hydraulic Fracturing and its Potential Impact on Drinking Water Resources.

U.S. Environmental Protection Agency, 2017a. Clean Water Act Methods.

U.S. Environmental Protection Agency, 2017b. Clean Water Act Section 404 and Agriculture. WWW Document, Clean Water Act. URL. https://www.epa.gov/cwa-404/clean-water-act-section-404-and-agriculture.

U.S. Environmental Protection Agency, 2017c. National primary drinking water regulations. WWW Document In: Ground Water and Drinking Water. URL. https://www.epa.gov/ground-water-and-drinking-water/national-primary-drinking-water-regulations.

van der Elst, N.J., Page, M.T., Weiser, D.A., Goebel, T.H.W., Hosseini, S.M., 2016. Induced earthquake magnitudes are as large as (statistically) expected. J. Geophys. Res. Solid Earth 121, 4575–4590 https://doi.org/10.1002/2016JB012818.

Veil, J., 2015. U. S. Produced Water Volumes and Management Practices in 2012. Ground Water Protection Council.

Walsh, F.R., Zoback, M.D., 2015. Oklahoma's recent earthquakes and saltwater disposal. Sci. Adv. 1, 1–9. https://doi.org/10.1126/sciadv.1500195.

Wang, Q., Chen, X., Jha, A.N., Rogers, H., 2014. Natural gas from shale formation- the evolution, evidence, and challenges of shale gas revolution in United States. Renew. Sust. Energ. Rev. 30, 1–28. https://doi.org/10.1016/j.rser.2013.08.065.

Wasylishen, R., Fulton, S., 2012. Reuse of Flowback & Produced Water for Hydraulic Fracturing in Tight Oil. The Petroleum Technology Alliance Canada, Calgary.

Williamson, T., Crawford, C., 2011. Estimation of suspended-sediment concentration from Total suspended solids and turbidity data for Kentucky, 1978-19951. JAWRA J. Am. Water Resour. Assoc. 47, 739–749. https://doi.org/10.1111/j.1752-1688.2011.00538.x.

Zhang, T., Bain, D., Hammack, R., Vidic, R.D., 2015. Analysis of radium-226 in high salinity wastewater from unconventional gas extraction by inductively coupled plasma-mass spectrometry. Environ. Sci. Technol. 49, 2969–2976. https://doi.org/10.1021/es504656q.

Zielinski, R.A., Otton, J.K., 1999. Naturally Occurring Radioactive Materials (NORM) in Produced Water and Oil-Field Equipment—An Issue for the Energy Industry. United States Geological Survey.

CHAPTER

10

Innovations in Monitoring With Water-Quality Sensors With Case Studies on Floods, Hurricanes, and Harmful Algal Blooms

Donna N. Myers*

U.S. Geological Survey, Scientist Emeritus, Denver, CO, United States
*Corresponding author: E-mail: dnmyers@usgs.gov

1 INTRODUCTION

Water-quality sensors serve an important role in monitoring, assessment, and research by describing high-temporal frequency changes at the sub-daily scale in basic properties and chemical characteristics of water such as salinity, dissolved oxygen (DO), nutrients, algal pigments, toxicants, and other quality measures. Measurements can be made on a frequency of seconds that are logged overtime intervals of months to years (Blaen et al., 2016; Glasgow et al., 2004).

Major applications of water-quality sensors included monitoring in aquaculture, drinking water, wastewater, fresh groundwater and surface water, estuaries, coastal waters, and open oceans (Technavio, 2016). Water users in the public and agricultural sectors use continuous water-quality data for watershed management and for early warning of chemical spills approaching water intakes. Water-quality sensor data also can be used to characterize water-quality impairments from water hazards such as floods, hurricanes, and harmful algal blooms (HABs). Scientists at research institutes and in academia gain fundamental insights into earth systems through observatories that measure and record high-frequency changes in hydrological, geochemical, and ecological processes associated with societally relevant environmental issues.

The terms continuous water-quality monitoring and continuous water-quality data are used throughout the chapter. Continuous water-quality monitoring refers to monitoring in situ at high-temporal frequencies at sub-daily time scales (Jones et al., 2014; Pellerin et al., 2014; Rode et al., 2016). Pellerin et al. (2014) define high frequency as measurement of changes as often as many times per hour or day. Blaen et al. (2016) define high-resolution monitoring as in situ monitoring at frequencies from one to hundreds of times per hour. Sub-daily is an appropriate time step to adequately describe a variety of natural processes and human activities that vary on time scales from seconds to hours and that may be repeated at diurnal, seasonal, and (or) annual intervals or at irregular intervals (Blaen et al., 2016; Chappell et al., 2017; Jones et al., 2014; Kraus et al., 2017; Pellerin et al., 2014). Continuous water-quality data and time-series data are terms often used synonymously with in situ high-frequency or high-resolution water-quality data. For purposes of this chapter, a monitoring platform at which continuous water-quality data are collected is called a water-quality monitor.

Tracing the development and application of continuous water-quality monitoring since its inception in the 1950s to the current time provides a long-term perspective from which progress can be measured. This chapter is a review and synthesis of major developments in continuous water-quality monitoring from its origins in the 1950s into the second decade of the 21st century. The chapter includes sections on sensor development, in situ monitoring platforms and deployments, data-collection methods, and the importance of quality control for the production of reliable data. Additional topics include data storage, transmission, and applications of continuous water-quality monitoring data. Consequently, the chapter briefly touches on many technical topics but does not deal with any one topic in great detail. The chapter describes several case studies of continuous water-quality monitoring as a tool to understand the immediate and longer-term impacts of large-magnitude floods and hurricanes on freshwater and estuarine systems.

The chapter concerns mainly with the application of water-quality sensor technology in the United States with relevant examples from other nations. The continuous water-quality monitor

network operated by the U.S. Geological Survey (USGS) in cooperation with its many state and local partners is highlighted in this chapter along with important efforts by other agencies and organizations within and outside the United States.

This chapter is organized around four major categories of technology innovations that have led to the successful development of continuous water-quality monitoring. These are (1) an instrument revolution that led to the development of water-quality sensors; (2) digital electronic data and communication systems to record and transmit data to monitoring organizations; (3) computerized databases to store, manage, and retrieve data; and (4) the Internet to deliver and disseminate data in a timely manner.

2 BACKGROUND AND HISTORY OF CONTINUOUS WATER-QUALITY MONITORING TO 1972

Water-quality monitoring began in the United States in the early 19th century primarily to describe the resource or in response to sanitary concerns and disease outbreaks. Most water-quality monitoring was carried out at weekly or monthly intervals typically for periods of a year or less at a given location or in short-term surveys lasting a few days to a few weeks (Myers, 2015).

By the first half of the 20th century, the development of water resources was a national priority and the objectives of many water-quality surveys were to characterize waters suitable for the development of public, industrial, and agricultural supplies. This objective was satisfied by daily sampling at a surface-water location for a year and sampling groundwater once or twice in total. At the time, daily sampling of surface water was considered high frequency. This frequency was assumed to adequately describe the variability in the mineral characteristics of rivers and lakes for water-supply and water-treatment purposes (Collins, 1910; Dole, 1909; Stabler, 1911).

However, laboratory analysis of daily samples was not typically undertaken for practical reasons; primarily because the time and cost of sample transport and laboratory analysis of daily samples was assumed to be unnecessary to satisfy the purpose of monitoring (Dole, 1909). To achieve an acceptable degree of temporal resolution, a strategy was developed in which equal volumes of 10 sequentially collected daily samples were combined into a single composite sample for laboratory analysis. This strategy resulted in 3 sets of results per month and about 36 sets of results per year (Dole, 1909). The composite method of sample analysis persisted in one form or another well into the 1950s because it was economical and captured semimonthly, monthly, and seasonal variations (Collins, 1910; Dole, 1909; Hem, 1985).

A time-series graph (Fig. 1) provides an early representation of temporal changes in streamflow and suspended and dissolved solids for the Cedar River near Cedar Rapids, Iowa. The time-series graph was hand drawn in the early 20th century from the tabular results of about 36 sets of 10-day composite sample results from 1906 to 1907 (Dole, 1909) (Fig. 1). The Cedar River near Cedar Rapids was one of a total of 98 stations on rivers and lakes sampled east of the one-hundredth meridian in 1906–07 for 14 variables including turbidity (TU) and concentrations of suspended and dissolved solids, calcium, carbonate, bicarbonate, chloride, iron, magnesium, nitrate, silica, and sodium plus potassium, and sulfate (Dole, 1909). This study was part of the first national water-quality survey of surface waters in the United States (Dole, 1909).

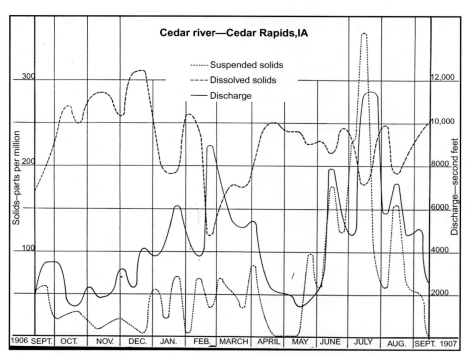

FIG. 1 Time-series graph of water discharge (streamflow) and concentrations of suspended and dissolved solids in parts per million from the Cedar River near Cedar Rapids, Iowa. Smoothed lines for solids concentrations were obtained from a series of 10-day composite samples collected from September 6, 1906 to September 17, 1907 (Dole, 1909).

Beginning around 1930, the development of advanced physical instrumentation to support chemical, biomedical, and agricultural analyses led to some of the first laboratory instruments for water analysis (Baird, 2002). The revolution in analytical instrumentation affected water analysis in three ways (Travis, 2002). First, new types of analytical instruments expanded the ability to detect a wide variety of chemical compounds using small amounts of sample with increased precision compared to visual readings. Second, automation of some or all parts of test methods reduced the time needed for analysis. Third, the detection-limit capabilities of new analytical instruments improved from milligrams per liter (mg/L) to <1 μg/L.

In 1934, the first commercial meter for measurements of hydrogen ion concentration (pH) in water became available (Gallwas, 2004). In 1937, a meter for measurements in water of specific electrical conductance (SC) became available (Collins et al., 1943; Durum, 1978; Hem, 1985). Technical details on the sensors are presented in the following section.

However, the revolution in chemistry that created advanced physical instruments did not have a major influence on methods of analysis in water laboratories until after the World War II. Most water analyses before 1950 were made using gravimetric, colorimetric, or volumetric methods also called wet-chemical methods (American Public Health Association, American Water Works Association, and Water Pollution Control Federation, 1971; Baird, 2002; Durum, 1978; Love, 1951).

Water pollution became an increasing concern in the late 1940s, but there was a lack of knowledge of its magnitude and extent. In response, the U.S. Congress passed a series of laws, the first being the Federal Water Pollution Control Act in 1948. Public agencies such as the U.S. Public Health Service, the Federal Water Pollution Control Administration, the USGS, state and local agencies, and academic researchers began to focus on ways to measure the extent and magnitude of water pollution on water resources and public health (Myers, 2015).

There was a growing recognition that in order to control water pollution, the fate of water pollutants was important to understand at the sub-daily scale. Wastewater discharges affecting diurnal concentrations of DO and 5-day biochemical oxygen demand (BOD_5) in rivers, streams, and estuaries became the subject of water-quality investigations (Schroepfer, 1942; Velz and Clark, 1950). Soon, factors that affected water quality at the sub-daily scale were identified, such as diurnal and seasonal cycles in solar radiation and water temperature (WT), tidal cycles, changing streamflow, time-variable waste inputs, and decomposition of pollutants with time and distance downstream from pollution sources (Velz and Clark, 1950).

Continuous water-quality monitoring began in the 1950s and started a new era in monitoring after more than 150 years of prior monitoring based solely on the collection and analysis of water samples. Continuous water-quality monitoring began to fill the previously identified need for high-temporal frequency water-quality data that was deemed critical by investigators like Schroepfer (1942) and Velz and Clark (1950). The first commercial thermistor for measurement of WT and the first commercial DO meter based on the polarographic electrode were introduced in 1951 and 1962, respectively (YSI, 2009, 2015a).

The first prototype water-quality monitors to continuously record SC and WT on a strip chart recorder were installed in 1955 at the existing USGS gaging stations measuring water levels and freshwater discharge on the Delaware Estuary near Philadelphia, Pennsylvania (Cohen and McCarthy, 1962) (Fig. 2). Innovative improvements to monitoring equipment on the Delaware River were made for the next several years by the USGS in cooperation with the City of Philadelphia and equipment manufacturers (McCartney and Beamer, 1962).

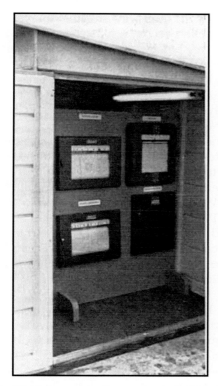

FIG. 2 The first water-quality monitor in the United States was deployed in 1955 on the Delaware River at Chester, Pennsylvania. The unit measured water temperature, dissolved oxygen, and specific electrical conductivity. Water-quality measurements were traced with a pen on a strip chart recorder. The water-quality monitor was housed in a large shelter with dimensions of about 8 ft. × 4 ft. × 7 ft. and placed on the banks of the Delaware River estuary (foot, ft.) (McCartney and Beamer, 1962). *Image from the U.S. Geological Survey.*

In 1960, pH and TU were added at one location to WT and DO (Fig. 2). Continuous measurement of DO was added to several monitors in the early 1960s to augment BOD_5 and other water-quality samples that were collected at weekly to monthly intervals (McCartney and Beamer, 1962). A network consisting of eight water-quality monitors was eventually established and operated for several years on the Delaware River in and near Philadelphia. The purpose of the water-quality monitors was to provide early notice to protect upstream freshwater-supply intakes on the Delaware River from encroachment of estuarine waters during low river levels and to monitor chemical and sanitary quality (McCartney and Beamer, 1962). Selected water-quality monitors established in this network in the 1950s and early 1960s continue to the current time serving a similar purpose nearly 70 years later.

The 1956 Federal Water Pollution Control Act ushered in an era of rapid development of new water-quality monitoring methods and approaches. Water pollution investigations began to track diurnal changes in stream conditions for pollution control planning purposes and to detect and warn water suppliers of approaching chemical spills. However, the continued reliance on wet-chemical methods limited the number of sample analyses per sampling station to about 15 over a period of several days (Kittrell, 1969). This frequency, while sub-daily, was acknowledged as being far from the optimal around-the-clock monitoring frequency needed to characterize pollutants in rivers, lakes, and estuaries.

In the early 1960s, the Ohio River Valley Water Sanitation Commission (ORSANCO) began to develop a network of water-quality monitors at three sites on the Ohio River near Cincinnati, Ohio recording WT, pH, SC, chloride, oxidation/reduction potential (ORP), and other water properties (Cleary, 1967; Ward, 1973). Potentiometric electrodes and other transducers were used as sensors in the water-quality monitors known as robot monitors (Cleary, 1967). A transducer is the part of a sensor that interacts with the environment to produce an electrical signal. The water-quality monitors were produced by the Schneider Instruments of Cincinnati, Ohio. (Any use of trade, firm, or product names is for descriptive purposes only and does not imply endorsement by the U.S. Government or by the publisher.) A few years later, the ORSANCO network of so-called robot monitors expanded to 11 sites and provided the first early-warning system regarding chemical and sewage spills to water providers with intakes along the urbanized and industrialized Ohio River (Cleary, 1967). This system has been sustained overtime, with updates as technology has improved, and is still serving its initial purpose to alert water suppliers of approaching spills.

The 1965 Water Quality Act established federal water-quality criteria for physical properties, chemical characteristics, and microbiological quality of recreational, public, industrial, and agricultural water supplies and for fish and wildlife protection. The 1965 Water Quality Act required states to establish their own enforceable water-quality standards based on the federally developed criteria. Numerical criteria were developed for water uses, including fresh and marine recreational waters, public water supplies, fish and wildlife propagation, and agricultural and industrial water supply (Federal Water Pollution Control Administration, 1968). Although not specifically required by the 1965 Water Quality Act, continuous monitoring provided high-frequency measurements to evaluate water-quality criteria based on the daily extremes such as daily minimum and daily average DO concentrations and maximum WT.

In 1965, a four-parameter, continuous-monitoring network measuring DO, pH, SC, and WT was operated by the USGS in cooperation with the State of Ohio. The network consisted of 42 water-quality monitors located on major rivers throughout the state at which 33 were measuring all four parameters (Durum and Blakey, 1965). Network data were used to evaluate water-quality criteria required by the 1965 Water Quality Act. Data were analyzed to determine percentiles as well as minimums, maximums, and daily average values to support development of water-quality standards for Ohio (Durum and Blakey, 1965). A special survey was made of DO concentrations in Ohio rivers and streams at low-streamflow conditions in September 1964. This information was conveyed to the state along with other supporting information on natural background levels in streams unaffected by water pollution. The results were used to inform the development of water-quality standards by the state of Ohio (Durum and Blakey, 1965).

By 1967, the number of water-quality monitors nationwide numbered 871 (Sayers, 1971). However, sites with continuous data from water-quality monitors represented a small fraction of the total number of water-quality monitoring sites nationwide. Samples were collected at the majority of sites at daily or less frequent intervals. At the national scale, continuous WT represented 11.4% of the total number of WT monitoring sites; continuous SC represented 4.6%; continuous DO represented 2.2%; and continuous pH represented only 1.1%. (Myers, 2015; Sayers, 1971) The principal suppliers of water-quality monitors in 1968 were Schneider Instrument Co., Cincinnati, Ohio; Honeywell, Inc., Fort Washington, Pennsylvania; and Fairchild Space and Defense Systems, Woodbury, New York. (Ward, 1973). In 1967, the continuous water-quality network of the USGS numbered 164 water-quality monitors in 26 states and the District of Columbia (Fig. 3).

The U.S. Environmental Protection Agency (EPA) was created in 1970 and the Clean Water Act (CWA) became law in 1972. These two drivers of water-quality monitoring set the stage for the next 50 years of monitoring technology developments. The EPA by authority of the CWA required water-quality monitoring of rivers, streams, and lakes as well as monitoring pollutant sources discharging to the waters of the United States. The CWA resulted in large expenditures of federal, state, and local funds for wastewater treatment with the intent of improving water quality in the United States. The CWA provided the legal basis to bring the waters of the United States. into compliance with water-quality standards.

Measurements of WT, pH, DO, and SC were being used, in part, to judge the effectiveness of water and wastewater treatment and to assess

3 INNOVATIONS IN WATER-QUALITY SENSORS

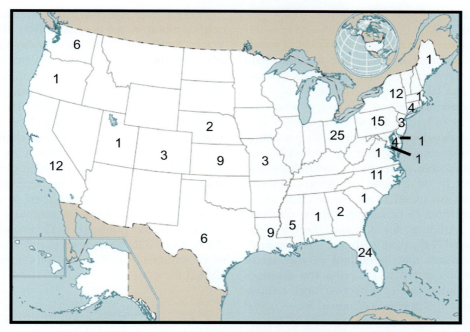

FIG. 3 In 1967, the U.S. Geological Survey (USGS) operated 164 water-quality monitors in 26 states and the District of Columbia (DC). Numbers on map represent number of water-quality monitors per state. *Image reproduced from the USGS.*

attainment of water-quality standards in rivers, streams, lakes, embayments, and estuaries. Another important use for continuous monitoring was for operational decision making in the control of water and wastewater-treatment processes. Placement of inline pH, TU, and DO sensors provided information for near immediate adjustments of treatment processes such as dosing of chemicals to assure that the proper quality resulted from treatment.

This section described the first 20 years of continuous water-quality monitoring as well as some limitations of monitoring at less frequent sampling intervals. Historical information helps to lay the groundwork with which to judge the importance of technological innovations that were made over the next 50 years to the current time. History also provides us with a long-term perspective from which progress can be measured. The remaining sections describe innovations that were major drivers for the technological improvements in continuous water-quality monitoring.

3 INNOVATIONS IN WATER-QUALITY SENSORS

A sensor is a device that detects and responds to input from the physical environment. The specific input could be light, heat, sound, moisture, pressure, or other environmental variables. Water-quality sensors evolved from four general categories of analytical measurements; electrical resistance, electrometric, optical, and acoustic. Public and academic laboratories in the physical and biomedical sciences, water analysis laboratories, research institutions, academia, and commercial vendors are essential sources for the development of instruments and sensors. The nexus from laboratory instruments to water-quality field sensors began with the placement of pH sensors in flow-through systems for process control in industry and in water and wastewater treatment (Ward, 1973). Process control stimulated the demand for handheld and inline sensors that were more durable and portable than laboratory instruments.

The following section discusses what are described as fully developed water-quality sensors. Fully developed is defined as providing "sensor measurements that are possible and readily applicable in terms of technology and deployment in unattended conditions." (Marcé et al., 2016). An excellent source of information on the performance of sensors can be obtained from the Alliance for Coastal Technologies (ACT). The ACT is a partnership of research institutions, resource managers, and private sector companies dedicated to fostering the development and adoption of effective and reliable sensors and platforms (Alliance for Coastal Technologies, 2018). The mission of ACT is to serve as an unbiased, third-party testbed for evaluating sensors and platforms for use in coastal and ocean environments (Alliance for Coastal Technologies, 2018). For example, recent reports are available from the ACT website (http://www.act-us.info/) on performance of sensors for continuously monitoring nutrients and DO, and a workshop report on sensors for monitoring harmful algae, cyanobacteria, and their toxins.

This section on sensor innovations includes general information on history of development, principles of measurement, common uses, and relevance to water-quality issues. Tables 1 and 2 describe the fully developed water-quality sensors discussed herein and their operational characteristics, strengths, and limitations.

Descriptions and manufacturers' specifications of specific instruments and equipment often are not specified in this chapter because currently (2018) available models are subject to change as new makes and models are frequently introduced. When mentioned, information on manufacturers and vendors is necessary to convey technical or historical information.

Publicly available data from continuous water-quality monitors are accessible from the website WaterQualityWatch (https://waterwatch.usgs.gov/wqwatch/). WaterQualityWatch is a USGS website that provides access to maps and data from water-quality monitors in surface waters of the United States. Maps are displayed at various scales from national, state, and site level. Water properties and chemical characteristics displayed on WaterQualityWatch include WT, SC, pH, TU, DO, fluorescent chlorophyll (FChl), and nitrate as nitrogen. In September 2017, WaterQualityWatch was accessed to determine the number of water-quality monitors with data available to the public (U.S. Geological Survey, 2017). These data provide an example of the scale of continuous water-quality monitoring in the United States.

3.1 Electrical Resistance Techniques

Electrical resistance instruments became available in the 1960s for monitoring continuous WT, wet and dry bulb air temperature, SC, and soil moisture (Smoot and Blakey, 1966).

3.1.1 Water Temperature

There are many types of temperature measurement devices ranging from liquid and gas thermometers to resistive and electrometric sensors and more recently, fiber optic cables. The most common types of water–temperature sensors rely on resistance measurements (Table 1). The two types of resistive temperature sensors are resistive temperature devices (RTDs) and thermistors (Wang et al., 2003). RTDs use a metal sensing element made of platinum, for example. The resistance of RTDs increases in a positive direction as temperature rises. Conversely, a thermistor uses a semiconductor such as a metal oxide as the sensing element. The electrical resistance of the semiconductor decreases sharply in a nonlinear manner as the temperature increases (Wang et al., 2003).

TABLE 1   General Performance Characteristics, Limitations, and Strengths of Fully Developed Resistance and Electrometric Water-Quality Sensors

Sensor type and reference	Type of water-quality measurement	Instrument range (by orders of magnitude or in units)	Precision or sensitivity	Specificity	Use setting(s): laboratory field site, or process control	Strengths	Limitations
Resistance sensors (Smoot and Blakey, 1966)							
Temperature thermistor	AT, WT	0–40°C	0.2°C	No known interferences	Robust in all settings	Battery powered, low maintenance, portable	Maintenance and calibration checks needed semimonthly to quarterly
Wheatstone bridge, carbon, or platinized nickel electrodes	SC	0–2000 or 2000–35,000 µS/cm	About 1% of full scale	Lacks specificity to ions	Robust in all settings	Battery powered, low maintenance, portable	Maintenance and calibration checks needed semimonthly to quarterly
Electrometric sensors							
Temperature thermocouple (Smoot and Blakey, 1966; U.S. Environmental Protection Agency, 1978)	WT	−40° to +70°C	0.2°C or about 1% of full scale	Few interferences	Robust in all settings	Unaffected by water color or TU	Similar to mechanical resistance sensors
Specific electrical conductance (Wilde, n.d.)	SC	0–2000 or 2000–35,000 µS/cm	About 1% of full scale	Lacks specificity to ions	Robust in all settings	Unaffected by water color or TU	Similar to mechanical resistance sensors
Hydrogen ion concentration (Wilde, n.d.)	pH	0–14 SU	0.01 SU	Highly selective for hydrogen ion	Robust in most settings, drift, and slow response in low ionic strength waters	Quick response, inexpensive and easy to use, no interferences from color or turbidity	Periodic calibration required in the field. Dependent on porous membrane that is fragile and can become clogged
Ion-selective electrodes (Radu et al., 2013; U.S. Environmental Protection Agency, 1978)	>60, monatomic ions (such as H^+) and polyatomic ions (such as ClO_4^{3-} and NO_3-)	−999 to 999 mV	<0.001 to >1 mg/L	Some cross reactivity among ions	Laboratory, onsite, ship board, and process control	Quick response. Inexpensive, easy to use, no interferences from color or TU	Frequent calibrations needed. Chemical interferences

Continued

TABLE 1 General Performance Characteristics, Limitations, and Strengths of Fully Developed Resistance and Electrometric Water-Quality Sensors—cont'd

Sensor type and reference	Type of water-quality measurement	Instrument range (by orders of magnitude or in units)	Precision or sensitivity	Specificity	Use setting(s): laboratory field site, or process control	Strengths	Limitations
Pt or Hg and Ag/AgCl electrodes (U.S. Environmental Protection Agency, 2017; Wilde, n.d.)	ORP	−999 to 999 mV	About 1% of full scale or ±20 mV	Lacks specificity to redox-sensitive ions	Laboratory, onsite, and process control	Unaffected by water color or TU	Biofouling, mechanisms available to clean probe
Polarographic electrodes (Wilde, n. d.; YSI, 2015a)	DO	0–20 mg/L	0.2 mg/L	Chemical interference from solvents, hydrogen sulfide, and certain other gases. Requires compensation for temperature, local barometric pressure or altitude, and salinity	Robust in most environmental settings	Unaffected by water color or TU; far less use of reagents compared to wet-chemical methods	Biofouling, mechanisms available to clean probe. Field calibrations needed weekly to biweekly. Membrane requires regular replacement

AT, air temperature; *WT*, water temperature; °C, degrees Celsius; *SC*, specific electrical conductance; *μS/cm*, microSiemens per centimeter at 25°C; *mg/L*, milligrams per liter; *μg/L*, micrograms per liter; *EPA*, U.S. Environmental Protection Agency; *ORP*, oxidation–reduction potential; *pH*, hydrogen ion concentration; *DO*, concentration of dissolved oxygen; *Pt*, platinum; *Hg*, mercury; *Ag/AgCl*, silver/silver chloride; *H⁺*, hydrogen ion; *Cl⁻*, chloride ion; *ClO₃⁻*, chlorate ion; *mV*, millivolts.

TABLE 2 General Performance Characteristics, Limitations, and Strengths of Fully Developed Optical and Acoustic Water-Quality Sensors

Sensor type and reference	Type of water-quality measurement	Instrument range (by orders of magnitude or in units)	Detection limit	Specificity	Use setting (s): laboratory field site, or process control	Strengths	Limitations
Optical sensors							
Turbidimeter, Photometer, VIS (Anderson, 2005)	TU	0–40, 0–1000, 0–4000, 10–10,000 NTU or FTU based on instrument type	0.05 NTU (lowest) FTU	Biased low to sand-sized particles, measures best at silt, and clay sizes smaller than 62.5 μm	Robust in all settings	Uniform use of standards made from formazin or polymer beads independent of instrument type	Biofouling, mechanisms available to clean probe. Interferences from scratches, color, colored particles, light leaks, air, or gas bubbles in sample
Colorimeters; VIS (Seabird Scientific, 2017; Systea SpA, n. d.; S::CAN, 2018)	Dissolved and total phosphorus as PO_4-P	0.0–0.3 mg/L as PO_4-P	0.007 mg/L as PO_4-P	Very selective and sensitive for specific ions	Laboratory, onsite, and process control. New microfluidic systems for in situ measurements	Newer systems are submersible, use filters, digestion reagents, and microfluidic tubes for sample preparation	In situ applications require frequent maintenance and calibrations, reagent disposal issues
	Ammonium ion NH_4^+	0.0–1.0 mg/L as NH_4-N	0.05 mg/L as NH_4-N				
	Nitrate+nitrite as nitrogen, NO_3+NO_2-N	0.0–0.7 mg/L NO_2+NO_3-N	0.01 mg/L NO_2+NO_3-N				
	Silicate ion SiO_4^{-2}	0.0–60 mg/L SiO_4^{-2}	0.3 mg/L SiO_4^{-2}				
Photometer, UV (Pellerin et al., 2013)	Nitrate as nitrogen NO_3-N	0.0–5.0 or 0.0–30 mg/L as NO_3	0.5 mg/L	Interference from TU >500 NTU, and CDOM 5 to >30 mg/L, compensated for water temperature	Robust in all settings; different instrument type for water vs wastewater monitoring	Quick response, inexpensive, easy to use, no chemical reagents	Biofouling, mechanisms available to clean probe, maintenance 3–4 weeks
Fluorometer UV/VIS (Turner	Rhodamine WT and other	0–10.0 μg/L, 0–100 μg/L, 0–1000 μg/L	0.1 μg/L	Dependent on excitation and	Robust in all settings	Background correction needed for water color and	Short-term deployments not subject to biofouling

Continued

TABLE 2 General Performance Characteristics, Limitations, and Strengths of Fully Developed Optical and Acoustic Water-Quality Sensors—cont'd

Sensor type and reference	Type of water-quality measurement	Instrument range (by orders of magnitude or in units)	Detection limit	Specificity	Use setting (s): laboratory field site, or process control	Strengths	Limitations
Designs, 2019d)	water-tracing dyes			emission spectra of sensor		TU. Temperature compensation required	
Fluorometer (VIS) (Hambrook Berkman and Canova, 2007; Turner Designs, 2019d)	FChl, phycocyanin, phycoerythrin accessory pigments	0–400 µg/L or 0.3 to >500 as in vivo in RFU or µg/L	0.03 µg/L (blue excitation) 0.3 µg/L (red excitation)	Dependent on excitation and emission spectra of sensor	Robust in all settings, no reagents needed	Background correction needed for water color and TU. Temperature compensation required	Biofouling, mechanisms available to clean probe. RFU is a relative measure. Collect and analyze samples for comparison of chl$a$ laboratory results to fluorometer measurements
Fluorometer, UV/VIS (Turner Designs, 2019d)	FDOM	0–1500 µg/L 0–3000 µg/L	0.05–0.4 µg/L	Dependent on excitation and emission spectra of sensor	Robust in all settings, no reagents needed	Background correction needed for water color and TU. Temperature compensation required	Biofouling, mechanisms available to clean probe
Luminometer, UV (Rounds et al., 2013)	DO concentration	0–20 mg/L	0.05 mg/L	High specificity for dissolved oxygen	Robust in all settings	Quick response, inexpensive, easy to use	Biofouling, mechanisms available to clean probe
Acoustic sensor							
ADVM (Landers et al., 2016)	Backscatter measurements of suspended-sediment concentration and particle size	0–35,000 mg/L Sample mass finer than 0.0625mm, sieve diameter (%)	About 50 mg/L	Requires concurrent streamflow data for surrogate statistical equation, relatively unbiased to particle size	Depends on frequencies of ADVM and water temperature	Complex set of equations to measure backscatter and convert to SSC	Biofouling, mechanisms available to clean probe

TU, turbidity; *NTU*, nephelometric turbidity units; *FTU*, formazin turbidity units; *VIS*, visible wavelengths; *UV*, ultraviolet wavelengths; *mg/L*, milligrams per liter; *µg/L*, micrograms per liter; $NO_3 – N$, nitrate ion as nitrogen; $PO_4 – P$, phosphate ion as phosphorus; $NO_2 + NO_3 – N$, nitrite plus nitrate as nitrogen; $NH_4 – N$, ammonium as nitrogen; *FChl*, fluorescent chlorophyll; *RFU*, relative fluorescence units; chl$a$, chlorophyll a; *FDOM*, fluorescent dissolved organic matter; *CDOM*, chromophoric dissolved organic matter; *ADVM*, acoustic Doppler velocity meter; *SSC*, suspended-sediment concentration.

Thermistors are one of the most commonly used devices for the measurement of water, soil, and air temperature and for temperature compensation for many other sensor measurements. As mentioned previously, commercial thermistors for WT measurements were introduced in 1951 (YSI, 2009). WT is a master variable, meaning that it governs other variables and processes such as pollutant transformation and degradation, chemical equilibria, dissolved-gas solubility and saturation, and thermal limitations of fish (U.S. Environmental Protection Agency, 1979, 1986). Because of their small size and low-energy requirements, thermistors, and RTDs are ideal for environmental temperature monitoring.

3.1.2 Specific Electrical Conductance

SC was one of the first instrumental methods used in water analysis and derived from the measurement of the electrical resistance generated between carbon and platinized nickel electrodes. SC is measured in milliSiemens per centimeter (mS/cm) or microSiemens per centimeter (μS/cm) at standard temperature of 25°C (Table 1). SC is the inverse of resistance, which is usually measured in Ohms. Laboratory measurement of SC in soils and water-soil mixtures to determine their salinity for irrigation quality purposes dates back to 1897 (Scofield, 1932). The Wheatstone bridge conductivity meter uses a configuration of parallel resistors to precisely determine the conductivity of a sample in a cup in one leg of the bridge circuit.

SC is a measure of the capacity of water (or other media) to conduct an electrical current. SC of water is a function of the types and quantities of dissolved substances in water, but there is no universal linear relation between total dissolved substances and conductivity (Radtke et al., 2005). SC serves as an inexpensive and direct measurement that indicates changing water-quality conditions, as a guide to water sampling, as an indicator of irrigation quality, and as a surrogate measurement for salinity and dissolved solids (Hem, 1985). From about

1937 to about 1946 and intermittently thereafter, daily measurements of SC were made by the USGS to determine the number of sequential daily water-quality samples to composite prior to analysis in order to better capture variability in chemical concentrations during periods of rapidly changing SC (Collins et al., 1943; Hem, 1985).

SC at 25°C is a useful measure of the salinity of an aquatic system and can be converted to salinity units (Weyl, 1964). In September 2017, 2657 WT and 1150 SC measurement stations in the United States were being routinely reported on the USGS WaterQualityWatch (U.S. Geological Survey, 2017).

3.2 Electrometric Techniques

Electrometric methods are a class of techniques in analytical chemistry that measure the electrical potential (volts) and/or current in an electrochemical cell containing the analyte and compare it to a reference cell with a known solution. Many types of water-quality sensors rely on electrometric methods for measurements of WT, SC, pH, and ion-selective electrodes (ISEs) that measure concentrations of monatomic and polyatomic anions and cations, ORP, and concentrations of DO.

3.2.1 Water Temperature

Thermocouples are electrometric instruments that are used to measure water and air temperatures (Table 1). A thermocouple is a relatively simple device made from two electrical conductors, typically two metallic wires of different composition that are connected at one point at their ends to make an electrical circuit (McGee, 1988; Wang et al., 2003). An electromotive force (EMF) is generated when current flows from the hot side to the cold side of the circuit. Temperature of water is sensed by a thermocouple that produces a direct-current (DC) milliVoltage output (Smoot and Blakey, 1966). A positive but nonlinear relation exists between the output voltage of the thermocouple and temperature.

Thermocouples were the first type of temperature probe or sensor. In 1826, the physicist A.C. Becquerel proposed that a thermocouple could be used for temperature measurement (McGee, 1988; Simpson et al., 1991). The thermocouple eventually was perfected in 1885 and accepted worldwide for temperature measurement in 1927 (McGee, 1988; Simpson et al., 1991). Prior to this time all temperature measurement was done with liquid or gas-filled thermometers (McGee, 1988).

Like thermistors, thermocouples can be placed inside a waterproof metal or ceramic shield to protect the sensor from exposure to harsh conditions. This makes the thermocouple useful in harsh environments such as near ocean vents or in corrosive effluents in treatment processes. The earliest known application of a thermocouple in a water-quality monitor was in the early 1960s (Smoot and Blakey, 1966). Because of their small size and low-energy requirements, thermocouples are ideal for environmental temperature monitoring.

3.2.2 Specific Electrical Conductance

General information on SC is included in Section 3.1.2. The potentiometric conductivity sensor is another type of sensor that operates with electrodes rather than on resistance. The electrometric SC sensor consists of six electrodes, two current electrodes, and two parallel pairs of voltage measuring electrodes. An internal temperature compensator automatically refers all measurements to 25°C, according to the standard potassium chloride–conductivity relationship. The output of this sensor is linearly proportional to SC (microOhms at 25°C) (Smoot and Blakey, 1966).

3.2.3 Hydrogen Ion Concentration (pH) and ISEs

Potentiometry, one type of electrometric method, is used to determine the concentration of a solute in solution. In potentiometric measurements, the potential between two electrodes is measured using a high impedance voltmeter.

Two types of potentiometric sensors are available for environmental applications, glass electrodes, and solid-state electrodes (Radu et al., 2013). Some examples of electrometric sensors include hydrogen ion concentration (pH), ISEs, and electrodes for measurement of ORP (Table 1).

Glass electrodes were the first introduced commercially beginning in the 1930s but are limited to measurement of a small number of ions (Frant, 1997). Glass electrodes are quick to respond in dilute natural waters but are fragile (Radu et al., 2013). Solid-state electrodes became available in the 1970s but may have a slower response time in dilute natural waters.

The concept of measuring hydrogen ion concentration (pH) using potentiometric devices dates back to 1906 (Frant, 1997). The first glass electrode for pH measurement was reported in 1928 (Frant, 1997). The first commercial laboratory instrument for pH measurement was the Beckman Model G pH meter, invented in 1934 by Arnold O. Beckman and produced until about 1950 (Gallwas, 2004). This invention started a revolution in the development of potentiometric sensors. The first pH meter included a glass electrode, reference electrode, and a current amplification system (Gallwas, 2004). The potentiometric pH meter remains the most common method of pH measurement in field and laboratory sensors (Table 1). In September 2017, the USGS was operating about 540 stations in the United States equipped with potentiometric sensors measuring pH (U.S. Geological Survey, 2017).

ISEs are able to detect the concentration of a single type of ion in a solution containing many other types of ions in solution (U.S. Environmental Protection Agency, 1978). ISEs measure voltage potentials between a measurement and reference electrode (U.S. Environmental Protection Agency, 1978). ISEs became commercially available for water analysis in the 1960s (Frant, 1997). By 1982, 24 different ISEs were available for water testing (Orion Research, 1982). Innovations since 2000 have expanded the types of ISEs and the capability to detect more than 60 monoatomic and polyatomic ions in solution at sufficiently low detection limits to make them useful for water analysis (Radu et al., 2013).

ISEs are available for measurement of monatomic and polyatomic ions (Table 1) including cations such as ammonium, calcium, magnesium, sodium, and potassium, and anions such as bromide, chloride, cyanide, fluoride, iodide, and nitrate (Orion Research, 1982; Radu et al., 2013; U.S. Environmental Protection Agency, 1978). ISEs are not subject to interferences from TU or water color (Blaen et al., 2016; Radu et al., 2013) (Table 1). However, ISEs are susceptible to interferences from other ions such as sulfide and to biofouling of the electrode surface (Radu et al., 2013).

ISEs are used most often in laboratories, onsite, on research vessels, and for process control in water and wastewater-treatment plants (Radu et al., 2013; U.S. Environmental Protection Agency, 1978). ISEs have had limited use as continuous water-quality sensors due to maintenance issues caused by biofouling, instrument drift, need for frequent calibration, and chemical interferences (Blaen et al., 2016; Radu et al., 2013; U.S. Environmental Protection Agency, 1978) (Table 1). Costs involved with frequent travel to remote locations for servicing the instrument can be prohibitively expensive.

In a survey of 75 drinking-water treatment plants in the United States, Belgium, the Netherlands, the United Kingdom, and Australia, 3 of the 10 most commonly monitored water properties in raw or treated public-drinking water are monitored using potentiometric sensors. In the four countries, pH is monitored in 79%–100% of the 75 drinking-water treatment plants, fluoride is monitored in 23%–81%, and total chlorine is monitored in 14%–100% (Raich, 2013).

3.2.4 ORP Sensors

ORP is a numerical index of the intensity of oxidizing or reducing conditions within a system (Hem, 1985). Water samples contain different types of oxidizing and reducing substances. Because the ion species cannot be identified by the ORP value, the measurement of ORP in natural water is in millivolts (mV) rather than concentration units (U.S. Environmental Protection Agency, 2017). Positive values for ORP indicate oxidizing conditions, whereas negative values indicate reducing conditions.

ORP electrodes were first used in the 1930s to determine the relation between an oxidizing environment and bactericidal properties of water containing chlorine (Schmelkes et al., 1939). ORP is currently measured using a solid-state system constructed with a reference electrode of silver/silver chloride and a platinum measurement electrode that measures the electrical potential of a solution (U.S. Environmental Protection Agency, 2017).

ORP is used to monitor drinking water and wastewater-treatment processes for oxidation state due to the addition of halogenated disinfectants such as chlorine. These and other studies showed a strong correlation of ORP and bactericidal activity of chlorine. ORP was suggested by the World Health Organization (WHO) for monitoring the oxidation levels of free chlorine in treated drinking water (World Health Organization, 1971). ORP also is used as an indicator of the levels of disinfectant in wastewater effluent. The presence of chlorine will result in a positive ORP value and the presence of hydrogen sulfide will result in a negative ORP value. ORP sensors were used in the ORSANCO water-quality monitoring network on the Ohio River in the early 1960s as a sanitary indicator (Cleary, 1967).

ORP also is commonly measured as a master variable in groundwater systems to characterize groundwater geochemistry. If the sample water contains a dominant oxidation/reduction system, the ORP value provides a clue about the ratio between the oxidizing substances and reducing substances. For example, iron ions are often the dominant pair of oxidizing and reducing ions in groundwater. In natural water, many redox-sensitive species contribute to the reading.

The ORP of solutions can be used to infer the form of the multivalent elements that dominate aquatic systems, such as ferrous ion (Fe^{2+}) in reducing conditions vs ferric ion (Fe^{3+}) in oxidizing conditions (Hem and Cropper, 1959). Reduction–oxidation (redox) processes affect the chemical quality of groundwater in all aquifer systems (McMahon et al., 2011). Beginning in the 1970s, continuous measurements of ORP in groundwater became routine with concurrent water-quality sampling as an indicator of the overall or governing geochemistry of a system (Wood, 1976).

3.2.5 Electrochemical Dissolved Oxygen Sensors

There are two types of electrochemical sensors for measurement of DO concentration in water, polarographic and galvanic sensors (YSI, 2009). The *Dissolved Oxygen Handbook—a practical guide to dissolved oxygen measurements* is a useful reference that describes principles of operation as well as similarities and differences in the two electrochemical methods (YSI, 2009). Both types of electrochemical sensors consist of an anode and a cathode that are confined in electrolyte solution by an oxygen permeable membrane. Oxygen molecules in the water sample diffuse through the membrane to the sensor at a rate proportional to the pressure difference across it (YSI, 2009). The oxygen molecules are reduced or consumed at the cathode producing an electrical signal that travels from the cathode to the anode and then to the instrument. The amount of oxygen diffusing through the membrane is proportional to the partial pressure of oxygen outside the membrane (YSI, 2009).

Major differences between polarographic and galvanic sensors are that the galvanic sensor does not require a polarizing voltage to be applied in order to reduce the oxygen that has passed through the membrane and so does not need a warm-up period as does the polarographic sensor (YSI, 2009). The cathode and anode of the galvanic sensor are made from silver and zinc, respectively. The cathode and anode for the polarographic sensor are gold and silver, respectively. Galvanic sensors have a shorter sensor life than polarographic sensors due to their electrode composition and that fact that they are always ready to make measurements (YSI, 2009).

The polarographic electrode for DO measurement was invented in 1956 by Dr. Leland Clark for biomedical applications (YSI, 2009). The technique was used to make measurements of DO concentrations in natural water as early as 1957 (Kanwisher, 1957). As mentioned previously, the first commercial field meter for measurement of DO in water was introduced in 1962 (YSI, 2015a).

Electrochemical DO sensors, due to their advantages for field use, replaced the traditional wet-chemical Winkler titration that had been used since 1915 for measurement of DO (American Public Health Association, 1915). Advantages of the electrochemical sensors are the ability to make inexpensive and instantaneous measurements, a large reduction in use of chemical reagents compared to the Winkler titration, and comparable sensitivity to the wet-chemical method (Table 1). The Winkler titration can be used as a periodic check on DO sensors (Rounds et al., 2013).

Disadvantages of electrochemical sensors include the need to recalibrate the sensor and perform sensor maintenance, such as changing the membrane and electrolyte filling solution, often on a weekly to biweekly basis (YSI, 2009) (Table 1). Sensors placed in organically enriched and biologically productive natural waters accumulate biofouling growths quickly and require more frequent routine maintenance than sensor deployed in dilute unproductive natural waters. In older models of electrochemical DO sensors, flow must be maintained across the surface of the membrane in order to make an accurate DO measurement. Flow at the membrane surface could be accomplished with a stirrer attachment.

Improvements in polarographic sensor technology created a sensor that uses rapid (millisecond) pulses of electricity across the electrodes rather than a constant voltage as in the older sensors. Rapid pulses of voltage and the addition of a third silver reference resulted in a polarographic that is electrode flow independent (YSI, 2009). Improvements in membrane composition and snap-on membranes have reduced the difficulty of changing the semipermeable membranes that are needed to separate the electrochemical sensor from the fluid to be measured (YSI, 2009).

The first analog DO field meters did not record or store data and could not be left unattended for any period of time to monitor and record water quality. Calibrations, maintenance, membrane changes, electrolyte refills, and water-quality measurements had to be recorded manually in log books. Portable field meters for DO are relatively small and lightweight with submersible sensors that are attached to the meter by cables with waterproof connections.

DO can change rapidly on a diurnal basis and is typically at its lowest levels overnight and during periods of seasonally high WTs as a result of the respiration of aquatic organisms (American Public Health Association, American Water Works Association, and Water Pollution Control Federation, 1971). Critically, low DO concentrations in surface water can be stressful or lethal for aquatic organisms such as macroinvertebrates and fish and capturing minimum concentrations is readily accomplished with unattended continuous monitoring.

By the 1980s, electrochemical DO sensors were being routinely installed in water-quality monitors at the USGS streamgages (Gordon and Katzenbach, 1983). In 2006, sensor technology for optical measurements of DO became available and replaced many electrochemical sensors in continuous-monitoring applications (YSI, 2015a).

3.3 Optical Techniques

Optical techniques measure the absorption, transmission, or light-scattering properties of substances in the visible (VIS), infrared (IR),

and ultraviolet (UV) ranges (Pellerin and Bergamasci, 2014). For water analyses, optical methods include but are not limited to spectrophotometry, colorimetry, photometry, fluorometry, and microscopy. The most common optical water-quality sensors measure basic water properties and characteristics such as TU and DO concentration. Notably, optical sensors also can measure nutrients such as ammonium, nitrate, nitrite, phosphate, and silicate. Nutrient measurements are relevant for understanding the importance of the quality of drinking water, groundwater, and natural aquatic systems.

The advantages of optical techniques are sensitivity and specificity for measurement of the constituent of interest, and minimal electronic drift of the instrument. Limitations include potential interferences from chromophoric or colored dissolved organic matter (CDOM), TU, and chemical compounds (Downing et al., 2012; Saraceno et al., 2017) (Table 2). Like other measurements of physical and chemical variables, optical measurements require temperature compensation. Biofouling is a common problem for optical sensors in natural waters. Innovate solutions include wipers, biofouling inhibitors and coatings, and compressed air blasts that are effective in automatic cleaning to remove biofouling and thereby extend the time interval of deployment between field visits (Table 2).

In a survey of a total of 75 drinking-water treatment plants in the United States, Belgium, the Netherlands, the United Kingdom, and Australia, TU was the most commonly measured optical property in raw or in treated water (Raich, 2013). TU was measured in raw or treated water in 89%–100% of the drinking-water treatments plants in the survey. Optical nitrate was measured at 57% of the drinking-water treatment plants surveyed in the United Kingdom (Raich, 2013).

3.3.1 Turbidity Sensors

Light-scattering properties of suspended particles in water are the basis for the optical measurement of TU. "TU is caused by the presence of suspended, colloidal, and dissolved matter, such as clay, silt, finely divided organic matter, plankton and other microscopic organisms, organic acids, and dyes" (Anderson, 2005). TU is an important measurement for water-treatment processes that remove particles to improve quality.

TU measurements date back to the use of the Jackson Candle turbidimeter. The Jackson candle turbidimeter consisted of a glass tube, open at the top and graduated in millimeters. The tube was held in place by a brass frame, open at the bottom and supported by a stand (Fig. 4). In the center of the stand was a standard candle positioned 3 in below the bottom of the glass tube (Fuller et al., 1900; Leighton, 1905) (Fig. 4). A water sample was slowly poured into the glass tube and the measurement recorded when the candle flame viewed from above had diffused to a uniform glow (Sadar, 1988). The result was reported in Jackson TU units (JTU). The practical limit of detection was relatively high, about 25 JTU, which was sufficient for TU measurements in most natural surface waters but insufficient for monitoring drinking water (Sadar, 1988).

A modern turbidimeter measures absorbance of light in relation to initial light intensity (Lawler, 2005). Turbidimetry is the measurement of TU by quantifying the degree of "attenuation" of a beam of light of known initial intensity (Lawler, 2005). Turbidimetry is usually measured in natural waters with fairly high TU in which the scattering particles are relatively large (Lawler, 2005). Starting in the 1940s, TU could be measured using spectrophotometers (Müller, 1954). The first commercial dedicated turbidimeter for water analysis was developed by the Hach Chemical Company in 1950 (Hach Chemical Company, 2014).

A nephelometer directly measures the light scattered, usually at 90 degree angle to the beam direction, by suspended particles (Lawler, 2005). Nephelometry is appropriate to water of lower TU in which the suspended particles are small such as in drinking water (Lawler, 2005). Most modern nephelometers are calibrated with known suspensions of formazin or polymer beads. Nephelometric methods offer precision and sensitivity. The EPA Method 180.1 and two other EPA approved methods define the optical geometry for TU measurements and specify the nephelometric method of analysis for TU regulatory reporting of TU in drinking water (Müller, 1954; U.S. Environmental Protection Agency, 1993).

There are eight basic types of instruments that are designed for different water conditions such as low, medium, and high levels of TU, ranges in particle sizes and shapes, and instruments that reduce or eliminate interferences such as color (Anderson, 2005). Concentration ranges

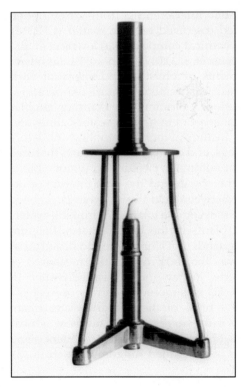

FIG. 4 The Jackson candle turbidimeter consisted of a glass tube closed at the bottom and graduated in centimeters and millimeters depth (Leighton, 1905). *Image from the U.S. Geological Survey.*

can vary widely among instruments by as much as four orders of magnitude (Table 2). Considerations of which instrument to use include type of light source, type of detector, and optical geometry and the angle of scattered light detection (Sadar, 1988). Selection of instrument depends on the water conditions to be monitored (Anderson, 2005).

Reporting units are dependent on the instrument and method used to measure TU (Anderson, 2005). The two most common measurements are reported as Nephelometric Turbidity Unit (NTU) and Formazin Nephelometric Unit (FNU). The main difference in the NTU and FNU measurements is the light source with the former being white light in the range of 400–680 nm and the latter being near IR in the range of 700–980 nm (Anderson, 2005).

In situ TU sensors became available in the early 2000s (YSI, 2015a). About the same time, the number of TU sensors began to increase at the USGS streamflow-gaging stations. In water year 2017, the USGS was reporting TU at approximately 577 stations in the United States. These sites are equipped with optical TU sensors on WaterQualityWatch (U.S. Geological Survey, 2017). A water year is the 12-month period October 1 through September 30 designated by the calendar year in which it ends.

3.3.2 Spectrophotometry, Colorimetry, and Chemical Analyzer Systems for Nutrient Analysis

Spectrophotometry is the quantitative measurement of the absorbance or transmission properties of a material as a function of wavelength (Laitinen and Ewing, 1977). Use of VIS and UV/VIS spectrophotometers in water analysis began in the 1940s. For example, the Coleman universal spectrophotometer was used in water laboratories for analysis of iron, manages, phosphate, silicate, and sulfate in the VIS range (Müller, 1954). The Coleman universal spectrophotometer could be expanded with accessories to serve as a fluorometer and as a nephelometer to measure TU (Müller, 1954) (Fig. 5). The Beckman DU spectrophotometer, the first commercial field instrument of its type for water analysis in the UV–VIS spectra, was introduced in 1941 and was used extensively for water analysis in the laboratory, onsite, or shipboard (Beckman et al., 1977; Simoni et al., 2003).

FIG. 5 A universal spectrophotomer used by the U.S. Geological Survey water laboratory in Waring, Wyoming, 1946. The instrument was manufactured by the Coleman Electric Company, Inc., Maywood, Illinois. The spectrophotometer was used to measure inorganic substances in water such as iron, manganese, phosphate, silicate, and sulfate in the visible range (Müller, 1954). *Image from the U.S. Geological Survey.*

However, spectrophotometers are typically insufficiently rugged for unattended field use.

A colorimeter is a light-sensitive device used for measuring the transmittance and absorbance of VIS light passing through a liquid sample at a specific wavelength. The colorimeter measures the concentration of the color that develops due to the reaction of specific reagents with the analyte to be measured in a solution. Automated systems using colorimetry for nutrients have been available in water analysis laboratories since the 1960s. These systems have been available only since mid-2000s for continuous water-quality measurements due to field maintenance challenges (Blaen et al., 2016) (Table 2). Major challenges of chemical analyzers is clogging of pumping and mixing tubes by small particles in water and biofouling and require maintenance to keep the microfluidic and microfiltration systems operational (Blaen et al., 2016) (Table 2). Biofouling, sample color, baseline drift, leakage, and buildup of the colorimetric complex in the sample lines also can lead to poor quality results (Ballinger, 1968).

Most commercially available bank-side and submersible chemical analyzers incorporate colorimeters and employ automated wet-chemical colorimetric methods (Fig. 6). The newer chemical analyzer systems can operate in an unattended mode for weeks; however, performance overtime is dependent on site conditions due to biofouling, for example. Chemical analyzer systems use microfluidic systems to pump and mix reagents in order to create colored complexes for measurement of the constituent of interest. Near-continuous measurements of nutrients in the form of their ions such as nitrate, nitrite, phosphate, ammonium, and silicate are possible (Seabird Scientific, 2017; Systea SpA, n.d.) (Table 2). Sample digestion steps are needed for colorimetric determinations of particulate forms of nutrients in water (Pellerin and Bergamasci, 2014). Prior to measurement, the sample may be prepared in several steps by the chemical analyzer before adding reagents for color development. These steps can include sample filtration and/or sample digestion to convert the substance of interest into the appropriate fraction such as dissolved or total prior to analysis. For example, total phosphorus can be measured as phosphate ion by colorimetry after chemical digestion and colorimetric reagent additions (Seabird Scientific, 2017).

In 2017, a microfluidic colorimetric chemical analyzer system, the WIZ in situ nutrient probe, became available at low cost and relatively low field maintenance requirements, while producing data of reliable quality (Systea SpA, n.d.; Chesapeake Biological Laboratory, 2017). In 2017, the WIZ optical nutrient measurement

FIG. 6 Innovative sensors and systems for continuous water-quality monitoring include (A) wet-chemical analyzers, (B) the optical nitrate sensor shown with antifouling brush, and (C) multiparameter sonde with antifouling wiper for measurement of five or more water-quality properties such as turbidity, dissolved oxygen, specific conductance, pH, and water temperature (Shoda et al., 2015).

3.3.3 Direct Optical Spectrophotometry

Direct optical spectrophotometry in the UV–VIS absorption range is the principle of measurement in several field sensors (Thomas et al., 2017). Direct optical spectrometry sensors use photometers to measure absorbance or fluorescence of various wavelengths of light specific to the constituent of interest. Direct optical sensors are simple to use, reagent free, require no preparation steps or disposables, have low operating costs, and can be applied in a broad range of water environments (Raich, 2013) (Table 2).

3.3.3.1 OPTICAL NITRATE SENSORS

Optical nitrate sensors have been used during the past few decades for wastewater monitoring as well as for freshwater, coastal, and oceanographic studies (Pellerin and Bergamasci, 2014). The first optical spectrophotometric method for analysis of nitrate in water, for example, was published in 1957 (Bastin et al., 1957) and eventually adopted as a screening method in the 19th edition of Standard Methods (American Public Health Association, American Water Works Association, and Water Environment Federation, 1995). A comparison of four types of optical nitrate sensors and their operational characteristics is described in Pellerin et al. (2013).

Sensors designed for freshwater and marine systems have lower measurement ranges compared to sensors designed to monitor wastewater-treatment processes owing to the comparatively large differences in media (Pellerin et al., 2013). Since 2010, optical nitrate sensors have gained broader usage in freshwater systems (Pellerin et al., 2013; Thomas et al., 2017) (Fig. 6).

Optical nitrate sensors are currently expensive for the initial investment, but on a "per measurement" basis as compared to colorimetric methods, they are cost effective. Most optical nitrate sensors measure in the UV range from 190 to 750nm with a peak wavelength of absorbance around 220nm for nitrate (Pellerin et al., 2013). Optical nitrate measurements are sensitive to variations in temperature and TU (Table 2). Additional interferences in nitrate measurements are caused by bromide ion and CDOM (Pellerin et al., 2013; Thomas et al., 2017).

Optical sensors measuring concentrations of nitrate have been increasing in number at water-quality monitoring locations operated by the USGS in the United States and territories since 2010. In September 2017, optical measurements of nitrate were being reported at 138 measurement stations on WaterQualityWatch (U.S. Geological Survey, 2017).

3.3.3.2 OTHER TYPES OF DIRECT OPTICAL SENSORS

Many chemical compounds in addition to nitrate can be detected using optical sensors in the UV and VIS spectra including but not limited to chemical oxygen demand; UV254; ozone, hydrogen sulfide, and monochloramine (Pellerin et al., 2013; Raich, 2013; S::CAN, 2018). Decades ago, correlations were found between the optical density of water in the UV region and the concentration of soluble organic matter in river waters and effluents (Palmer, 1970). UV absorbance at the wavelength of 254nm (UV254) is a qualitative measure of the aromatic organic content of natural water (Thomas et al., 2017). UV254 can indicate the potential during drinking-water treatment to form disinfection byproducts such as chlorinated and brominated organic compounds. UV–VIS spectrophotometers, for example, are capable of screening for algal pigments in water using nine wavelengths from UV (365nm) to VIS red (676 nm) enabling absorption over a wide spectrum (Turner Designs, 2019a).

3.3.4 Fluorescence Spectroscopy

Fluorescence spectroscopy is used in a wide array of water-quality sensors to detect compounds like water-tracing dyes, chlorophyll and accessory pigments in algae and cyanobacteria, tryptophan, fluorescent dissolved organic matter (FDOM), optical brighteners, and heavy and light oils from petroleum (Turner Designs, 2019b). Another fluorescence-based technique is luminescence spectroscopy used since 2006 to determine concentrations of DO in water (YSI, 2015a).

Fluorescence is defined as the emission of electromagnetic waves of characteristic energy when atoms or molecules decay from an excited state to a lower energy state (Wilson et al., 1986). The excitation is the result of subjecting the atom or molecule to radiation of a particular wavelength, followed almost immediately by emission of radiation at a different characteristic wavelength. In most cases, the emitted light has a longer waver length and lower energy that the absorbed light or energy. The fluorescence ceases when the external source is removed (Wilson et al., 1986). Molecules that fluoresce when excited by photons are known as fluorophores (Laitinen and Ewing, 1977; Smart and Laidlaw, 1977). Knowledge of the fluorescent properties of natural substances dates back to 1565 with the first recorded observation of a bluish opalescence in a water infusion from the wood of a small Mexican tree, *Lignum nephriticum* (Laitinen and Ewing, 1977).

Reliable laboratory-grade spectrofluorometric instruments were developed in the 1940s (Laitinen and Ewing, 1977). However, laboratory instruments with diffraction gratings were not always practical for field applications. This led to the development of filter fluorometers (Smart and Laidlaw, 1977). A filter fluorometer is an instrument containing a lamp or other means of exciting fluorescent radiation in a sample, filters to limit the light to certain ranges of wavelengths, and a detector to measure relative fluorescent intensities that vary with the concentration of the substance under examination (Wilson et al., 1986). Filter fluorometers use primary and secondary filters and other optical components to target the characteristic wavelengths of the substance being measured (Wilson et al., 1986). Filter fluorometers are sufficiently sensitive for many applications in water analysis.

Measurements with filter fluorometers need to be temperature compensated (Wilson et al., 1986). Modern field fluorometers and in situ and handheld sensors include temperature compensation algorithms. Many fluorometer measurements need correction for optical interferences in the sample such as light-scattering effects of particles, background fluorescence, and water color (Downing et al., 2012; Hambrook Berkman and Canova, 2007; Pellerin and Bergamasci, 2014; Saraceno et al., 2017; Wilson et al., 1986) (Table 2).

3.3.4.1 SENSORS FOR RHODAMINE WT AND OTHER FLUORESCENT DYES

Fluorescent dyes, such as rhodamine WT, are added to water to trace the movement of water and solutes in rivers and groundwater. The first known hydrologic study using dye tracers was made in 1877 when fluorescein dye was added to the Danube River to trace water and solute exchanges between the Danube River and a tributary of the Rhine River (Lakowicz, 2006). About 60 h after addition of the dye to the Danube River, fluorescent dye was measured using a fluoroscope in the tributary to the Rhine. Soon after, other hydrologic studies were conducted near Paris, France using fluorescein and a laboratory fluoroscope (Dole, 1906).

In the early to mid-1960s, filter fluorometers were used to help determine the streamflow and travel time of solutes for modeling the transport and fate of pollutants in streams (Smart and Laidlaw, 1977; Wilson et al., 1986). At first, many samples were collected manually over short periods of time to detect the leading edge, rising limb, peak, and falling limb of the dye-concentration curve in water (Wilson et al., 1986; YSI Environmental, 2001). Automated pumping and flow-through detection systems in the 1980s and 1990s reduced the need for frequent manual collection of samples (Wilson et al., 1986; YSI Environmental, 2001). Handheld sensor systems capable of detecting various fluorescent dyes became available in the early 2000s and replaced the large filter fluorometers for dye-concentration measurement in water (Turner Designs, 2003; YSI Environmental, 2001). Temperature compensation and background correction for water color and TU are needed for fluorometer measurements (Table 2).

3.3.4.2 FLUORESCENT CHLOROPHYLL AND ACCESSORY PIGMENTS

Chlorophyll is the molecule involved in photosynthesis and is present in all plants and in cyanobacteria. Chlorophyll *a* is commonly measured as an indicator of algal biomass in fresh and marine waters (Hambrook Berkman and Canova, 2007). The first recorded use of a field fluorometer for in situ optical fluorescence measurement of FChl concentrations in the ocean off Baja, California was in 1966 (Lorenzen, 1966). Improvements overtime in field fluorometers include ability to measure UV and VIS spectra at wavelengths indicative of different algal groups, various types of chlorophyll molecules, and accessory pigments (Turner Designs, 2019c). For example, detection of potentially toxin-producing and (or) bloom-forming algae and cyanobacteria can be evaluated from in vivo (obtained from a living organism) sensor measurements of FChl and the nontoxic accessory pigments phycocyanin in freshwaters and phycoerythrin in marine waters (Graham et al., 2017; Hambrook Berkman and Canova, 2007) (Table 2).

Benefits of the in vivo sensor measurement of FChl and other pigments include simplicity of use, selectivity and sensitivity to the pigment molecules, and rapid nondestructive measurement (Zeng and Li, 2015). In situ sensors do not require pretreatment of samples and are near reagent free except for calibration solutions. Unless corresponding samples are collected for laboratory analysis, in situ sensors have no sample shipping, preservation, or container costs. Another benefit is the ability to observe trends in phytoplankton and cyanobacteria concentrations at sub-daily time scales that provide insights into community production and dynamics (Hambrook Berkman and Canova, 2007).

Limitations of the in vivo FChl measurement are the lack of comparability overtime at the same location and among sites in different locations. Therefore, in vivo FChl and other pigments are relative measures and are reported as relative fluorescence units (RFU). The measurement is affected by cell structure, particle size, physiological state of the cell, and environmental conditions. In vivo FChl measurements also can vary among algal or cyanobacterial species. Cells may show different fluorescence intensities even with similar chlorophyll content (Hambrook Berkman and Canova, 2007). Variability between field and laboratory measurements can depend on the distribution of photosynthetic organisms in the water column and seasonal differences in algal or cyanobacterial species, their abundance, and physiological condition (Hambrook Berkman and Canova, 2007). Variation was found to be nearly 10-fold in the ratio of the in vivo measurements of fluorescence to extractable chlorophyll from analysis in the laboratory (Loftus and Seliger, 1975). Therefore, in vivo measurements of FChl provide an indirect measurement of concentration of chlorophyll *a* that is often used as a surrogate for algal abundance and biomass (Hambrook Berkman and Canova, 2007; Loftus and Seliger, 1975; Zeng and Li, 2015). Collection of periodic samples for direct analysis of extractable chlorophyll *a* collected before and after or periodically during the time in situ sensor measurements are made provides a basis for adjustment of in vivo FChl measurements to chlorophyll *a* concentration (Hambrook Berkman and Canova, 2007; Zeng and Li, 2015).

HABs have been indicated indirectly using fluorescence-based sensors for measurement of FChl and phycocyanin in freshwater. The sensor measures FChl and phycocyanin, the accessory pigment associated solely with cyanobacteria. These measurements can be combined to warn of potentially harmful levels of algal toxins in freshwater systems, particularly those that serve as source water to drinking-water treatment plants (Graham et al., 2017; Turner Designs, 2019d). Cyanobacteria that produce toxins are common to freshwater streams, lakes, reservoirs, and ponds. Under certain conditions of high nutrient concentrations and high light intensity, cyanobacteria can proliferate and form blooms in a water body. These blooms may produce toxic compounds that can be harmful to the health of the environment, animals, and humans (U.S. Environmental Protection Agency, 2015).

In September 2017, the USGS was reporting FChl at 63 water-quality monitors on Water-QualityWatch (U.S. Geological Survey, 2017). These measurements may be adjusted for comparability and consistency against results from laboratory-analyzed samples using a standardized fluorometric method (Hambrook Berkman and Canova, 2007).

3.3.4.3 FLUORESCENT DISSOLVED ORGANIC MATTER SENSOR

Measurements of particulate and dissolved organic matter (DOM) provide valuable information about the carbon cycle in fresh and marine waters with implications for drinking-water quality, contaminant transport, and environmental quality (Cole et al., 2007; Jones et al., 2014; Pellerin and Bergamasci, 2014). Organic matter in aquatic systems includes a variety of simple to complex animal- and plant-based organic molecules and associated particles that are released by living and dead organisms (Pellerin and Bergamasci, 2014).

Measuring the fraction of DOM that fluoresces is diagnostic of DOM type and amount (Pellerin and Bergamasci, 2014). The component of chromophoric (colored) DOM that fluoresces when excited by UV light is known as FDOM. Starting in the mid-1990s, monitoring of FDOM as a surrogate measurement for terrestrial inputs of carbon to the coastal ocean and to freshwater rivers and lakes have been in use (Blaen et al., 2016; Downing et al., 2012; Hudson et al., 2007; Pellerin and Bergamasci, 2014). Fluorometric sensors measuring FDOM provide key information at relevant timescales to help better understand the sources, transport, and fate of carbon in freshwater systems (Downing et al., 2012).

Sensors measuring FDOM require temperature compensation and correction for interferences from light-scattering particles and other dissolved substances over a range of conditions in field settings (Downing et al., 2012) (Table 2).

3.3.4.4 OPTICAL DISSOLVED OXYGEN SENSOR

A luminescent sensor for optical detection of DO concentration was introduced for water measurements in 2006 (YSI, 2009). Oxygen diffusing across the membrane of the sensor quenches luminescent chemical dyes in the sensor (YSI, 2009). The luminescence is measured by the sensor and compared against a reference solution. Temperature and salinity are compensated in commercially available sensor systems (YSI, 2009). Optical DO sensors have several advantages compared to polarographic electrodes for measurement of DO concentration in water (Tables 1 and 2). Compared to the electrochemical sensors for measurement of DO, optical sensors are relatively more stable with less drift overtime and do not require frequent changes of fragile membranes or additions of electrolyte filling solution and do not require flow across the sensor surface. Optical DO sensors require less field maintenance than electrochemical sensors (Wilde, n.d.) (Table 2).

Optical sensors measuring concentrations of DO are increasing in number at water-quality monitors operated by the USGS in the United States and territories. For example, in September 2017, the USGS was reporting 646 measurement stations for DO on WaterQualityWatch (U.S. Geological Survey, 2017).

3.4 Acoustic Sensors for Suspended-Sediment Concentration

Acoustic Doppler technology for measurement of suspended-sediment concentration (SSC) and particle-size distributions in flowing water can be obtained from acoustic Doppler velocity meters (ADVMs) designed to measure water velocity (Wood, 2014). Acoustic measurements are based on pulses of sound transmitted by the ADVM through water at a known frequency along two or more beams perpendicular or angled to streamflow (Wood, 2014). ADVM sensors output a pulse strength indicator called backscatter, which is the acoustic energy scattered by suspended-sediment particles back toward the signal source (Gray and Landers, 2014; Landers et al., 2016). The backscatter pulse can serve as an indicator of the concentration and particle sizes of sediment in a volume of water at a cross section of a flowing river (Wood, 2014). SSC and particle-size distributions determined by acoustic backscatter methods are influenced by the acoustic frequencies of the sensor, distance from the transducer to the particles, the volume of sediment and water within the range of the sensor, WT and water density, sediment density, and sediment shape (Gray and Landers, 2014) (Table 2).

An acoustic sediment method was first described for use in flowing water systems to estimate SSC and particle-size distributions in the mid-1970s (Urick, 1975). The method has been increasing in use since the mid-1990s for measurement of suspended sediment in the United States. A thorough discussion of principles of measurement can be found in Gray and Landers (2014).

3.5 Innovations in Handheld Meters and Water-Quality Sondes

Portable handheld meters (meters) were initially developed in the 1950s and 1960s for making rapid water-quality measurements in the field and laboratory. Portable meters have evolved overtime to be versatile instruments. Innovations include the addition of EDLs for automated data recording, data storage, and data analysis with connection to a laptop using a universal serial bus (USB) cable. Portable meters can carry global positioning system (GPS) receivers for location tracking and software for displaying measurement locations on a map. Another innovation that provides flexibility in the use of handheld meters is the universal port for interchangeable sensors.

Until the 1990s, multiple instruments had to be taken to the field to make spot measurements. For example, five instruments were needed in the field to measure SC, pH, TU, FChl, and DO. A sonde is an instrument probe that automatically logs and transmits information about its underwater surroundings (YSI, 2015a). The innovation that sondes provide for water-quality applications is the integration of multiple sensors and an EDL into one submersible instrument thus reducing the need to purchase and transport multiple instruments to the field. Sondes are designed for unattended monitoring for variable periods of time depending on the application, instrument, battery life, and other factors.

In the late 1960s, a fully submersible multiparameter sonde was developed by the Plessey Co. Ltd., Marine Systems Division and was tested in the Canadian Great Lakes by the Ministry of Ontario, Canada (Palmer, 1970). The portable sonde was capable of measuring DO, WT, suspended solids, SC, and depth (D);

was battery operated: and data were stored on magnetic tape (Palmer, 1970). The first submersible water-quality sondes could be deployed unattended in the environment for only a few days (Ballinger, 1968; Palmer, 1970). Challenges of using sondes include protecting the sonde from debris, freezing in cold weather, drying out if water levels drop, sediment clogging, geochemical fouling from shifts in redox conditions, and biofouling (Wagner et al., 2006).

Batteries were the primary power supplies in the first water-quality sondes. In the mid-1990s, extended battery life and higher electronic stability allowed unattended monitoring to increase from 2 to 3 days to 45 days or longer depending on the number of sensors and accessories. Increased battery life extended the time between service visits and allowed deployments in areas not serviced by alternating current (AC) power (YSI, 2015a). Solar panels began to be used at water-quality monitors in the late 2000s to provide power to recharge batteries in the sondes, and also to power EDLs, modems, Wi-Fi and Bluetooth receivers, computers, and pumps. Most sondes contain an internal battery as a backup to prevent loss of data from the EDL in case of a power outage.

In the 1990s, four-parameter sondes became commercially available for WT, SC, DO, and pH. The four sensor ports were not interchangeable with other available sensors such as for ORP, TU, FChl, or FDOM (YSI, 2015a). The first miniaturized fluorescence-based probes for FChl and TU were integrated into multiparameter sondes in the early 2000s (Turner Designs, 2019b; YSI, 2015a). In the early 2000s, expansion from four to six sensor ports provided expanded capacity for additional types of measurements and further reduced the need to deploy multiple instruments at field sites. Another notable feature of the newer generation of sondes, like portable meters, is the universal sensor port which allows a sonde to accept fluorometric sensors for FChl and FDOM, and other types of optical sensors such as those measuring optical nitrate.

Portable sondes can be deployed for in situ measurements at permanent water-quality monitors and have the benefit of eliminating the need to pump water to the sensors for measurements to be made. Elimination of the pump reduces power consumption and allows for remote deployments. Unless telemetry is provided, data must be manually retrieved from the remotely deployed instrument necessitating a field visit (Wagner et al., 2006; YSI, 2015a).

A new generation of water-quality sondes that became available after 2012 are smaller, lighter, and have reduced power consumption compared to those of the prior generation (YSI, 2015a). Smart sensor technology allows calibration data to be stored in the sensor memory and used to recalibrate automatically (YSI, 2015a). Newer instruments employ a wide variety of antifouling measures including a single sensor plane, copper-based antifouling materials, pump-operated flow-through systems that can turn on for measurement and off to reduce biofilm growth, and chemical injections into pump tanks, shutter systems, brushes, and wipers (Fig. 6). These innovations further extend the time interval between service visits to address biofouling (YSI, 2015a).

Information provided in the report Global Hydrological Monitoring Industry Survey describes trends in many aspects of continuous water monitoring from 2002 to 2012 (Aquatic Informatics, 2015). The report documents the responses to a questionnaire from 700 water-monitoring professionals in 90 countries representing federal, regional, state, provincial, and local governments; academic institutions; consulting firms; and industry (Aquatic Informatics, 2015). Responses to the Survey indicated that the use of multiparameter sondes increased by 35% from 2002 to 2012. In 2002, 26% of Survey respondents reported using multiparameter sondes, whereas in 2012, 61% reported using multiparameter sondes (Aquatic Informatics, 2015). Growth in water-quality monitors is taking place on a global scale (Technavio, 2016).

4 STANDARDIZED PROCEDURES FOR CONTINUOUS WATER-QUALITY MEASUREMENTS

The production of high-quality and reliable data is achieved in part by use of well-described and published methods, standardized practices and protocols, and quality-control procedures. Standardized and reliable methods are documented and published to assure that data users have the information they need to produce data that are accurate, representative of natural water conditions, reproducible, consistently collected, and citable.

4.1 Manuals and Standardized Methods for Water-Quality Sensors

Sensor manufacturers provide manuals that contain essential information about the sensors and software systems including operational instructions and requirements. More recently, manufacturers have been broadening the scope of manuals to include information on how to deploy and maintain sensors and water-quality monitors.

The USGS independently develops guidelines for continuous water-quality monitoring and makes those guidelines available in peer reviewed and published manuals. Overtime, the USGS manuals have included information on the selection, installation, operation, and maintenance of monitoring systems, quality control, and evaluation of continuous water-quality data. The USGS has a long history of producing consistent high-quality information from its multipurpose national network of water-quality sensors. Examples are drawn from the USGS and governmental and academic publications to provide examples of documented methods and quality-control procedures.

In 1966, the USGS published a technical memorandum with instructions for the USGS field personnel for processing multiparameter water-quality data recorded on a single 16-channel tape (Durum, 1978). Instructions included visiting each station regularly such as biweekly to every 3 weeks to clean and, if necessary, to recalibrate water-quality sensors. Information collected during site visits was used to quality assure the data and to provide a data quality rating for each time period. Importantly, representative water-quality samples and cross-sectional water-quality profiles were required to be collected, analyzed, and used to verify that water-quality sensors measuring at one location accurately portray the stream, lake, estuary, or embayment conditions. These instructions also were documented in publications describing the first water-quality monitors operated by the USGS (Durum, 1978; McCartney and Beamer, 1962). In 1966, a USGS report described two types of water-quality monitoring systems deployed at streamflow gage and their operational characteristics (Smoot and Blakey, 1966).

In 1983, the USGS published its first manual on quality-assurance practices for water-quality sensors (Gordon and Katzenbach, 1983). Building upon the previous guidelines, the purpose of this manual was to provide instructions for the selection, installation, and use of monitoring systems and evaluation of continuous water-quality data for WT, pH, DO, and SC. Guidelines were provided for various situations regarding site and sensor selection, installation, calibration, operation, and maintenance. Hardware and engineering principles provided by this manual ensured that the data represented the environment to be monitored and met standards of the time (Gordon and Katzenbach, 1983).

At a time when major changes were taking place in sensor monitoring, the USGS published a manual in 2000 (Wagner et al., 2000) and again in 2006 in the report series Techniques and Methods (TM) as Chapter 1-D3 (Wagner et al., 2006). This report provided guidelines on five sensors; WT, pH, DO, SC, and TU. Information in Wagner et al. (Wagner et al., 2006) encompassed a comprehensive list of topics including site- and monitor selection, sensor inspection and calibration methods, field procedures, data evaluation, drift corrections, and data computation. Also included were data record-review and data-reporting processes.

The USGS National Field Manual (NFM) for the Collection of Water-Quality Data is a compendium of methods, techniques, and procedures in 10 chapters covering a broad swath of topics in field measurement (Wilde, n.d.). Chapter A6 of the NFM contains seven sections on routine field measurements of WT, DO, SC, pH, ORP, TU, and use of multiparameter instruments (Wilde, n.d.). Chapter 7.4.2 contains a set of instructions for measuring and quality assuring in vivo FChl and phycocyanin pigments inclusive of data management and interpretation (Hambrook Berkman and Canova, 2007). Additional information on calibration, secondary standards, and quantitation of FChl is available, for example, from Turner Designs (2019d).

The USGS published a comprehensive manual on using optical nitrate sensors (Pellerin et al., 2013). The manual includes topics on selection, operating principles, sensor design and key features, sensor characterization, typical interferences, and approaches for sensor deployment (Pellerin et al., 2013). The report also includes information about maintaining sensor performance in the field, calibration protocols, quality-assurance techniques, data formats, and reporting. Although the focus of this report is UV nitrate sensors, many of the principles can be applied to other optical sensors for water-quality studies (Pellerin et al., 2013).

Other noteworthy manuals on continuous water-quality sensors and related topics are available from academic institutions such as the Virginia Institute of Marine Science (Miles, 2009). State and provincial governments provide several notable examples of continuous water-quality manuals (Government of Newfoundland and Labrador, 2013; Province of British Columbia, 2015; Shull and Lookenbill, 2015; Texas Commission on Environmental Quality (variously dated), n.d.). Of special note are the sections on groundwater monitoring and monitoring considerations under winter conditions of snow, ice, and low temperatures (Government of Newfoundland and Labrador, 2013). The variables of interest for continuous groundwater monitoring differ somewhat from surface-water monitoring with the addition of ORP to other commonly measured variables like pH, SC, and WT (Government of Newfoundland and Labrador, 2013). In addition to manuals on operation, the Texas Commission on Environmental Quality (Texas Commission on Environmental Quality, 2018) published a Quality Assurance Project Plan for the Texas continuous water-quality monitoring network.

4.2 Approaches and Standardized Methods for Surrogate Water-Quality Models

A water-quality surrogate is a continuous sensor measurement used as a proxy to statistically estimate the concentration of a water-quality constituent that cannot be readily measured or is impractical to collect at high-temporal frequency (Loving et al., 2014). The development of surrogate water-quality models began in the early- to mid-2000s using linear and log-linear regression approaches to estimate concentrations, loads, and yields of water-quality constituents from continuous water-quality data and streamflow measurements (Christensen et al., 2000; Rasmussen et al., 2005; Rasmussen and Ziegler, 2003). Surrogate measurements can be converted into values for the variable of interest using specific statistical practices and procedures. Statistical methods have been developed to correct for bias in regression estimates that is introduced from log transformations (Helsel and Hirsch, 2002).

The use of surrogate models for computing time-series values of water-quality constituents is increasing. The growth in the number of publications relying on the surrogate statistical approach for estimating concentrations and loads of water-quality variables led to the development of standardized methods. Standardization provides a mechanism to promote increased efficiency in activities performed routinely. Standardization also provides a level of consistency to assure that statistical surrogate models are routinely applied, are appropriately developed, and are statistically valid.

Many organic chemical constituents that could be modeled from surrogate measurements have statistical issues associated with values below the detection limit that require special statistical consideration. Other types of statistical models beyond linear regression may be worthwhile to include in standardized methods based on readily available examples from current publications. The need for additional standardized methods to address surrogate modeling of a broad range of water-quality constituents other than SSC is apparent.

4.2.1 Approaches for Developing Surrogate Water-Quality Models

Continuous water-quality measurements such as WT, pH, DO, TU, SC, FDOM, and other measurements including continuous streamflow and meteorological measurements can serve as proxy variables in regression analysis to determine concentrations, flux, and loads of water-quality constituents of interest. The predicted values from the surrogate regression multiplied by the concurrent measurements of streamflow can be further analyzed to compute constituent stream flux and load. Flux is sometimes referred to as instantaneous load, which has units of mass per unit time (Kraus et al., 2017). Constituent load is calculated as the integrated flux over a specified period of time and has units of mass (Kraus et al., 2017). The time over which the flux is integrated must be specified. For example, an annual load is the flux integrated over a year (Kraus et al., 2017).

Surrogate models published by the USGS and by others predict concentrations of suspended sediment, suspended solids, chloride, dissolved solids, indicator bacteria, total nitrogen (TN), total phosphorus, trace elements, and selected organic compounds (Baldwin et al., 2013; Christensen et al., 2000; Galloway, 2014; Jastram, 2014; Johnson et al., 2007; Miller et al., 2007; Rasmussen et al., 2008; Rode et al., 2016). Measurements of FDOM have been successfully used as surrogate measurements for other constituents such as methyl mercury (Bergamasci et al., 2012) and DOC (Snyder et al., 2018). Fig. 7 shows

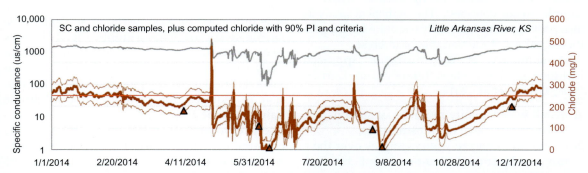

FIG. 7 Computed and measured chloride concentrations and 90% prediction intervals based on a surrogate statistical relation of specific electrical conductance (SC) for the Little Arkansas River, Kansas. Chloride concentrations and 90% prediction intervals are indicated by thick and thin orange lines, respectively. Specific electrical conductance measurements (SC) are indicated by the gray line and the secondary maximum contaminant level for chloride concentration is indicated by the red line. Triangles indicate sample measurements to confirm surrogate statistical relations (Rasmussen et al., 2016) (https://nrtwq.usgs.gov/).

a time series of SC and computed and measured chloride concentrations in 2014 for the Little Arkansas River in Kansas from a previously published statistical surrogate relation (Christensen et al., 2000; Rasmussen et al., 2016).

Once the sensor is installed and begins collecting data, a statistically relevant number of samples for laboratory analysis of the variable of interest are collected concurrently with the surrogate measurements over a range of conditions. The conditions to consider for sample collection include a range of hydrologic, weather, and climatic conditions. Sampling and measurements also need to cover the range of temporal changes related to seasonal and diurnal and interannual patterns. Other continuous measurements can include streamflow; ranges in wind speed, direction, and wave height in lakes; and tidal cycles in estuaries. Human-influenced conditions, for example, can include ranges in patterns of water use, water management, and runoff from various upstream land uses.

Results from sample analyses are used to develop a statistical regression equation between the surrogate measurements and the variable of interest. The regression equation provides the basis to estimate concentrations of the variable of interest from the high-temporal frequency surrogate measurements during periods when no samples were collected. The statistical analysis also provides an uncertainty estimate for the predicted values of the variable of interest.

A summary of the surrogate approach includes six basic steps (Francy and Darner, 2006; Loving et al., 2014):

1. install water-quality monitors at stations, store, and retrieve or transmit data,
2. collect and analyze water samples over the range of hydrologic, temporal, and chemical conditions,
3. develop site-specific regression models using sample results and in stream sensor values,
4. estimate modeled concentrations and if desired, fluxes, loads, and uncertainty,
5. continue sampling to verify models, and
6. publish modeled results and display surrogate values on the Web, if desired.

4.2.2 Standardized Methods for Computing Suspended-Sediment Concentration (SSC) and Loads From Surrogate Measurements

In 2009, the EPA reported that suspended sediment was one of the leading causes of water-quality impairment affecting beneficial uses of water in the United States (Wood, 2014). Monitoring suspended sediment and its impacts on water resources is an important activity given the scope of the sedimentation problem. Historically, samples for SSC were collected daily or with automatic samplers due its highly variable particle size and concentrations. SSC is affected by seasonal, annual, climatic, and hydrologic factors. However, cost and other factors led to a 45% decline in long-term monitoring stations for SSC from 710 in 1975 to 390 stations in 2014 (Myers and Ludtke, 2017).

The search for more efficient methods of providing reliable SSC concentration and load data led to the development of a surrogate method. A comprehensive manual for developing and using surrogate statistical models using linear regression for estimating SSC and loads from continuous TU and streamflow measurements was published by Rasmussen et al. (2009). The surrogate approach has a number of advantages over previous computational methods from daily samples (Rasmussen et al., 2009), as follows: (1) no subjective interpolation or estimation is required for periods without sample values; (2) the computational procedure is precisely reproducible; (3) the approach is less time consuming than historical interpolation techniques; and (4) estimates of uncertainty can be computed for the SSC time series.

For example, using time-series log-transformed TU and streamflow measurements, concentrations and loads of SSC were developed for the Red River of the North at Fargo, North Dakota

(Galloway, 2014). Continuous TU and stream-flow were statistically significant explanatory variables for estimating log-transformed SSC, and samples were collected and analyzed over a range of hydrological and TU conditions. The regression equation provided a satisfactory statistical relation between SSC, log-transformed TU, and log-transformed streamflow, with a standard deviation of 0.10, an adjusted R^2 of 0.95, and an average relative percent difference between measured and predicted concentrations of SSC of 12.7%. The regression equation was used to estimate SSCs with 90% prediction intervals from the TU and streamflow data for the period of record.

A manual for developing and using surrogate statistical models for estimating SSC and loads from time-series ADVM backscatter measurements and streamflow was published by Landers et al. (2016). ADVM measurements are another form of surrogate measurement of a proxy for SSC; therefore, samples for laboratory analysis must be collected periodically to develop a statistical relation to backscatter measurements (Landers et al., 2016; Wood, 2014).

Landers et al. (2016) identified advantages of the acoustic backscatter approach over other sediment-surrogate methods: (1) better representation of river cross-sectional conditions from large measurement volumes, compared to other surrogate instruments that measure data at a single point; (2) high-temporal resolution of collected data; and (3) data integrity when biofouling is present. An additional advantage of this technique is the potential expansion of monitoring SSC at sites with existing ADVMs used in streamflow-velocity monitoring.

Wood and Teasdale (2013) compared the surrogate models for SSC and loads obtained from three different sensors—acoustic backscatter, turbidimeter, and laser diffraction sensors—for two rivers in Idaho. High-frequency ADVM sensor backscatter measurements provided superior statistical surrogate models of suspended sediment at both stations, explaining more than 90% of the variability in concentrations (Wood and Teasdale, 2013).

4.2.3 Surrogate Methods Forecasting Water Quality at Recreational Beaches

A statistical modeling approach was developed to estimate the concentrations of intestinal bacteria such as *Escherichia coli* (*E. coli*) or enterococci at bathing beaches. Concentrations of *E. coli* and enterococci in relation to an established water-quality criteria or standard are used as indicators of illness risk from water-contact recreation. Modeled output from a surrogate relation is further analyzed to compute a daily forecast of the probability that *E. coli* or enterococci numbers will exceed recreational water-quality standards (Francy and Darner, 2006).

The need for the modeling approach arose from the problem of lag time of 24h or more between collecting, analyzing, and finally reporting sample results. Results are needed sooner because concentrations of *E. coli* and other indicator bacteria can change on the scale of hours. Using laboratory methods, collection, analysis, and reporting of sample results takes from 18 to 24h. Therefore, timely reporting to protect public health on the same morning as a water sample is collected is not possible. A same-day laboratory method of analysis and reporting of *E. coli*, although available, is expensive and requires a highly trained analyst. These testing requirements are beyond the reach of most beach managers.

Consequently, a step-by-step procedure was developed in 2006 and refined in 2013 describing how continuous data such as wave height, weighted 48-h antecedent rainfall, continuous TU, and WT, for example, can be used to develop models that predict bathing water quality at Lake Erie beaches (Francy et al., 2013; Francy and Darner, 2006). The procedures can be used to develop and test predictive models at other beaches. Forecasts of recreational quality from surrogate models, or Nowcasts, have been provided on the Internet since 2006 to the public for use in planning recreational beach activities (Francy et al., 2013).

In 2012, the EPA released a standardized modeling program called Virtual Beach (VB) based on the surrogate water-quality measures including TU, WT, and hydrologic and meteorological data (U.S. Environmental Protection Agency, 2018). In 2017, VB and other surrogate models were used to forecast recreational quality at 21 Lake Erie beaches and 1 recreational river. Forecasts were provided on the Nowcast website (U.S. Geological Survey, 2018). At 20 of 22 recreational areas in 2017, the VB model performed better than the previous day's laboratory result to forecast *E. coli* exceedances of the bathing water standard (U.S. Geological Survey, 2018). Tools for data management, data processing, and guidance on the use of predictive models are available for the development of daily beach water-quality forecasts (U.S. Geological Survey, 2013).

5 SENSOR PLATFORMS AND DEPLOYMENTS

The application of continuous water-quality monitoring into an ever wider range of natural environments has been made possible by innovations in sensors, how they are integrated into their support systems, and how they are deployed in the environment. These innovations include constructed platforms upon which or within which the equipment is housed. Equipment may include water level, water velocity, and current sensors, meteorological equipment, water-quality sensors and sondes, computer software, hardware such as EDLs, communication systems, GPS, batteries, AC power, and solar-power systems.

Deployment encompasses all the processes involved in getting monitoring systems up and running properly in the environment and maintaining the systems to provide reliable data on a continuous basis. Two types of sensor platforms and their methods of deployment are discussed herein: (1) stationary sensor platforms on land or in water and (2) mobile and autonomous vehicle platforms. Organizations such as USGS, other government agencies, and academic institutions operate networks of water-quality monitors with similar objectives. Data from these networks provide data that are representative of current conditions at regional and national scales and contribute information for water-quality research and assessments.

5.1 Stationary Platforms

Stationary platforms on land, such as gaging stations, are housed in shelters that protect equipment and sensors to measure, record, and transmit hydrologic, water quality, and meteorological data (Sauer and Turnipseed, 2010). Shelters are located on the banks of streams, rivers, reservoirs, and lakes; on bridges and causeways; and on other permanent structures.

Stationary platforms in open water, such as moored observatories, support equipment above and below water to make observations of hydrologic and water-quality properties. Moored observatories are secured to the bottom of the water body by wires, buoys, weights, and floats (Weller et al., 2000).

5.1.1 Stationary Platforms on Land

As previously mentioned, the pioneering work to develop and deploy water-quality monitors was initiated in 1955 on the Delaware River estuary in Pennsylvania (McCartney and Beamer, 1962) (Fig. 2). The first sensor systems typically consisted of a lower unit that contained a flow chamber with flow-through cells for pumping lake, estuary, or river water to the electronic sensors that were submerged in individual flow chambers; a middle unit that contained a servo controller to condition sensor signals to analog or digital signals and send them to the top unit that contained either an analog strip chart or a digital punch-tape paper recorder (McCartney and Beamer, 1962; Smoot and Blakey, 1966) (Fig. 8). In the ORSANCO system, the top unit also contained a data telemeter transmitter linked by 0–15 Hz teletype-grade lines linked to the sensor(s). This arrangement allowed the system to be interrogated on an hourly basis from a central location (Cleary, 1967). These systems had to be connected to AC power with an AC/DC converter for the analog sensors (McCartney and Beamer, 1962).

The first land-based water-quality sensors were housed in large shelters with dimensions of about 8 ft. × 4 ft. × 7 ft. and placed on the banks or shores of streams, lakes, and estuaries (Cleary, 1967; McCartney and Beamer, 1962; Palmer, 1970) (Figs. 2 and 8). Sensor systems were placed in the same shelters as recording water-level gages. Shelters for recording gaging stations were introduced in 1894 long before the invention of water-quality sensors (Follansbee, 1994). Colocation conveniently provided the same data logging and telemetry systems for water-quality sensor systems as used for recording and transmitting water stage and/or tidal levels. Additional inputs for meteorological variables such as rainfall, air temperature, and photosynthetically active radiation (PAR) were accommodated on 10-channel recorders. By the mid-1960s, a system measuring WT and SC became available that operated on battery power with submersible in situ sensors and no pump (Smoot and Blakey, 1966).

By the 1980s, submersible in situ sensors were being placed at stationary installations eliminating the need for pumps and flow chambers. Sensors were protected by placement in a long polyvinyl chloride (PVC) plastic sleeve that extended from the gage house to the

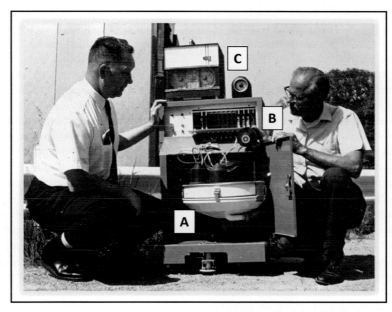

FIG. 8 One of several types of water-quality monitors developed by the U.S. Geological Survey in the 1960s. This system consisted of (A) a lower unit with submersible pump and flow-through chamber containing the sensors. Water was to be pumped to the sensors that were submerged in individual flow chambers. A middle unit (B) contained a programmed servo controller device to issue commands to record data from a selected channel. The top of the unit (C) was comprised of a digital punch-tape recorder and timer (Smoot and Blakey, 1966). *Image from the U.S. Geological Survey.*

measurement position in the stream, lake, or river. The sensors inside the sleeve were protected from damage by debris, low water levels, and ice. The lower end of the pipe was perforated with sufficient 1-in holes to allow the water to flow around the sensors. The ability to retrieve sensors for servicing under all conditions is an important aspect of installation and deployment (Gordon and Katzenbach, 1983; Wagner et al., 2000). To protect the shelter and its contents from floods, the structure is typically located on land above the anticipated 200-year (0.5% annual exceedance probability) flood stages of the river (Sauer and Turnipseed, 2010).

Overtime, the variety of shelters has increased to fit expanding capabilities. Many modern instrument shelters are built to house instruments measuring water stage, velocity, and water-quality instruments. The shelters also accommodate externally placed antenna for data transmission and solar panels for power. Shelters internally house EDLs, modems, computers, and microprocessor-based instruments for sensor control. Additional information on shelters including requirements and considerations for deployment in a variety of situations are found in Sauer and Turnipseed (Sauer and Turnipseed, 2010). Traditional 4 ft. × 4 ft. × 7 ft. shelters have been emplaced at thousands of water-quality monitors and streamflow measurement stations nationwide by the USGS.

An example of a newer type of installation is located on the Kankakee River near Davis, Indiana (Shoda et al., 2015) (Fig. 9). The gage is one of the first to be constructed by the USGS without a shelter that can accommodate some of the most modern equipment and instruments. The gage is instrumented with a water-level sensor for streamflow measurement, a rain gage, a submersible six-parameter water-quality sonde, a nitrate sensor, and a separate microfluidic colorimetric phosphate sensor. Continuous streamflow and water-quality data are collected at 5–15 min intervals, stored, transmitted, and made publicly accessible on the USGS websites such as WaterQualityWatch within 1 h. The site also provides a satisfactory location for representative stream sampling of selected water-quality constituents for laboratory analysis. Laboratory results are used to develop a time series for analyzed constituents based on the surrogate measurements of streamflow and selected continuous water-quality data (Shoda et al., 2015).

5.1.2 Stationary Platforms in Open Water

Moored observatories are data-collection platforms (DCPs) that are anchored on the bottom in open waters of lakes, estuaries, embayments, and the ocean (LaCapra, 2017). A mooring is an anchor on the seafloor, attached to a line, chain, cable, or some combination that is held taut by a flotation buoy on or near the surface (LaCapra, 2017). Subsurface and surface buoys keep the mooring upright and support instruments (Weller et al., 2000). Scientific instruments attached to a mooring line and buoy can make continuous measurements from above the water surface to the bottom. Buoys mounted above water support meteorological sensors, satellite or radio transmitters and receivers, and solar panels (Weller et al., 2000). An example of a surface buoy for water-quality monitoring in the San Francisco Bay Delta is shown in Fig. 10. Buoys located under water hold various instruments such as water-quality sensors, current meters, power supplies, and data recorders (Weller et al., 2000).

The first observatories on the open ocean were weather ships that from 1940 to 1980 recorded ocean temperatures and provided weather conditions for navigation (Dinsmore, 1996; Weller et al., 2000). Often weather ships hosted oceanographic researchers with scientific missions. This network of ocean weather stations was eventually replaced by satellite systems serving the same purpose (Weller et al., 2000). The end of the weather ship network left

FIG. 9 The U.S. Geological Survey's sentinel gage on the Kankakee River at Davis, Indiana (05515500). The sentinel gage is one of the first gages to be constructed without a shelter. The platform is equipped with a rain gage, streamflow gage, and a submersible water-quality sonde for measurement of pH, turbidity, water temperature, specific electrical conductance, and concentrations of dissolved oxygen. A separate optical sensor is provided for measurement of nitrate. Data are transmitted to the GOES satellite and subsequently displayed within 1 h on the National Water-Information System website (Shoda et al., 2015).

researchers and scientists without a platform from which to make ocean observations (Weller et al., 2000). In the 1980s, the National Oceanic and Atmospheric Administration (NOAA) and research organizations such as the Woods Hole Oceanographic Institute (WHOI) have been deploying subsurface moorings for weather and climate prediction and for oceanographic research (Weller et al., 2000). In the 1990s, two test-bed moorings provided some of the first interdisciplinary deepwater platforms for oceanographic monitoring and research since the weather ship network (Dickey et al., 2006; Weller et al., 2000).

Surface moorings face challenges from exposure to the harsh ocean environment that includes corrosive levels of salinity, storms, wind, and waves. Ocean moorings are built to

FIG. 10 Deployment of monitoring buoy from which multiparameter water-quality sondes are suspended (Kraus et al., 2017). *Photograph by Bryan Downing, U.S. Geological Survey, September 9, 2014.*

withstand extreme conditions such as category 2 or higher hurricanes. Moorings and instruments are subject to biofouling from a variety of ocean organisms that require maintenance to remove (LaCapra, 2017). Subsurface moorings are subject to less stressful conditions and now routinely collect information for as many as 2 years without servicing (Weller et al., 2000).

From 1993 to 2008, the Bermuda Test-bed Mooring (BTM) and from 1997 to 2000, the HALE-ALOHA (H-A) test-bed mooring, were anchored at depths exceeding 4500 m. The BTM was located about 80 km southeast of Bermuda (Dickey et al., 2006; Glasgow et al., 2004). The H-A mooring was located about 100 km northeast of Hawaii, USA. One goal of the BTM and H-A platforms was to develop and test new deep-sea technologies, interdisciplinary sensors, antifouling techniques, hardware, and data telemetry systems. Another goal was to facilitate scientific studies requiring high-temporal frequency, long-term geophysical, geochemical, optical, and biological data (Dickey et al., 2006).

Instruments above the ocean surface of the BTM and H-A moorings collected meteorological and spectral radiometric measurements from a buoy tower. Subsurface measurements included ocean currents, continuous measurements of WT, SC, light absorption and scattering properties of seawater, other optical properties, and concentrations of nitrate and trace elements. The BTM profiler system measured these properties and characteristics at various depths from 2 to 500 m. Data were telemetered to ship or to shore.

Using BTM data, detailed studies were published on nutrient fluxes due to eddies, effects of hurricanes on physical, optical, and biogeochemical properties of seawater, and ground-truth data for satellite-derived ocean color. Continuous, long-term data collected from the BTM and H-A mooring have led the way for detailed studies of a variety of other physical, chemical, biooptical, and ecological processes on time scales from minutes to years (Dickey et al., 2006).

The test-bed moorings have served as models for more than a dozen additional ocean, coastal, and coral reef research moorings in the United States and internationally that host investigations of the ocean environment (Dickey et al., 2006). The test-bed moorings were especially important to the design of the National Science Foundation's multi-institutional Ocean Observatories Initiative (OOI). OOI is a system of observatories such as the Coastal Pioneer Array managed by the WHOI located offshore on the continental shelf of New England (LaCapra, 2017).

Another example of a water-quality monitor on a stationary floating platform is located on Lake Mead, a reservoir on the Colorado River in Arizona and Nevada in the southwestern United States. A study was undertaken to better understand reservoir inflow, circulation, and ecosystem processes that influence one of the most important water resources in the region (Veley and Moran, 2012). The study collected data from 2004 to 2009 and assessed reservoir quality during a time when water levels fell 17 ft. due to long-term drought conditions in the Colorado River basin (Veley and Moran, 2012). Automated variable-depth profiling systems mounted on a floating platform were installed to collect continuous water-quality data at five locations in the reservoir (Veley and Moran, 2012).

The five floating platforms on Lake Mead included the following equipment: YSI variable-depth-winch assembly for lowering and raising the YSI 6000 multiparameter sonde to profile six variables at specific depths in the water column (Veley and Moran, 2012). The sonde was equipped with sensors for WT, SC, DO, pH, TU, and D. The platform also featured a Campbell Scientific CR10X EDL/sensor control module for all the systems onboard the platform, Wavecom GSM cellular modem package for communication, and a 12-V, 95 A battery charged by a 30-W solar panel. The solar panel powered the instruments, communication systems, and winch for making profile measurements. Data were transmitted to a secured base station for daily data review, processing, programming updates, and troubleshooting, as needed.

5.1.3 Monitoring and Research Networks

The National Ecological Observatory Network (NEON) is the newest long-term observatory network established in the United States in 2010 (National Ecological Observatory Network, 2018). The NEON provides open-source data that characterize terrestrial, freshwater, and atmospheric components of ecosystems across the United States over a 30-year timeframe (National Ecological Observatory Network, 2018).

The NEON network hosts integrated observatories of terrestrial, aquatic, and atmospheric measurement platforms. As of 2018, the NEON is collecting water data, such as water properties and chemical and biological characteristics, on stationary platforms at 34 sites; 24 streams, 3 rivers, and 7 lakes (National Ecological Observatory Network, 2018). Land-based stationary platforms are used for streams and buoy-mounted sensors are used for lakes and large rivers. Sensor platforms continuously collect data that characterize hydrologic conditions and surface-water quality, including DO, pH, SC, TU, FChl, nitrate, FDOM, and underwater PAR (National Ecological Observatory Network, 2018). Groundwater wells, installed in multiple aquatic sites, collect groundwater elevation, WT, and SC (National Ecological Observatory Network, 2018). As of 2018, the NEON network was still completing its build-out phase of ecological observatories.

As of December 2016, the USGS operated a network of 2685 water-quality monitors in the United States and Puerto Rico; an increase from 1555 in December 2006. This represents a 73% increase overtime (U.S. Geological Survey, n.d.) (Fig. 11). The USGS network of water-quality monitors has multiple objectives and is cooperatively supported by the USGS and a large number of state and local partners. The number of water-quality monitors is about four times greater in surface water than in groundwater. The number of water-quality monitors in springs has remained small yet relatively constant overtime ranging from 16 to 29 water-quality monitors.

According to an industry survey, in 2012, the number of water-quality, streamflow, and other types of monitors per hydrologic network globally was 269, an 18% increase in network size since 2002 (Aquatic Informatics, 2015). By 2022, hydrologic networks are likely to increase on average by 53% according to the survey respondents due to unmet needs for monitoring. Continuous water-quality monitoring was identified by a majority of survey respondents, 44% and 41%, respectively, as a primary or secondary purpose of monitoring (Aquatic Informatics, 2015).

5.2 Mobile Platforms

A wide variety of mobile platforms have been developed overtime, some operate as stand-alone monitors like pontoon boats, while others are associated with mooring observatories in the open ocean like autonomous underwater vehicles (AUVs). Mobile platforms have the advantage of being able to characterize three-dimensional (3D) patterns in ocean circulation and chemical gradients overtime. AUVs are programmed to traverse a path while collecting data. AUV capabilities are well beyond those of a mooring or research vessel to provide spatial surveys overtime. AUV paths are repeated overtime to track temporal changes in the water environment (Eriksen et al., 2001). When deployed in warm, sunlit, and biologically productive waters, AUVs are vulnerable to biofouling and require more frequent maintenance than if deployed in cold, nutrient-poor waters without sunlight (LaCapra, 2017).

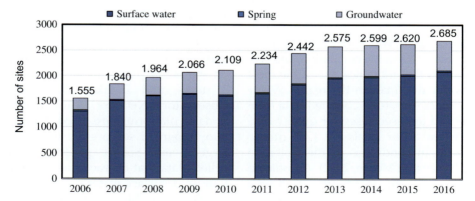

FIG. 11 The number of water-quality monitors operated by the U.S. Geological Survey steadily increased by 73% from 2006 to 2016. In December of calendar year 2016, water-quality monitors numbered 2685; whereas, in December of calendar year 2006 water-quality monitors sites numbered 1555 (U.S. Geological Survey, n.d.).

5.2.1 *Pontoon Boats*

Pontoon boats are mobile platforms that can be anchored for periods of time as a stationary platform. A pontoon boat can be trailered to the sampling location, launched from a boat ramp, and then driven to the study area. The pontoon boat can be used for transects and longitudinal studies. In 2003, the NOAA's Monitoring and Event Response for Harmful Algal Blooms (MERHAB) Program funded development of the pontoon boat named MER-HAB Autonomous Research Vessel In Situ or MARVIN (Glasgow et al., 2004).

The MARVIN was designed and built by the AMJ Environmental in St. Petersburg, Florida, which was affiliated with Glasgow et al. (2004) and YSI (2007). Research onboard the MARVIN addresses the health of humans and marine animals at risk from HABs. In 2006, two additional MARVIN pontoon monitoring systems were built. From 2005 to 2007, MARVIN pontoon boats were deployed in three locations in Florida, in the St. Johns River, the Indian River Lagoon near Cape Canaveral, and the Caloosahatchee River, to measure and track HABs (YSI, 2007).

Continuous water-quality sensors and chemical analyzers onboard the MARVIN recorded indicators and drivers of HABs. A multiparameter sonde measured pH, DO, WT, SC, FChl, and TU. Nitrate and phosphate concentrations were monitored with an automated nutrient analyzer using colorimetric methods. Underwater quantum radiation sensors provided information on ambient light and PAR. Testing of additional instrumentation in 2006 included Turner Designs Phytoflash (later Aquaflash) for accurate estimations for total chlorophyll *a* and photosynthetic efficiency, a FDOM sensor, and an optical nitrate sensor. A water-velocity meter measured water level and current speed and direction. Above deck, an instrument tower contained sensors for meteorological measurements. A GOES satellite transmitter, cellphone modem, and cellphone provided communication and data transmission. Power for the entire MARVIN system was supplied by two 12-V deep-cycle marine batteries charged by solar panels (YSI, 2007).

5.2.2 *Autonomous Underwater Vehicles*

Autonomous and robotic vehicles provide a means to perform 3D synoptic surveys of aquatic areas relatively quickly and inexpensively with minimum effort. AUVs are tetherless and can carry several electronic and optical sensors for measuring gradients in water-quality properties as well as currents and other physical properties (Curtin et al., 1993). AUVs provide detailed spatial information through programmed survey paths.

There are two types of AUVs; one type uses a battery-powered propeller system for propulsion and the other type uses a battery operated buoyancy controller for propulsion and moves horizontally and vertically on wings. AUVs need software programs to control rendezvous and docking maneuvers, survey paths, gradient detection and following, obstacle avoidance, adaptive sampling, terrain following, and fault detection and recovery (Curtin et al., 1993). AUVs are designed to operate in marine and freshwater environments but are limited by battery life and currents traveling faster than their operational speed. AUVs are programmed to travel a survey path at constant depth, constant height above the bottom, or using an undulating path (Jackson, 2013). AUVs can be launched from a ship or from land.

AUVs perform integrated synoptic surveys collecting a variety of hydrometric, bathymetric, and water-quality data (Jackson, 2013). AUVs can carry instruments to create high-resolution bathymetry, side-scan sonar imagery, sensors for measurement of multiparameter water-quality properties, water velocity, and currents. Data are collected simultaneously at similar spatial and temporal resolution (Jackson, 2013).

5.2.2.1 PROPELLER-DRIVEN AUVs

Propeller-driven AUV development began in the 1980s (Curtin et al., 1993). The Odyssey was a prototype propeller-driven AUV designed to complete sampling tasks autonomously (Curtin et al., 1993). The Odyssey was 2.1 m long, had a maximum diameter of 0.6 m, and was rated to a depth of 6700 m. A variety of subsystems such as water-property sensors, propulsion motor, control-surface actuators, and sonar transducers were located inside the unit (Curtin et al., 1993). In the early 2000s, propeller-driven AUVs came into broad use in coastal shelf areas and in the open ocean. Currently, there are at least nine manufacturers of propeller-driven ocean-going AUVs (UST, 2018).

YSI introduced a propeller-driven AUV in 2008 for high-resolution water quality and bathymetry mapping of inland and coastal waters (YSI, 2015a). The AUV has flexible options for water-quality monitoring, bottom mapping, and water current profiling and can be deployed by one person (YSI, 2015a). Applications include baseline monitoring, bottom mapping, event response, and point- and nonpoint-pollutant source characterization (YSI, 2015a). Instruments onboard the AUV system for conducting synoptic surveys include acoustic Doppler current profiler, Doppler velocimetry log, single-beam echo sounder, side-scan sonar, differential GPS receiver, and a 6-port sonde for measurement of WT, SC, pH, DO, and TU (Jackson, 2013). Additional sensors for FChl, phycocyanin, and rhodamine WT dye can be added or are interchangeable with other sensors.

The newest model AUV has forward object avoidance, built in GPS, Wi-Fi, continuous data streaming, and operates at speeds from 1 to 4 knots or 1.2 to 4.6 mph. At 2.5 knots (3 mph) mission time of the AUV is about 14 h on battery charge. The AUV can operate at depths as much as 100 m (YSI, 2015a). The dimensions of the various AUV models range from 152 to 216 cm in length, have a tube diameter of 14.7 cm, and weigh from 27 to 39 kg (YSI, 2018).

Jackson (2013) describes several applications of EcoMapper. One application was mapping near-surface SC levels in Milwaukee Harbor in Wisconsin, USA to identify the location of an effluent outfall and areal extent of the plume of wastewater effluent within a river plume of high SC. Another survey was made along the nearshore area of a public bathing beach on Lake Erie, Ohio, USA to understand the effects of the influent of a nearby contaminated tributary stream on beach-water quality. The EcoMapper AUV also has been used for depth and transect profiling of water-quality variables in stratified lakes and reservoirs (Jackson, 2013).

An example of a propeller-driven AUV for deployment in marine coastal areas is the REMUS 600 AUV, manufactured by Hydroid-Kongsberg (Kongsberg, 2018). The REMUS 600 AUV is deployed at the coastal pioneer array (array) operated by the WHOI (LaCapra, 2017). The array was built from 2011 to 2016 as a network of 10 stationary observatory moorings at seven locations with AUVs to travel paths spanning 160 miles^2 (414 km^2). AUVs are complimentary additions to the stationary moorings at oceanographic observatories like the Array. The location of the Array, 90 miles (145 km) offshore near the edge of New England's continental shelf, is optimal for investigations of currents, salinity, and nutrients that affect marine life and fisheries on the shelf (LaCapra, 2017). The array is part of the National Science Foundation's OOI.

The array uses a stationary mooring for vertical profiles and propeller-driven AUVs and gliders for spatial surveys. See Section 5.2.2.2 for glider description. Two propeller-driven REMUS 600 AUVs and six Slocum gliders are programmed to collect measurements beyond the 160-miles^2 Array extending across about 9300 miles^2 (24,087 km^2) near the shelf edge (LaCapra, 2017). The AUVs collect an integrated set of physical, biological, and chemical properties, as well as the spatial and temporal variability.

The REMUS 600 AUV is battery powered with a propeller system that can traverse a path at 3.0 knots (3.5 mph), over a period of 2 days on a path about 161 km long (Kongsberg, 2018). The REMUS 600 is 3.25 m long, 32.4 cm in diameter, and weighs 240 kg and can dive to a depth of 600 m. As part of the array, the REMUS 600 AUVs collect continuous measurements of current velocity, DO, nitrate concentration, and other water properties on its programmed surveys of the continental shelf. When submerged, a REMUS AUV communicates with a research vessel using an acoustic modem. When at the surface, the REMUS 600 communicates by satellite at preprogrammed locations. Scientists on the ship can monitor a vehicle's status and communicate any needed adjustments.

5.2.2.2 GLIDERS

AUV gliders passively travel the ocean by changing their buoyancy by adjusting their volume to be either slightly smaller or larger than that of an equal mass of seawater (Eriksen et al., 2001). Gliders have wings that provide hydrodynamic lift to propel the vehicle forward as it sinks or rises (Eriksen et al., 2001; LaCapra, 2017). Power for the buoyancy system and sensors is supplied by lithium or alkaline batteries. Gliders sample the ocean in a saw-tooth flight path (Rudnick et al., 2004). Gliders also can perform as virtual profilers by maintaining the same horizontal position and changing vertical position in the water column (Rudnick et al., 2004). Buoyancy-driven gliders travel slowly and consume small amounts of power allowing them to be deployed autonomously for as much as a year. Due to their slow speeds, gliders can be pushed off course by ocean currents that exceed their forward velocity.

There are three basic types of gliders that were developed in the 1990s (Rudnick et al., 2004). All three types have approximately the same (1) lengths from 150 to 200 cm; (2) wing spans from 100 to 120 cm; (3) diameters from 20 to 30 cm, and (4) weights from 51 to 52 kg. All three carry similar payloads from 3.5 to 5.0 kg (Rudnick et al., 2004). The Seaglider (Eriksen et al., 2001) was built at the University of Washington and was designed to operate most efficiently in the open ocean in missions of several months duration and paths of several thousand km. The Slocum Battery was manufactured by the Webb Research Corp. and is optimized for shallow and coastal operations. The Spray (Sherman et al., 2001) was built at the Scripps Institute of Oceanography and designed for efficient deepwater surveys. Initially, gliders could carry only a few sensors for measurement of SC, D, and WT (CDT). Recently, gliders have provided options to add fluorescence sensors to detect FChl, phycoerythrin, and FDOM. An optical nitrate sensor can be mounted on the outside of some of the gliders (Jones et al., 2011).

The coastal pioneer array supports six Slocum gliders. The Slocum gliders are 1.5 m long, 22 cm in diameter, travel at a speed of about 0.6 knots (0.7 mph), and weigh 64 kg. The Slocum gliders travel a path as much as 1500 miles long. The gliders can dive to depths of either 200 m or 1000 m to collect data on WT, salinity, and depth.

6 DATA ACQUISITION, STORAGE, AND TELEMETRY

This section and Section 7 draw upon information from the continuous water monitoring, electronic data collection, and telemetry systems of the USGS over the last 7 decades. The USGS uses basically the same software and hardware systems for time-series precipitation, water, and water-quality data; therefore, the following sections do not differentiate among data types. The section also provides some examples of the systems in place at other organizations and relevant results from a 2012 survey of data storage and telemetry equipment and software systems used by water-monitoring agencies.

6.1 Data Acquisition and Transfer

A major challenge before the 1950s was retrieval of water data from field stations. The manual extraction of large amounts of data from paper strip charts on which measurements were recorded with an ink pen greatly limited hydrologic science (Blakey, 1970). An important innovation was the introduction of the digital-punch paper tape so that data from analog instrument signals could be recorded digitally and then machine processed for computer entry (Blakey, 1970). The same punch paper tapes that were introduced to record water stage were used for water-quality sensors (McCartney and Beamer, 1962; Smoot and Blakey, 1966). By 1965, digital-punch paper tapes became routine equipment for recording time-series data on water levels, precipitation, and water quality (Blakey, 1970). By 1968, streamgages and water-quality monitors were equipped with transmitters for telemetering data to a base station using exclusively leased 0–15 Hz teletype-grade phone lines (Blakey, 1970).

In the 1970s, new technologies, especially in the field of electronics, led to several innovations in the digital recording and storage of time-series data. Digital-data storage came into wide-scale use starting in the mid-1970s (Sauer and Turnipseed, 2010). An electronic data logger or EDL is a generic term for digital-data storage device located on a sensor or sonde either internally on a memory module or on a removable memory card. Data are retrieved either by downloading directly from the EDL to a field computer or by removing the memory card and transferring the data from the card to a computer. Modern EDLs can store relatively large amounts of data from instruments without frequent servicing (Sauer and Turnipseed, 2010).

EDLs and DCPs could be programmed to electronically record measurements on a specific, regular time interval, or on a user-defined schedule. Electronic timers in data-acquisition systems are typically very accurate, a critical need in time-series data collection (Sauer and Turnipseed, 2010). Data on DCPs were transferred by a landline or cellular telephone modem, or a land-based radio such as a spread-spectrum radio that sends data using low power across many different frequencies. DCPs store relatively large amounts of data received from individual instrument recorders.

The use of digital multichannel EDLs and DCPs by water-monitoring organizations has increased globally from 37% in 2002 to 71% in 2012 with the remaining 29% using analog strip chart recorders (Aquatic Informatics, 2015).

6.2 Data Delivery

The ability to transmit water data in a timely manner began well before the advent of digital telemetry systems. Starting in 1921, a system for emergency transmission of water data used a landline telephone to deliver data from the USGS streamgages to selected governmental agencies involved in water management (Hirsch and Fisher, 2014). By 1931, data could be transmitted by radio technologies (Hirsch and Fisher, 2014). These communication systems were limited to use by the USGS, U.S. Bureau of Reclamation, National Weather Service, or U.S. Army Corps of Engineers (Hirsch and Fisher, 2014).

Major advances that began in the 1970s and accelerated after 2000 have greatly improved the timeliness of water data delivery. These advances have allowed water data to be made available to users very quickly relative to the time of data collection (Wong and Kerkez, 2016). Real-time (RT) data refer to data collected and delivered to data users at similar frequencies. "RT data are data made available as soon as they are collected in order to make a decision within a constrained time window" (Wong and Kerkez, 2016). For example, RT data are necessary to manage hydroelectric generating stations and to adjust process controls in water and wastewater plants. Use of RT data in environmental monitoring is growing but has not yet been widely adopted (Wong and Kerkez, 2016).

Near-RT data refers to data delivered rapidly relative to frequency of collection but at a slower rate relative to RT data. Almost all continuous water-quality measurements from field sensors are collected at 5-, 15-, or 60-min frequencies.

6 DATA ACQUISITION, STORAGE, AND TELEMETRY

Data are processed, transmitted, and disseminated at frequencies typically from 1 to 4 h (U.S. Geological Survey, 1998; U.S. Geological Survey, 2002). Near-RT data are satisfactory for most water resources decision-making purposes and so is the more common form of delivery of water data (Wong and Kerkez, 2016).

6.2.1 Data Transmission by GOES Satellite Systems

A large improvement in water data transmission and delivery came about as a result of the development of earth-orbiting satellites. Beginning in 1975, the Geostationary Operational Environmental Satellite-2 (GOES-2) became an important method of telecommunication for water data for the USGS and weather data for the National Weather Service (Glasgow et al., 2004). The GOES satellites are now available to other governmental and nongovernmental agencies. The GOES system uses geosynchronous satellites that orbit the Earth in the equatorial plane, approximately 23,000 mi above the Earth's surface. Currently, transmissions are sent to GOES-15 which was launched in 2010 (Sauer and Turnipseed, 2010).

Data from the GOES satellite are received by the Data Collection and Automated Processing System (DAPS) in Wallops, Virginia. DAPS is a centralized computer-based processing unit that monitors the GOES satellite transmissions. Each DCP has a scheduled transmission interval that ranges from 1 to 4 h, although during flood events, the DCP can be programmed to transmit as often as the recording interval of the data, for example, every 15 min. DAPS distributes the new transmissions to the USGS receiving stations through the domestic communications satellite (DOMSAT). Raw (uncorrected) transmissions from the DOMSAT are translated into a computer-readable format by using a decoding program. Since 1990, delivery times of continuous measurements have decreased to hourly or more frequently on the Internet (U.S. Geological Survey, 1998, 2002). A schematic of the flow of data from the land surface to communication and environmental satellites and then to the Internet is provided in Fig. 12.

6.2.2 Cellular, Wireless, and Other Communication Methods

As of 2018, other communications methods are available such as Bluetooth or Wi-Fi for wireless broadcast of data to nearby tablets or laptops, or WaterLOG modems and satellite transmitters for long-range communications (Sauer and Turnipseed, 2010). Current-generation field sensors can be linked together through wireless sensor networks and can deliver data to the Internet via an Ethernet router with 4G/3G/2G interfaces or to a base station on landline telephone or within a short range by wireless (Wi-Fi) connection (Hart and Martinez, 2006; Libelium, 2018; YSI, 2015a). Web-enabled communication systems provide efficiency and improve reliability of water-monitoring data in two ways because managers of continuous-monitoring networks can interrogate monitoring stations from the office. The length of time between field maintenance trips may be extended if it is determined in the office that operational performance of the monitor is acceptable. Conversely, if a water monitor has malfunctioned, a maintenance field trip can be promptly dispatched to reduce the loss or corruption of data.

The Urban Coastal Institute (UCI) at the Monmouth University supports 11 long-term, near-RT water-quality monitors located in the northern estuaries of New Jersey (Tiedemann et al., 2009). Data are collected by YSI 6600 EDS sondes and multiparameter data logger systems, designed for long term, in situ monitoring. Each seven-parameter YSI 6600 sonde reports data every 15 min. The YSI 6200 Data-Acquisition System is programmed to execute sensor sampling, biofouling maintenance, data buffering, and cellular and radio telemetry to a base station at the Monmouth University (Nichols, n.d.; Tiedemann et al., 2009). The data acquisition

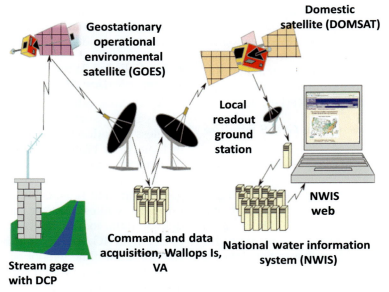

FIG. 12 Schematic of data transmission from a streamflow gage to the internet. Data from the GOES satellite are received every 1–4 h from the Data-Collection Platform (DCP) located at the streamflow gage and transmitted to the Command and Data Acquisition (CDA) system in Wallops Island, Virginia. The CDA system distributes the new transmissions to the U.S. Geological Survey (USGS) local readout ground stations through the domestic communications satellite (DOMSAT). Raw transmissions from the DOMSAT are translated into a computer-readable format by using a decoding program and are uploaded to the USGS's National Water-Information System web page (NWISweb) (Sauer and Turnipseed, 2010).

and telemetry system then provides data to the UCI public website in near RT.

The use of digital technologies for transmitting time-series data is increasing (YSI, 2015a). In 2012, EDLs were reported as the most frequently used data storage and retrieval technology, installed at 60% of monitoring stations worldwide. Next most frequently used were telephone (33%), satellite (27%), radio (20%), and web-enable sensors (20%). There was a steep uptrend in the adoption of web-enabled sensors from 4% of stations in 2002 to 20% of stations in 2012 (YSI, 2015a). By 2022, it is estimated that approximately 40% of stations will use either web-enabled sensors, satellite, and/or telephone to transmit water-monitoring data. Less than a quarter, about 22%, of stations are projected to continue to rely on analog systems for data retrieval by 2022. Percentages in the industry report can sum to greater than 100% due to redundant communications technologies used at a single station (YSI, 2015a).

7 DATA-PROCESSING SOFTWARE, WATER-INFORMATION SYSTEMS, AND WEB APPLICATIONS

Sensors collect large amounts of data overtime. For example, WT sensor observations collected every 15 min for a year will result in more than 35,000 observations. In 2014, there were 11,264 water-monitoring stations operated by the USGS at which one or more water properties or chemical characteristics were being measured at high frequency, typically every 15 min (Hirsch and Fisher, 2014). Starting in 2007 and onward, the USGS began providing quality-assured high-frequency values online (Hirsch and Fisher, 2014).

The number of water-quality, streamflow, and other types of water monitors per hydrologic network globally was 269 in 2012 and the size of networks is anticipated to increase by 2022 (YSI, 2015a). The contributions of large datasets to research, data analysis, and scientific understanding can be offset by data management challenges in what has become known as the sensor data deluge (Porter et al., 2012). The sensor data deluge is a condition in which more data are collected than can be effectively and efficiently managed resulting in missed opportunities to analyze and interpret data and to develop new conceptual understandings of the earth system and its ecological processes (Porter et al., 2012). Given the large amount of time-series data being collected by water-monitoring agencies and organizations, specialized computer software and hardware are needed for data management.

Data from water-quality monitors delivered in near RT are raw data or data that are only partially processed and corrected. These data are often needed in RT or near RT by data users for purposes of daily operational decision making or emergency response. Detailed review and approval of data after delivery from hydrologic platforms remains a time intensive but critically important process. A process to review data after initial arrival from a water-quality monitor or other instrument is facilitated by data management software.

In addition to specialized software for data processing, databases serve as data repositories for many millions of high-frequency data values overtime. Computer software and hardware also are needed to provide users with a method to select and retrieve the desired data for analysis as efficiently as possible. Innovations overtime in national computerized databases have been important for sharing these data with water managers, academic researchers, private companies, and the public.

This section describes the evolution of software, hardware, and computer database systems in use at the USGS from the 1950s onward. The USGS solutions described herein represent a national scale example and are well documented in publications. The USGS uses the same software and hardware systems for time-series precipitation, water, and water-quality data; therefore, this section uses applicable examples of all types.

7.1 Data-Processing Software and Water-Information Systems

In the late 1960s, the USGS began to provide data processing for its state offices and began to create a national database for water data. In 1971, the first of such database systems was the National Water Data Storage and Retrieval System (WATSTORE). Time-series data were an important feature of WATSTORE. WATSTORE was operated and maintained on the central mainframe computer facility at the USGS National Center in Reston, Virginia, USA and at approximately 50 terminals in state and field offices throughout the United States. Direct access was possible only through terminal equipment capable of interfacing with an IBM 370/155 computer (Bartholoma et al., 2003). WATSTORE had only minor software upgrades and limited capacity for processing the types and amounts of data that could be stored within the system.

At the time of WATSTORE, the processing of time-series data by digital computer was in a state of flux because of continual improvement in data storage and processing procedures (Rantz, 1982). Typically, records of time-series data were manually checked, reviewed, and approved prior to publication. Annual dissemination of water data reports as paper publications had been the primary method of data release. The national WATSTORE database was updated annually from state mainframes on the order of 1–2 years after the data were collected. Paper publications were useful for purposes that were not time sensitive such as water resources planning and hydrologic research, but not for purposes of immediate need such as flooding, operational decision making at dams and levees, and water hazard alerts (Hirsch and Fisher, 2014). Also, paper reports were limited to reporting summarized data such as daily values rather than individual unit values of high-frequency measurements.

WATSTORE was eventually replaced beginning in the mid-1980s with a new database that was installed on a distributed network of minicomputers and fileservers across the country for the storage and retrieval of water data. The National Water-Information System I (NWIS-I) replaced WATSTORE in the USGS offices throughout the United States. The NWIS system is composed of four subsystems; one of which was the Automated Data-Processing System (ADAPS) (U.S. Geological Survey, 1998). ADAPS consisted of a collection of computer programs and databases for the processing, review, storage, and retrieval of time-series data (U.S. Geological Survey, 1998). The USGS developed ADAPS in-house and began using it in 1985. Once the data were processed locally in ADAPS, they were uploaded to the national WATSTORE database for archival.

From 1985 to 2016, ADAPS served as the primary data-processing software in the USGS for management and review of time-series data. Documentation of the ADAPS system is found in user manuals, for example Bartholoma et al. (2003). The water data stored in ADAPS resulted from the processing of data collected from automated recorders and by observations and manual measurements nationwide. The data in ADAPS were retrieved manually or were relayed through telephones or by satellites from gaging stations to computers in USGS offices hosting ADAPS (U.S. Geological Survey, 1998).

NWIS-I and WATSTORE were replaced by NWIS-II (Edwards et al., 1987; Mathay, 1991). The NWIS-II system became fully functional around 1994 and is currently used in the USGS. NWIS-II marked a major improvement in the amount and quality of metadata that were stored. Metadata is information about a measurement or an analysis. Metadata increases the value of data for reuse. By storing the metadata with the actual data results, future users of the data can make more accurate assessments of its appropriateness to their needs. A geographic information system (GIS) is an integral part of the NWIS-II system. GIS can be used to establish or verify the location information of a data site in NWIS-II.

In 2012, the USGS began a project to replace ADAPS with the commercial software system AQUARIUS Time-Series. The AQUARIUS Time-Series project was completed in 2017. The USGS migrated more than 100 years of historical unit-value and daily streamflow data and other time-series data from ADAPS into AQUARIUS Time-Series (Aquatic Informatics, 2017). More than 50,000 measurement locations, 16,500 active gaging sites, and 3 million time-series datasets are managed in the new centralized system. These time-series data represent more than 100 billion high-frequency values at the sub-daily timescale (Aquatic Informatics, 2017).

Outside the USGS, a variety of commercial software solutions, spreadsheets, and custom solutions have been developed and are available to process time-series data. Recently, options for cloud data storage have become available (Libelium, 2018; YSI, 2015a). In the report on global hydrological monitoring trends, about 35% of respondents reported using Microsoft Office Software such as Excel as their primary tool to process data (Aquatic Informatics, 2015). About the same number of respondents, 34%, reported that they use a custom solution; either built in-house or by a contractor (Aquatic Informatics, 2015). Use of commercial data management systems was reported by 28% of respondents.

The rapid adoption of continuous-monitoring technologies is producing growing volumes of data with increased data complexity and metadata. Meeting data user expectations for uninterrupted flows of quality assured, reliable, and timely data are challenges. Software systems are needed by all water-quality monitoring agencies, researchers, and water utilities to handle time-series data.

7.2 Web Applications

Public dissemination and reuse of data on the Internet has positive public benefit (Cho et al., 2017). In 1994, the first web server was installed on a newly formed World Wide Web (Internet) at a USGS State office. Soon, other USGS State and local offices began deploying web servers and distributing near-RT data through separate USGS websites (U.S. Geological Survey, 2002). In 2001, the NWISweb system, an application of NWIS, was created to improve that service by aggregating all publically available data into one national database accessible through one website, NWISweb (U.S. Geological Survey, 2002). NWISweb integrates and makes publicly available many types of water data, including near-RT surface and groundwater levels, streamflows, precipitation, and water-quality data.

WaterQualityWatch is an application that imports near-RT data from NWISweb to summarize current conditions in a nationally consistent and useful way on a map that is scalable from the national to the local scale. The map can be updated instantaneously as data are retrieved from NWISweb. For example, gradients in WT are readily apparent from north to south (Fig. 13). Cool to warm symbol colors on the map indicate a gradient of cold to warm WTs.

The USGS provides additional web applications for data delivery to the public. WaterAlert (https://maps.waterdata.usgs.gov/mapper/wateralert/) began providing NWISweb data users with the ability to select a monitoring site and a

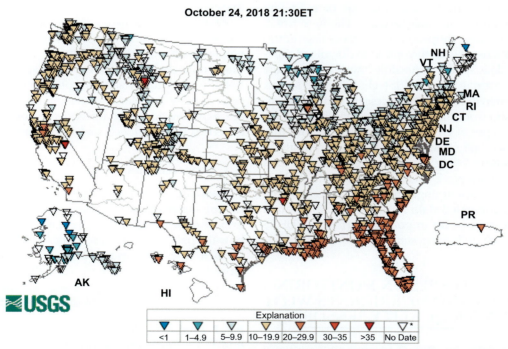

FIG. 13 A national scale map of water temperature from the U.S. Geological Survey (USGS) website WaterQualityWatch retrieved on October 24, 2018 at 21:30 Eastern Standard Time. These data are updated at intervals from 1 to 4 h and made available to the public on the National Water-Information System web page (NWISweb). Other water-quality properties and characteristics on the WaterQualityWatch website include streamflow, specific electrical conductance, pH, dissolved oxygen, turbidity, nitrate, and fluorescent chlorophyll (U.S. Geological Survey, 2017) (https://waterwatch.usgs.gov/wqwatch/).

water-quality property, set a threshold value, and sign up to receive a text or email message when a value meets or exceeds the threshold. In 2013, WaterNow (https://water.usgs.gov/waternow/) software enabled NWISweb data users to subscribe to a text or email report of current conditions at a selected monitor. NWISweb mapping tools (https://maps.waterdata.usgs.gov/mapper/index.html) provide users with options to filter and select data and display those data on maps for data exploration purposes. In 2013, the USGS released MobileWaterData (https://m.waterdata.usgs.gov/), a mobile-friendly version of NWISweb that delivers current conditions in a mobile-friendly format to smart phones.

Data services also are available from NWISweb. Rather than retrieving data from the NWIS website, an innovative approach to data access directly into the body of analysis or modeling code is known as data services. The new USGS WaterServices (https://waterservices.usgs.gov/) provides users with the option to download time-series data in the extensible markup language (XML) format from NWISweb. Software scripts are written to contain the necessary code to download data from the USGS data services, check for certain types of errors, and organize the information into data structures that are appropriate to the analyses being made by the particular data analysis package. Data retrieval and data analysis can be an efficient single integrated process. In the future, integrated data retrieval and application systems will be an important part of the USGS water data delivery system.

8 CASE STUDIES ON MONITORING FLOODS AND HURRICANES WITH WATER-QUALITY SENSORS

Increasing air and WTs, and more intense precipitation and runoff can result in climatic extremes that negatively affect river and lake-water quality (Georgakakos et al., 2014).

Delivery of excessive loads of storm-derived pollutants such as nitrogen, phosphorus, and suspended sediment combined with warm WTs are increasing the frequency and severity of HABs and hypoxia in fresh and marine waters (Michalak, 2016). A number of published studies document the use of water-quality sensors to investigate extremes in water quality resulting from impacts of hurricanes, high-intensity precipitation events, and floods (Marcé et al., 2016).

Large storms can lead to large increases in TU, suspended sediment, and nutrient loading to receiving waters. These can be tracked by meteorological and water-quality monitors. Measuring and understanding the effects on water quality of seasonal floods, tropical storms, and hurricanes requires high-temporal frequency monitoring at the event scale of minutes to hours as well as continued monitoring of the long-term impacts that can last from days to years (Marcé et al., 2016).

Water-quality sensors record daily extreme values such as minimum, maximum, and other values with low probability of observation. These data are difficult to capture using conventional water-quality sample collection and laboratory analysis. Many water-quality extremes are expected to occur infrequently but can be observed during floods and hurricanes.

8.1 Seasonal Floods

Two case studies are presented to describe the use of water-quality monitors to measure the impact of seasonal floods on water quality and how the patterns of flood response are repeatable in each case. The first case study involves the largest river in North America and the third largest river in the world and the second case study involves the 11th largest lake in the world.

8.1.1 Floods in the Mississippi River Basin and Lake Pontchartrain, Louisiana

The effects of extreme seasonal flooding in the Mississippi River on water quality of Lake Pontchartrain in Louisiana demonstrate the

8 CASE STUDIES ON MONITORING FLOODS AND HURRICANES WITH WATER-QUALITY SENSORS

benefit of using water-quality monitors combined with sampling to document the immediate and longer-term effects that persist well beyond the event scale of a few hours to a few days. Studies also demonstrate that patterns in response of the impacted systems are repeatable and, therefore, can be modeled and predicted, albeit with some uncertainty in a changing climate.

Flood studies in 2008 and 2011 by the USGS and the U.S. Army Corps of Engineer measured the response of Lake Pontchartrain to relatively large diversions of fresh nutrient-rich flood waters from the Mississippi River through the Bonnet Carré Spillway into the relatively nutrient-poor estuarine waters of Lake Pontchartrain (Mize and Demcheck, 2009).

The Bonnet Carré Spillway (Spillway) is located 28 miles northwest of New Orleans, Louisiana and is part of an integrated flood-control system for the lower Mississippi River system. Lake Pontchartrain is a 629-miles^2 (1629 km^2) estuarine embayment that is typically closed-off to the Mississippi River. Lake Pontchartrain has two outlets on the southeastern shore of the lake that connect it to Mississippi Sound which is a part of the northern Gulf of Mexico. Diversions are made during large floods to reduce stress on the downstream levee system of the Mississippi River.

Heavy rains in the valleys of the Mississippi and Ohio Rivers in April 2008 and in the valleys of the Mississippi, Missouri, and Ohio Rivers in May 2011 caused flooding that threatened the downstream levees in the City of New Orleans. In 2008 and 2011, the U.S. Army Corps of Engineers opened the Bonnet Carré Spillway (Spillway) to varying degrees to release flood waters into Lake Pontchartrain (Lake) and to ease pressure on downstream levees (U.S. Army Corps of Engineers and Mississippi River Commission, 2012). The previous release from the Spillway was in 1997 (Mize et al., 2013).

Diversion of Mississippi River flood water to Lake Pontchartrain through the Spillway in 2011 was of much greater magnitude and duration than in 2008. Total volume of river water discharged into the Lake through the Spillway in 2008 was about 145% of the estimated lake volume (Mize et al., 2013). The total volume discharged through the Spillway into the Lake in 2011 was about 400% of the estimated lake volume (Mize et al., 2013). In 2008, the river diversion began in April 2008 and lasted approximately 28 days, whereas the diversion in 2011 began in May 2011 and lasted approximately 44 days (Mize et al., 2013).

The influence of large volumes of nutrient-rich freshwater from the Mississippi River on the relatively nutrient-poor estuarine waters and on the phytoplankton communities in Lake Pontchartrain was documented in field investigations that spanned a short period before Spillway inflows, during the inflows, and after the inflows had passed through the lake. In both years, a water-quality sonde measuring WT, pH, SC, salinity, and DO and a flow-through nitrate analyzer were deployed at a single location on the Lake Pontchartrain Causeway. The water-quality monitor location was about 15 miles east of the Spillway, 3.5 miles (5.6 km) from the south shore of the Lake, and in the direct path of the Spillway inflow. Continuous measurements were made hourly in 2008 and 2011 starting just before and lasting several weeks when Spillway inflows were passing through the Lake.

As part of the same study, weekly water-quality samples were collected at six sites in 2008 and at seven sites in 2011 (Mize et al., 2013; Mize and Demcheck, 2009). Weekly samples were collected just before, for several weeks during, and for several weeks after the Spillway inflows had flushed through the Lake. Samples were collected at the Spillway, at three sites on the Causeway Bridge that approximately bisects Lake Pontchartrain from north to south, and at the two outlets to Mississippi Sound at the southeastern side of the Lake. An additional site at Grand Pass in Mississippi Sound near the

EVALUATING WATER QUALITY TO PREVENT FUTURE DISASTERS

outflow of the Lake was added in the 2011 study. One sampling site on the Causeway Bridge was collocated with the water-quality monitor in both years. Water-quality samples were analyzed for concentrations of various forms of nitrogen and phosphorus, chlorophyll *a*, major inorganic ions, and for characterization of phytoplankton taxa and relative abundance by cell numbers and biovolumes.

Although specific spatial patterns of nutrient, sediment, and phytoplankton response to the Spillway diversion varied between the two floods, general patterns emerged. In both floods, concentrations of all forms of nitrogen and phosphorus were elevated in Spillway inflows compared to the Lake. Due to the larger magnitude and longer duration of inflows in 2011, load estimates for TN, nitrite plus nitrate-N, dissolved orthophosphate and total phosphorus, suspended sediment, and dissolved silica delivered to the Lake through the Spillway were two to four times larger than those estimated for 2008 (Mize et al., 2013).

Measurements of WT, nitrate-N, and SC from the water-quality monitor on the Causeway in 2008 and 2011 provided a high-temporal frequency time series showing influence of the river water as it passed through the Lake. At the water-quality monitor, changes were recorded such as nitrate minimums prior to arrival of the Spillway water, and then an increase, peak, and decline in nitrate concentrations as the river water passed by the water-quality monitor and out of the Lake into Mississippi Sound (Mize et al., 2013). The inverse pattern in WT and SC was observed over the same time period as colder and fresher water passed by the water-quality monitor during the diversion (Mize and Demcheck, 2009). Weekly water-chemistry sample results were in general agreement with sensor measurements.

In 2008, phytoplankton at all sites in the Lake responded to the influx of nutrient-rich freshwater from the river with changes in the phytoplankton community structure and abundance. The phytoplankton shifted from a dominance of diatoms in the river water and green algae and diatoms in the lake prior to the Spillway diversion to a community dominated by cyanobacteria in the latter phases of the study after the diverted river water had passed out of the lake (Mize et al., 2013). The taxa of cyanobacteria that were identified in 2008 were capable of toxin production. An independent study of the 2008 spillway diversion documented the cyanobacterial bloom and levels of microcystin, a cyanobacteria toxin, at concentrations in Lake Pontchartrain as much as 1.7 µg/L later in the summer of 2008 (Bargu et al., 2011).

In 2011, however, owing to the longer duration and larger magnitude of Spillway inflows, the phytoplankton communities at all sampling locations in the Lake later in the study were not dominated by cyanobacteria as in 2008 (Mize et al., 2013). Nutrient-rich water and phytoplankton communities more like those in the Spillway inflow passed through Lake Pontchartrain into Mississippi Sound as measured at the sampling site at Grand Pass. Rather than developing in the Lake as in 2008, the cyanobacterial bloom developed in Mississippi Sound in 2011 due to the timing and magnitude of the influx of nutrient-rich water (Mize et al., 2013).

Cyanobacterial blooms are a repeatable pattern after seasonal Spillway inflows. In addition to the 2008 and 2011 blooms, a cyanobacterial bloom closed the Lake to water recreation and fishing after the Spillway inflow in 1997 (Mize et al., 2013).

8.1.2 Seasonal Floods and Western Lake Erie Algae Blooms

An example of the repeatable and long-term impacts of seasonal floods on nutrient delivery from watersheds to receiving waters comes from the Lake Erie watershed. Lake Erie (Lake) is the 11th largest freshwater lake in the world and is located in the United States and Canada and is bordered by the States of Michigan, Ohio, Pennsylvania, and New York and the Province of Ontario, Canada. The majority of Lake Erie

HABs develop within the shallow and warm western basin of the Lake (Wynne et al., 2015). Western Lake Erie suffers from seasonal blooms of cyanobacteria in the late summer that are typically more widespread and abundant in wetter than average years due to springtime rains that wash freshly applied fertilizers and manure into the lake from the predominantly agricultural Maumee River watershed in Indiana, Michigan, and Ohio (Stow et al., 2015). The Maumee River is the single largest tributary source of nitrogen and phosphorus to the Lake and enters the western end of the Lake at Toledo, Ohio (Stow et al., 2015). The frequency and severity of annual cyanobacterial blooms in Lake Erie appear to be increasing after 2008 (Wynne et al., 2015).

The City of Toledo obtains its drinking water directly from western Lake Erie. Winds and water currents can transport cyanobacterial blooms within proximity to drinking-water intakes. During these times, HABs and associated toxins can be drawn into the water intake and into the treatment plant. If not treated, blooms can cause odor, taste, and color problems in treated drinking water and can be harmful to human and animal health. During a cyanobacteria bloom in western Lake Erie in August 2014, microcystins were drawn into the drinking-water intake of Toledo, Ohio. Microcystin concentrations were found greater than the WHO's provisional guideline of 1 μg/L at the time (World Health Organization, 2003; Wynne et al., 2015). Bottled water was made available to nearly 500,000 residents of Toledo as a temporary remedy until the toxins were removed by enhancing the drinking-water treatment process (Wynne et al., 2015).

In response to the HABs that affected Toledo's drinking water in 2014, several lake-shore communities in northern Ohio, Ohio universities, and the NOAA deployed water-quality sensors in western Lake Erie on 9 buoys in open water, at the Toledo intake, and at 10 other water intakes (Chaffin et al., 2018). The high-temporal frequency water-quality monitors detect changes that routine weekly sampling could miss.

Near-RT water-quality sensors transmit data and information around the clock to characterize HAB-related water quality and to alert water providers to the magnitude and location of HABs in western Lake Erie (Chaffin et al., 2018). Western Lake Erie buoys are equipped with sensors for measuring, recording, and transmitting near-RT data for WT, pH, SC, FChl, phycocyanin, TU, FDOM, and concentrations of DO, nitrate, and phosphate. Near-RT data are provided on a public website (Chaffin et al., 2018; World Health Organization, 2003).

In 2015, the EPA established nonregulatory 10-day health advisory limits in treated drinking water for two cyanotoxins; microcystin and cylindrospermopsin (National Oceanic and Atmospheric Administration, 2018a). Health advisories describe nonregulatory concentrations of drinking-water contaminants at or below which adverse health effects are not anticipated to occur over specific exposure durations (National Oceanic and Atmospheric Administration, 2018a).

In 2017, NOAA deployed the first Environmental Sample Processor (ESP) in freshwater on a stationary mooring near the Toledo, Ohio water intake in Lake Erie (U.S. Environmental Protection Agency, 2015). Prior to 2017, the ESP mainly had been deployed for oceanographic-scale research and monitoring in marine environments (Greenfield et al., 2008; National Oceanic and Atmospheric Administration, 2018b).

In the early 2000s, the Monterey Bay Aquarium Research Institute (MBARI) began development of the ESP (Herfort et al., 2016; Monterey Bay Aquarium Research Foundation, 2018). The ESP is an autonomous robotic instrument that provides near-RT detection of HABs and their toxins. The ESP collects and filters water samples to concentrate particulate material like cyanobacterial cells and colonies and run molecular diagnostic tests using molecular probe technology for genetic analysis to determine if potentially harmful cyanobacteria are present. Results can be accessed or provided to water managers and the public in near RT. Samples also can be collected and preserved for microscopic identification of specific cyanobacteria taxa and for testing of the presence and concentrations of microcystins in the laboratory.

The ESP is a useful addition to the NOAA's system of tracking, reporting, and forecasting HABs in Lake Erie. The NOAA uses ESP and buoy data, satellite images, and lake modeling to issue semiweekly HAB Bulletins and HAB forecasts for the Lake. Microcystin concentrations from Lake Erie samples are analyzed by enzyme-linked immunosorbent assay (ELISA) within 48h of collection and this information is uploaded to the NOAA's western Lake Erie toxin tracking website (U.S. Environmental Protection Agency, 2015). The NOAA posts near-RT data from the ESP and buoys on their public website that provides warnings to water suppliers and the public of the development of HABs in western Lake Erie (U.S. Environmental Protection Agency, 2015).

8.2 Hurricanes and Tropical Cyclones

Several research investigations of the effects of tropical cyclones and hurricanes on estuarine-water quality were published in the early 2000s (Paerl et al., 2001; Walker, 2001). The consequences of excessive rainfall associated with tropical storms include elevated freshwater input and associated downstream salinity depression (Walker, 2001) along with accumulations of bottom materials (Peierls et al., 2003) often from damaged vegetation and/or nonpoint and point sources of pollutants in watersheds affected by the storms. More recently, human health and environmental concerns have been raised about preventing chemical releases during heavy rains, storm surges, and flooding caused by hurricanes Harvey in 2017 and Florence in 2018 (Miner et al., 2018).

Water-quality monitors in estuaries provided high-frequency data on the immediate impacts of hurricanes that would have been difficult to collect using water-quality sampling campaigns. Coastal networks of hydrometeorological and water-quality monitors (coastal networks) in the York River estuary of the Chesapeake Bay National Estuarine Research Reserve (CBNERR) in Virginia (Reay and Moore, 2005) and in Grand Bay National Estuarine Research Reserve (GBNERR) in Mississippi (YSI, 2015b) showed similar patterns in water-quality responses to tropical storms.

Water-quality monitors were instrumented with YSI data sondes at several locations in both estuaries in the CBNERR and GBNERR along gradients in salinity and depth (Reay and Moore, 2005; YSI, 2015b). Continuous water levels and continuous water-quality measurements were collected before, during, and after the storms in both studies. Weather stations provided data on wind speed. Water-quality monitors at all study sites in both NERRs displayed some effects of the passage of the storms in measurements of hydrological, physical properties, and chemical characteristics. Water-quality monitors describe the passage of the storms with far greater temporal resolution than sampling studies recording, for example, minimum and maximum values for each day before, during, and after (Reay and Moore, 2005; YSI, 2015b).

Water-quality monitors were able to remain active during and after both storms despite extreme conditions. Results showed, to varying degrees depending on location, that water levels and TU changed in both estuaries with the arrival of winds and storm surges that caused mixing and resuspension of bottom materials (Reay and Moore, 2005; YSI, 2015b). Depending on location, salinity initially increased with storm surges in both estuaries during the first hours of the storms and then decreased due to precipitation and tributary inflows over several days afterward (Reay and Moore, 2005; YSI, 2015b). At one site in the CBNRR, turbidities during the storm exceeded 0.1% of recorded TU values observed that season. In both studies, DO levels were depressed after the storms at

8 CASE STUDIES ON MONITORING FLOODS AND HURRICANES WITH WATER-QUALITY SENSORS

shallow estuarine locations. No anoxia was observed in the CBNRR study but one site in the GBNRR study remained anoxic for 2 days. Depressed DO values were inferred to be from watershed material loadings of vegetation and debris during the storms rather than other factors (Reay and Moore, 2005; YSI, 2015b). During Hurricane Katrina, one water-quality monitor in GBNERR continued to record measurements despite being completely submerged for about 3.5 h during the storm surge (YSI, 2015b).

8.3 Storm-Hardened Platforms

A challenge of characterizing water quality during floods, hurricanes, and other water hazards is the severe physical conditions that damage or destroy water-quality monitors on land, in coastal areas, and on moorings at sea that are not built to withstand such conditions. For example, Cho et al. (2012) describe the partial loss of salinity monitors in Chesapeake Bay and its major tributaries resulting from hurricane-force winds and storm surges of Hurricanes Floyd in 1999 and Isabel in 2003. Both storms were category 2 hurricanes at landfall and tracked north along Chesapeake Bay. Measurements of salinity, water level, depth, water velocity, and wind speed recorded at observational buoys and weather stations are needed to evaluate the impact of hurricanes and storm surges in estuaries (Cho et al., 2012).

Storm hardening is a method that physically strengthens the stationary installation to make it less susceptible to damage from the high winds and storm surges associated with hurricanes and with swift and high water levels associated with floods. Hardening usually requires significant investment and can take several years to complete. Storm hardening is especially important for installations placed on river banks and in shallow coastal waters.

The 2005 hurricane season in the northern Gulf of Mexico destroyed many existing USGS gaging stations, some with water-quality sensors.

In response, a program to storm harden more than 120 damaged or destroyed Gulf-Coast streamgages was undertaken with funding from the U.S. Congress (Rebich et al., 2015). As part of this effort, the USGS storm-hardened 10 monitoring platforms in the coastal waters of the northern Gulf of Mexico from the Mississippi Sound to Vermilion Bay, Miss (Rebich et al., 2015). This area is subject to high winds and storm surge during hurricanes. The hardened gages, constructed by the U.S. Army Corps of Engineers, are 10 ft. × 10 ft. platforms on the top of a 90-ft. long, 3-ft. diameter steel pipe driven 60 ft. into the Gulf bottom in approximately 10 ft. of water (Rebich et al., 2015). This configuration elevates the monitoring equipment about 30 ft. above the normal water surface. The instruments at these 10 gages collect water level, SC, salinity, water and air temperatures, wind speed and direction, precipitation, and barometric pressure. Data are telemetered in near RT to the Internet. In addition, TU and DO are collected at two sites, one of which is Mississippi Sound at Gulfport Light (301912088583300) (Rebich et al., 2015) (Fig. 14).

The benefits of storm hardening include acquisition of uninterrupted near-RT water level, wind speed, and water-quality data including extreme values. Storm hardening also reduces equipment replacement and repair costs (Rebich et al., 2015). Where previous installations were not storm hardened and were destroyed, the new storm-hardened installations were able to withstand direct exposure to Hurricane Gustav, a 2008 category 2 storm and in subsequent storms that followed (Rebich et al., 2015).

In the open ocean, moorings are exposed to extreme conditions that can cause damage or destruction of the platform and its instruments. The Pioneer Array, located 90 miles offshore on the continental shelf of New England, is designed to withstand a 100-year (1% annual exceedance probability) storm with winds of 65 mph and waves heights from 14 to 18 m (LaCapra, 2017). Storm conditions can stress

FIG. 14 Example of storm-hardened water-quality and water-level monitor operated by the U.S. Geological Survey in Mississippi Sound. (A) Platform before storm hardening and (B) platform after completion of storm hardening (Rebich et al., 2015).

the cables that attach the moorings to the sea bed. To help the moorings endure the extreme tension created by powerful winds, waves, and currents, Pioneer's designers created, tested, and installed highly elastic hoses to carry the sensitive electrical wires. The stretchable hoses are engineered to withstand turbulent coastal waters (LaCapra, 2017).

9 ADVANTAGES AND LIMITATIONS OF WATER-QUALITY SENSORS

Advances and innovations since 1950 have paved the way for continuous water-quality monitoring to become an increasingly viable means of quickly and accurately collecting water-quality data at the sub-daily scale over periods of months to years. Continuous water-quality monitoring has advantages and benefits relative to discrete water-quality sampling but those benefits and advantages come with challenges and limitations (Table 3).

Continuous water-quality sensors are providing scientific insights into sub-daily patterns in water quality by providing high-temporal frequency observation that are leading to new and improved understanding of geochemical and biophysical processes in rivers, streams, lakes, wetlands, estuaries, and oceans. Continuous water-quality sensors produce large datasets, however, that can result in data management challenges. Large datasets require data management and software system(s) for review, quality control, storage, and retrieval.

Sensors can be deployed in a wide range of aquatic environments from the smallest streams

TABLE 3 Advantages and Limitations of Continuous Water-Quality Monitoring

Advantages	Limitations
Ability to collect high-temporal-frequency data over a wide range of conditions reduces the likelihood that changes in water quality are missed or obscured	Large datasets require a data management plan and software system(s) for review, quality control, storage, and retrieval
Sensors can be deployed in a wide range of aquatic environments from the smallest streams, groundwater, to the open ocean and over a wide range of environmental conditions from floods and hurricanes to droughts and in extreme environments	Risk of equipment damage or loss during hurricanes and floods, freezing conditions, ice cover, or low water levels. Corrosion can be a challenge from constant deployment in chemically adverse environments
High-frequency measurements and short lag time in transmitting and receiving results can inform decisions that are vital for protection of life, property, health, and the environment and operational decision making	Meeting expectations of data users for uninterrupted flows of quality assured, reliable, and timely data are a challenge. Satellite telemetry systems, phone lines, cellular, or Wi-Fi networks are needed to transmit and receive data from remote autonomous sensors in real time or near-real time. Internet and or intranet websites are needed for data display and dissemination
Fully developed sensor technologies provide data with high resolution, high precision, and low bias	Field calibration and checks are periodically needed to control instrument drift, physical and chemical interferences, and biofouling
Sensors provide a method for early trend detection, help identify monitoring gaps, and ensure timely data for science-based decision support related to ecosystems and human health	Additional standardized methods are needed to support scientific and management needs and for regulatory decision making
Sensor data are comparably lower cost on a per measurement basis due to reduced need to collect and transport water samples to laboratories for analysis	Substantial costs can be associated with initial equipment purchase, field installation, power requirements, operation, and ongoing maintenance
Advances in technology are providing new types of water-quality sensors to measure variables previously unavailable as continuous water-quality data. Surrogate techniques statistically predict unmeasured water-quality constituents that are otherwise impractical or too expensive to collect at high frequencies	Fewer types of water-quality analyses available from water-quality sensors compared to laboratory methods

and groundwater to the open ocean and over a wide range of environmental conditions from floods and hurricanes to droughts and chemical spills or other extreme environments such as the deep ocean. Water-quality monitors deployed in the environment are at risk of damage or loss during hurricanes and floods, freezing conditions, ice cover, or low water levels. Corrosion of sensors and other parts of water-quality monitors can be an issue in chemically adverse conditions in the environment.

High-frequency sensor measurements and short lag times in transmitting and receiving data can inform decisions that are vital for protection of life, property, the environment, and operational decision making. Meeting expectations of data users for uninterrupted flows of reliable and timely data can be a challenge. Reliable water-quality monitors that provide uninterrupted data and communication networks are needed to transmit and receive data from autonomous sensors in a timely manner.

Fully developed sensor technologies provide data with high resolution, high precision, and low bias. Field calibration and checks are periodically needed to control instrument drift, physical and chemical interferences, and biofouling. Time and resources are needed to install, service, and maintain the optimal operation of water-quality sensors to produce reliable data.

Sensors provide a method for early trend detection, help identify monitoring gaps, and ensure timely data for science-based decision support related to ecosystems and human health. Additional standardized methods are needed to support scientific and management needs and for regulatory decision making.

Continuous water-quality data are collected at comparably lower cost on a per measurement basis compared to water-quality samples due to reduced collection, transportation, and laboratory analysis costs. However, substantial costs can be associated with initial equipment purchase, field installation, power requirements, operation, and ongoing maintenance.

Advances in technology have provided new types of optical water-quality sensors and chemical analyzer systems to measure variables previously unavailable as in situ continuous water-quality data. Surrogate techniques statistically predict unmeasured water-quality constituents that are otherwise impractical or too expensive to collect at high frequencies. Surrogate techniques help fill gaps in the capabilities of continuous water-quality monitors. However, continuous water-quality monitors continue to be limited in the types of physical properties and chemical characteristics that can be monitored in comparison to the wide variety of test methods available in water analysis laboratories.

10 SUMMARY AND CONCLUSIONS

Innovations in technology overtime have made continuous water-quality monitoring an increasingly viable way of quickly and accurately making in situ water-quality measurements at high frequency. Water-quality monitors continue to serve their original functions as assessment tools in support of water-quality regulations and policies and as early-warning systems for alerting water suppliers of spills approaching their water-supply intakes. Additional functions include alerting the public to the possibility of microbiological contamination in areas used for water recreation. As in the past, water-quality monitors continue to support the protection of human health and the environment from unanticipated chemical and microbiological water hazards.

Continuous water-quality measurements are providing scientific insights into sub-daily patterns and extremes in water quality that are leading to new and/or improved understanding of geochemical processes in rivers, streams, lakes, estuaries, and oceans. Water-quality monitoring is revealing patterns in the immediate and longer-term effects of HABs and hypoxia brought about by nutrient-enriched flood waters discharging to lakes and estuaries. Deployments of water and water-quality monitors to capture the impacts of floods, hurricanes, and other water hazards will necessitate hardened platforms that can withstand damaging winds, storm surges, and floods.

Continuous water-quality monitoring has several advantages and benefits but those benefits and advantages come with challenges and limitations. Provision of water-quality data using traditional field sampling and laboratory analysis at frequencies from daily, weekly, monthly, or less often can miss important natural and human-influenced phenomena that occur at high-temporal frequency. The need to measure water quality at high-frequency warrants critical evaluation in relation to goals and objectives of data to assure reliable and timely data at a reasonable cost.

New types of optical sensors and improvements in chemical analyzer systems are increasing the types of water properties and characteristics that can be measured in situ at

EVALUATING WATER QUALITY TO PREVENT FUTURE DISASTERS

10 SUMMARY AND CONCLUSIONS

high-temporal frequency. Multiparameter sondes, smart sensors, and universal sensor ports have made monitoring at high frequency more flexible and efficient overtime. However, users of water-quality monitors need guidelines for sensor selection, installation, operation, and quality-control procedures to ensure data quality.

Water-quality sensors are housed in stationary platforms such as gage houses, buoys, moorings, pontoon boats, and AUVs, and in integrated coastal and ocean observatories. A trend toward small and lighter sensors, EDLs, and other electronic equipment, new antifouling mechanisms, extended battery life, submersible sensors, and solar-power cells have allowed for remote deployments of water-quality monitors in the smallest streams to the deep ocean. Consequently, autonomous operation of water-quality monitors has increased from a few days to a few weeks or several months, thereby reducing the cost of operation.

Data storage and transmission innovations have enhanced continuous water-quality monitoring possibly more than other factor. EDLs and DCPs have become standard equipment in water-quality monitors. Data transmission and communication systems have advanced from teletype, wire radio, and land lines to satellite telemetry, cellular, and Wi-Fi networks. Delivery of continuous water-quality data has improved from annual or less often in paper reports to RT or near RT in a matter of an hour or less relative to the time of data collection.

Data from water-quality monitors delivered in RT or near RT are raw data or data that are only partially processed and corrected. These data are often needed in RT or near RT for purposes of daily operational decision making or emergency response. However, after data are initially received, a process to review data quality, reliability, and completeness in a timely manner is essential for ensuring that the needs of the user community are met. Detailed review of data remains a time intensive but critically important process. Most organizations are using commercial spreadsheets, homegrown or consultant developed software, or commercial systems for data management.

Data-processing software and water-information systems are essential for management of large amounts of continuous data. Innovations overtime in water-information systems have been important for secure data storage, archival, retrieval, use, and reuse. Databases and computer systems serve as data repositories for many millions to billions of high-frequency data values overtime.

Public dissemination and reuse of data on the Internet has positive public benefit. Data availability on the Internet is resulting in greater use of data and information leading to better informed decision making. Web applications provide options for data retrieval by laptop, tablet, or smart phone. Geospatial mapping and customization of data retrievals also contribute to enhanced data retrievals. Data services provide flexible retrieval formats that have made data analysis more efficient and less costly in term of time and human resources for water managers, academic researchers, private companies, and the public. By the early 2000s, water recreationists, fishermen, surfers, beach-goers, and others in the public became frequent users of streamflow and water-quality data from the Internet to plan outdoor water-related activities.

The number of water-quality monitors has been steadily increasing and this trend is expected to continue into the 2020s. Continuous RT water-quality data are used for decisions regarding drinking water and wastewater treatment, regulatory programs, recreation, public safety, and scientific research. The USGS network of water-quality monitors has experienced a 73% increase from 2006 to 2016. In calendar year 2016, stations in the USGS network numbered more than 2600 nationwide. Continuous water-quality networks operated by other organizations also are experiencing growth in their continuous-monitoring networks. The innovations and technical improvements have resulted in the growth in continuous water-quality monitoring.

EVALUATING WATER QUALITY TO PREVENT FUTURE DISASTERS

References

Alliance for Coastal Technologies, 2018. Supporting innovation to better understand, predict, and manage coastal, ocean and great lakes environments. http://www.act-us.info/. (Accessed 14 December 2018).

American Public Health Association, 1915. Standard Methods for the Examination of Water and Sewage, second ed. American Public Health Association, Boston, MA. 144 pp.

American Public Health Association, American Water Works Association, and Water Environment Federation, 1995. Standard Methods for the Examination of Water and Wastewater, 19th ed. American Public Health Assoc., American Water Works Assoc., and Water Environment Fed, Washington, DC. pp. 4:85–86.

American Public Health Association, American Water Works Association, and Water Pollution Control Federation, 1971. Standard Methods for the Examination of Water and Wastewater, 13th ed. American Public Health Assoc., American Water Works Assoc., Water Pollution Control Fed, New York, NY. 874 pp.

Anderson CW (2005), Alkalinity and acid neutralizing capacity (ver. 2.1). U.S. Geological Survey Techniques of Water-Resources Investigations, book 9, chap. A6., sec. 6.7, Sept. 2005, accessed 9.25.2018, from http://pubs.water.usgs.gov/twri9A6/.

Aquatic Informatics, 2015. Global Hydrologic Monitoring, Industry Trends. Aquatic Informatics, Vancouver, BC. 30 p.

Aquatic Informatics, Big water data collected by the U.S. geological survey is now managed in AQUARIUS time-series across 25 states. 2017, accessed 10.23.2018 at https://aquaticinformatics.com/news/usgs-big-water-data/.

Baird, D., 2002. Analytical chemistry and the "big scientific instrumentation revolution" In: Morris, P.J.T. (Ed.), From Classical to Modern Chemistry, the Instrumental Revolution. London, UK, Royal Society for Chemistry and the Chemistry Museum, pp. 29–56.

Baldwin, A.K., Robertson, D.M., Saad, D.A., Magruder, C., 2013. Refinement of regression models to estimate real-time concentrations of contaminants in the Menomonee River drainage basin, southeast Wisconsin, 2008–11. U.S. Geological Survey Scientific Investigations Report 2013-5174, 113 p. http://pubs.usgs.gov/sir/2013/5174/.

Ballinger, D.G., 1968. Automated water-quality monitoring. Environ. Sci. Technol. 2 (8), 606–610.

Bargu, S., White, J.R., Li, C., Czubakowski, J., Fulweiler, R.W., 2011. Effects of freshwater input on nutrient loading, phytoplankton biomass, and cyanotoxin production in an oligohaline estuarine lake. Hydrobiologia 661, 377–389. https://doi.org/10.1007/s10750-010-0545-8.

Bartholoma SD, Kolva JR, Nielsen JP, User's manual for the national water information system of the U.S. Geological Survey, automated data processing system (ADAPS), 2003. U.S. Geological Survey Open-File Report 03-123 Version 4.3, 407 p.

Bastin, R., Weberling, T., Padilla, F., 1957. Ultraviolet spectrophotometric determination of nitrate. Anal. Chem. 29, 1795–1797.

Beckman, A.O., Gallaway, W.S., Kaye, W., Ulrich, W.F., 1977. History of spectrophotometry at Beckman Inc. Anal. Chem. 49 (3), 280A–300A.

Bergamasci, B.A., Fleck, J.A., Downing, B.D., Boss, E., Pellerin, B.A., Ganju, N.K., Schoellhamer, D.H., Byington, A.A., Heim, W.A., Stephenson, M., Fujii, R., 2012. Mercury dynamics in a San Francisco estuary tidal wetland: assessing dynamics using in situ measurements. Estuaries Coast 35 (4), 1036–1048.

Blaen, P.J., Khamis, K., Lloyd, C.E.M., Bradley, C., Hannah, D., Krause, S., 2016. Real-time monitoring of nutrients and dissolved organic matter in rivers: capturing event dynamics, technological opportunities and future directions. Sci. Total Environ. 569–570, 647–660.

Blakey, J., 1970. Design of an automatic monitoring system. In: Kerrigan, J.E. (Ed.), Proceedings of the National Symposium on Data and Instrumentation for Water-Quality Management. University of Wisconsin, Madison, pp. 243–247.

Chaffin, J.D., Kane, D.D., Stanislawczyk, K., Parker, E.M., 2018. Accuracy of data buoys for measurement of cyanobacteria, chlorophyll, and turbidity in a large lake (Lake Erie, North America): implications for estimation of cyanobacterial bloom parameters from water quality sonde measurements. Environ. Sci. Pollut. Res. 25, 25175–25189. https://doi.org/10.1007/s11356-018-2612-z.

Chappell, N.A., Jones, T.D., Tyche, W., 2017. Sampling frequency for water quality variables in streams: systems analysis to quantify minimum monitoring rates. Water Res. 123, 49–57. https://doi.org/10.1016/j.waters.2017.06.047.

Chesapeake Biological Laboratory, Nutrient sensor challenge winners announced at ASLO conference, 2017. accessed 9.22.2018 at https://www.umces.edu/news/nutrient-sensor-challenge-winners-announced-aslo-conference.

Cho, K.-H., Wanga, H.V., Shen, J., Valle-Levinson, A., Teng, Y., 2012. A modeling study on the response of Chesapeake Bay to hurricane events of Floyd and Isabel. Ocean Model. 49 (50), 22–46.

Cho A, Fischer A, Doyle M, Levy M, Kim-Blanco P, Webb R, The Value of Water Information: Overcoming the Global Data Drought. Xylem White Paper, August 2017. accessed 10.24.2018 at http://xylem.com/waterdata.

Christensen, V.G., Xiaodong, J., Ziegler, A.C., 2000. Regression analysis and real-time water-quality monitoring to

REFERENCES

estimate constituent concentrations, loads, and yields in the Little Arkansas River, South-central Kansas, 1995-99. U.S. Geological Survey Water-Resources Investigations Report 00-4126, 36 p. http://ks.water.usgs.gov/Kansas/pubs/reports/wrir.00-4126.html.

Cleary, E.J., 1967. The ORSANCO story. The Johns Hopkins Press, Baltimore, MD, pp. 197–210.

Cohen, B., McCarthy, L.T., 1962. Salinity of the Delaware Estuary. U.S. Geological Survey Water Supply Paper 1856B, 47 p.

Cole, J.J., Prairie, Y.T., Caraco, N.F., McDowell, W.H., Tranvik, L.J., Striegl, R.G., Duarte, C.M., Kortelainen, P., Downing, J.A., Middelburg, J.J., Melack, J., 2007. Plumbing the global carbon cycle: integrating inland waters into the terrestrial carbon budget. Ecosystems 10, 171–184. https://doi.org/10.1007/s10021-006-9013-8.

Collins, W.D., 1910. Quality of the surface waters of Illinois. U.S. Geological Survey Water-Supply Paper 239, 102 pp.

Collins, W.D., Howard, C.S., Love, S.K., 1943. Quality of surface waters of the United States, 1941. U.S. Geological Survey Water-Supply Paper 942, 73 pp.

Curtin, T.B., Bellingham, J.G., Catipovic, J., Webb, D., 1993. Autonomous oceanographic sampling networks. Oceanography 16 (3), 86–94.

Dickey T, Chang G, Moore C, Hanson A, Karl D, Manov D, Spada F, Peters D, Kemp J, Schofield O, Glenn S (2006), The Bermuda testbed mooring and HALE-ALOHA mooring programs: innovative deep-sea global observatories. Presented at the Oceans 2006 Conference Sept. 18–21, 2006, Boston, MA, 4 p. DOI: https://doi.org/10.1109/OCEANS.2006.307099.

Dinsmore, R.P., 1996. Alpha, bravo, charlie—ocean weather ships 1940–1980. Oceanus 39 (2), 9–10.

Dole, R.B., 1906. Use of fluorescein in the study of underground waters. U.S. Geological Survey Water Supply and Irrigation Paper 160. pp. 73–85.

Dole, R.B., 1909. The quality of surface waters in the United States, part 1. Analyses of waters east of the one hundredth meridian. U.S. Geological Survey Water Supply Paper 236, 118 pp.

Downing, B.D., Pellerin, B.A., Bergamaschi, B.A., Saraceno, J.F., Kraus, T.E.C., 2012. Seeing the light: the effects of particles, dissolved materials, and temperature on in situ measurements of DOM fluorescence in rivers and streams. Limnol. Oceanogr. 10, 767–775.

Durum, W.H., 1978. Historical profiles of quality of water laboratories and activities, 1879–1973. U.S. Geological Survey Open-File Report 78-432, 235 pp.

Durum, W.H., Blakey, J.F., 1965. Time-weighted analysis as an indicator of natural stream quality. In: Prestented to the Association of State and Interstate Water Pollution Control Administrators, Dayton, Ohio. December 7–8, 1965: 11 p. (four map attachments).

Edwards, M.D., Putnam, A.L., Hutchison, N.E., 1987. Conceptual design for the National Water Information System. U.S. Geological Survey Bulletin 1792, 22 p.

Eriksen, G.C., Osse, T.J., Light, R.D., Wen, T., Lehman, T.W., Sabin, P.L., Ballard, J.W., Chiodi, A.M., 2001. Seaglider: a long-range autonomous underwater vehicle for oceanographic research. IEEE J. Ocean. Eng. 26 (4), 424–436.

Federal Water Pollution Control Administration, 1968. Water quality criteria. A Report of the National Technical Advisory Committee to the Secretary of Interior. Washington, DC, 234 pp.

Follansbee, R., 1994. A History of the Water Resources Branch, U.S. Geological Survey; Volume I, From Predecessor Surveys to June 30, 1919. U.S. Geological Survey Unnumbered Series, 286 p.

Francy, D.S., Darner, R.A., 2006. Procedures for developing models to predict exceedances of recreational water-quality standards at coastal beaches. U.S. Geological Survey, Techniques and Methods 6-B5, 34 p. http://pubs.usgs.gov/tm/2006/tm6b5/.

Francy, D.S., Brady, A.M.G., Carvin, R.B., Corsi, S.R., Fuller, L.M., Harrison, J.H., Hayhurst, A., Lant, J., Nevers, M.B., Terrio, P.J., Zimmerman, T.M., 2013. Developing and implementing predictive models for estimating recreational water quality at great lakes beaches. U.S. Geological Survey Scientific Investigations Report 2013-5166, 68 p. https://doi.org/10.3133/sir20135166/.

Frant, M.S., 1997. Where did ion selective electrodes come from? The story of their development and commercialization. J. Chem. Educ. 74 (2), 159–166.

Fuller, G.W., Whipple, G.C., Clark, H.W., Gehrmann, A., Wyatt, J., Jordan, E.O., 1900. Standard methods of water analysis. Science 12 (311), 906–915.

Galloway, J.M., 2014. Continuous water-quality monitoring and regression analysis to estimate constituent concentrations and loads in the Red River of the North at Fargo and Grand Forks, North Dakota, 2003–12. U.S. Geological Survey Scientific Investigations Report 2014-5064, 37 p. https://doi.org/10.3133/sir20145064.

Gallwas, J., 2004. Beckman Arnold Orville, 1900–2004. Anal. Chem. 76 (15). 264A–265.

Georgakakos, A., Fleming, P., Dettinger, M., Peters-Lidard, C., Richmond, T.C., Reckhow, K., White, K., Yates, D., 2014. Climate change impacts in the United States: chap. 3. In: Richmond, T.C., Yohe, G.W. (Eds.), Third National Climate Assessment, U.S. Global Change Research Program, pp. 69–112. https://doi.org/10.7930/J0G44N6T. http://nca2014.globalchange.gov/report/sectors/water.

Glasgow, H.B., Burkholder, J.M., Reed, R.E., Lewitus, A.J., Kleinman, J.E., 2004. Real-time remote monitoring of water quality: a review of current applications, and advancements in sensor, telemetry, and computing technologies. J. Exp. Mar. Biol. Ecol. 300, 409–448.

Gordon AB, Katzenbach M (1983), Guidelines for use of water-quality monitors. U.S. Geological Survey Open-File Report 83-681, 94 p.

Government of Newfoundland and Labrador, 2013. Protocols Manual for Real Time Water-Quality Monitoring in Newfoundland and Labrador. Dept. of Environment and Conservation, Water Resources Management Division. November 2013, 91 p.

Graham, J.L., Dubrovsky, N.M., Eberts, S.M., 2017. Cyanobacterial harmful algal blooms and U.S. Geological Survey science capabilities. U.S. Geological Survey Open-File Report 2016–1174, 12 p. https://doi.org/10.3133/ofr20161174.

Gray, J.R., Landers, M.N., 2014. Measuring suspended sediment. In: Ahuja, S. (Ed.), Comprehensive Water Quality and Purification. In: vol. 1. Elsevier, United States of America, pp. 157–204.

Greenfield, D., Marin III, R., Doucette, G.J., Mikulski, C., Jensen, S., Roman, B., Alvarado, N., Scholin, C.A., 2008. Field applications of the second-generation environmental sample processor (ESP) for remote detection of harmful algae: 2006-2007. Limnol. Oceanogr. 6, 667–679.

Hach Chemical Company, 2014. History and heritage. http://www.hach.com/history. (Accessed 5 March 2018).

Hambrook Berkman, J.A., Canova, M.G., 2007. Algal biomass indicators (ver. 1.0). U.S. Geological Survey Techniques of Water-Resources Investigations, book 9, chap. A7, section 7.4, accessed at http://pubs.water.usgs.gov/twri9A/.

Hart, J.K., Martinez, K., 2006. Environmental sensor networks: a revolution in the earth system science? Earth Sci. Rev. 78, 177–191.

Helsel, D., Hirsch, R., 2002. Statistical methods in water resources. U.S. Geological Survey, Techniques of Water-Resources Investigations Book 4, Chapter A3, (522 p).

Hem, J.D., 1985. Study and interpretation of chemical characteristics of natural water. U.S. Geological Survey Water-Supply Paper 2254, 264 pp.

Hem JD, Cropper WH (1959) Survey of ferrous-ferric chemical equilibria and redox potentials; U.S. Geological Survey; Water Supply Paper 1459, pp. 1–32.

Herfort, L., Seaton, C., Wilkin, M., Roman, B., Preston, C., Marin, R., Seitz, K., Smith, M., Haynes, V., Scholin, C., Baptista, A., Simon, H., 2016. Use of continuous, real-time observations and model simulations to achieve autonomous, adaptive sampling of microbial processes with a robotic sampler. Limnol. Oceanogr. 14, 50–67. https://doi.org/10.1002/lom3.10069.

Hirsch, R.M., Fisher, G.T., 2014. Past, present, and future of water data delivery from the U.S. Geological Survey. J. Contemp. Water Res. Ed. 153, 4–15.

Hudson, N., Baker, A., Reynolds, D., 2007. Fluorescence analysis of dissolved organic matter in natural, waste, and polluted waters—a review. River Res. Appl. 23, 631–649.

Jackson, P.R., 2013. Integrated synoptic surveys using an autonomous underwater vehicle and manned boats. U.S. Geological Survey Fact Sheet 2013-3018, 4 p.

Jastram, J.D., 2014. Streamflow, water quality, and aquatic macroinvertebrates of selected streams in fairfax county, Virginia, 2007–12. U.S. Geological Survey Scientific Investigations Report 2014-5073, 68 p. https://doi.org/10.3133/sir20145073.

Johnson, K.S., Needoba, J.A., Riser, S.C., Showers, W.J., 2007. Chemical sensor networks for the aquatic environment. Chem. Rev. 107 (2), 623–640.

Jones, C., Webb, D., Glenn, S., Schofield, O., Kerfoot, J., Kohut, J., Aragon, D., Haldeman, C., Haskin, T., Kahl, A., Hunter, E., 2011. Slocum glider, expanding the capabilities. Proc. International Symposium on Unmanned Untethered Submersible Technology, Portsmouth, New Hampshire, USA, August 21–24, 2011, pp. 165–173.

Jones, T.D., Chappell, N.A., Wlodek, T., 2014. First dynamic model of dissolved organic carbon derived directly from high-frequency observations through contiguous storms. Environ. Sci. Technol. 48, 13289–13297.

Kanwisher, J., 1957. Polarographic oxygen electrode. Limnol. Oceanogr. 4 (2), 210–217.

Kittrell, F.W., 1969. A practical guide to water quality study of streams. Report CWR-5. Federal Water Pollution Control Admin, 135 p.

Kongsberg (2018), AUV/marine robots and autonomous vehicles. accessed 10.19.2018 at https://www.km.kongsberg.com/ks/web/nokbg0240.nsf/AllWeb/A4B26E3B2000D960C12580C3004D6014?OpenDocument.

Kraus, T.E.C., Bergamaschi, B.A., Downing, B.D., 2017. An introduction to high-frequency nutrient and biogeochemical monitoring for the Sacramento–San Joaquin Delta, northern California: U.S.Geological Survey. Scientific Investigations Report 2017–5071, 41 p. https://doi.org/10.3133/sir20175071.

LaCapra, V., 2017. A pioneering vision. Oceanus 53 (1), 34–40.

Laitinen, H.A., Ewing, G.W., 1977. A History of Analytical Chemistry. American Chemical Society, York, PA. 358 pp.

Lakowicz, J.R., 2006. Principles of Fluorescence Spectroscopy, third ed. Springer. 954 p.

Landers, M.N., Straub, T.D., Wood, M.S., Domanski, M.M., 2016. Sediment acoustic index method for computing continuous suspended-sediment concentrations. U.S. Geological Survey Techniques and Methods, book 3, chap. C5, 63 p. https://doi.org/10.3133/tm3C5.

Lawler, D.M., 2005. Spectrophotometry: turbidimetry and nephelometry. In: Brown, C., Vega-Montoto, L., Wentzell, P. (Eds.), Encyclopedia of Analytical Science. Elsevier, pp. 343–351. https://doi.org/10.1016/B0-12-369397-7/00718-4.

Leighton, M.O., 1905. Field assay of water. U.S. Geological Survey Water Supply and Irrigation Paper 151, 77 pp.

REFERENCES

Libelium, 2018. Waspmote plug & sense quick overview. Document version: v8.0, Libelium Comunicaciones Distribuidas S.L. 48 p.

Loftus, M.E., Seliger, H.H., 1975. Some imitations of the in vivo fluorescence technique. Chesapeake Sci. 16 (2), 79–92.

Lorenzen, C.J., 1966. A method for the continuous measurement of in vivo chlorophyll concentration. Deep Sea Res. 13, 223–227.

Love, S.K., 1951. Water analysis. J. Am. Water Works Ass. 23 (2), 253–257.

Loving, B., Putnam, J., Turk, D., 2014. Continuous real-time water information—an important kansas resource. U.S. Geological Survey Fact Sheet 2014-3003. 4 p. https://doi.org/10.3133/fs20143003.

Marcé, R., George, G., Buscarinu, P., Deidda, M., Dunalska, J., de Eyto, E., Flaim, G., Grossart, H.P., Istvanovics, V., Lenhardt, M., Moreno-Ostos, E., Obrador, B., Ostrovsky, I., Pierson, D.C., Potužák, J., Poikan, S., Rinke, K., Rodríguez-Mozaz, S., Staehr, P.A., Šumberová, K., Waajen, G., Weyhenmeyer, G.A., Weathers, K.C., Zion, M., Ibelings, B.W., Jennings, E., 2016. Automatic high-frequency monitoring for improved lake and reservoir management. Environ. Sci. Technol. 50, 10780–10794. https://doi.org/10.1021/acs.est.6b01604.

Mathay SB, System requirements specification for the U.S. Geological Survey's National Water Information System II, 1991, U.S. Geological Survey Open-File Report 91-525, 622 p.

McCartney, D., Beamer, N.H., 1962. Continuous recording of water quality in the Delaware Estuary. J. Am. Water Works Assoc. 54 (10), 1193–1200.

McGee, T.D., 1988. Principles and Methods of Temperature Measurement. Wiley, New York. 581 p.

McMahon, P., Chapelle, F.H., Bradley, P., 2011. Evolution of Redox Processes in Groundwater in Aquatic Redox Chemistry. ACS Symposium Series, vol. 1071. American Chemical Society, pp. 581–597. chap. 26.

Michalak, A.M., 2016. Study role of climate change in extreme threats to water quality. Nature 535, 549–550.

Miles, E.J., 2009. Shallow water-quality monitoring— continuous monitoring station selection, assembly, and construction. Virginia Institute of Marine Science Special Report in Applied Marine Science and Ocean Engineering no. 412, 223 p.

Miller, C., Gutiérrez-Magness, A.L., Feit Majedi, B.L., Foster, G.D., 2007. Water quality in the Upper Anacostia River, Maryland: continuous and discrete monitoring with simulations to estimate concentrations and yields, 2003–05. U.S. Geological Survey Scientific Investigations Report 2007-5142, 43 p.

Miner, K., Wayant, N., Ward, H., 2018. Preventing chemical releases in hurricanes. Science 362 (6411), 166.

Mize, S.V., Demcheck, D.K., 2009. Water quality and phytoplankton communities in Lake Pontchartrain during and after the Bonnet Carré Spillway opening, April to October 2008, in Louisiana, USA. Geo-Mar. Lett. 29, 431–440. https://doi.org/10.1007/s00367-009-0157-3.

Mize, S.V., Demcheck, D.K., Rivers, B.W., 2013. Water-quality and phytoplankton communities in Lake Pontchartrain before, during, and after the Bonnet Carré Spillway openings in 2008 and 2011, Louisiana. Mississippi River and Tributaries 2011 Post-Flood Report, Appendix F, Environmental and Cultural Resources, U.S. Army Corps of Engineers, 21 p.

Monterey Bay Aquarium Research Foundation (2018), The Environmental Sample Processor (ESP). accessed 12.14.2018 at https://www.mbari.org/technology/emerging-current-tools/instruments/environmental-sample-processor-esp/.

Müller, R.H., 1954. Test shows Coleman is best buy. Anal. Chem., 1954 26 (4), 43A–46A. https://doi.org/10.1021/ac60088a747.

Myers, D.N., Ahuja, S., 2015. Foundations of water-quality monitoring and assessment in the United States. In: Food, Energy and Water: The Chemistry Connection. Elsevier, pp. 21–79.

Myers, D.N., Ludtke, A.S., 2017. Progress and lessons learned from water-quality monitoring networks. In: Ahuja, S. (Ed.), Chemistry and Water: The Science Behind Sustaining the World's Most Crucial Resource. In: 2017, Elsevier, pp. 23–112.

National Ecological Observatory Network, The Observatory, 2018, National Ecological Observatory Network accessed 10.16.2018 at https://www.neonscience.org/observatory.

National Oceanic and Atmospheric Administration (2018a), Great Lakes meteorological real-time coastal observation network (ReCON) accessed 10.16.2018 at https://www.glerl.noaa.gov/metdata/.

National Oceanic and Atmospheric Administration, 2018b. Environmental sample processer. 10.14.2018. https://www.glerl.noaa.gov//res/HABs_and_Hypoxia/esp.html.

Nichols n.d. (undated), Real-Time Water Quality Monitoring in New Jersey Estuaries. Fact Sheet, Monmouth University, 1 p., accessed 3.11.2019 at https://www.monmouth.edu/uci/documents/2018/10/real-time-water-quality-monitoring-in-nj-estuaries.pdf/.

Orion Research, 1982. Handbook of Electrode Technology. Orion Research 45 p.

Paerl, H.W., Bales, J.D., Ausley, L.W., Buzzelli, C.P., Crowder, L.B., Eby, L.A., Fear, J.M., Go, M., Peierls, B.L., Richardson, T.L., Ramus, J.S., 2001. Ecosystem impacts of three sequential hurricanes (Dennis, Floyd, and Irene) on the United States' largest lagoonal estuary, Pamlico Sound, N.C. Proc. Natl. Acad. Sci. 98 (10), 5655–5660.

Palmer, M.D., 1970. Monitoring of water quality. Water Res. 4, 765–770.

Peierls, B.L., Christian, R.R., Paerl, H.W., 2003. Water quality and phytoplankton as indicators of hurricane impacts on a large estuarine ecosystem. Estuar. Coasts 26 (5), 1329–1343.

Pellerin, B.A., Bergamasci, B.A., 2014. Optical sensors for water quality. Lakeline, 34 (1), 13–17. May 2014.

Pellerin, B.A., Bergamaschi, B.A., Downing, B.D., Saraceno, J.F., Garrett, J.A., Olsen, L.D., 2013. Optical techniques for the determination of nitrate in environmental waters: guidelines for instrument selection, operation, deployment, maintenance, quality assurance, and data reporting. U.S. Geological Survey Techniques and Methods 1-D5, 37 p.

Pellerin, B.A., Bergamaschi, B.A., Gilliom, R.J., Crawford, C.G., Saraceno, J., Frederick, C.P., Downing, B.D., Murphy, J.C., 2014. Mississippi River nitrate loads from high-frequency sensor measurements and regression-based load estimation. Environ. Sci. Technol. 48, 12612–12619. https://doi.org/10.1021/es504029c.

Porter, J.H., Hanson, P.C., Lin, C.C., 2012. Staying afloat in the sensor data deluge. Trends Ecol. Evol. 27 (2), 121–129.

Province of British Columbia, 2015. Continuous Water-Quality Sampling Programs: Operating Procedures. Resources Information Standards Committee, Victoria, BC. 115 p.

Radtke DB, Davis JV, Wilde FD, Specific electrical conductance (ver. 1.2) , 2005, U.S. Geological Survey Techniques of Water-Resources Investigations, book 9, chap. A6, sec. 3, August 2005, accessed 9.25.2018, from http://pubs.water.usgs.gov/twri9A6/.

Radu, A., Radu, T., McGraw, C., Dillingham, P., Anastasoa-Ivanova, S., Diamond, D., 2013. Ion selective electrodes in environmental analysis. J. Serb. Chem. Soc. 78 (11), 1729–1761.

Raich, J., 2013. Review of Sensors to Monitor Water Quality. Publications Office of the European Union, European Commission, Luxembourg. 33 pp.

Rantz SE, Discharge ratings using simple stage-discharge relations, 1982, U.S. Geological Survey Water-Supply Paper 2175, Chap. 10 v. 2 pp. 544–600.

Rasmussen, P.P., Ziegler, A.C., 2003. Comparison and continuous estimates of fecal coliform and *Escherichia coli* bacteria in selected Kansas Streams, May 1999 through April 2002. U.S. Geological Survey Water-Resources Investigations Report 03-4056, 87 p. http://ks.water.usgs.gov/Kansas/pubs/reports/wrir.03-4150.html.

Rasmussen, T.J., Ziegler, A.C., Rasmussen, P.P., Stiles, T.C., 2005. Estimation of constituent concentrations, densities, loads, and yields in Lower Kansas River, northeast Kansas, using regression models and continuous water-quality monitoring, January 2000 through December 2003. U.S. Geological Survey Scientific Investigations Report 2005-5165, 117p. http://pubs.usgs.gov/sir/2005/5165/.

Rasmussen, T.J., Lee, C.J., Ziegler, A.C., 2008. Estimation of constituent concentrations, loads, and yields in streams of Johnson County, northeast Kansas, using continuous water-quality monitoring and regression models, October 2002 through December 2006. U.S. Geological Survey Scientific Investigations Report 2008-5014, 103 p. at http://pubs.usgs.gov/sir/2008/5014/.

Rasmussen, P.P., Gray, J.R., Glysson, G.D., Ziegler, A.C., 2009. Guidelines and procedures for computing time-series suspended-sediment concentrations and loads from in-stream turbidity-sensor and streamflow data. U.S. Geological Survey Techniques and Methods, book 3, chap. C4, 52 p.

Rasmussen, P.P., Eslick, P.J., Ziegler, A.C., 2016. Relations between continuous real-time physical properties and discrete water-quality constituents in the little Arkansas River, South-central Kansas, 1998-2014. U.S. Geological Survey Open-File Report 2016-1057, 20 p.

Reay, W.G., Moore, K.A., 2005. Impacts of tropical cyclone Isabel on shallow water quality of the York River estuary. In: Sellner, K.G. (Ed.), Hurricane Isabel in Perspective. Chesapeake Research Consortium, pp. 135–144. CRC Publication 05-160.

Rebich, R.A., Wilson, D., Runner, M., 2015. In: Monitoring network design to assess potential water-quality improvements associated with the Mississippi Coastal Improvement Program in the Mississippi Sound. Presented at 2015 Mississippi Water Resources Conference, April 7, 2015. http://wrri.msstate.edu/conference/pdf/rebich_richard2015.pdf. (Accessed 10 January 2018).

Rode, M., Wade, A.J., Cohen, M.J., Hensley, R.T., Bowes, M.J., Kirchner, J.W., Arhonditsis, G.B., Jordan, P., Kronvang, B., Halliday, S.J., Skeffington, R.A., Rozemeijer, J.C., Aubert, A.H., Rinke, K., Seifeddine, J., 2016. Sensors in the stream: the high-frequency wave of the present. Environ. Sci. Technol. 50, 10297–10307. https://doi.org/10.1021/acs.est.6b02155.

Rounds SA, Wilde FD, Ritz GF, Dissolved Oxygen (ver. 3.0): U.S. Geological Survey Techniques of Water Resources Investigations, 2013, book 9, chap. A6, sec. 6.2, 19 p. http://water.usgs.gov/owq/FieldManual/Chapter6/6.2_v3.0.pdf.

Rudnick, D.L., Davis, R.E., Eriksen, C.C., Fratantoni, D.M., Perry, M.J., 2004. Underwater gliders for ocean research. Mar. Technol. Soc. J. 38 (1), 48–59.

S::CAN (2018), Spectrometer probes. accessed 12.14.2018 at https://www.lab-environ.com/categories/Spectrometer-probes/7.

Sadar, M.J., 1988. Turbidity Science. Technical Information Series—Booklet No. 11, Hach Chemical Company. 26 p.

Saraceno, J.F., Shanley, J.B., Downing, B.D., Pellerin, B.A., 2017. Clearing the waters: evaluating the need for site-specific field fluorescence corrections based on turbidity

REFERENCES

measurements. Limnol. Oceanogr. 15, 408–416. https://doi.org/10.1002/lom3.10175.

Sauer, V.B., Turnipseed, D.P., 2010. Stage measurement at gaging stations. U.S. Geological Survey Techniques and Methods book 3, chap. A7, 45 p. http://pubs.usgs.gov/tm/tm3-a7/.

Sayers, W.T., 1971. Water quality surveillance: the federal-state network. Environ. Sci. Technol. 5 (2), 114–119.

Schmelkes, F., Horning, E., Campbell, G., 1939. Electro-chemical properties of chlorinated water. J. Am. Water Works Ass. 31 (9), 1524–1537.

Schroepfer, G.T., 1942. An analysis of stream pollution and stream standards. Sewage Work. J. 14 (5), 1030–1063.

Scofield, C.S., 1932. Measuring the salinity of irrigation waters and of soil solutions with the wheatstone bridge. U.S. Department of Agriculture; Circular 232, 20 pp.

Seabird Scientific, 2017. HydroCycle PO_4 in situ dissolved phosphate. 2 p., accessed 3.11.2019 at. https://www.seabird.com/hydrocycle-po/product-downloads?id=54721314201.

Sherman, J., Davis, R.E., Owens, W.B., Valdes, J., 2001. The autonomous underwater glider 'Spray'. IEEE Ocean. Eng. 26, 437–446.

Shoda, M.E., Lathop, T.R., Risch, M.R., 2015. Real-time, continuous water-quality monitoring in Indiana and Kentucky. U.S. Geological Survey Fact Sheet 2015-3041, 4 p.

Shull, D., Lookenbill, J., 2015. Bureau of Clean Water Continuous Instream Monitoring Protocol. Pennsylvania Department of Environmental Protection, Harrisburg, PA. 66 p.

Simoni, R.D., Hill, R.L., Vaughan, M., Tabor, H., 2003. A classic instrument: the Beckman DU spectrophotometer and its inventor, Arnold O. Beckman. J. Biol. Chem. 278 (49), 79–81. http://www.jbc.org/content/278/49/e1.full.html#ref-list-1.

Simpson, J.B., Pettibone, C.A., Kranzler, G.A., 1991. Temperature. In: Henry, Z.A., Zoerb, G.C., Birth, G.S. (Eds.), Instrumentation and Measurement for Environmental Sciences, third ed. Am. Soc. Agric. Engs. pp. 6-1 to 6-17.

Smart, P.M., Laidlaw, I.M.S., 1977. An evaluation of some fluorescent dyes for water tracing. Water Resour. Res. 13 (1), 15–35. https://doi.org/10.1029/WR013i001p00015.

Smoot, G.F., Blakey, J.F., 1966. Systems for monitoring and digitally recording water quality parameters. U.S. Geological Survey Open-File Report, 14 p.

Snyder, L., Potter, J.D., McDowell, W.H., 2018. An evaluation of nitrate, fDOM, and turbidity sensors in New Hampshire streams. Water Resour. Res. 54, 2466–2479. https://doi.org/10.1002/2017WR020678.

Stabler, H., 1911. Some stream waters of the western United States with chapters on sediment carried by the Rio Grande and the industrial application of water analyses. U.S. Geological Survey Water-Supply Paper 274, p. 188.

Stow, C.A., Cha, Y.K., Johnson, L.T., Confesor, R., Richards, R.P., 2015. Long-term and seasonal trend decomposition of Maumee River nutrient inputs to Western Lake Erie. Environ. Sci. Technol. 49, 3392–3400.

Systea SpA n.d., WIZ portable in-situ probe for water analysis. WIZ_05E, 4 p.

Technavio, 2016. Global water quality sensors market 2016–2020. 70 p. https://www.technavio.com/report/global-semiconductor-equipment-global-water-quality-sensors-market-2016-2020. (Accessed 7 July 2018).

Texas Commission on Environmental Quality, 2018. Quality assurance project plan for continuous water-quality monitoring network program. Austin, TX, 125 p.

Texas Commission on Environmental Quality (variously dated), n.d. Standard operating procedures: continuous water-quality monitoring network. accessed 9.20.2018 at https://www.tceq.texas.gov/waterquality/monitoring/cwqm_sops.html.

Thomas, O., Caussel, J., Jung, A.V., Thomas, M.F., 2017. Natural water. In: Thomas, O., Burgess, C. (Eds.), UV-Visible Spectrophotometry of Water and Wastewater, second ed. In: vol. 2017. Elsevier, pp. 225–257. https://doi.org/10.1016/B978-0-444-63897-7.00007-X.

Tiedemann, J., Nickels, J., Witty, M., 2009. Real-time water quality monitoring shows diurnal variation of dissolved oxygen that contributes to unsolved fish kills. Proc. Water Environ. Fed. 2009, 3282–3291.

Travis, A.S., 2002. Instrumentation in environmental analysis, 1935–1975. In: Morris, P.J.T. (Ed.), From Classical to Modern Chemistry, the Instrumental Revolution. In: 2002. Royal Society for Chemistry and the Chemistry Museum, London, UK, pp. 285–308.

Turner Designs, 2003. Affordable fluorescence solutions in the palm of your hand. Turner News. v. 3, unpaginated.

Turner Designs, 2019a. ICAM, in situ absorption. Publication S-0159, 2 p., accessed 3.11.2019 at. http://www.comm-tec.com/Docs/Brochure/Turner/ICAM_S-0159.pdf.

Turner Designs, 2019b. Optical specification guide: cyclops submersible sensors. accessed 3.11.2019 at http://www.comm-tec.com/Docs/Brochure/Turner/ICAM_S-0159.pdf.

Turner Designs, 2019c. Phyto find—in situ algal classification. Publication S-0212 Rev. E, 2 p., accessed 3.12.2019 at http://docs.turnerdesigns.com/t2/doc/brochures/S-0212.pdf.

Turner Designs, 2019d. Application note: fluorometer calibration for in vivo detection of cyanobacterial pigments. undated. Publication S-0024 Rev. A, 3 p., accessed 3.12.2019 at http://docs.turnerdesigns.com/t2/doc/appnotes/S-0024.pdf.

U.S. Army Corps of Engineers and Mississippi River Commission, 2012. Room for the river. Summary Report of the 2011 Mississippi River Flood. Vicksburg. 32 pp.

U.S. Environmental Protection Agency, 1978. Ion selective electrodes in water analysis. U.S. Environmental Protection Agency, EPA 600/2-78-106, 37 pp.

U.S. Environmental Protection Agency, 1979. A national compendium of fish and water temperature data. U.S. Environmental Protection Agency, EPA 600/3-79-056, 208 pp.

U.S. Environmental Protection Agency, 1986. Quality Criteria for Water 1986, Temperature. U.S. Environmental Protection Agency, EPA 440/5-86-001, 395 pp.

U.S. Environmental Protection Agency, 1993. Methods for the determination of inorganic substances in environmental samples: Cincinnati, Ohio. EPA/600/R-93/100, 178 p.

U.S. Environmental Protection Agency, 2015. 2015 drinking water health advisories for two cyanobacterial toxins. Office of Water 820F15003, 3 p.

U.S. Environmental Protection Agency, 2017. Field measurement of oxidation-reduction potential (ORP). SESD-PROC-113-R2, 22 pp.

U.S. Environmental Protection Agency (2018), Exposure Assessment Models—Virtual Bbach: Center for Exposure Assessment Modeling, accessed 10.08.2018 at http://www2.epa.gov/exposure-assessment-models/virtual-beach-vb\.

U.S. Geological Survey, 1998. National water information system (NWIS). U.S. Geological Survey Fact Sheet 98-027, 2 p.

U.S. Geological Survey, 2002. NWISWeb: new site for the nation's water data. U.S. Geological Survey Fact Sheet 02-128, 2 p.

U.S. Geological Survey, 2013. Tools for beach health data management, data processing, and predictive model implementation. U.S. Geological Survey Fact Sheet 2013-3068, 6 p.

U.S. Geological Survey (2017), WaterQualityWatch accessed September 1, 2017 at http://waterwatch.usgs.gov/wqwatch/.

U.S. Geological Survey, NowCast—A daily nowcast of recreational water quality conditions. U.S. Geological Survey, 2018, Accessed 10.8.2018 at https://ny.water.usgs.gov/maps/nowcast/.

U.S. Geological Survey n.d. (variously dated), Budget justifications and performance information. accessed 12.11.2018. https://www.usgs.gov/about/organization/science-support/budget/usgs-budget-archives.

Urick, R.J., 1975. Principles of Underwater Sound, second ed. McGraw Hill, New York. 384 p.

UST, 2018. Autonomous underwater vehicle manufacturers (AUV; UUV; ROV). https://www.unmannedsystemstechnology.com/category/supplier-directory/platforms/uuv-manufacturers/. (Accessed 17 October 2018).

Veley, R.J., Moran, M.J., 2012. Evaluating lake stratification and temporal trends by using near-continuous water-quality data from automated profiling systems for water years 2005–09, Lake Mead, Arizona and Nevada. U.S. Geological Survey Scientific Investigation Report 2012–5080, 25 p. Available at http://pubs.usgs.gov/sir/2012/5080/.

Velz, J., Clark, R.N., 1950. Sampling for effective evaluation of stream pollution. Sewage Ind. Waste. 22 (5), 666–684.

Wagner, R.J., Mattraw, H.C., Ritz, G.F., Smith, B.A., 2000. Guidelines and standard procedures for continuous water-quality monitors: site selection, field operation, calibration, record computation, and reporting. U.S. Geological Survey Water Resources Investigation Report 00-4252, 53 p.

Wagner, R.J., Boulger Jr., R.W., Oblinger, C.J., Smith, B.A., 2006. Guidelines and standard procedures for continuous water-quality monitors—station operation, record computation, and data reporting. U.S. Geological Survey Techniques and Methods 1–D3, 51 p. + 8 attachments; accessed at http://pubs.water.usgs.gov/tm1d3.

Walker, N.D., 2001. Tropical storm and hurricane wind effects on water level, salinity, and sediment transport in the river-influenced Atchafalaya-Vermilion Bay system, Louisiana USA. Estuar. Coasts 24 (4), 498–508.

Wang, S., Tang, J., Younce, F., 2003. Temperature measurement. In: Encyclopedia of Agricultural, Food, and Biological Engineering. Marcel Dekker, New York, pp. 987–993.

Ward, R.C., 1973. Data Acquisition Systems in Water-Quality Management. EPA-R5-73-014, U.S. Environmental Protection Agency, 260 p.

Weller, R., Toole, J., McCartney, M., Hogg, N., 2000. Outposts in the ocean. Oceanus 42 (1), 20–23.

Weyl, P.K., 1964. On the change of electrical conductance of seawater with temperature. Limnol. Oceanogr. 9 (1), 75–78. https://doi.org/10.4319/lo.1964.9.1.0075.

Wilde FD., n.d. ed. (variously dated), Field measurements. U.S. Geological Survey Techniques of Water-Resources Investigations, book 9, chap. A6, with sec. 6.0–6.8, accessed 10.7.2018 at http://pubs.water.usgs.gov/twri9A6/.

Wilson Jr., J.F., Cobb, E.D., Kilpatrick, F.A., 1986. Fluorometric procedures for dye tracing. U.S. Geological Survey Techniques of Water-Resources Investigations, book 3, ch. A12, 34 p.

Wong, B.P., Kerkez, B., 2016. Real-time environmental sensor data: an application to water quality using web services. Environ. Model. Software 84, 505–517.

Wood, W.W., 1976. Guidelines for collection and field analysis of groundwater samples for selected unstable constituents. U.S. Geological Survey Techniques of Water Resources Investigations, book 1, chap. D-2, 24 p.

Wood, M.S., 2014. Estimating suspended sediment in rivers using acoustic doppler meters. U.S. Geological Survey Fact Sheet 2014-3038, 4 p.

Wood, M.S., Teasdale, G.N., 2013. Use of surrogate technologies to estimate suspended sediment in the Clearwater River, Idaho, and Snake River, Washington, 2008–10. U.S. Geological Survey Scientific Investigations Report 2013-5052, 30 p.

REFERENCES

World Health Organization, (Ed.), 1971. International Standards for Drinking Water. third ed World Health Organization, Geneva. 70 pp.

World Health Organization, 2003. Cyanobacterial toxins: microcystin-LR in drinking-water; background document for development of WHO guidelines for drinking-water quality. WHO/SDE/WSH/03.04/57, 14 p.

Wynne, T.T., Davis, T.W., Kelty, R., Anderson, E.J., Joshi, S.J., 2015. NOAA forecasts and monitors blooms of toxic cyanobacteria in Lake Erie. In: Clear Waters. New York Water Environment Association, Syracuse, NY. 21–23 pp. http://www.glerl.noaa.gov/pubs/fulltext/2015/20150041.pdf.

YSI, 2007. Pontoon-Mounted Autonomous Monitoring System Keeps Constant Watch on Harmful Algal Blooms. YSI. Application Note, 3 p.

YSI, 2009. The Dissolved Oxygen Handbook—A Practical Guide to Dissolved Oxygen Measurements. YSI. 43 p.

YSI, 2015a. Evolution of Water-Quality Monitoring. YSI. eBook E112, 54 p.

YSI. Using real-time telemetry for ecological monitoring of coastal wetlands, 2015b, YSI Environmental Application Note A578, (2 p).

YSI, i3XO EcoMapper™ autonomous underwater vehicle, specifications, 2018, YSI Fact Sheet #E50 Rev. A, 2 p.

YSI Environmental, Water Tracing, in situ dye Fluorometry and the YSI 6130 Rhodamine WT Sensor, 2001, YSI Environmental, White Paper 1006 E46-01, 7 p.

Zeng, L., Li, D., 2015. Development of in situ sensors for chlorophyll concentration measurement. J. Sensors 2015, 1–16. https://doi.org/10.1155/2015/903509.

CHAPTER
11

Biosensors for Monitoring Water Pollutants: A Case Study With Arsenic in Groundwater

Jason Berberich[a,], Tao Li[b], Endalkachew Sahle-Demessie[b]*

[a]Department of Chemical, Paper and Biomedical Engineering, Miami University, Oxford, OH, United States [b]National Risk Management Research Laboratory, Office of Research and Development, U.S. Environmental Protection Agency, Cincinnati, OH, United States
*Corresponding author: E-mail: berberj@miamioh.edu

1 INTRODUCTION

The demand for safe freshwater has been increasing globally mainly due to population growth, the increase in municipal, agricultural, and industrial activities, and changing land use. In the United States, about 1.5 billion cubic meters of water is used each day (Maupin et al., 2014). According to the United Nations Department of Economics and Social Affairs (UNDESA), water quality is one of the main challenges that societies have faced during the last 50 years (Neary, 2008). Increase in human activities has a negative impact on the quality of water. Water-quality issues are related to human health, sanitation, and biodiversity. Despite much progress in cleaning-up waterways in many areas, water pollution remains a serious global problem. Declining water quality could threaten human health, limit food production, reduce ecosystem functions, and hinder economic growth. A sustainable supply of safe water is needed for maintaining people's livelihoods, healthy ecosystems, and robust economies. There are many challenges including biological and chemical contaminants in source water and water distribution systems. Aging water system infrastructure also threatens the safety and sustainability of many water resources. Dynamics of water quality is dependent on both human interventions of water systems and natural causes such as detention in reservoirs, channelization as water diversions, and changes in rainfall and snowmelt (Kraus et al., 2011). Water-quality changes may also be caused by more subtle transformations in soil temperature, atmospheric deposition, and shifting vegetation patterns where the effects are complex and interconnected. Detecting short-term water-quality changes are vital, in the events such as a chemical spill or a wastewater lagoon discharging following a heavy rain. Monitoring long-term changing trends is also important for resource management. New technologies that include in situ sensors, data platforms, and analysis have provided improved opportunity to monitor and manage large-scale water system.

1.1 The Changing Profile Water Pollution

Water pollution consists of a wide range of factors that are linked to chemical, physical, and microbial components. In most industrialized countries, the types of legacy and emerging contaminants continue to be a concern. However, over the years, the balance of pollutants has shifted markedly (Tanabe and Ramu, 2012). The quality of rivers and lakes has been improved as the result of reduced flows of untreated sewage and industrial pollutants to rivers. A study on lake water by the United Nations Environment Program (UNEP) identified four main problems causing pollution in freshwater lakes and water. These include pathogens (which are caused by human and animal waste), organic matter (including plant nutrients from agricultural runoff such as nitrogen or phosphorus), chemical pollution, and salinity (from irrigation, domestic wastewater, and runoff of mines into rivers) (Waste, 2010). Nonpoint source pollutions increased as the result of the growing industrial and agricultural activities, increase in population, living standards in developing countries. Economic growth correlated to increases of suspended sediments, nutrients, pesticides, and oxygen-consuming substances in water (Mainstone and Parr, 2002; Ricciardi and Rasmussen, 1999). Many anthropogenic activities generate polluting by-products including microbial pathogens, nutrients, oxygen-consuming materials, heavy metals, and persistent organic matter (Sutton et al., 2013). Fig. 1 summarizes the primary causes of water-quality reduction and their impact. Minor changes in the complex ecosystems impact the balance of water chemistry and alter the quality of water (Karr and Dudley, 1981). The presence of multiple contaminants synergistically causes amplified or different impacts than the cumulative effects of pollutants (Vinebrooke et al., 2004). The continued input of pollutants can ultimately exceed an ecosystem's resilience, leading to dramatic and possibly irreversible losses. Groundwater systems contaminated by anthropogenic pollutants are difficult and costly to restore.

Environmental laws such as the Clean Water Act, in the United States, and accompanying

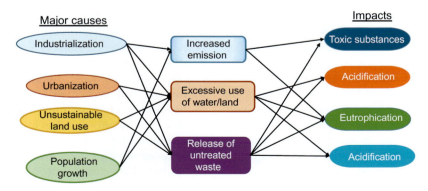

FIG. 1 Factors which cause water-quality problems in surface waters.

1 INTRODUCTION

regulatory measures, have succeeded in cutting water pollutions from concentrated "point sources" like factories and municipal sewage treatment plants. In the United States, the volume of treated municipal wastewater increased by 30% from 1972 to 1992 while the organic pollution [measured as the biological oxygen demand (BOD)] from these plants dropped by 36% (Burian et al., 2000). However, the national efforts in cleaning-up waterways have not been as effective in reducing "nonpoint" pollutants such as toxins, nutrients, sediments in the form of runoff from agriculture, urban and suburban stormwater, oil and gas operations, and mining industry (Carpenter et al., 1998). A growing number of legacy and emerging contaminants from domestic, commercial, and industrial activities find their ways to waste and wastewater plants. In other areas, new pollutants such as pesticides, personal care products (PPCPs), pharmaceuticals, and nanomaterials are disposed into sewer systems and reach regional wastewater treatment plants. There has been an increase in the concentrations of degradation by-products from disposed organic compounds and caused accumulation of persistent metabolites (Sørensen et al., 2007) in water treatment. Many of these plants are not designed to treat emerging contaminants such as disposed pharmaceuticals. Once released to the environment these chemicals may have hormonal effects and toxicity in aquatic animals. The impact of chronic exposure to these chemicals is not fully understood. The conventional pollution prevention approaches may not be fully effective to prevent the impacts of these pollutants (Boxall et al., 2003).

1.2 Sources of Water Pollution

1.2.1 Nutrient Pollution

The most prevalent water-quality problem is eutrophication, which is the result of enrichment of nutrients in an ecosystem. A significant fraction of high-nutrient loads is phosphorus and nitrogen from urban and farmlands. Large bodies of water were also threatened by eutrophication. World Wildlife Fund reported that in the past century the phosphorus concentration had increased by a factor of 10 in the Baltic Sea (Vandeweerd et al., 2006). A study by the United Nations Environmental Program (UNEP) has indicated that about 40% of the lakes and reservoirs globally have been affected by eutrophication due to nutrient enrichment on coastal hypoxia, creating algal blooms that are harmful to human health (Ricciardi and Rasmussen, 1999). The mechanism of eutrophication is not fully understood, but the two primary nutrients to blame are phosphorus and nitrogen. The levels of nutrients in many lakes and rivers increased significantly over the past 50 years due to increased discharge of domestic wastes and nonpoint pollution from agricultural practices and urban development (Mainstone and Parr, 2002). The impact in water-quality reduction is noticeable in shallow bodies of water that are surrounded by land such as the Baltic Sea, where marine biotopes are threatened by loss of air or reduction in quality from eutrophication, contamination, unregulated fishery activities, and human settlements. By 2000, water-quality degradation in many lakes of industrialized countries stopped or slowed down due to the increased use of wastewater treatment technologies (Schindler, 2012). However, eutrophication of lakes and rivers is on the fast rise in many countries, where there is excessive fertilizer application in agriculture and at places which lack pollution reduction due to economic reasons. More than 80% of sewage in developing countries is discharged untreated, containing everything from human waste, farm waste to highly toxic industrial discharges. The sewage discharges have polluted rivers, lakes, and coastal areas. Both the composition and quantity of nutrients impact the formation of harmful algal blooms (HABs) (Hallegraeff, 2003). HAB may be caused and sustained by exogenous chronic low-level nutrient delivery or by episodic events. Hence, the detection and prediction of HABs and their toxins have become critical for water-quality management.

Pollution of freshwater ecosystems impacts the habitat and quality of aquatic life and other wildlife. The type of contaminants and the levels of contamination determine the suitability of water for many uses such as drinking, bathing, and agriculture. Removing pollutants from water is usually difficult, costly, and often impossible. The impacts of pollution on drinking water sources of small communities are more significant than in larger metropolitan areas, since small towns may not afford expensive treatment technologies. Extreme- or hyper-eutrophication of drinking water sources, for example, Lake Dianchi, and Lake Taihu in China, has resulted in ecological and human health issues. The problem in these lakes was so severe that all the native water plants and many species of fish were killed. Anoxic conditions kill snails in the bottom water. Due to the poor quality of the water, it has been challenging to supply water for domestic use that meets regulatory standards. Many polluting industries such as leather tanning and chemicals manufacturing are moving from developed countries to developing countries. Although there have been some regions that have shown improvements in water quality, water pollution is on the rise globally. The effects of eutrophication in the Florida Everglades are causing a shift in its native flora and fauna. The total phosphorus concentration in Lake Okeechobee was 69 µg/L despite the use of over 17,000 ha of stormwater treatment areas since 2004 (Richardson et al., 2007).

Water contamination from agricultural livestock operations has been a persistent concern. However, the growth of concentrated animal feeding operations (CAFOs) in the past 3 decades presents a greater risk to water quality because of the increased volume of waste and the contaminants in the waste. The manure or sludge from intensified agriculture may leach to surface and groundwater and cause health problems (Richardson et al., 2007; Ruley and Rusch, 2002). CAFOs have been a concern as a "point source" for a variety of potential contaminants. The waste streams contain nutrients, growth hormones, antibiotics, chemicals additives in the manure or in equipment cleaners, pathogens such as *Escherichia coli*, animal blood, silage leachate from corn feed, or copper sulfate used in footbaths for cows. For example, a large farm with 800,000 pigs may produce over 1.6 million tons of waste a year, which is one and a half times more than the annual sanitary waste produced by the city of Philadelphia, Pennsylvania, a city with a population of more than 1.5 million (Mittal, 2009). Thus, CAFOs require large waste treatment plant next to the operation. Many times, the wastewater is stored in lagoons prior to the treatment. Floods following rains can cause waste storage lagoons to overflow, causing large discharges to nearby water bodies (Field and Struzeski Jr, 1972). Contaminants also travel over land or through surface drainage systems to nearby creeks or through man-made ditches.

1.2.2 Accidental Spills

In this section, water-quality problems caused by chemical micropollutants from accidental spills are discussed. Fig. 2 shows the most significant pathways by which emerging contaminants are introduced into the aquatic environment and end up in drinking water. One study pointed out that in 2015 one in four Americans lived in places where the water systems were in some violation of safety regulations, including the 1974 Safe Drinking Water Act (Percival et al., 2017). Although major chemical spills that make it into the news seem to be rare, chemical spills happen often. According to the Toxic Release Inventory prepared by the US EPA, more than 190 million pounds of a toxic

1 INTRODUCTION

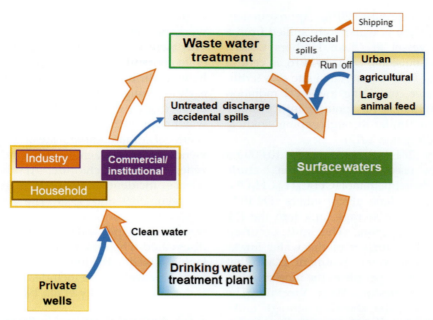

FIG. 2 Schematic representation of pathways of legacy and emerging contaminants introduced in the surface waters and drinking water sources.

substance were released into surface or groundwaters in 2016 (Blackman Jr, 2016). In addition to source water contaminations, the aging water distribution system also caused water-quality problems (Sartor and Boyd, 1972).

There are many episodes of accidental contaminations of rivers, lakes, and groundwaters, which are sources of raw drinking water supplies. Contamination of drinking water supplies by harmful bacteria and other microorganisms is a major concern. For example, Aeromonads are ubiquitous in aquatic environments and are readily isolated from both nutrient-rich and nutrient-poor environments (Holmes et al., 1996). *Aeromonas* species are regarded as pathogenic microorganisms involved in gastroenteritis. They are widely distributed in the aquatic environment, including raw and processed drinking water. Concern over the presence of mesophilic *Aeromonas* in public drinking water supplies has been expressed in recent years (Igbinosa et al., 2012). The contamination of a drinking water distribution system due to mixing with sewage water and other pollutants has happened in many parts of the world. This has led to contamination of drinking water by the mesophilic *Aeromonas* in Scotland, and in Finland (Ward et al., 1996).

There are many routes for groundwater contamination, but the major ways are flooding, improper use and disposal of contaminants, nonpoint source, and problems related to well integrity. For example, The Texas Commission on Environmental Quality (2015) was dealing with more than 3400 total groundwater contamination cases documented 276 new cases of groundwater contamination. One-third of these contaminations originate from petroleum storage tanks, Authorities had to notify private well owners that their drinking water might be contaminated (Bregman and Edell, 2016). As another example, on February 2018, approximately 400,000–450,000 gal of water were spilled from a containment vessel at Clean Harbors Colfax, Grant Parish Louisiana, USA. In 2016, in central Florida's rural Polk County, at a site of a fertilizer plant, water containing low-level radiation and other pollutants has poured into primary drinking water aquifer through gaping sinkhole 45 ft. wide (Klingener, 2017).

In January 2014, approximately 10,000 gal of chemicals used to process coal, crude 4-methylcyclohexylmethanol ($CH_3C_6H_{10}CH_2OH$) mixed with propylene glycol ethers (DiPPH), were spilled from a storage tank into the Elk River in West Virginia. The spill occurred 2.4 km (1.5 miles) upriver of the water intake of the municipal water treatment plant that serves about 300,000 people in the nine counties surrounding Charleston, West Virginia. The company owning the chemical storage tanks failed to inspect and repair corroding tanks. The contaminated water overwhelmed the Kanawha Valley water treatment activated carbon bed adsorption system (Jeter et al., 2016; Whelton et al., 2017). Officials issued do-not-use order 2 h after MCHM was detected at the treatment plant (Omur-Ozbek et al., 2016). By that time, the West Virginia Poison Center started getting phone calls from the public about a severe reaction from bathing and drinking the water. The Elk River is a tributary of the Ohio River. The spill plume traveled downstream at least 390 miles to the Ohio River at Louisville, Kentucky. The impacted population involved more than 2 million people relying on water from the polluted part of the Ohio River (Whelton et al., 2015, 2017). In September 2017 Eastman Chemical Company settled a lawsuit for $151 M.

Although limited in number, there have been incidents of intentional water contamination. In 1972, a local terrorist group acquired 30–40 kg of typhoid bacteria cultures with the intention to use it against water supplies in Chicago. In 2000, workers at the Cellatex chemical plant in France dumped 5000 L of H_2SO_4 into a tributary of the Meuse River because they were denied workers' benefits (Carus, 2002; Gleick, 2006). This increasing concern of accidental and deliberate contamination of drinking water supplies have contributed to the idea of integrating early-warning systems (EWSs) for water managers (States et al., 2003). Other incidents of water supply contamination are Camp Lejeune chemical contamination caused by trichloroethylene (TCE), tetrachloroethylene (PCE), benzene, and vinyl chloride. Another industrial release incident happened at Woburn, MA between 1969 and 1978. The residents were exposed to chlorinated solvents and arsenic from a municipal water supply. The event was discovered when the number of childhood leukemia spiked in the town 4 years after the water from new wells were included in the municipal water supply (Rinsky et al., 1987).

1.2.3 Natural Pollution Sources: The Case of Arsenic Pollution of Drinking Water Sources

Some water contaminants produce harmful health effects only when their concentrations are above certain thresholds. Nitrates (NO_3) produce methaemoglobinaemia that can cause "blue baby" syndrome (Knobeloch et al., 2000). A second group consists of elements that are essential to human health, such as fluoride and arsenic, but exposure to excess quantities can create health risks. Many metal ions such as selenium, copper, and zinc are essential to health in low concentrations. However, metals tend to accumulate in tissues, and exposure to them over a long time or at high levels can lead to illness. Exposure to elevated concentrations of trace metals can have negative consequences for both wildlife and humans. Contaminants in the third group have very low thresholds of hazards. These include genotoxic substances such as pesticides and arsenic. Heavy metals occur naturally. High concentration of heavy metals can enter food and water as the result of environmental discharge from mining and heavy industry.

1 INTRODUCTION

Arsenic is distributed in the earth's crust at an average concentration of $2\,mg\,kg^{-1}$. It occurs in trace quantities in all rock, soil, water, and air. The natural processes such as volcanic action and low-temperature volatilization are the most important natural source of arsenic, which also produces one-third of the atmospheric flux of arsenic. Schwarzenbach et al. studied the pollution of drinking water from natural arsenic and found that it affects as many as 140 million people in 70 countries. Fig. 3 shows estimates of the global levels of arsenic in drinking water supplies (Schwarzenbach et al., 2010) and Fig. 4 shows arsenic concentration in at least 25% of the groundwater samples in the United States. Although arsenic is present in more than 200 mineral species, the most common one is arsenopyrite, which is an iron arsenic sulfide (FeAsS). As deposits of arsenopyrite become exposed to the atmosphere, usually through mining, the mineral slowly oxidizes, converting the arsenic into oxides that are more soluble in water, leading to acid mine drainage (AMD) (Akai et al., 2004; Appelo and Postma, 2004). Other anthropogenic industrial activities such as smelting of nonferrous metals and the burning of fossil fuels are the major processes that contribute to anthropogenic arsenic contamination of air, water, and soil (Bhattacharya et al., 1997, 2002; Cullen and Reimer, 1989). The use of arsenic-containing pesticides has left vast tracts of agricultural land contaminated and leads to arsenic accumulation in crops such as rice (Abedin et al., 2002).

The use of the waterborne wood preservative *ammoniacal copper zinc arsenate* (ACZA) has also led to arsenic contamination of the environment. Chromated arsenicals-treated wood is used to produce commercial wood shakes, shingles, permanent foundation support beams, and other wood products permitted by approved labeling. Copper-chromated arsenicals (CCA), is a waterborne wood preservative that has been used for pressure treatment of lumber since the 1930s. Pressure treatment of lumber has been used to protect wood against termites, fungi, and other pests that can degrade the integrity of wood products. The unintended environmental release of heavy metals from CCA-treated timber can occur at many points along the life cycle of the product, from manufacture, to handling and use, and to disposal (Khan et al., 2006). Although elemental arsenic is not soluble in water, arsenic salts exhibit a wide range of solubility depending on pH and the ionic environment in water. Arsenite [As(III)] is the dominant form under reducing

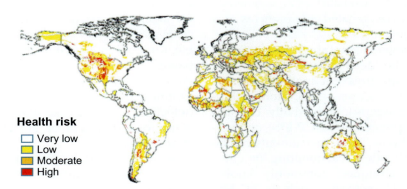

FIG. 3 Levels of arsenic in drinking water at significant levels. *Source: From Schwarzenbach, R.P., et al., 2010. Global water pollution and human health. Annu. Rev. Environ. Resour. 35, 109–136; United Nations Environment Program (UNEP).*

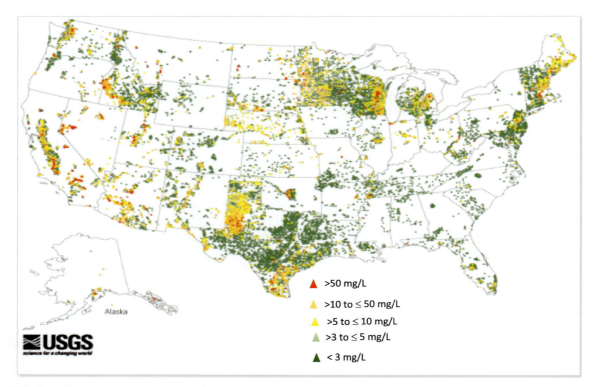

FIG. 4 A map of the United States showing available data on arsenic concentrations (μg/L or ppb) as of 2001 from samples of groundwater. Note that *orange or red dots* denote concentrations higher than that allowed in drinking water by US EPA regulations. *Source: US geological survey.*

conditions; however, arsenate [As(V)] is generally the stable form in oxygenated environments such as surface waters.

There are several laboratory instrumental methods for the determination of arsenic. These include spectroscopic methods (atomic absorption spectroscopy—AAS, atomic fluorescence spectroscopy), inductively coupled plasma–mass spectrometry (ICP-MS), and voltammetry methods. These instruments can be integrated with other fractionating or chromatographic techniques, commonly known as "hyphenated" methods, such as LC-MS, LC-ICP-MS to determine species of metallic pollutants, such as arsenic. These hyphenated methods have increased sensitivity range for arsenic compounds. Although laboratory systems are highly sensitive, they are unsuitable for field measurements. A field test kit based on the color reaction of arsine with mercuric bromide had been used for blanket groundwater testing in Bangladesh, and the detection limit was 50–100 μg/L under field conditions (Rahman et al., 2002; van Geen et al., 2005).

Arsenic is naturally present in rocks and sediments that form aquifers tapped for drinking water (see Fig. 5). However, the arsenic found in rock and sediment is immobile; thus, only trace levels of arsenic are found in groundwater.

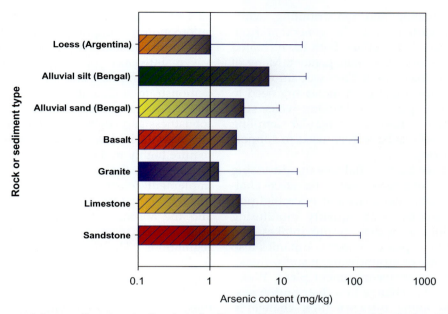

FIG. 5 Some arsenic is naturally present in all rocks and sediments that form aquifers tapped for drinking. Bars show average and lines show the range arsenic contents.

Certain natural geochemical conditions and processes can lead to the release of arsenic from the rocks and into the groundwater that is subsequently used for drinking. Monitoring metals in surface and groundwater supplies provides background information needed to determine the suitability of water resources for human consumption. Evidence suggests that levels of arsenic in groundwater aquifers in many parts of the world are below the World Health Organization (WHO) and the US Environmental Protection Agency (EPA) drinking water guideline (10 μg/L). However, arsenic remains a serious threat to health in some parts of the world such as Bangladesh and Cambodia, where shallow aquifer tube wells are abundant (Akai et al., 2004). Up to half of the estimated 10 million tube wells in Bangladesh might be contaminated with arsenic (Smith et al., 2000). As a result, up to 70 million people in Bangladesh were exposed to water with arsenic at greater than 10 μg/L (Berg et al., 2001). The effects of high levels of arsenic exposures, such as occurred in Bangladesh, have been well documented. Although well water reduced the high levels of cholera and other waterborne diseases, they have led to high rates of arsenic poisoning (Bhattacharjee, 2007; Bhattacharya et al., 2007).

2 MONITORING OF ARSENIC IN GROUNDWATER

2.1 Managing Groundwater Contamination by Arsenic is Limited by Analysis

In the United States, monitoring of surface and groundwater is done as part of the Clean Water Act and Safe Drinking Water Act. Water-quality monitoring is used to alert us to current, ongoing, and emerging problems and to provide the science-based information for sound decisions by local, state, tribal, and national stakeholders. Legacy pollutants have been accumulating and redistributing, while emerging pollutants may be generated from new chemicals or materials. Both have created challenges for pollution management. Many of the pollutants have made their way into drinking water. Thus, there is an increasing need to monitor water quality in drinking water supplies. Online monitoring and regular sampling for off-line analysis have been used to support decision-making.

Sensors play an essential role in addressing current and long-term water resource challenges for many chemicals and microbial pollutants. Currently, water-quality monitoring occurs mainly at water supply intakes and water treatment plants. Water companies also monitor surface or groundwater near the intake of the drinking water treatment plant. Water quality may also change in a distribution system due to aging infrastructure, change in water chemistry, and generation of disinfection by-products. For many water systems, there are few or no sensors along water distribution lines or at the point of use. Currently, the online monitoring of drinking water systems is used for basic physicochemical parameters includes measuring flow rate, physical characteristics such as turbidity, particle count, temperature, and chemical characteristics such as pH, conductivity, oxygen, and chlorine. Apart from general parameters, specific parameters such as fluoride, nitrate, or total organic carbon can also be monitored online. The current water monitoring system may not be adequate to warn the public and water treatment managers of water-quality issue, such as pollution by toxic metals.

The rising concern of the possible accidental or deliberate release of contaminants into water supply system has increased the need to for early-warning methods to assess drinking water at a system level (EWSs). Such water management system integrates online monitoring, data collection, interpretation, and communication of measured data. Online sensors can provide a rapid response and become an integral part of an EWS. This requires a significant degree of automation in sampling, data acquisition, and data transmission. The sensor should also be low maintenance and upgradeable, have low power consumption, and low cost. They should demonstrate sufficient sensitivity, allow minimal false positives/false negatives; exhibit robustness and ruggedness, be able to continually operate in a water environment, and not require highly trained operators.

Currently, the main obstacles to effectively implement sensors are the lack of standards for contamination testing in drinking water and the absence of the links between available sensor technologies and water-quality regulations. The International Organization for Standardization (ISO) is preparing globally applicable guidelines—the future ISO 11830 on the crisis management of water utilities. Biosensors will play a vital role in monitoring organic and inorganic priority pollutants and other toxins as an environmental quality-monitoring tool.

Groundwater quality is becoming a more urgent issue due to intensified human activity and global climate change. According to a U.S. Census American Housing Survey in 2015, about 13 million people in the United States obtained drinking water from private wells. In general, well water requires more testing and treatment than the water from the treatment plant because the condition of groundwater can be highly variable. Typical well water test includes bacteria, chlorine, hardness, pH, alkalinity, iron, nitrate/nitrite, copper, chlorine, and fluorine. The tests are carried out periodically (yearly in general) to monitor the effectiveness of treatment and track changes of water quality over time. The test needs to be more comprehensive in special circumstances. For example, when a family has pregnant women or infants, more stringent tests should be carried out to ensure water safety. Many more special tests are needed for well water in the proximity of human activities. Intensive agriculture may

EVALUATING WATER QUALITY TO PREVENT FUTURE DISASTERS

cause pesticide pollution in the groundwater. Gas drilling can raise the concentration of salt, drilling chemicals, heavy metals, and naturally occurring low-level radiation. Landfilling hazardous materials may cause organic pollution and mobilize heavy metals in sediments. For end users, groundwater testing is mainly for risk mitigation. For public agencies, the pollution of groundwater escalates to resource conservation. Both will benefit from the field use sensors.

Groundwater pollution by arsenic is a prime example of water-quality management for both consumers and public agencies. The scale and complexity of arsenic pollution have been intensively studied. Experience has been built globally for arsenic risk management and groundwater resource conservation. This knowledge is the most valuable reference for pollution control and remediation. It also provides important framework for analytical method development for practical application. These uses should be designed for low-income regions of the world to be low cost, practical, and reliable for monitoring individual wells on a regular basis. These sensors must be easy to use in addition to being affordable and sufficiently reliable.

2.1.1 Groundwater as the Most Important Resource

Groundwater is an essential resource for human activity. It constitutes two-thirds of the freshwater resources in the world (Chilton, 1996). Groundwater is generally free from significant microbial pollution and adequate for potable water supply with little or no treatment. In the rural areas of the United States, 96% of domestic water is supplied from groundwater. Groundwater quality is largely dependent on the dissolved inorganics. Among all the inorganics found in groundwater, arsenic causes the most severe concern because exposure to arsenic has caused many health crises all over the world (Mandal and Suzuki, 2002). Arsenic is either geogenic or from human activities.

Groundwater pollution appeared to be the primary pathway of human exposure (Mandal and Suzuki, 2002; Smedley and Kinniburgh, 2002). Consumption of polluted well water caused most of the arsenic poisoning episodes. Livestock health was also impacted by consuming arsenic-contaminated water from wells. In addition to drinking water, arsenic may also enter the food chain via crops when polluted groundwater is used for irrigation (Khan et al., 2009). The pollution can be more severe in the future because the demand for groundwater has been growing and increasing as the population grows (Chilton, 1996). Because of the risk from arsenic exposure, in 2001 the US EPA set the maximum permissible concentration (MCL) of arsenic for public water supplies for drinking water at a concentration of 10 µg/L (Frey and Edwards, 1997; Mineral Commodity Summaries, 2017).

Arsenic is the 20th most abundant element in the earth's crust. It can occur in 200 different mineral forms (Mandal and Suzuki, 2002; Smedley and Kinniburgh, 2002). Contamination of aquifers is related to geographical characters and mining. The most severe pollution is in South and Southeast Asia, including Bangladesh, West Bengal, Cambodia, and Vietnam. These places have Holocene alluvial and deltaic sediment aquifers. A similar type of pollution is also found in the Danube Basin in Romania and Hungary. Polluted volcanic sediment or geothermal aquifers of large areas have been found in Lagunera in Mexico, Antofagasta in Chile, Carson Desert in the United States, and Chaco Pampean Plain in Argentina. Aquifers of lacustrine sediment and old-basin-filled sediment are polluted in many areas. These areas include the San Joaquin Valley (CA), Southern Carson desert (NV), and Basin and Range (AR) in the United States, and Inner Mongolia in China. Mining activity also caused arsenic pollution in groundwater in Lavrion in Greece, Fairbanks (AK), Coeur d'Arlene (ID), and Lake Oahe (SD) in the United States, Ron Phibun in Thailand,

and Moira Lake (ON) in Canada (2 Kumar) (Mandal and Suzuki, 2002). In the United States, disposal of arsenic pesticides and warfare agents also contribute to arsenic pollution. Out of the 1191 Superfund Sites, about 30% listed arsenic as a contaminant of concern (Welch et al., 2000).

Since human exposure happens during arsenic migration and transport, the knowledge of its mobilization and sequestration is critical for pollution management. The general pathway can be simplified to a diagram with three nodes (Fig. 6) (Bondu et al., 2016). Arsenic is released from the primary source in bedrock to enter groundwater. Precipitation can sequester the dissolved arsenic with other minerals in the sediment. The sediment can serve as the secondary source of arsenic through highly complicated geochemical reactions. It is challenging to prevent human exposure to arsenic through groundwater because many factors can trigger arsenic release and influence transportation of dissolved arsenic (Smedley and Kinniburgh, 2002). It is unreliable to conduct risk assessment by extrapolation with a small quantity of data (Khan et al., 2009). Importantly, the interconversion between sequestered arsenic in a secondary source and dissolved arsenic can be erratic. Arsenic forms a mineral complex with Fe(OH)$_3$ in sediment. A modeling study showed that the sediment that is initially in equilibrium with 2 μg/L water may release arsenic under certain conditions, leading to [As] > 200 μg/L in water (Welch et al., 2000). Apparently, close monitoring should be an essential aspect to manage the pollution.

2.2 Factors That May Influence the Concentration of Arsenic in Water

2.2.1 Arsenic Concentration in Groundwater Is Dependent on the Dynamics of Mobilization and Sequestration

Arsenic has been found in high concentrations in certain rock-forming minerals, sediment rocks, and surficial deposits. Notable examples include sulfide minerals, iron-rich sediments, coals, bituminous shale, or mining-contaminated sediments (Smedley and Kinniburgh, 2002). The most notable primary source of arsenic is the oxidative dissolution from pyrite and arsenopyrite (Welch et al., 2000). The overall reactions of pyrite and arsenopyrite with oxygen can be summarized in Eqs. (1) and (2), respectively. Ferric hydroxide is the product in both cases. Both As(III) and As(V) can form complex with Fe(OH)$_3$ by adsorption. In addition, the arsenic released by oxidative dissolution is typically As(V), which can form salts with Ca(II) and Fe(III). The dissociation constants (K_d) of these salts are typically around 10^{-5} mol/L. Therefore, arsenic can be sequestered in the precipitation of these salts or through adsorption to ferric hydroxide (Magalhaes 8) (Magalhaes, 2002).

$$4FeS_2 \text{ (Pyrite)} + 15O_2 + 14H_2O = 4Fe(OH)_3 + 8H_2SO_4 \quad (1)$$

FIG. 6 General pathway for human exposure to arsenic through groundwater.

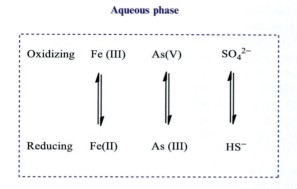

FIG. 7 Redox interconversions and solid formation in As-S-Fe system.

$2FeAsS$ (Arsenopyrite) $+ 7O_2 + 8H_2O = 2Fe(OH)_3$
$+ 2H_3AsO_4 + 2H_2SO_4$ (2)

Other chemicals, such as nitrate or Fe(II) can also release the arsenic in pyrite by sulfide oxidation (Herath et al., 2016). Arsenopyrite can also be oxidized with nitrate as the final electron acceptor. The oxidations usually involve many steps and slow, but can be facilitated by microbial metabolism. In water, many redox couples may participate in the transformations. They include O_2/H_2O, Mn(IV)/Mn(II), NO_3^-/N_2, NO_3^-/NO_2^-, Fe(III)/Fe(II), SO_4^{2-}/HS^-, and CO_2/CH_4 (Gorny et al., 2015). Many microorganisms can catalyze this oxidative dissolution. The type of oxidative microorganism and their growth depend on the pH and redox potential of the water, and the available nutrients (Herath et al., 2016).

The concentration of arsenic and their speciation are dependent on the redox state of the groundwater, chemical composition, microbial activity, absorption and adsorption processes. The interconversion between As(III) and As(V) are dependent on the redox potential as well as the pH. However, oxidation of As (III) with oxygen is kinetically slow. Therefore, other pathways are more significant. As(III) oxidation with manganese oxide is highly efficient. In nature, bacterial oxidation may play an important role in arsenic redox interconversion (Gorny et al., 2015).

The mobilization and sequestration of arsenic in groundwater are closely interconnected with iron and sulfur (Fig. 7) (Gorny et al., 2015; Hering et al., 2011; Welch et al., 2000). Under intensively oxidizing and strong acidic conditions, arsenate can form scorodite with Fe(III). The sulfate is easy to mobilize. More commonly in an oxic environment, ferric hydroxide precipitates out of water under neutral pH. Both As(V) and As(III) can be removed from water by coprecipitation or adsorption to $Fe(OH)_3$ or on mixed-valence iron oxides such as mackinawite. Although As(III) has less affinity to iron oxide (Welch 3) (Welch et al., 2000), it can be oxidized to As(V) by Fe(III) in the presence of UV light or by oxygen with Fe(II) catalysis (Gorny et al., 2015). Both reactions may contribute to arsenic sequestration (Hering et al., 2011). The sorption of arsenic depends on the surface property of iron oxide, the ionization of arsenate or arsenite, and competing anions in the water matrices. The pH dictates the ionization. The pK_a values of arsenate are 11.53, 6.97, and 2.20; while the values for arsenite [As(III)] are 13.4,

12.13, and 9.22. Therefore, arsenate is anionic, while arsenite is mostly neutral in groundwater. The efficacy of competing anions has been tested with isotherm experiment. For As(V) adsorption to Mg-Al-CO$_3$ hydroxide surface, the order of competing for anion is $HPO_4^{2-} > SO_4^{2-} > CO_3^{2-} > Cl^- > NO_2^- > F^-$. The order of anions for As(III) adsorption to Fe-Mn-oxide surface follows $HPO_4^{2-} > SiO_3^{2-} > CO_3^{2-} > SO_4^{2-}$ (Gorny et al., 2015).

Microbial respiration turns groundwater into a suboxic environment or an anoxic environment when organic carbon is supplied (Gorny et al., 2015; Hering et al., 2011). The Fe(III) is reduced to Fe(II), causing the dissolution of iron oxide mineral and release of arsenic into water (Smedley and Kinniburgh, 2002). The concentration of arsenic in water correlated to those of dissolved iron and ammonium ion in a reductive environment (Anawar et al., 2003) suggesting arsenic was released after oxygen and nitrate was consumed by microbial activity (Nickson et al., 2000). Based on a considerable amount of evidence, reductive dissolution has been widely accepted as the primary mechanism to release arsenic from sediment to groundwater in many parts of the world (Fendorf et al., 2010; Herath et al., 2016; Smedley and Kinniburgh, 2002). However, the sources of organic carbon are still uncertain. It can be from sedimentary carbon such as peat (Mailloux et al., 2013; McArthur et al., 2001) or anthropogenic carbon such as landfill (deLemos et al., 2006; Hering et al., 2011).

Under an anoxic condition, pyrite formation becomes significant when sulfide is generated through sulfate reduction. The authigenic pyrite was found enriched with arsenic in Bengal Basin sediment, Bangladesh (Lowers et al., 2007). The concentration of arsenic in the pyrite depends on the rate of sulfate reduction, the concentration of arsenic, and the ratio of dissolved arsenic vs sulfide. Under a highly reducing condition, arsenic was found to form realgar, α-As$_4$S$_4$ when sulfide activity was high (O'Day et al., 2004). Apparently, the formations of pyrite and realgar contribute to arsenic sequestration under reducing conditions.

Water transportation further complicates the dynamics of arsenic mobilization and sequestration during the recharge and discharge of groundwater. Redox perturbation can mobilize arsenic in sediment as the result of groundwater seepage or water table fluctuation (Hering et al., 2011). Water transportation contributes to the dissolution by delivering organic carbon and changing the chemical environment around the arsenic sediment (Mailloux et al., 2013). For dissolved arsenic, the migration and adsorption in water are strongly influenced by the interchange between As(III) and As(V) (Smedley and Kinniburgh, 2002). Sediment accumulation takes place in a river basin with the incoming arsenic. Groundwater flow is facilitated by seasonal shift, such as the change from Monsoonal rains and dry season; or human activities such as irrigation pumping (Fendorf et al., 2010).

2.2.2 Analysis of Arsenic for Nonpoint Source Pollution in Groundwater

There have been numerous studies of arsenic in groundwater. The objectives of these studies are:

- Survey of the occurrence of arsenic in local, regional, or national groundwater resources (Focazio et al., 1999).
- Evaluation of arsenic in drinking water supplies to estimate how a drinking water standard would affect compliance (Frey and Edwards, 1997)
- Assessment of groundwater in response to health crisis incidents (Mandal et al., 1996).
- Survey of well water and variability characterization for hazard assessment and guidance for risk mitigation (Zheng and Zheng, 2017).
- Study of arsenic dynamics in groundwater in an anthropogenically polluted area such as landfills and arsenic waste disposal areas (Clancy et al., 2013; deLemos et al., 2006).

- Geochemistry and hydrology study of in aquifers concerning arsenic dynamics in groundwater (Chakraborty et al., 2015; Fendorf et al., 2010; Harvey et al., 2006).
- Survey and characterize arsenic pollution from groundwater irrigation (Brammer and Ravenscroft, 2009).

In most cases, the arsenic pollution was widely spread in sediments with no identifiable point sources. Hence, a comprehensive survey and point-of-use check were dominant approaches for risk management.

Blanket survey of arsenic occurrence in U.S. groundwater was carried out when reliable instrument became available in 1970s. This is an early example of nationwide evaluation of groundwater resources. Between 1973 and 1998, the USGS compiled arsenic data for approximately 20,000 groundwater sites all over the United States (Focazio et al., 1999). The samples were collected with consistent protocols mostly from wells and springs, and some from public supply systems. The samples were filtered and acidified in the field, then analyzed with atomic-adsorption spectrometry (detection limits $\leq 1\,\mu g/L$) in labs. A database for groundwater arsenic point was created and used for several types of risk evaluations at the national scale. The map of arsenic in groundwater shows that most states in the western US had a high frequency of high arsenic ($>10\,\mu g/L$) regions (Fig. 4). These states include Oregon, California, Arizona, New Mexico, Idaho, Utah, and Colorado. In the central US, South Dakoda, North Dakoda, and Oklahoma also have many arsenic hot spots. In the northeast, high frequency of arsenic hot spots was found in Michigan, New Hampshire, Massachusetts, and New Jersey. The arsenic point database was also combined with US EPA safe drinking water information system to evaluate the impact on specific population for each county. Overall, 8% of the area exceeded the minimum concentration level of $10\,\mu g/L$. However, this database was not comprehensive. There was no information on the depth of the sample locations. The samplings were not frequently enough to conclude temporal variation.

For compliance assessment, additional surveys focused on public water supply systems (Frey and Edwards, 1997). The likelihood of arsenic occurrence was characterized by natural occurrence factor (NOF), a score derived from USGS water-quality database. The sampling plan was designed base on public water system size, NOF, and source water type. A total of 800 sites were selected in three surveys. Each was sent with sampling bottles with instructions for acidification or filtration. Out of 809 sampling kits sent, 517 were returned. The samples were analyzed with inductively coupled plasma–mass spectrometry (ICP-MS), which as the detection limit at $0.1\,\mu g/L$. It was found arsenic occurrence in U.S. public water supply was mostly $<2\,\mu g/L$. 53%–71% of groundwater supplies met this level, and 61%–88% of surface water supplies. Water finishing significantly reduced arsenic of surface water. Only 1%–3% of the water supplies were estimated to violate the arsenic standard of $20\,\mu g/L$.

The most serious arsenic pollution is in southern Asia. The eruption of arsenic health crisis took place in late 1980s before any surveys were feasible due to limitation of analytical tools and lack of resources. Research in 1990s revealed that 34 million people in West Bengal in India were exposed to arsenic from groundwater, causing a high rate of skin lesions and cancers (Mandal et al., 1996). The scale and severity of the crisis underscored the enormous risk of groundwater pollution by arsenic. In the coming years, local and international teams carried out many studies to understand the distribution of arsenic, the risk in water use, and the health hazard. The relationship of arsenic fluctuation and environmental events was studied to support water resource management. The research in this area provided the most valuable knowledge about the occurrence and mobilization of arsenic in groundwater. The field experience also revealed many unmet needs for effective risk mitigation in groundwater consumption.

In 1987, there was an outbreak of arsenical dermatosis involving 197 patients in five districts of West Bengal, India (Mandal et al., 1996). A survey was conducted to evaluate the groundwater in seven districts covering an area of 37,493 km² in this area between 1989 and 1996. In total, 20,000 tube well water samples were analyzed. Arsenic levels were determined by injection-hydride generation-atomic spectrometry (FI-HG-AAS) with a limit of detection of 3 µg/L. About 45% of the water had >50 µg/L, and the highest concentration was 3.7 mg L⁻¹. It was estimated that more than 1 million people in the area were affected by the contamination, and 20% of the population developed arsenic skin lesions. Additional analyses of hair, nail, and urine arsenic were expected to characterize the hazard better.

In 1992, arsenic skin lesion cases were identified in the people from Bangladesh. A blanket survey was initiated in 1996 to evaluate groundwater pollution in all districts of Bangladesh (Chakraborti et al., 2010). Between 1996 and 2010, 52,202 samples were collected from hand tube wells all over the nation and analyzed in the labs with FI-HG-AAS. The lower detection limit of the method was 3 µg/L. Overall, 27.2% of the tube wells had As above 50 µg/L. The samples were from 50 out of the 64 districts, suggesting the pollution was nationwide. The impacted people were 22 million by the high As tube wells. Therefore, a safe water source had to be designated for each community. The depth of tube wells correlated to the frequency of arsenic pollution. High-frequency pollutions were found in the wells of 15–25 m depth. Deeper wells with depth > 100 m or surface water were recommended to reduce the risk of arsenic pollution. Meanwhile, there was an effort to label every hand tube well regarding meeting the local standard of 50 µg/L (Rahman et al., 2002). Field-testing kits were used to evaluate 1.3 million out of 4.3 million wells in Bangladesh (Chakraborti et al., 2010). However, validating the field test with FI-HG-AAS showed that the field-testing kits were highly unreliable. In a validation with 290 wells, false negatives were 68% and false positives were 35%. The false negative is a serious issue as it generates additional risk for arsenic exposure.

In early 2000, Columbia University and partner institutions in Bangladesh launched a long-term study of the health effects of As exposure on a cohort of 12,000 people in relation to the As in groundwater supply. Central Bangladesh, the Araihazar area, was selected for the study because the site had a wide range of exposure within a limited space (Van Geen et al., 2007). Araihazar is a transitional area located ~30 km northeast of Dhaka on the boundary between the uplifted Pleistocene Madhupur Terrace in the northwest and a large lower area with Holocene aquifer in the south. Before the site was selected, it had been found that the well water in the northwest was generally safe, while the well water in the south was highly polluted by arsenic (van Green et al., 2003).

The selected site has an area of 25 km² with approximately 6000 tube wells around 2001 when a survey was carried out to characterize the spatial variability of well water (van Green et al., 2003). A total of 5971 tube well samples were collected between March 2000 and December 2001 with information on location, depth, and year of installation. They were analyzed with graphite furnace atomic absorption (GFAA, detection limit 1 µg/mL). The range of arsenic spanned from <5 to 860 µg/L. In all, 52% of the surveyed wells exceeded the local standard. Location data showed that most of the wells were clustered in villages. The depths of these water-wells ranged from 8 to 90 m. Most of the wells had depths between 12 and 24 m. Most of the wells in this range had arsenic concentrations >50 µg/L. For wells with a depth of 20–30 m, 80% were contaminated with As above 50 µg/L. They typically were associated with reducing Holocene alluvial aquifer with gray sand sediment. The distribution of polluted wells was patchy. Safe wells could be found in many clusters in walking distance (presumably <100 m). Therefore, well switching was proposed to mitigate exposure risk. Wells with 30–90 m depth represented 15% of the total. Only 8% of them were polluted. They typically drew water from oxic pleistocene aquifer associated with orange sand sediment. A regression analysis showed that As concentration might increase over time, with a rate of $16\pm2\,\mu g/L$ per decade. For those clusters that had no save wells, deeper community wells were installed to mitigate the risk. In all, 51 community wells with depths ranging from 14 to 164 m were installed between 2001 and 2013 (Van Geen et al., 2007). The depth was principally guided by the sand color of the aquifer. Over 1000 L of water were typically pumped from a community well each day. These wells were monitored over 5 years with sampling every 2 weeks or 1 month. Most wells provided safe water over this period. Only four (7.8%) of them experienced arsenic increase after 2 years. The increases were either gradual or erratic on a monthly scale.

Temporal variability of arsenic in the "safe wells" was evaluated in a study from January 2001 to January 2004 in the same area in Araihazar, Bangladesh (Cheng et al., 2005). The 20 tube wells were selected based on characters representing location, depth, age, and ownership. They all met the local standard at the beginning. Samples were taken biweekly or monthly and analyzed with HR-ICP-MS. For 17 of the wells, the standard deviation was <10 μg/L and over the 3 years. The large fluctuation was found in three wells. Two were shallow wells (with 8 and 10 m in depth) with short excursions or seasonal trend for arsenic. One was a deep public well (with 60 m in depth) with a sudden arsenic increase at the end of the study. Although most wells remained safe over the 3 years, the incidents of large fluctuation were hard to predict. A period analysis is still

necessary for all wells for risk management (Ravenscroft et al., 2006; Sengupta et al., 2006; van Geen et al., 2005). Meanwhile, additional evaluation was carried out to study temporal variability of water chemistry in deep and shallow aquifers in the same area (Dhar et al., 2008). Over the 3 years, groundwater was sampled periodically from 37 monitoring wells with depths from 5 to 91 m. The concentrations of arsenic in these wells were from <5 to 600 μg/L based on previous studies. The evaluation includes measuring water level, arsenic, phosphorus, major cation, and chloride, and redox-sensitive elements including Fe, Mn, and S. The data on time scale were analyzed regarding fluctuation, seasonality, and excursions. The 26 shallow wells in the set had <19% RSD over the time, with the largest decrease of −41 μg/L per year for wells with >200 μg/L of As and a more significant increase of 19 μg/L per year with a well containing 150 μg/L of As. The fluctuation of other redox-sensitive species, including Fe, Mn, and S was much higher ($\pm90\%$) than that of As. The arsenic fluctuation in deep aquifer had <10% RSD, although variations of Fe, Mn, and S were also higher.

In 2012–13, a second blanket survey was carried out to evaluate the tube wells in Araihazar (van Geen et al., 2014). The samples were collected and analyzed in the field with a kit based on Gutzeit method. In total, 48,790 wells were covered by a team of 10 women with 2 male supervisors. Colored placards were put on wells to inform users of the As concentration, with red for As >50 μg/L, green for As >10 and ≤50 μg/L, and blue for As ≤10 μg/L. The survey found the number of tube wells increased to 10,879 in the area, almost doubled since the first blanket survey in 2000–01. The survey represented 48% of the total; including 29% of the wells meeting the local standard and 19% of the wells found unsafe. The unsafe wells in the sample set dropped from 52% in the first blanket survey to 40% in the second blanket survey. However, 66% of the unsafe wells in the survey were still used for drinking and cooking, suggesting arsenic exposure was still high after hazard had been proven and safer alternatives were proposed. Using field kit enabled more comprehensive evaluation. The quality of the field analysis was evaluated with a random subset of 502 wells. The samples were analyzed with HR-ICP-MS to find the accuracy of field assay. There were 9 false negatives in the blue group of 118 wells, 27 false negatives in the green group of 173 wells. There were 13 false positives in the green group, and 33 false negatives in the red group of 211 wells. Importantly, the false negatives were 7.2% of the total, indicating the benefit of using a field kit to inform users of exposure risk.

The study in Araihazar demonstrated that health risk mitigation of diffuse pollution in groundwater heavily relies on analysis. For the blanket survey, spatial and temporal knowledge is still sketchy because of the lack of resolution. Arsenic concentrations were widely different between wells that were 10–100 m apart (van Green et al., 2003). Although arsenic level was found relatively stable in both deep and shallow aquifers, a sharp rise of arsenic was found at monthly scale, and a gradual increase was also observed more frequently in aging wells (Cheng et al., 2005; Van Geen et al., 2007). The dynamics of arsenic in groundwater was the result of many factors, including local geological and geomorphologic features, hydrological events and associated water chemistry, and human intervention such as waste disposal and water pumping (Chakraborty et al., 2015). Although deep wells were recommended to reduce exposure risk, arsenic could increase at any time due to pollution by water from shallow aquifers through defects of installation or failed parts (Choudhury et al., 2016; Van Geen et al., 2007). Seasonality was commonly used as a measure to characterize temporal variability (Cheng et al., 2005; Dhar et al., 2008), but it could not explain the short excursions of As concentration fluctuation. This suggests the incidents might be related to more specific

hydrology events, which are highly variable in terms of time and spatial scales (Skøien et al., 2003). The concern of the vulnerability of deep aquifer is also legitimate because recharge of a deep aquifer might involve downward migration of As on a large scale (Fendorf et al., 2010). Some prudent approaches to address these issues involve identification of safe aquifer for digging wells, comprehensive evaluation of well water in polluted area, well switching to mitigate exposure risk, and periodical check of arsenic concentration in "safe wells," and using biomarkers to monitor exposure (Ravenscroft et al., 2006; Sengupta et al., 2006; Van et al., 2002; Van Geen et al., 2007; Zheng and Zheng, 2017). Ideally, the analysis is carried out on demand, which is dependent on the population, number, and quality of the wells, water use, hydrology events, and time to get results for decision-making. Apparently, demand at this level can only be met by field assay; because the cost, throughput, and implement efficiency of lab analysis are all prohibitive.

2.2.3 Analysis of Arsenic for Point Source Pollution in Groundwater

Metal ore mining has been recognized as the most significant source of environmental pollution due to its scale and the characteristics of the waste (Dudka and Adriano, 1997; Lottermoser, 2010). Following material extraction from the ground, wastes are generated in every subsequent step on the mining site. Beneficiation, the mineral processing step prior to metallurgical extraction, generates a large amount of waste called *tailing*. *Tailings* are a heterogeneous mixture of sedimentary rocks and soils. The physical and chemical characteristics of tailing depend on mineralogy and geochemistry, the mining practices, and moisture content. They can be used in backfilling of the mine, or applications such as construction, but they are generally deposited in a tailings dam or pond on or near the mine site. In metallurgical treatment, the residues from leaching or smelting processes are called *slags*. The metallurgical process also generates additional waste including ash and flue ash. Copper, zinc, and lead are mined as their corresponding sulfide ores, including chalcopyrite, galena, and sphalerite. Global arsenic emission from copper smelting was estimated to be 12,800 tons year^{-1}, the largest source of anthropogenic source (Matschullat, 2000). The tailings and slags for metal mining, either from monometallic or polymetallic ores, were found to have a very high concentration of arsenic ranging from 10^2 to 10^4 mg kg^{-1} (Garelick et al., 2008). The erosion of the waste is dramatically faster due to increased exposure to air and rainfall. The leachate from mining waste erosion is typically rich in toxic metals or metalloids such as arsenic, lead, cadmium, etc. Although gold exists in free elemental form, gold mining generates a large amount of process water containing arsenic rich tailings (Acheampong et al., 2013; Garelick et al., 2008). In the gold mining area of Takab, Iran, the sediment from Zashuran River contained arsenic ranging from 125 to 125,000 mg kg^{-1}. The AngloGold Ashanti mine in Obuasi area of Ghana processed 470,000 tons of ore per month. The waste from the mine included tailing discharged at 286.2 tons h^{-1}, and 400–500 m^3 h^{-1} wastewater. The arsenic concentration of the wastewater was 21.6 mg L^{-1} (Acheampong et al., 2013). Other metal or metalloid mines, such as tin and tungsten, were also created arsenic pollution through uncontrolled tailing and slags (Garelick et al., 2008; Liu et al., 2010).

The primary mechanism for arsenic mobilization from mine waste involves releasing arsenic from sulfide mineral by oxidation to generate AMD, and redox reaction of arsenic. The released arsenic undergoes adsorption and desorption of arsenic species, and precipitation or coprecipitation with other metals as the pH, redox potential, and concentration of other components in the matrices change along the pathway of mobilization (Cheng et al., 2009).

Sulfur and Fe(II) oxidation are two main reactions in AMD generation from pyrite in exposed mine waste. The process involves both geochemistry and microbial activity in the oxidations and release of other elements. This is best illustrated in a case study with Richmond Mine of Iron Mountain superfund site in California (Edwards et al., 2000; Nordstrom and Alpers, 1999). Richmond was one of the major mines in Iron Mountain area, which is a massive sulfide ore body in Northern California. Between 1860s and 1960s, it was mined for Ag, Au, Cu, Fe, Zn, and pyrite. The site was the largest Cu producer in California. Multiple runoff took place during sudden surges of acid mine water despite waste dams had been built. The discharge into the Sacramento River not only caused massive fish-kill but also threatened the people in Redding, a downstream town with 10,000 residents. In 1983, the Iron Mountain area was listed as a Superfund site. For Richmond mine alone, the oxidative erosion of pyrite was estimated to be 2500 tons year^{-1}. The resulting acid mine water was acidic (down to pH -3.5) and metal-rich (up to 200 g L^{-1}), with arsenic as high as 850 mg L^{-1}. The remediation started in 1986 with mine plugging, partial capping, and surface water diversion to minimize the generation of AMD and reduce the runoff; and lime neutralization, metal ion precipitation to reduce the hazard of AMD.

Pyrite-rich waste in mining site is unstable and can become the point source of pollution. The abandoned Carnoulès Pb–Zn mine in southeastern France has long been identified as a point source for arsenic pollution (Casiot et al., 2003). A flood in 1976 flushed the mine tailing by several kilometers and polluted the entire Amous river valley along the way. A total of 1.5 million tons of pyrite-rich tailing stock was later collected and formed a 5500 m^2 impoundment behind a dam in the upstream of religious creek. The water input of this impoundment was either from natural spring or rainwater. Resulting from pyrite oxidation, arsenic concentration could reach as high as 160 mM (5280 ppm) and the pH was ~3.1. The output of this

EVALUATING WATER QUALITY TO PREVENT FUTURE DISASTERS

impoundment was the AMD at the bottom of the dam. The concentration of arsenic in the AMD was 0.7–4.5 mM (23.1–148.5 ppm), therefore causing downstream pollution. The mine tailing water also polluted groundwater underneath through downward migration. The As concentration in the shallow groundwater strongly correlated to the dissolved oxygen and reached as high as 162 mM. When depth increased from 2 to 14 m, the pH gradually increased to 6, and arsenic was mostly sequestered through adsorption by iron sulfate minerals.

The dissolution and migration of arsenic from mine waste to groundwater are strongly dependent on weather and geological features. Zimapan mining area is in Hidalgo State, Mexico. The area has been mined for Ag, Au, Zn, and Pb since the 17th century, leaving a large amount of tailing in the vicinity of Zimapan Township and widespread waste from smelting fume in the entire area (Armienta and Segovia, 2008). The 35,000 inhabitants in the town used deep well water for drinking and shallow well water for irrigation (Labastida et al., 2013). A survey with hair and skin samples from the local people showed that they suffered arsenic pollution. The background arsenic concentration in soil and non-mineralized zones was 74–151 mg kg^{-1}, the soil arsenic in the mineralized zone was 294–2580 mg kg^{-1} (Armienta et al., 2000), and the tailing arsenic reached as high as 22,000 mg kg^{-1}. Potable water was drawn from deep well with arsenic as high as 1.0 mg L^{-1}. The local aquifer is formed by fractured limestone, with groundwater flows from north to south. The groundwater is recharged by precipitation of less than 400 mm year^{-1}. The sulfide mineral in tailing was oxidized during the dry season when rock was exposed to air. In the rainy season, arsenic was mobilized when the water table rose to the tailing waste (Rodríguez et al., 2004). The distribution of arsenic during mobilization exhibited spatial and temporal variability, depending on the regional geological feature. Water chemistry evaluation showed that the groundwater directly from mine tailing had high sulfate and low sodium, while the groundwater from the diffuse source had low sulfate and high sodium (Sracek et al., 2010). Some of the wells in the volcanic area, away from the arsenic point source, produce low arsenic water (<0.05 mg L^{-1}) persistently. However, the yield of the wells was not adequate for the residents. Mobile water potabilization plant based on ferric sulfate adsorption was proposed for drinking water supply (Armienta et al., 2000). Passive treatment system by neutralization with local limestone was also explored to mitigate AMD from the tailing (Labastida et al., 2013).

Incidents of massive pollution due to tailings dam collapse highlighted the importance of waste remediation (Guerra et al., 2017; Queiroz et al., 2018). Although mine site is a point source, the cost-effective remediation still requires tremendous amount investigation on the site (Nordstrom, 2011). The goal of remediation depends on natural background condition. Material balance constraints need to be estimated based on the occurrence and abundance of the pollutants. Studies on the reactivity of minerals, regional hydrology are also essential to modeling contaminant dissolution, transport, and redistribution.

Superfund site associated with industrial arsenic waste represents an important type of point source (Welch et al., 2000). The Vineland Superfund site has a history of arsenic pollution in soil and groundwater resulting from arsenical biocide manufacture between 1950 and 1994 (Keimowitz et al., 2005a, 2017). The arsenic migrated via the surface water system from Blackwater Branch to Maurice River and ended up in Union Lake. Over the pathway, the total dissolved arsenic concentrations increased from 22 ± 4 nM (1.65 µg/L) to 272 ± 36 nM (20.4 µg/L) and stream discharge increased from 207 ± 5.5 to 950 ± 25 L s^{-1}. The arsenic content in the sediment of Union Lake reached as high as 15 mmol As kg^{-1} on a dry weight basis (1125 mg kg^{-1}). The pollution posed a serious risk because Union Lake was a man-made lake for

EVALUATING WATER QUALITY TO PREVENT FUTURE DISASTERS

recreational activity in the residential area. For remediation, groundwater pump and treatment started in 2000. The remediation reduced the total arsenic flux and the concentration of arsenic in the surface water, but the pollution remained as of 2004 (Keimowitz et al., 2005a). To end offsite transport of arsenic, the pollution was evaluated by characterizing the source, partition, and dissolution of arsenic along the pollution pathway. Porewater and sediment samples were taken from different depths. They were analyzed for As, Fe, Mg, and S. The total arsenic in Blackwater Branch was estimated to be $\sim 22 \pm 3$ mol (1.64 kg) day^{-1}. Minor inputs, including those from surface soil arsenic mobilization, groundwater treatment plant effluent, and remobilization of arsenic in the sediment of Blackwater Branch, contributed only <3% of this total. The arsenic was mainly groundwater discharge from the Superfund site. The chemical compositions of sediment and porewater were different in Blackwater Branch and Union Lake. The Fe/S ratio was low in Blackwater Branch and had high As(III) content; therefore, arsenic was mobilized by forming thioarsenite (AsS_3^{3-}). In Union Lake, sulfide concentration was controlled by high Fe and most of the arsenic was As(V), therefore arsenic was mobilized by reductive dissimilation via Fe(III) reduction. The seasonal change caused an anoxic condition in Union Lake. Under the condition, manganese was released prior to the dissolution of iron and arsenic. As the result of buffering effect from manganese, arsenic concentration only increased to ~ 30 ppb despite the high concentration in sediment (Keimowitz et al., 2017).

Non-arsenic landfills are also potential point sources of arsenic pollution (deLemos et al., 2006; Ford et al., 2011; Keimowitz et al., 2005b). A former landfill in Kennebec County, Maine, was a quarry of sand and gravel production in 1920s (Keimowitz et al., 2005b). It became a landfill site between 1930 and 1970, with wastes including volatile and semi-volatile organic chemicals. Due to organic chemical contamination of surrounding area, it was listed as a Superfund site in 1983. Water monitoring with wells on the flow path of leachate showed that arsenic concentrations were all >100 μg/L between 1987 and 1998, despite of remediation efforts including landfill capping, vapor extraction, and injection of oxidants. A 2005 survey showed that the groundwater directly under the landfill had redox potential of −95.4 mV and arsenic concentration of ~ 300 μg/L. The arsenic in the peripheral region was only 2 μg/L. The sediment beneath the landfill had arsenic content at 5 mg kg^{-1}, like that of the peripheral sediment. No arsenic accumulation or hot spots were found in the sediment beneath the landfill. Therefore, the authors proposed that the high arsenic in groundwater was the result of reductive dissolution by landfill leachate. Another similar case was in the Coakley Landfill that caused benzene pollution in North Hampton, New Hampshire (deLemos et al., 2006). The arsenic in landfill leachate was 23 μg/L, but the concentration of As in the water from a nearby spring was 170 μg/L. Mineralogy study of the sediment and lab evaluation of the clay samples showed that arsenic release correlated well with Fe increase. The release of arsenic was attributed to a microbial reduction of Fe(III) in the clay.

In general, arsenic mobilization in groundwater is possible wherever redox environment is disrupted by additional carbon source. In 1979, an oil pipeline rupture left approximately 2675 barrels of unrecoverable crude oil in Bemidji, Minnesota. In 1983, the site was enrolled in the Toxic Substances Hydrology Program sponsored by the U.S. Geological Survey (USGS). It was found dissolved As reached 230 mg L^{-1} in the centerline of a contaminant plume, suggesting arsenic in sediment was mobilized by the organic plume (Ziegler et al., 2017a,b). Over 35 years, the organic plume had been expanding in the direction of groundwater flow, which had an average linear velocity of 0.06 m day^{-1}. Microbial oxidation of the hydrocarbon created alternative anoxic-suboxic transition zones as the plume expanded. Fe(III) was also reduced to Fe(II) in the anoxic environment, releasing As from the sediment. The oxidation state of As was not determined, but its concentration always correlated with Fe in Fe-reducing, anoxic-suboxic transition, and oxic sediment conditions. At the leading edge of the hydrocarbon plume, the groundwater underwent a transition from anoxic to suboxic condition. The transition caused As sequestration by Fe(III) precipitant. Based on the mass balance principle, a total of 78% of the arsenic was mobilized by the hydrocarbon plume. Between 1993 and 2013, arsenic migrated 30 min down gradient in the redistribution. Although this was a localized arsenic redistribution and had not caused health risk, the organic plume is still expanding. Close monitoring is necessary because if the pollution extends and natural attenuation is inadequate, where the point source may eventually turn into a diffuse source with significant spatial and temporal variabilities. The same type of As mobilization was also observed in enhanced reductive dechlorination (ERD), which involved the injection of molasses to promote mineralization of chlorinated hydrocarbons (He et al., 2010). Arsenic mobilization needs to be taken into consideration when ERD is planned for remediation.

The understanding of arsenic transportation is critical for pollution point source management. A survey was carried out to delineate arsenic transportation at Shepley's Hill Landfill at Fort Devens, Massachusetts (Ford et al., 2011). Wells, shallow tube wells, and push-point transects were used for sampling and flow measurement in a space ($60 \times 60 \times 15\,m^3$) adjacent and under Red Cove, a pond that received the polluted groundwater. The objectives included identification of plume origin, mapping contaminant plume, and estimating the flux of arsenic in the hydraulic pathway from pollution point to surface water. Water chemistry analysis linked groundwater to landfill leachate,

confirming landfill might constitute a pollution point source by arsenic mobilization. High arsenic concentration ($\sim$600–1000 µg/L) was found in wells on the northwest side of the pond, while relatively lower As (<10–400 µg/L) was found from wells on the southeast side. The concentration distribution of arsenic correlated to groundwater flow. The As was mainly discharged from the southeast side into the pond, where the arsenic was sequestered in sediment.

The interface between groundwater and surface water (GSI) is critical to the transportation of pollutants into surface water (Ping et al., 2006). GSI is defined chemically as the zone in which porewater concentrations of a conservative tracer are between 10% and 90%. The thickness of the zone is from a few centimeters to several meters. Water chemistry changes significantly in GSI due to many geochemical reactions, energy exchange, and physical events. Comprehensive characterization of these events at GSI was expected to better plan for surface water management (Smith et al., 2008). The objectives are to characterize the pathways of water exchange and chemical transformations, quantify the flux of water and concerned chemicals, and project the dynamics of concerned pollutants in stream water. Many sampling methods are available for the study in GSI. Multiple samplings are needed to address the needs for scale and resolution. Large-scale sampling is needed to characterize the fate and transport of pollutants, while high-resolution sampling is needed to investigate temporal and spatial variability in the water exchange at GSI (Kalbus et al., 2006). In the sampling methodology study, freeze-core and bead column sampling were employed for arsenic GSI evaluation at Auburn Road Landfill site, Londonderry, New Hampshire (Ping et al., 2006). The GIS was identified in the seepage area along the leading edge of the arsenic plume with a field kit. Freeze-core sampling enabled sediment analysis with cm-scale resolution. It was found iron and arsenic changed significantly over a 10 cm range in the sediment. Arsenic in surface water just above the sediment in the seepage area was much lower. Bead column captures arsenic and iron accumulation and is used to evaluate oxidative sequestration of arsenic. It was found that the levels of surface water might have a strong impact on arsenic sequestration at GSI. The sampling methods were to be used to study the impact of seasonal change on the transport of As across GIS.

Arsenic analysis for pollution point source characterization is largely dependent on laboratory spectroscopic methods such as ICP-MS, ICP-AES, or graphite furnace AAS, because of their high sensitivity and since they afford quantitative results. Sampling plan needs to balance between scale and resolution. Temporal resolution is usually inadequate to characterize variability. Speciation of arsenic can be a very tedious task because sample preservation or separation of As(V) with ion exchange at the spot is needed for lab analysis (Ping et al., 2006). Stripping voltammetry or Keimowitz et al. (2005a) was used to determine the speciation on site. The sampling plan can be optimized based on onsite mapping of arsenic distribution with a field kit (Ping et al., 2006).

2.2.4 Current Field Test Tools and Applications

A couple of basic approaches exist for monitoring arsenic in the field (Melamed, 2005). The simplest available is a colorimetric test based on variations of the Gutzeit method (Sanger and Fisher Black, 1908). These tests start with the addition of zinc powder (or sodium borohydride) and acid (such as sulfamic acid, tartaric acid, or hydrochloric acid) to a sample of groundwater. The reaction will reduce inorganic arsenic to arsine gas (AsH_3) which is stripped from the sample and reacts with a paper test strip that is impregnated with mercuric chloride. The reaction of arsine with mercuric chloride produces the colored product AsH_2HgBr. Depending on the amount of arsenic present in the sample, the test strip will go from white (no arsenic) to yellow (small amounts or arsenic) to brown (at higher concentrations). For quantitative analysis, the color of the test strips is compared with a color chart to estimate inorganic arsenic to the concentration near the MCL level (Spear et al., 2006). Most tests require incubation from 15 to 30 min according to manufacturer's recommendations (Spear et al., 2006), although there have been reports of improved results at lower concentrations when test times are increased (van Geen et al., 2005). These colorimetric test kits have found application in large field testing with mixed results (Rahman et al., 2002; van Geen et al., 2005). Reliability of the test results is complicated due to the qualitative nature of comparing colored sample papers with colored charts making it difficult to determine arsenic at the lower concentrations perfectly (van Geen et al., 2005). The Gutzeit test is also prone to interference by selenium, tellurium, antimony, and sulfides (Melamed, 2005; Spear et al., 2006). Some test kits include approaches to remove sulfides, such as by reaction with lead acetate (van Geen et al., 2005). The significant advantage of the colorimetric test kits are the small footprint and easy storage. In addition, the simplicity to perform the test makes them straightforward to use by someone who is not a skilled technician. The typical test costs less than $5 per sample. However, a drawback of the tests is the risk of exposure to arsine gas; the user should use protective equipment when performing the test. In addition, the tests contain mercury bromide so the tests need to be disposed of with hazardous waste.

There are other tests that can potentially be used in the field including anodic stripping voltammetry (ASV) and portable X-ray fluorescence; however, these tests require transport of fragile instrumentation and require some basic training to operate the equipment effectively. In ASV, an electrode is placed into a water sample containing arsenic where a portion of the arsenic is collected (reduced) onto the

EVALUATING WATER QUALITY TO PREVENT FUTURE DISASTERS

electrode surface. Next, the arsenic is stripped off the electrode by oxidation and the amount of current required to strip off the arsenic is determined to quantify the amount of arsenic present. ASV is capable of measuring arsenic from 0.1 to 300 µg/L (Melamed, 2005). Antimony, bismuth, and copper are known interferents (Melamed, 2005). X-ray fluorescence is a spectroscopic technique where a sample is irradiated with X-rays or gamma rays; the sample will then emit characteristic X-rays that can be used to identify the atoms in the sample. One advantage of this technique includes the possibility of characterizing elemental composition of a solid media such as soil directly; however, the detection limit is higher than other techniques requiring sample preconcentration before analyzing water samples (Melamed, 2005).

3 BIOSENSORS FOR MONITORING ARSENIC IN WATER

The monitoring of pollutants in water sources is needed for managing risks related to human exposure and protection of ecosystems. As described above, many tools are available for monitoring pollutants but they are often cumbersome and complex requiring trained personnel in a laboratory setting. Although there has been a significant improvement in approaches for field sampling and analysis, many of the tools are still costly or cumbersome making them difficult to apply for monitoring applications in the field. For over 30 years, there has been an interest in using biosensors, chemical sensors with a biologically based detection element, for analysis in environmental applications. Biosensors have been successfully commercialized for medical and many industrial applications such as in the food and drink industries (Mello and Kubota, 2002). Biosensors have been developed for single-use, disposable

applications, such as the glucose biosensor developed for diabetics (Yoo and Lee, 2010). These robust sensors have relatively simple electronics that operate electrochemical sensors coupled with disposable enzyme electrodes. Biosensors have also been developed for continuous operations such as fermentation monitoring and analysis of multiple blood samples for bedside monitoring in hospitals (Collison and Meyerhoff, 1990). The development of biosensors for arsenic monitoring could significantly improve our ability to manage remediation operations and monitor the spread of arsenic in contaminated sites. In addition, simple single-use biosensors for arsenic could be used for individuals to check water quality at home; this would be especially useful for individuals that rely on well water in areas that are known to have high arsenic levels in their groundwater. This section introduces the basic types of biosensors and discusses developments in biosensors for arsenic detection. In addition, we will discuss the application of arsenic biosensors to field monitoring and barrier that need to be overcome before arsenic biosensors are successfully used in the field.

3.1 Biosensor Technology

Chemical sensors and biosensors need to perform two main functions: recognition and transduction (Fig. 8). The signal recognition component is important in providing the sensing function and responds to the presence of the analyte of interest. For biosensors, the recognition element is biologically based, typically a protein, DNA or a living cell, and detection is driven by specific affinity interactions between the biorecognition component and the analyte of interest. Once an analyte has bound the biological recognition element, a quantifiable signal must be relayed (or transduced) to indicate that the binding event occurred. Signal transduction may take place in many forms

EVALUATING WATER QUALITY TO PREVENT FUTURE DISASTERS

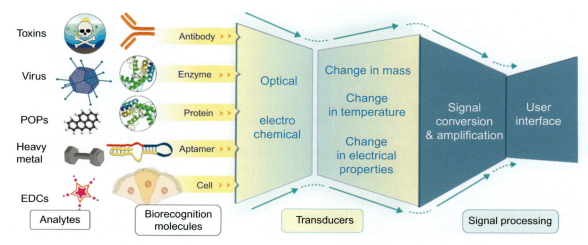

FIG. 8   Basic components of a biosensor.

including thermal, mechanical, resistive, capacitive, electrochemical, or optical. This section introduces common biorecognition elements and transduction methods that are frequently encountered in biosensor design. A thorough discussion on biosensors is beyond the scope of this work and a number of recent books and reviews discuss many of the basic details (Banica, 2012; Evtugyn, 2013; Moretto and Kalcher, 2014; Mulchandani et al., 1998).

3.1.1 Common Biorecognition Elements

The biorecognition component of the biosensor can generally be lumped into three main categories. The first is affinity-binding elements that include antibodies, aptamers, and nucleic acids. The second category is biocatalytic sensors that use enzymes and the third category is biocatalytic sensors that use whole microorganisms.

3.1.1.1 AFFINITY-BINDING ELEMENTS

Antibodies are proteins produced by the immune system in response to a pathogen or foreign substance. Antibodies are of special interest for diagnostics due to their very high selectivity and low binding constants. The ability to produce antibodies in animals in response to different antigens and advances in genetic and protein engineering has made antibodies critical for the immunodiagnostic application. There are other proteins, like antibodies, that cells use to interact with and respond to their environment that can bind very tightly to many analytes of environmental interest and these may also be used in biosensing applications. The three-dimensional structure and distribution of amino acid functional groups of a protein influence how tightly they bind to ligands of interest. Affinity interactions of proteins and antibodies are used in enzyme-linked immunoassays with surface plasmon resonance (SPR) for different environmental applications.

DNA (deoxyribonucleic acid) and RNA (ribonucleic acid) are composed of nucleotides that tightly bind to complementary nucleic acids to form double-bonded strands through a process referred to as hybridization. This hybridization process is very specific. A sequence of nucleic acids that is specific and complementary to a nucleic acid sequence for a specific microorganism can be synthesized and used as a probe to identify the specific microorganisms. Nucleic acid-based biosensors find use in environmental monitoring for closely monitoring pathogenic microorganism and viruses (Kumar et al., 2018).

Aptamers are single-stranded oligonucleotides (DNA or RNA) that fold into a specific three-dimensional structure like proteins, and selectively bind to low and high molecular weight organic and inorganic molecules with high affinity. Their binding constants are in the micromolar and picomolar ranges making them comparable to antibody/antigen affinity interactions. Aptamers can be designed for specific analytes using a selection procedure referred to as systematic evolution of ligands by exponential enrichment (SELEX) (Gopinath, 2007; Sampson, 2003). There has been a significant effort to develop biosensors for a range of analytes of interest in environmental applications using optical, electrochemical, and other techniques (Hayat and Marty, 2014; Willner and Zayats, 2007).

3.1.1.2 ENZYMES

Enzymes are catalytic proteins that have been used for the development of many different biosensors for biomedical, food, and environmental applications. Enzymes typically have very high specificity for substrates making them very selective sensors. If an enzyme can be found with high selectivity for an analyte of interest, they can be used as sensor recognition elements. Typically, the challenge of developing an enzyme-based biosensor is finding a sensitive and selective method for monitoring the product of an enzymatic reaction. Enzymes that have been used in the development of biosensors include glucose oxidase which oxidizes glucose in the presence of oxygen to gluconic acid and hydrogen peroxide, alcohol oxidase which oxidizes methanol to formaldehyde and hydrogen peroxide and nitrate reductase which reduces nitrate to nitrite (Almeida et al., 2010). Some enzymes can be inhibited by several analytes that are of interest for environmental sensing

applications (Amine et al., 2016; Arduini and Amine, 2014; El Harrad et al., 2018). For example, the enzyme acetylcholinesterase is inactivated by low concentrations of organophosphate pesticides. By monitoring the enzymatic activity change of acetylcholinesterase, it is possible to determine the concentration of organophosphate pesticides in a sample. Many heavy metals are also potent inhibitors of enzymes and have been detected using enzymatic biosensors (Malitesta and Guascito, 2005; Verma and Singh, 2005).

3.1.1.3 WHOLE CELLS

Enzymes and proteins have long been used for development of biosensors; however, over the last 2 decades, significant advances in genetic engineering have made it possible to design whole cells with complex metabolic cascades for sensing and signaling changes in their environment. In the coming years, whole cell biosensors are likely to have a large impact on environmental monitoring (Gui et al., 2017). The typical whole cell biosensor incorporates a gene circuits with a regulatory gene and a reporter gene. The regulatory gene responds to the analyte of interest and the primary function is to turn off/on the expression of the reporter gene in the presence of the analyte. The reporter gene expresses a protein that can be monitored using a transducer. Common reporter elements include the expression of green fluorescent protein (GFP) for fluorescent monitoring, lucifier/luciferase for luminescent detection, or the enzyme β-galactosidase for generation of color or a redox active product that can be monitored electrochemically (Bereza-Malcolm et al., 2015).

3.1.2 Transduction Methods

The biosensor transducer is responsible for monitoring the biorecognition binding event and creating a signal that can be actively monitored and quantified. Transduction methods broadly fit into two main categories, those based on monitoring some change in physical property

3 BIOSENSORS FOR MONITORING ARSENIC IN WATER

of the material and those based on monitoring a change in composition. Some of the most common methods are described below.

3.1.2.1 CHANGE IN MASS

One approach for monitoring a binding event between a biological compound such as an antibody and an analyte is to monitor the change in mass that occurs on binding. Sensitive acoustic biosensors, such as piezoelectric biosensors, lead to a change in vibrational frequency on binding of an analyte. A typical example is the quartz crystal microbalance which leads to a change in vibrational frequency upon binding of an analyte. This transduction approach has often been used with antibodies.

3.1.2.2 CHANGE IN TEMPERATURE

Although not as common, biosensors have been developed that monitor the change in thermal properties on binding of an analyte to the biorecognition element. Typically, this type of transduction method is used with enzyme-catalyzed processes, which develop a larger change in temperature due to the catalyzed reaction.

3.1.2.3 CHANGE IN ELECTRICAL PROPERTIES

Many biosensors have been developed that monitor changes in electrical properties. These transduction methods often involve miniaturized biosensors that lead to a change in resistance or a change in dielectric properties on binding of an analyte to the biorecognition element.

3.1.2.4 ELECTROCHEMICAL METHODS

One of the most common approaches for biosensor design involves electrochemical procedures for biosensor transduction. Potentiometric biosensors use potentiometric methods to monitor changes in ions such as hydronium, ammonium, or fluoride ions. Potentiometric sensors that monitor changes in pH from enzyme-catalyzed hydrolytic reactions have

found many applications. Amperometric transduction techniques measure the electric current produced from oxidation/reduction reactions that take place at an electrode surface. Amperometric biosensors are commonly used with oxidative enzymes that produce hydrogen peroxide as a product. A third common electrochemical transduction approach involves electrochemical impedance. This approach measures the change in AC current through an electrochemical cell. This approach has generated a lot of interest recently for label-free affinity biosensors for monitoring affinity binding (Daniels and Pourmand, 2007).

3.1.2.5 OPTICAL METHODS

There are many varieties of spectroscopic methods that are commonly used as transduction methods for biosensors. Many of these rely on changes in UV, visible, or infrared radiation associated with the production of or consumption of chemical species. Optical transduction methods can monitor the absorption of or the emission of electromagnetic radiation. These approaches have also been used to monitor change in refractive index of a sensing layer or light scattering of nanoparticle systems. Colorimetric sensors are one of the most basic biosensors relying on changing of color to monitor a transduction event.

3.2 Biosensors for Arsenic Detection

Over the last few decades, different types of biosensors have been developed for arsenic detection. Many of these are specific to arsenite or arsenate providing the potential opportunity to differentiate between both species. A short review of the major types of arsenic biosensors developed is provided below. Reviews that are more detailed have been published over recent years (Chen and Rosen, 2014; Kaur et al., 2015; Merulla et al., 2013; Upadhyay et al., 2018).

EVALUATING WATER QUALITY TO PREVENT FUTURE DISASTERS

3.2.1 Bioaffinity Biosensors

Many aptamer-based biosensors have been developed for As(III) (Song et al., 2016; Wu et al., 2012a,b). Song et al. developed a biosensor using aptamers with Au@Ag core-shell nanoparticles as a substrate for surface-enhanced Raman scattering (Song et al., 2016). Their technique had a linear range from 0.5 to 10 ppb and a detection limit of 0.1 ppb. No tested ions interfered with the biosensor and the sensor was shown to work well with lake water samples. Wu et al. developed an aptamer-based sensor composed of nanoparticles prepared from crystal violet and aptamers for As(III) (Wu et al., 2012b). This group measured changes in Resonance Rayleigh Scattering of the nanoparticles due to changes in size on addition of As(III). They found their assay had a detection limit of 0.2 ppb with a linear range of 0.1–200 ppb. No interference by other metal ions was observed. The same group developed a similar assay using aptamers with gold nanoparticles (Wu et al., 2012c).

Phytochelatins and metallothioneins are cysteine-rich peptides and proteins that are believed to play an important role in regulating metal ions and scavenging toxic metals in eukaryotes and some prokaryotes (Grill et al., 1987; Palmiter, 1998). Biosensors with phytochelatins and metallothioneins have been developed for detection and monitoring of metal ions using techniques such as SPR (Wu and Lin, 2004) and ASV (Irvine et al., 2017). By immobilizing metallothionein to paper discs on screen-printed electrodes, the metallothioneins selectively concentrate arsenic (III) allowing detection using ASV down to a concentration of 13 ppb (Irvine et al., 2017).

Whole-cell sensors are frequently used for biosensor development since the transcriptional regulators are often very sensitive to their chemical environment. The transcriptional regulators can be cloned into cells using genetic engineering techniques and can be used to switch on and off reporter genes. However, the response time of whole-cell sensors can vary from a few hours to a few days. In order to reduce the response times, Kawakami cloned the ArsR DNA-binding transcriptional regulator from *Escherichia coli* and developed a fusion protein with GFP to give ArsR-GFP. ArsR will bind to DNA containing a particular promoter-operator sequence (P_{ars}–O_{ars}) if there is no As(III) present. In the presence of As(III), the affinity of ArsR for P_{ars}–O_{ars} is reduced. They attached oligonucleotides containing P_{ars}–O_{ars} to the surface of microplate wells to develop what is effectively a competitive binding assay for arsenite. If no arsenite is present in a sample, ArsR-GFP binds to the P_{ars}–O_{ars} sequence present on the plate and a fluorescent signal is observed. If arsenite is present in a sample, a reduction in ArsR-GFP binding occurs leading to a reduced fluorescent signal. They reported a detection limit of 5 µg/L with As(III) and an assay time of 40 min. Rosen's group protein engineered the transcriptional regulator AfArsR from *Acidithiobacillus ferrooxidans* to develop a protein that was insensitive to inorganic arsenic but could bind to methylarsenite and phenylarsenite (Chen et al., 2014a). They developed an assay method based on fluorescence anisotropy of the AfArsR-DNA interactions and were able to vary their ability to detect inorganic and organic arsenicals by protein engineering.

While not being precisely affinity biosensors, electrochemical biosensors have been developed for monitoring damage to DNA by monitoring changes in guanine-DNA oxidation signals on damage to DNA by radiation or chemicals (Wang et al., 1997). This basic approach has been applied to monitoring the oxidative degradation of DNA by arsenic (Ferancová et al., 2007; Labuda et al., 2005; Ozsoz et al., 2003). The basic mechanism of these sensors involves measuring the oxidation of guanine residues in DNA electrochemically using differential pulse voltammetry and comparing the signal to DNA that has been oxidized with arsenic. If arsenic is present in the sample and oxidizes the guanine residues, the DPV signal was reduced (Ozsoz et al., 2003). This form of sensor is relatively straightforward but has limited selectivity since anything that oxidizes guanine will potentially be detected. Improvements to this sensor approach were demonstrated using DNA oxidation catalysts composed of Ru(II) complexed with bipyridine (Labuda et al., 2005).

3.2.2 Enzymatic Biosensors

Enzyme-based biosensors are a very promising approach for the development of simple single-use and continuous biosensors for field monitoring. The success of single-use glucose biosensors used by diabetics demonstrates the possibility. There are two major approaches to using enzymes for arsenic detection. The first approach uses arsenite as a substrate and monitors arsenic concentration by measuring the oxidation current. The second approach uses arsenic as an inhibitor and measures changes in the enzyme activity with a time of exposure.

Only one report exists of a biosensor that directly uses arsenic as a substrate (Male et al., 2007). The report uses the molybdenum containing enzyme arsenite oxidase which was isolated from a chemolitoautotrophic bacterium isolated from a gold mine. The enzyme arsenite oxidase oxidizes As(III) to As(V) using oxygen. The authors in this paper immobilized the enzyme to a multiwalled carbon nanotube-modified electrode allowing direct electron transfer between the oxidized/reduced form of the enzyme and the electrode (mediator free). Amperometric tests were performed at 300 mV and under these conditions, a linear response was observed up to 500 ppb arsenite with a detection limit of 1 ppb. The authors demonstrated that the sensor was not subjected to interference by copper a common ion present in groundwater that causes interferences with electrochemical stripping techniques. Interference

was observed by humic acid which is oxidized at the electrode surface; however, this could be corrected by using a dual-electrode setup where one electrode had arsenite oxidase and the second electrode does not. Similar electrode designs are used in clinical biosensors to correct for electroactive components found in blood.

The second approach to develop enzymatic biosensors for arsenic takes advantage of the fact that arsenic can bind tightly to proteins and in some cases can be a potent inhibitor of enzyme activity (Shen et al., 2013). Enzyme inhibition-based biosensors have been studied for a number of years and have been used successfully for monitoring of pesticides and chemical warfare agents (Amine et al., 2016; Arduini and Amine, 2014; El Harrad et al., 2018; Luque de Castro and Herrera, 2003). Biosensor design based on enzyme inhibition can be more complicated than sensors designed to analyze for the substrate. To use an inhibition-based biosensor, the substrate (without inhibitor) must be applied to the sensor and the amount of enzyme activity on that sensor must be determined. Next, the sample containing the inhibitor is applied to the sensor and often allowed to incubate for some period of time. After incubation with the inhibitor, the enzyme activity is remeasured. A decrease in enzyme activity is indicative of the presence of an inhibitor. The disadvantage of this approach is that some sensors can be inhibited by many types of compounds or even classes of compounds. However, in general, if an enzyme is inhibited by a compound, it is likely to be toxic. If the correct enzyme and inhibitor can be identified, extremely sensitive biosensors can be developed. For example, protein engineering was used to design a *Drosophila melanogaster* acetylcholinesterase that was used to develop an amperometric biosensor with a detection limit down to attomolar levels for the organophosphate pesticide dicholorvos (Sotiropoulou et al., 2005).

11. BIOSENSORS FOR MONITORING WATER POLLUTANTS

A number of enzymes have been reported that are inhibited by arsenic at very low concentrations (Shen et al., 2013). Many of them are inhibited by As(III) or As(V) to varying degrees. For example, the binding constant, $\underline{K}_d$, of acetylcholinesterase from electric eel with arsenite is 15 μM depending on the pH (Wilson and Silman, 1977). Acetylcholinesterase is not inhibited by As(V). The enzyme acid phosphatase is reversibly inhibited by arsenate with a K_d of <1 mM but is inhibited by arsenite with a K_d of >1 17 mM (Shen et al., 2013). Table 1 lists many enzymes that have been used for the development of inhibition-based biosensors. Many have these sensors have detection limits well below the EPA/WHO guidelines of 10 ppb (133 nM). Almost all of the inhibition-based biosensors developed to date are amperometric biosensors.

3.2.3 *Whole-Cell Biosensors*

A whole-cell biosensor utilizes a whole cell, typically a bacterium, with a physical transducer to develop a signal that can be easily quantified (typically electrochemical, colorimetric, luminescent, or fluorescent). The simplest whole-cell biosensor consists of an arsenic sensitive promoter and a reporter gene that will produce a protein that will generate a quantifiable signal. The first whole-cell biosensors for arsenic used the genes from Plasmid R773 from *E. coli* that provides resistance to the bacterium to arsenic (Corbisier et al., 1993). The basic machinery includes an arsenite sensitive promotor (P_{arsR}), an arsenite efflux pump (ArsAB), and an arsenite reductase (ArsC). The expression of the *ars* operon is controlled by the binding of the repressor protein ArsR to the ArsR binding site (ABS) upstream of the promotor. Binding of

TABLE 1   Examples From the Literature of Enzymatic Biosensors for Arsenic Detection

Enzyme	Sensing Method	Substrate	Reported Arsenic Detection Limit
Acetylcholinesterase (Stoytcheva et al., 1998a)	Amperometric	Acetylthiocholine iodide	0.2 nM (Arsenite)
Acetylcholinesterase (Stoytcheva et al., 1998b)	Amperometric	Acetylthiocholine iodide and acetylthiocholine perchlorate	0.2 nM (Arsenite)
Acetylcholinesterase (Gumpu et al., 2018)	Amperometric	Acetylthiocholine chloride	30 nM
Acid phosphatase ad polyphenol oxidase (Cosnier et al., 2006)		Phenyl phosphate	2 nM (Arsenate)
Arsenite oxidase (Male et al., 2007)	Amperometric	Arsenite	1 ppb (Arsenite)
Acetylcholinesterase (Sanllorente-Méndez et al., 2010)	Amperometric	Acetylthiocholine iodide	11 nM (Arsenite)
Acid phosphatase (Sanllorente-Méndez et al., 2012)	Amperometric	2-Phospho-L-ascorbic acid	0.11 μM (Arsenate)
Acid phosphatase and acetylcholinesterase (del Torno-de Román et al., 2015)	Amperometric	2-Phospho-L-ascorbic acid and acetylthiocholine iodide	2.0 μM (Arsenate) 35.9 μM (Arsenite)
Laccase (Wang et al., 2016)	Amperometric	2,2'-Azino-bis(3-ethylbenzothiazoline-6-sulfonic acid	13 μM (Arsenite) 132 μM (Arsenate)

EVALUATING WATER QUALITY TO PREVENT FUTURE DISASTERS

ArsR to the promotor prevents transcription; however, in the presence of arsenite, the affinity of ArsR to ABS is reduced allowing transcription of the downstream elements. The first reports of whole-cell sensors for arsenic utilized the *ars* operon from *E. coli* (Corbisier et al., 1993); however, arsenic and other metal sensitive operons have been found in other microorganisms including eukaryotes (Daunert et al., 2000; Merulla et al., 2013).

Over the last 2 decades, a number of arsenite sensitive whole-cell biosensors have been developed by building different arsenite sensitive genetic circuits using the ArsR As(III)-responsive transcriptional repressor (Chen and Rosen, 2014). In the absence of arsenite, the ArsR repressor binds to the ABS operator site of the promotor preventing transcription (Fig. 9); however, in the presence of arsenite, the affinity of ArsR for the promotor is reduced and expression of the reporter gene downstream of the promotor can take place. Many approaches have been utilized to synthesize these reporter genes together to reduce background expression of the reporter genes (Chen et al., 2014b; Merulla et al., 2013). Most of the whole-cell biosensors for arsenic have been developed for arsenite; however, Rosen's group demonstrated the ability to tune the selectivity of the whole-cell sensors for organic arsenic derivatives (Chen et al., 2014b; Chen and Rosen, 2014).

By optimizing the arsenic detection gene circuits and the reporter genes the detection limits, background responses, and response times can be tuned (Stocker et al., 2003). The typical whole-cell biosensor utilizing the ArsR-P_{arsR} system has been reliably shown to have detectable range of 10–50 μg/L of arsenite and responds within a few hours (Trang et al., 2005). The whole-cell biosensors have been packaged in a number of single-use and continuous detection formats and have been miniaturized (Roggo and van der Meer, 2017). Whole-cell biosensors appear to be the most mature of the biosensors for arsenic detection (Chen and

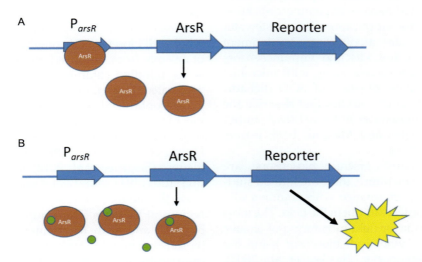

FIG. 9 Basic sensing mechanism of whole-cell arsenic biosensors based on ArsR gene. (A) If no arsenite is present, background amounts of ArsR are synthesized which binds to the operator of the ArsR promoter (P_{arsR}) preventing transcription of the reporter genes. (B) If present, arsenite will bind to the ArsR repressor reducing the affinity for P_{arsR}. Since ArsR is no longer bound to P_{arsR}, transcription of the reporter gene can take place.

Rosen, 2014; van der Meer, 2016) and have even been used in field testing in Bangladesh and Vietnam (Siegfried et al., 2012; Trang et al., 2005). Shelf-life studies and reliability studies have been performed (Date et al., 2010).

3.3 Biosensors for Field Monitoring

Over 20 years ago, there was significant growth in biosensor research and it was predicted that there would be many commercial biosensors for environmental monitoring applications (Rogers, 1995; Rogers and Gerlach, 1996, 1999). Although that optimism has not gone away, there are still a limited number of commercial biosensors for environmental monitoring compared with biomedical and food applications (Bahadir and Sezginturk, 2015; Rodriguez-Mozaz et al., 2005). There has been significant progress in basic research to develop biosensors as judged by the number of published journal articles and journals that exist for biosensors. While there has been a considerable success in the development of biosensors for biomedical applications, there has been a limited success in commercialization of biosensors for applications in wastewater and drinking water monitoring. Development of immunoassay test kits and immunosensors have found some success along with microbial biosensors for determination of BOD (Bahadir and Sezginturk, 2015). Studies that describe the application of biosensors in formal field studies are limited (Rogers and Mascini, 1998; Rogers and Williams, 1995).

Although many arsenic biosensors are described in the literature, there are limited studies comparing performance with commercially available arsenic test kits out in the field. The most studied biosensor for field monitoring of arsenic is the ArsR-P_{ars} whole-cell biosensor using the luminescent bioreporter (Stocker et al., 2003). A number of studies have been performed with these whole-cell sensors in simulated groundwater in order to optimize sensitivity, reduce assay complexity, and shorten required assay time (Harms et al., 2005; Stocker et al., 2003). Studies were also performed to determine components of simulated groundwater that could impact the performance of the biosensor which might lead to false positive or false negative responses (Harms et al., 2005). For simulated groundwater, Harms and coworkers found that the common groundwater components phosphate, Fe(II) and silicate could have an impact on sensor performance. They discovered that phosphate at a concentration of 0.25 g/L would cause a slight increase in the background signal, but did not impact the As(III) signal. Thus, high concentrations of phosphate could be interpreted as a false positive. Silicate and Fe(II) were both observed to reduce light production by the sensors which could be potentially interpreted as a false negative. They found that silicate could potentially polymerize and sequester some of the arsenic. Fe(II) likely oxidized to form insoluble Fe(III) which may also sequester arsenite by adsorption to the iron precipitate. They found that the addition of 100 μM of EDTA to samples kept the arsenic soluble and available for the whole-cell biosensors. They demonstrated the utility of the whole-cell biosensors by monitoring the removal of arsenite from simulated groundwater using a water filtration column composed of sand and zero-valent iron granules (Harms et al., 2005).

The optimized whole-cell luminescent arsenic bioreporters were validated in two large field tests; one of the tests was in Vietnam (Trang et al., 2005) and the second was in Bangladesh (Siegfried et al., 2012). In the first validation study, 194 groundwater samples were taken from the Red River and Mekong River Delta regions of Vietnam. The samples were analyzed by AAS and with the luminescent whole-cell biosensor using the ArsR-P_{ars} system. The water samples from the region were highly variable in concentrations of arsenic, iron, phosphate, bicarbonate, ammonium, and chloride. The bioreporter tested had a detection limit of 7.5 μg/L and

EVALUATING WATER QUALITY TO PREVENT FUTURE DISASTERS

3 BIOSENSORS FOR MONITORING ARSENIC IN WATER

a linear range of 0–75 µg/L and had a 90 min incubation time. A solution of decanal needed to be added to the cell vials as a substrate for the luciferase catalyzed luminescent reaction. After 3 min, the luminescence was monitored. In order to prevent iron precipitation, EDTA was added to the samples. Results from the tests were excellent; of the 194 samples, 8% of the samples gave false negative (these were samples in the 10–19 µg/L range) and 2.4% of the samples were false positives. These results were quite favorable when compared to test kits based on the Gutzeit method (Rahman et al., 2002).

In a follow-up validation in Bangladesh using a similar bioreporter (Siegfried et al., 2012), Jan Roelof van der Meer and collaborators worked to simplify the bioreporter assay so that it could be used by a single person with minimal training where they could collect 150 samples, test, analyze, and distribute the results within an 8 h period. The bioreporter was similar to that used in the study in Vietnam, except that the plasmid contained a tetracycline resistance gene, and a bacterial luciferase gene that could regenerate its substrate. Tests were performed on more than 40 water samples and were verified by ICP-MS and tested using commercially available arsenic test kits that use the Gutzeit method. Tests with the bioreporter cells were performed using different sample preparation methods and times. As before, it was found that it was important to treat test samples with EDTA to prevent precipitation of iron and arsenic in the samples. When compared to results from ICP-MS, the bioreporter tests gave 2 false positives and no false negative out of 41 samples when using EDTA. The authors noted that the Gutzeit tests matched well with the results from ICP-MS but did not give any details on performance. When compared to the Gutzeit test kits, the bioreporter assays were more amenable to testing a large number of samples. Additionally, the bioreporter test kits had fewer components, generated less waste, and did not produce toxic

chemical waste that required disposal (mercury). However, the potential disadvantage of the bioreporter cells is the requirement to use genetically modified organisms. The authors designed the bioreporter test method to isolate the bioreporter cells in sealed vials to prevent release into the environment and included antibiotic resistance genes so that the cell would only grow in the presence of tetracycline.

The future of biosensors for water monitoring in the field and monitoring arsenic in particular shows great promise. In a biosensor, a bioreceptor is integrated with the transducer in one unit; the selectivity of the bioreceptor enables direct pollutant analysis in complex matrices. The simplicity of the biosensor makes it possible to manufacture almost entirely with renewable materials that are nontoxic, low cost, and therefore disposable. These features are highly desirable to characterize water pollution and to ensure water safety in the field. As shown in the case of arsenic pollution in groundwater, spatial variability has been characterized with blanket surveys using laboratory-based methods, but the spatial resolution and scale were not adequate for water resource management. Field kits based on Gutzeit reaction were used in the field to label the wells in Bangladesh (Chakraborti et al., 2010). It is more challenging to monitor temporal variability because arsenic mobilization and sequestration are related to weather events, human activities, and incidents that may change the chemical environment of arsenic deposits. The analytical need can only be met with a field test kit. The Gutzeit test kit has been the only commercial product for field analysis. Although it is semi quantitative, the field experience provided important guidance for biosensors in field application (Feldmann, 2008). Several critical requirements for a fieldable biosensor include:

- good accuracy with sufficient sensitivity and precision;

EVALUATING WATER QUALITY TO PREVENT FUTURE DISASTERS

- practical and straightforward to use by non-trained users;
- short analysis time; and
- low cost.

The Gutzeit method suffers from many issues such as interference by sulfide, poor reproducibility, and the requirement for toxic chemicals. Nevertheless, it was still selected as the most suitable field-testing method by Chemists Without Borders in 2011 (Lizardi, 2016). Two whole-cell biosensors were evaluated in the pool of options, but they suffered from lack of accuracy or poor stability.

4 CONTAMINANTS OF EMERGING CONCERN AND FUTURE PRIORITIES

4.1 Sources, Fate, and Transport of CECs

A large array of chemicals that are used in our everyday lives including PPCP (e.g., soaps and disinfectants) and health-products (such as prescription and nonprescription drugs) and fragrances, surfactants, and pesticide are frequently detected in drinking water supplies.

These chemicals called collectively, contaminants of emerging concern (CECs), are now ubiquitous in the aquatic environment. These chemicals are released from nonpoint sources through excretion, bathing, or direct disposal and survive the wastewater treatment process and are discharged to surface and groundwaters (Daughton and Ternes, 1999; Díaz-Cruz and Barceló, 2004; Glassmeyer et al., 2008; Halling-Sørensen et al., 1998; Heberer, 2002; Kostich et al., 2010; Pal et al., 2014; Petrie et al., 2015). There are thousands of potential emerging contaminants, with more continually being introduced. Toxicological research and the development of water-quality standards and treatment technologies have not kept up with the rate at which we are finding them in the environment.

CEC can be classified into three categories:

- The first consists of compounds that are newly introduced into the environment, including industrial compounds that have only recently been developed.
- The second category comprises compounds that have only recently been detected in the environment although they are possibly around for longer times. The release of these chemicals to the drinking water source has been happening for decades. What is new is the ability of the environmental analytical chemists to detect extremely low concentrations and inform the public of their risk. Hence the appropriate name the chemical is "contaminants of emerging concerns" (CEC). These chemicals are found at very low concentrations in the environment. The current analytical techniques, such as LC/MS/MS technique (Gros et al., 2006a), and biological methods used bioassays allowed the detection and a wider array of persistent, nonpolar organic compounds in environmental samples, and animal tissues (Van der Oost et al., 2003).
- The third category emerging contaminants consist of compounds that are known and have been in use for a long time as benign compounds (e.g., hormones) but they are recently recognized to have adverse effects on ecosystems or humans (Gómez et al., 2006; Gros et al., 2006b).

Quantifying the fate and transport of CECs in natural systems is critical to understanding their risk to human health and the environment. Wastewater treatment plants, along with landfills, and feedlots are commonly cited as a major source of CECs (Fig. 10) (Caliman and Gavrilescu, 2009; Daughton and Ternes, 1999; Halling-Sørensen et al., 1998).

CECs are significant because the risk they pose to human health and the environment has not yet been fully understood. The most common emerging contaminants originate from

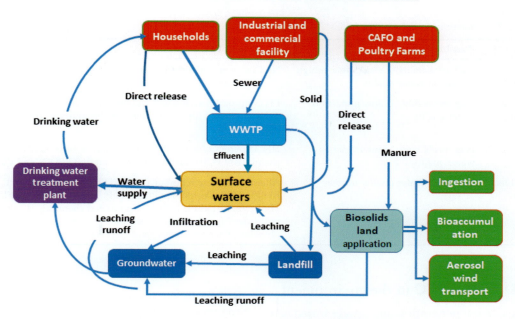

FIG. 10 Pathways of exposure for contaminants of emerging concern. *Modified from Percival, R.V., et al., 2017. Environmental Regulation: Law, Science, and Policy. Wolters Kluwer Law & Business.*

diffuse sources, including both organic and inorganic chemicals in drinking water sources. Examples include pharmaceuticals, PPCPs, nanomaterials, and endocrine disrupting compounds (EDC) and their metabolites. The presence of PPCP/EDC in water supplies has been known for over 30 years. Much of the original concerns were based on physiological abnormalities of the aquatic organisms caused by high concentrations of EDCs near wastewater treatment facilities and drinking water supplies. Because EDCs are lipophilic and persist in the environment, they could accumulate in sediments and in living organisms. Many emerging contaminants are from nonpoint sources such as elimination from the body, flushing of unused medication or expired products, leachate from landfills, rinse off from showering or bathing, and agricultural runoff. There is little toxicological information for chronic, low-level exposure to most of the EDCs. Hence, one of the vital problems is the identification of future risks, since the available information on prognosis of exposure and toxicity of complex mixtures of CEC is lacking. Several compounds in the EPA's contaminant candidate list III (CCL III) are not regulated because they have no maximum exposure threshold.

Thousands of chemicals are used in consumer products. Only a handful of CECs reach level of concern for impacts on human health and the environment. Assessing the environmental hazard posed by CEC requires information on their fate, transport, and persistence in the environment, their potency as ecological toxicants and measurements of their occurrence in ecosystems. CEC can be grouped into different classes depending on the potential risks they pose as shown in Table 2. Table 2 provides the environmental risk tiers and associated monitoring strategies and management actions. The risk-based hierarchy tiers may be used as a guide for sensor development, monitoring strategies, and management actions.

11. BIOSENSORS FOR MONITORING WATER POLLUTANTS

TABLE 2 Prioritization Scheme for CECs

Tier	Possible Impact	Example CECs
4-High concern	High probability of a moderate or high impact on water quality	No CECs currently in this tier
3-Moderate concern	High probability of low impact on water quality	PFOS fipronil nonylphenol and nonylphenol ethoxylates polybrominated diphenyl ethers polybrominated diphenyl ethers (PBDEs)
2-Low concern	High probability of no impact on water quality	HBCD pyrethroids (14 chemicals) pharmaceuticals (100+ chemicals) PPCP ingredients (10 chemicals) PBDDs and PBDFs
1-Possible concern	Impact on water quality unclear	Alternative flame retardants (BEH-TEBP, EH-TBB, DBDPE, PBEB, BTBPE, HBB, Dechlorane Plus, TPhP, TDCPP, TCPP, TCEP, TBEP, TBPP, V6, EBTEBPI, TBECH) fluorinated chemicals (17 chemicals) pesticides (dozens of chemicals) plasticizers (bisphenol A, phthalates) nanomaterials short-chain chlorinated paraffins

4.2 Detecting CEC in the environment

The current generation and newly tested sensors will improve the temporal resolution for measurements of nutrients, organic pollutants, metals, and CEC. Selectivity and robustness are both important in field use. Water sensors have to recognize the contaminants, and the transducers have to report the signals in complex water matrices.

Sensor technologies provide important tools that enable us to monitor water quality and make associated water resource management decisions. Sensors have been increasingly used for monitoring water quality such as temperature, salinity, dissolved oxygen, pH, turbidity, fluorescent dissolved organic matter (a surrogate for chlorophyll, and therefore phytoplankton, abundance), and nitrate. Several biosensors show significant promise for in situ, long-term monitoring family of pollutants, such as nutrients, or specific contaminants. Sensors developed to determine several analytes in parallel are useful tools in environmental monitoring and screening. Online monitoring of emerging contaminants provides accurate and timely information that helps to minimize risk to humans and manage the underlying drivers of water-quality

challenges. For example, continuous and real-time monitoring of nutrients provide daily, seasonal, and event based changes in water-quality and vital information that provide for conditions high-nutrient concentrations that can lead to HABs, which can produce toxins that further contaminate drinking water and imperil wildlife. The real-time data collected from an array of sensors can be transmitted via cell phone and a public website. While continuous monitoring has many advantages, issues with sensor accuracy and precision, developing protocols for sensor maintenance and minimizing biofouling, are some of the challenges.

A wide variety on emerging chemicals such as alkylphenolic surfactants, steroid sex hormones, and pharmaceuticals are used in increasing volumes and have been suspected as endocrine disruptors or as causative agents of bacterial resistance (Lopez de Alda et al., 2003). New biosensors have been developed lately for the determination of some of these compounds as, for instance, sensors for detection of alkylphenol ethoxylates, estrone, and endocrine disruptors (Rodriguez-Mozaz et al., 2004).

There is a need for increased field deployments of sensors for legacy contaminants and

EVALUATING WATER QUALITY TO PREVENT FUTURE DISASTERS

CECs. Although laboratory methods such as LC/MS/MS have shown significant progress in the past decade, there are still major technical challenges in monitoring and managing CECs. These include the increasing number of new chemicals in commerce that has the potential to cause adverse environmental impacts, the lack of water-quality standards or adverse effects thresholds to allow for interpretation of exposure levels, and the absence of practical field monitoring and analytical methods to measure many CECs. Once the challenges are overcome, a network of sensors could be used to gather information at a plant level, water distribution systems, or even for watershed management as discussed in Section 1.2.3. In addition to real-time detections, these sensors will detect and provide early-warning allowing for adaptive sampling and increasing public awareness.

A biosensor can detect the presence of a substrate by using biological components, which then provides a quantifiable signal for increasing confidence in the risk assessment of both regulated and emerging chemical contaminants. Biomonitoring tools (e.g., bioassays, biomarkers, microbial community analyses) have great potential for increasing confidence in the risk assessment of both regulated and emerging chemical pollutants. The network of sensors used for environmental monitoring, and screening that can determine several analytes simultaneously, based on real-time parallel monitoring of multiple species. The data from a sensor is matched with molecular biomarkers as detection tools to understand the fate-, transport, and transformation of emerging pollutants including nanoparticles in environmental matrices. Innovative tools in chemical, optical, and biomonitoring of current and emerging pollutants provide water-quality information across the continuum for a watershed, lakes, and river basins, and inform regulatory decisions and the development of management options.

5 CONCLUSIONS

With the global population expected to reach 10 billion by 2050, unsustainable agricultural and industrial activities will continue to lead to issues with water quality and availability. The changing profile of water pollution and sources of regulated and emerging contaminants will continue to be a challenge for drinking water treatment plants. Increasingly stringent regulatory limits will require new robust sensors for trace detection. Sensing techniques that are available today for monitoring known water contaminants are typically expensive and require trained technicians to use them correctly. There are simple low cost, portable test kits available to detect many contaminants, such as arsenic, but they are typically unreliable detecting As at the low concentrations that are relevant to drinking water sources and human health. Advances in biosensors provide the opportunity for simple to use, disposable or continuous tests, for monitoring many of the common and emerging contaminants that water-quality personnel are facing today. Although many biosensors have been developed for regulated and emerging contaminants of concern, additional research and development is needed to prove their utility in the field. More work is also vital to get robust biosensors into the hands of non-trained end-users and to demonstrate that fieldable biosensors can be developed with good accuracy, sensitivity, and precision, with short analysis times and at a low cost.

DISCLAIMER AND ACKNOWLEDGMENT

The views expressed in this chapter are those of the authors and do not reflect the official policy or position of the Unites States Environmental Protection Agency. Mention of trade names, products, or services does not convey official EPA approval, endorsement, or recommendation. This manuscript has been subjected to the Agency's review and has been approved for publication.

References

Abedin, M.J., et al., 2002. Arsenic accumulation and metabolism in rice (Oryza sativa L.). Environ. Sci. Technol. 36 (5), 962–968.

Acheampong, M.A., Paksirajan, K., Lens, P.N.L., 2013. Assessment of the effluent quality from a gold mining industry in Ghana. Environ. Sci. Pollut. Res. 20 (6), 3799–3811.

Akai, J., et al., 2004. Mineralogical and geomicrobiological investigations on groundwater arsenic enrichment in Bangladesh. Appl. Geochem. 19 (2), 215–230.

Almeida, M.G., et al., 2010. Nitrite biosensing via selective enzymes—a long but promising route. Sensors 10 (12), 11530–11555.

Amine, A., et al., 2016. Recent advances in biosensors based on enzyme inhibition. Biosens. Bioelectron. 76, 180–194.

Anawar, H.M., et al., 2003. Geochemical occurrence of arsenic in groundwater of Bangladesh: sources and mobilization processes. J. Geochem. Explor. 77 (2–3), 109–131.

Appelo, C.A.J., Postma, D., 2004. Geochemistry, Groundwater and Pollution. CRC Press.

Arduini, F., Amine, A., 2014. Biosensors based on enzyme inhibition. Adv. Biochem. Eng. Biotechnol. 140, 299–326.

Armienta, M.A., Segovia, N., 2008. Arsenic and fluoride in the groundwater of Mexico. Environ. Geochem. Health 30 (4), 345–353.

Armienta, A., et al., 2000. Health risk and sources of arsenic in the potable water of a mining area. In: Interdisciplinary Perspectives on Drinking Water Risk Assessment and Management. vol. 260. IAHS Publishers, pp. 9–16.

Bahadir, E.B., Sezginturk, M.K., 2015. Applications of commercial biosensors in clinical, food, environmental, and biothreat/biowarfare analyses. Anal. Biochem. 478, 107–120.

Banica, F.-G., 2012. Chemical Sensors and Biosensors: Fundamentals and Applications. John Wiley & Sons, p. 834.

Bereza-Malcolm, L.T., Mann, G., Franks, A.E., 2015. Environmental sensing of heavy metals through whole cell microbial biosensors: a synthetic biology approach. ACS Synth. Biol. 4 (5), 535–546.

Berg, M., et al., 2001. Arsenic contamination of groundwater and drinking water in Vietnam: a human health threat. Environ. Sci. Technol. 35 (13), 2621–2626.

Bhattacharjee, Y., 2007. A Sluggish Response to Humanity's Biggest Mass Poisoning. American Association for the Advancement of Science.

Bhattacharya, P., Chatterjee, D., Jacks, G., 1997. Occurrence of arsenic-contaminated groundwater in alluvial aquifers from Delta Plains, Eastern India: options for safe drinking water supply. Int. J. Water Resour. Dev. 13 (1), 79–92.

Bhattacharya, P., et al., 2002. Arsenic in the environment: a global perspective. In: Handbook of Heavy Metals in the Environment. Marcell Dekker, New York, pp. 147–215.

Bhattacharya, P., et al., 2007. Arsenic in the Environment: Biology and Chemistry. Elsevier.

Blackman Jr., W.C., 2016. Basic Hazardous Waste Management. CRC Press.

Bondu, R., et al., 2016. A review and evaluation of the impacts of climate change on geogenic arsenic in groundwater from fractured bedrock aquifers. Water Air Soil Pollut. 227 (9), 1–14.

Boxall, A.B., et al., 2003. Peer Reviewed: Are Veterinary Medicines Causing Environmental Risks? ACS Publications.

Brammer, H., Ravenscroft, P., 2009. Arsenic in groundwater: a threat to sustainable agriculture in South and South-east Asia. Environ. Int. 35 (3), 647–654.

Bregman, J.I., Edell, R.D., 2016. Environmental Compliance Handbook. CRC Press.

Burian, S.J., et al., 2000. Urban wastewater management in the United States: past, present, and future. J. Urban Technol. 7 (3), 33–62.

Caliman, F.A., Gavrilescu, M., 2009. Pharmaceuticals, personal care products and endocrine disrupting agents in the environment–a review. CLEAN–Soil, Air, Water 37 (4–5), 277–303.

Carpenter, S.R., et al., 1998. Nonpoint pollution of surface waters with phosphorus and nitrogen. Ecol. Appl. 8 (3), 559–568.

Carus, W.S., 2002. Bioterrorism and Biocrimes: The Illicit Use of Biological Agents Since 1900. The Minerva Group.

Casiot, C., et al., 2003. Geochemical processes controlling the formation of As-rich waters within a tailings impoundment (Carnoulès, France). Aquat. Geochem. 9 (4), 273–290.

Chakraborti, D., et al., 2010. Status of groundwater arsenic contamination in Bangladesh: a 14-year study report. Water Res. 44 (19), 5789–5802.

Chakraborty, M., Mukherjee, A., Ahmed, K.M., 2015. A review of groundwater arsenic in the Bengal Basin, Bangladesh and India: from source to sink. Curr. Pollut. Rep. 1 (4), 220–247.

Chen, J., Rosen, B.P., 2014. Biosensors for inorganic and organic arsenicals. Biosensors 4 (4), 494–512.

Chen, J., et al., 2014a. Biosensor for organoarsenical herbicides and growth promoters. Environ. Sci. Technol. 48 (2), 1141–1147.

Chen, B., et al., 2014b. Therapeutic and analytical applications of arsenic binding to proteins. Metallomics 7 (1), 39–55.

Cheng, Z., et al., 2005. Limited temporal variability of arsenic concentrations in 20 wells monitored for 3 years in

REFERENCES

Araihazar, Bangladesh. Environ. Sci. Technol. 39 (13), 4759–4766.

Cheng, H., et al., 2009. Geochemical processes controlling fate and transport of arsenic in acid mine drainage (AMD) and natural systems. J. Hazard. Mater. 165 (1), 13–26.

Chilton, J., 1996 Ground water, in Water Quality Assessments— A Guide to Use of Biota, Sediments and Water in Environmental Monitoring, D. Chapman, Ed. 1996, UNESCO/WHO/UNEP: Great Britain. http://apps.who.int/iris/bitstream/handle/10665/41850/0419216006_eng.pdf;jsessionid=73122E70BB90C7102EBD5E533C6C116A?sequence=1

Choudhury, I., et al., 2016. Evidence for elevated levels of arsenic in public wells of Bangladesh due to improper installation. Ground Water 54 (6), 871–877.

Clancy, T.M., Hayes, K.F., Raskin, L., 2013. Arsenic waste management: a critical review of testing and disposal of arsenic-bearing solid wastes generated during arsenic removal from drinking water. Environ. Sci. Technol. 47 (19), 10799–10812.

Collison, M.E., Meyerhoff, M.E., 1990. Chemical sensors for bedside monitoring of critically Ill patients. Anal. Chem. 62 (7), A425.

Corbisier, P., et al., 1993. Luxab gene fusions with the arsenic and cadmium resistance operons of Staphylococcus aureus plasmid-Pi258. FEMS Microbiol. Lett. 110 (2), 231–238.

Cosnier, S., et al., 2006. Specific determination of As(V) by an acid phosphatase – polyphenol oxidase biosensor. Anal. Chem. 78 (14), 4985–4989.

Cullen, W., Reimer, K., 1989. Arsenic speciation in the environment. Chem. Rev. 89, 713–764. Find this article online.

Daniels, J.S., Pourmand, N., 2007. Label-free impedance biosensors: opportunities and challenges. Electroanalysis 19 (12), 1239–1257.

Date, A., Pasini, P., Daunert, S., 2010. Fluorescent and bioluminescent cell-based sensors: strategies for their preservation. In: Whole Cell Sensing Systems I: Reporter Cells and Devices. vol. 117, Springer Nature Switzerland AG, pp. 57–75.

Daughton, C.G., Ternes, T.A., 1999. Pharmaceuticals and personal care products in the environment: agents of subtle change? Environ. Health Perspect. 107 (Suppl 6), 907.

Daunert, S., et al., 2000. Genetically engineered whale-cell sensing systems: coupling biological recognition with reporter genes. Chem. Rev. 100 (7), 2705–2738.

del Torno-de Román, L., et al., 2015. Dual biosensing device for the speciation of arsenic. Electroanalysis 27 (2), 302–308.

deLemos, J.L., et al., 2006. Landfill-stimulated iron reduction and arsenic release at the Coakley superfund site (NH). Environ. Sci. Technol. 40 (1), 67–73.

Dhar, R.K., et al., 2008. Temporal variability of groundwater chemistry in shallow and deep aquifers of Araihazar, Bangladesh. J. Contam. Hydrol. 99 (1), 97–111.

Díaz-Cruz, S., Barceló, D., 2004. Occurrence and analysis of selected pharmaceuticals and metabolites as contaminants present in waste waters, sludge and sediments. In: Series Anthropogenic Compounds. Springer, pp. 227–260.

Dudka, S., Adriano, D.C., 1997. Environmental impacts of metal ore mining and processing: a review. J. Environ. Qual. 26 (3), 590–602.

Edwards, K.J., et al., 2000. Geochemical and biological aspects of sulfide mineral dissolution: lessons from Iron Mountain, California. Chem. Geol. 169 (3), 383–397.

El Harrad, L., et al., 2018. Recent advances in electrochemical biosensors based on enzyme inhibition for clinical and pharmaceutical applications. Sensors 18 (1), 164.

Evtugyn, G., 2013. Biosensors: Essentials. Springer Science & Business Media, p. 274.

Feldmann, J., 2008. Onsite testing for arsenic: field test kits. Rev. Environ. Contam. Toxicol. 197, 61–75.

Fendorf, S., Michael, H.A., van Geen, A., 2010. Spatial and temporal variations of groundwater arsenic in South and Southeast Asia. Science 328 (5982), 1123–1127.

Ferancová, A., et al., 2007. Interaction of tin(II) and arsenic(III) with DNA at the nanostructure film modified electrodes. Bioelectrochemistry 71 (1), 33–37.

Field, R., Struzeski Jr., E.J., 1972. Management and control of combined sewer overflows. J. Water Pollut. Control Fed., 1393–1415.

Focazio, M.J., et al., 1999. A Retrospective Analysis on the Occurrence of Arsenic in Ground-Water Resources of the United States and limitations in Drinking-Water-Supply Characterizations. U.S. Geological Survey, pp. 1–21. p. i–vi.

Ford, R.G., et al., 2011. Delineating landfill leachate discharge to an arsenic contaminated waterway. Chemosphere 85 (9), 1525–1537.

Frey, M.M., Edwards, M.A., 1997. Surveying arsenic occurrence. J. Am. Water Works Assoc. 89 (3), 105–117.

Garelick, H., et al., 2008. Arsenic pollution sources. Rev. Environ. Contam. Toxicol. 197, 17–60.

Glassmeyer, S.T., et al., 2008. Environmental presence and persistence of pharmaceuticals: an overview. In: Aga, D.S. (Ed.), Fate of Pharmaceuticals in the Environment and in Water Treatment Systems. CRC Press, Boca Raton, FL.

Gleick, P.H., 2006. Water and terrorism. Water Policy 8 (6), 481–503.

Gómez, M.J., et al., 2006. Determination of pharmaceuticals of various therapeutic classes by solid-phase extraction and liquid chromatography–tandem mass spectrometry analysis in hospital effluent wastewaters. J Chromatogr. A 1114 (2), 224–233.

Gopinath, S.C.B., 2007. Methods developed for SELEX. Anal. Bioanal. Chem. 387 (1), 171–182.

Gorny, J., et al., 2015. Arsenic behavior in river sediments under redox gradient: a review. Sci. Total Environ. 505, 423–434.

Grill, E., Winnacker, E.-L., Zenk, M.H., 1987. Phytochelatins, a class of heavy-metal-binding peptides from plants, are functionally analogous to metallothioneins. Proc. Natl. Acad. Sci. U. S. A. 84 (2), 439–443.

Gros, M., Petrović, M., Barceló, D., 2006a. Multi-residue analytical methods using LC-tandem MS for the determination of pharmaceuticals in environmental and wastewater samples: a review. Anal. Bioanal. Chem. 386 (4), 941–952.

Gros, M., Petrović, M., Barceló, D., 2006b. Development of a multi-residue analytical methodology based on liquid chromatography–tandem mass spectrometry (LC–MS/MS) for screening and trace level determination of pharmaceuticals in surface and wastewaters. Talanta 70 (4), 678–690.

Guerra, M.B.B., et al., 2017. Post-catastrophe analysis of the Fundão Tailings Dam Failure in the Doce River System, Southeast Brazil: potentially toxic elements in affected soils. Water Air Soil Pollut. 228 (7), 252.

Gui, Q., et al., 2017. The application of whole cell-based biosensors for use in environmental analysis and in medical diagnostics. Sensors 17 (7), 1623.

Gumpu, M.B., et al., 2018. Amperometric determination of As(III) and Cd(II) using a platinum electrode modified with acetylcholinesterase, ruthenium(II)-tris(bipyridine) and graphene oxide. Microchim. Acta 185 (6), 297.

Hallegraeff, G., 2003. Harmful algal blooms: a global overview. In: Manual on Harmful Marine Microalgae. vol. 33, UNESCO, Paris, France, pp. 1–22.

Halling-Sørensen, B., et al., 1998. Occurrence, fate and effects of pharmaceutical substances in the environment—a review. Chemosphere 36 (2), 357–393.

Harms, H., et al., 2005. Effect of groundwater composition on arsenic detection by bacterial biosensors. Microchim. Acta 151 (3–4), 217–222.

Harvey, C.F., et al., 2006. Groundwater dynamics and arsenic contamination in Bangladesh. Chem. Geol. 228 (1), 112–136.

Hayat, A., Marty, J.L., 2014. Aptamer based electrochemical sensors for emerging environmental pollutants. Front. Chem. 2, 41. https://doi.org/10.3389/fchem.2014.00041.

He, Y.T., et al., 2010. Geochemical processes controlling arsenic mobility in groundwater: a case study of arsenic mobilization and natural attenuation. Appl. Geochem. 25 (1), 69–80.

Heberer, T., 2002. Occurrence, fate, and removal of pharmaceutical residues in the aquatic environment: a review of recent research data. Toxicol. Lett. 131 (1–2), 5–17.

Herath, I., et al., 2016. Natural arsenic in global groundwaters: distribution and geochemical triggers for mobilization. Curr. Pollut. Rep. 2 (1), 68–89.

Hering, J.G., et al., 2011. Role of coupled redox transformations in the mobilization and sequestration of arsenic. ACS Symp. Ser. 1071 (Aquatic Redox Chemistry), 463–476.

Holmes, P., Niccolls, L.M., Sartory, D., 1996. The ecology of mesophilic Aeromonas in the aquatic environment. In: The Genus Aeromonas, Wiley & Sons, Hoboken, NJ, pp. 127–150.

Igbinosa, I.H., et al., 2012. Emerging Aeromonas species infections and their significance in public health. Sci. World J. 2012. https://doi.org/10.1100/2012/625023.

Irvine, G.W., Tan, S.N., Stillman, M.J., 2017. A simple metallothionein-based biosensor for enhanced detection of arsenic and mercury. Biosensors 7 (1), 14.

Jeter, T.S., et al., 2016. 4-MCHM sorption to and desorption from granular activated carbon and raw coal. Chemosphere 157, 160–165.

Kalbus, E., Reinstorf, F., Schirmer, M., 2006. Measuring methods for groundwater—surface water interactions: a review. Hydrol. Earth Syst. Sci. 10 (6), 873–887.

Karr, J.R., Dudley, D.R., 1981. Ecological perspective on water quality goals. Environ. Manag. 5 (1), 55–68.

Kaur, H., et al., 2015. Advances in arsenic biosensor development—a comprehensive review. Biosens. Bioelectron. 63, 533–545.

Keimowitz, A.R., et al., 2005a. Arsenic redistribution between sediments and water near a highly contaminated source. Environ. Sci. Technol. 39 (22), 8606–8613.

Keimowitz, A.R., et al., 2005b. Naturally occurring arsenic: mobilization at a landfill in Maine and implications for remediation. Appl. Geochem. 20 (11), 1985–2002.

Keimowitz, A.R., et al., 2017. Manganese redox buffering limits arsenic release from contaminated sediments, Union Lake, New Jersey. Appl. Geochem. 77, 24–30.

Khan, B.I., et al., 2006. Release of arsenic to the environment from CCA-treated wood. 1. Leaching and speciation during service. Environ. Sci. Technol. 40 (3), 988–993.

Khan, N.I., et al., 2009. Human arsenic exposure and risk assessment at the landscape level: a review. Environ. Geochem. Health 31 (Suppl. 1), 143–166.

Klingener, N., 2017. Next Water Quality Project for Keys: Clean Up Canals. http://www.wlrn.org/post/next-water-quality-project-keys-clean-canals. [(Accessed 25 February 2019)].

Knobeloch, L., et al., 2000. Blue babies and nitrate-contaminated well water. Environ. Health Perspect. 108 (7), 675.

Kostich, M.S., et al., 2010. Predicting variability of aquatic concentrations of human pharmaceuticals. Sci. Total Environ. 408 (20), 4504–4510.

Kraus, T.E., et al., 2011. How reservoirs alter drinking water quality: organic matter sources, sinks, and transformations. Lake Reservoir Manage. 27 (3), 205–219.

Kumar, N., et al., 2018. Emerging biosensor platforms for the assessment of water-borne pathogens. Analyst 143 (2), 359–373.

Labastida, I., et al., 2013. Treatment of mining acidic leachates with indigenous limestone, Zimapan Mexico. J. Hazard. Mater. 262, 1187–1195.

REFERENCES

Labuda, J., et al., 2005. Voltammetric detection of damage to DNA by arsenic compounds at a DNA biosensor. Sensors 5 (6), 411–423.

Liu, C.-p., et al., 2010. Arsenic contamination and potential health risk implications at an abandoned tungsten mine, southern China. Environ. Pollut. 158 (3), 820–826.

Lizardi, C., 2016. Penny per test—low cost arsenic test kits. In: Abstracts of Papers of the American Chemical Society, p. 252.

Lopez de Alda, M.J., et al., 2003. Liquid chromatography–(tandem) mass spectrometry of selected emerging pollutants (steroid sex hormones, drugs and alkylphenolic surfactants) in the aquatic environment. J. Chromatogr. A 1000 (1–2), 503–526.

Lottermoser, B., 2010. Mine Wastes, Characterization, Treatment and Environmental Impacts, third ed Springer-Verlag Berlin Heidelberg, p. 400.

Lowers, H.A., et al., 2007. Arsenic incorporation into authigenic pyrite, Bengal Basin sediment, Bangladesh. Geochim. Cosmochim. Acta 71 (11), 2699–2717.

Luque de Castro, M.D., Herrera, M.C., 2003. Enzyme inhibition-based biosensors and biosensing systems: questionable analytical devices. Biosens. Bioelectron. 18 (2), 279–294.

Magalhaes, M.C.F., 2002. Arsenic. An environmental problem limited by solubility. Pure Appl. Chem. 74 (10), 1843–1850.

Mailloux, B.J., et al., 2013. Advection of surface-derived organic carbon fuels microbial reduction in Bangladesh groundwater. Proc. Natl. Acad. Sci. U. S. A. 110 (14), 5331–5335.

Mainstone, C.P., Parr, W., 2002. Phosphorus in rivers—ecology and management. Sci. Total Environ. 282, 25–47.

Male, K.B., et al., 2007. Biosensor for arsenite using arsenite oxidase and multiwalled carbon nanotube modified electrodes. Anal. Chem. 79 (20), 7831–7837.

Malitesta, C., Guascito, M.R., 2005. Heavy metal determination by biosensors based on enzyme immobilised by electropolymerisation. Biosens. Bioelectron. 20 (8), 1643–1647.

Mandal, B.K., Suzuki, K.T., 2002. Arsenic round the world: a review. Talanta 58 (1), 201–235.

Mandal, B.K., et al., 1996. Arsenic in groundwater in seven districts of West Bengal, India—the biggest arsenic calamity in the world. Curr. Sci. 70 (11), 976–986.

Matschullat, J., 2000. Arsenic in the geosphere—a review. Sci. Total Environ. 249 (1), 297–312.

Maupin, M.A., et al., 2014. Estimated Use of Water in the United States in 2010. US Geological Survey.

McArthur, J., et al., 2001. Arsenic in groundwater: testing pollution mechanisms for sedimentary aquifers in Bangladesh. Water Resour. Res. 37 (1), 109–117.

Melamed, D., 2005. Monitoring arsenic in the environment: a review of science and technologies with the potential for field measurements. Anal. Chim. Acta 532 (1), 1–13.

Mello, L.D., Kubota, L.T., 2002. Review of the use of biosensors as analytical tools in the food and drink industries. Food Chem. 77 (2), 237–256.

Merulla, D., et al., 2013. Bioreporters and biosensors for arsenic detection. Biotechnological solutions for a worldwide pollution problem. Curr. Opin. Biotechnol. 24 (3), 534–541.

Mineral Commodity Summaries, 2017. U.D.O. Interior and U.G. Survey. Reston, VA.

Mittal, A., 2009. Concentrated Animal Feeding Operations: EPA Needs More Information and a Clearly Defined Strategy to Protect Air and Water Quality From Pollutants of Concern. DIANE Publishing.

Moretto, L.M., Kalcher, K., 2014. Environmental Analysis by Electrochemical Sensors and Biosensors: Applications. Springer, p. 455.

Mulchandani, A., Rogers, K.R., Rogers, K., 1998. Enzyme and Microbial Biosensors: Techniques and Protocols. Humana Press, p. 280.

Neary, G.M.C.a.J.P., 2008. Water Quality for Ecosystem and Human Health, second ed. UNEP GEMS/Water Programme.

Nickson, R., et al., 2000. Mechanism of arsenic release to groundwater, Bangladesh and West Bengal. Appl. Geochem. 15 (4), 403–413.

Nordstrom, D.K., 2011. Hydrogeochemical processes governing the origin, transport and fate of major and trace elements from mine wastes and mineralized rock to surface waters. Appl. Geochem. 26 (11), 1777–1791.

Nordstrom, D.K., Alpers, C.N., 1999. Negative pH, efflorescent mineralogy, and consequences for environmental restoration at the Iron Mountain Superfund site, California. Proc. Natl. Acad. Sci. U. S. A. 96 (7), 3455–3462.

O'Day, P.A., et al., 2004. The influence of sulfur and iron on dissolved arsenic concentrations in the shallow subsurface under changing redox conditions. Proc. Natl. Acad. Sci. U. S. A. 101 (38), 13703–13708.

Omur-Ozbek, P., Akalp, D., Whelton, A., 2016. Tap water and indoor air contamination due to an unintentional chemical spill in source water. In: Proc. International Structural Engineering and Construction Society EURO-MED-SEC-01, May, pp. 24–29.

Ozsoz, M., et al., 2003. Electrochemical biosensor for the detection of interaction between arsenic trioxide and DNA based on guanine signal. Electroanalysis 15 (7), 613–619.

Pal, A., et al., 2014. Emerging contaminants of public health significance as water quality indicator compounds in the urban water cycle. Environ. Int. 71, 46–62.

Palmiter, R.D., 1998. The elusive function of metallothioneins. Proc. Natl. Acad. Sci. U. S. A. 95 (15), 8428–8430.

Percival, R.V., et al., 2017. Environmental Regulation: Law, Science, and Policy. Wolters Kluwer Law & Business.

Petrie, B., Barden, R., Kasprzyk-Hordern, B., 2015. A review on emerging contaminants in wastewaters and the environment: current knowledge, understudied areas and recommendations for future monitoring. Water Res. 72, 3–27.

Ping, G., et al., 2006. Sampling methods to determine the spatial gradients and flux of arsenic at a groundwater seepage zone. Environ. Toxicol. Chem. 25 (6), 1487–1495.

Queiroz, H.M., et al., 2018. The Samarco mine tailing disaster: a possible time-bomb for heavy metals contamination? Sci. Total Environ. 637-638, 498–506.

Rahman, M.M., et al., 2002. Effectiveness and reliability of arsenic field testing kits: are the million dollar screening projects effective or not? Environ. Sci. Technol. 36 (24), 5385–5394.

Ravenscroft, P., Howarth, R.J., McArthur, J.M., 2006. Comment on "limited temporal variability of arsenic concentrations in 20 wells monitored for 3 years in Araihazar, Bangladesh". Environ. Sci. Technol. 40 (5), 1716–1717.

Ricciardi, A., Rasmussen, J.B., 1999. Extinction rates of North American freshwater fauna. Conserv. Biol. 13 (5), 1220–1222.

Richardson, C.J., et al., 2007. Estimating ecological thresholds for phosphorus in the everglades. Environ. Sci. Technol. 41 (23), 8084–8091.

Rinsky, R.A., et al., 1987. Benzene and leukemia. N. Engl. J. Med. 316 (17), 1044–1050.

Rodríguez, R., Ramos, J.A., Armienta, A., 2004. Groundwater arsenic variations: the role of local geology and rainfall. Appl. Geochem. 19 (2), 245–250.

Rodriguez-Mozaz, S., et al., 2004. Biosensors for environmental monitoring of endocrine disruptors: a review article. Anal. Bioanal. Chem. 378 (3), 588–598.

Rodriguez-Mozaz, S., et al., 2005. Biosensors for environmental monitoring—a global perspective. Talanta 65 (2), 291–297.

Rogers, K.R., 1995. Biosensors for environmental applications. Biosens. Bioelectron. 10 (6), 533–541.

Rogers, K.R., Gerlach, C.L., 1996. Peer reviewed: environmental biosensors: a status report. Environ. Sci. Technol. 30 (11), 486A–491A.

Rogers, K.R., Gerlach, C.L., 1999. Peer reviewed: update on environmental biosensors. Environ. Sci. Technol. 33 (23), 500A–506A.

Rogers, K.R., Mascini, M., 1998. Biosensors for field analytical monitoring. Field Anal. Chem. Technol. 2 (6), 317–331.

Rogers, K.R., Williams, L.R., 1995. Biosensors for environmental monitoring: a regulatory perspective. Trends Anal. Chem. 14 (7), 289–294.

Roggo, C., van der Meer, J.R., 2017. Miniaturized and integrated whole cell living bacterial sensors in field applicable autonomous devices. Curr. Opin. Biotechnol. 45, 24–33.

Ruley, J.E., Rusch, K.A., 2002. An assessment of long-term post-restoration water quality trends in a shallow, subtropical, urban hypereutrophic lake. Ecol. Eng. 19 (4), 265–280.

Sampson, T., 2003. Aptamers and SELEX: the technology. World Patent Inf. 25 (2), 123–129.

Sanger, C.R., Fisher Black, O., 1908. The quantitative determination of arsenic according to the method by Gutzeit. Zeitschrift Fur Anorganische Chemie 58 (2), 121–153.

Sanllorente-Méndez, S., Domínguez-Renedo, O., Arcos-Martínez, M.J., 2010. Immobilization of acetylcholinesterase on screen-printed electrodes. Application to the determination of arsenic(III). Sensors 10 (3), 2119–2128.

Sanllorente-Méndez, S., Domínguez-Renedo, O., Arcos-Martínez, M.J., 2012. Development of acid phosphatase based amperometric biosensors for the inhibitive determination of As(V). Talanta 93, 301–306.

Sartor, J.D., Boyd, G.B., 1972. Water Pollution Aspects of Street Surface Contaminants. vol. 81. US Government Printing Office.

Schindler, D.W., 2012. The dilemma of controlling cultural eutrophication of lakes. Proc. R. Soc. B. rspb20121032.

Schwarzenbach, R.P., et al., 2010. Global water pollution and human health. Annu. Rev. Env. Resour. 35, 109–136.

Sengupta, M.K., et al., 2006. Comment on "limited temporal variability of arsenic concentrations in 20 wells monitored for 3 years in Araihazar, Bangladesh". Environ. Sci. Technol. 40 (5), 1714–1715.

Shen, S., et al., 2013. Arsenic binding to proteins. Chem. Rev. 113 (10), 7769–7792.

Siegfried, K., et al., 2012. Field testing of arsenic in groundwater samples of bangladesh using a test kit based on lyophilized bioreporter bacteria. Environ. Sci. Technol. 46 (6), 3281–3287.

Skøien, J.O., Blöschl, G., Western, A.W., 2003. Characteristic space scales and timescales in hydrology. Water Resour. Res. 39 (10), 1304.

Smedley, P.L., Kinniburgh, D.G., 2002. A review of the source, behaviour and distribution of arsenic in natural waters. Appl. Geochem. 17 (5), 517–568.

Smith, A.H., Lingas, E.O., Rahman, M., 2000. Contamination of drinking-water by arsenic in Bangladesh: a public health emergency. Bull. World Health Organ. 78 (9), 1093–1103.

Smith, J.W.N., et al., 2008. Groundwater-surface water interactions, nutrient fluxes and ecological response in river corridors: translating science into effective environmental management. Hydrol. Process. 22 (1), 151–157.

Song, L., et al., 2016. A novel biosensor based on Au@Ag core–shell nanoparticles for SERS detection of arsenic (III). Talanta 146, 285–290.

Sørensen, S.R., et al., 2007. Degradation and mineralization of nanomolar concentrations of the herbicide dichlobenil

REFERENCES

and its persistent metabolite 2, 6-dichlorobenzamide by Aminobacter spp. isolated from dichlobenil-treated soils. Appl. Environ. Microbiol. 73 (2), 399–406.

Sotiropoulou, S., Fournier, D., Chaniotakis, N.A., 2005. Genetically engineered acetylcholinesterase-based biosensor for attomolar detection of dichlorvos. Biosens. Bioelectron. 20 (11), 2347–2352.

Spear, J.M., et al., 2006. Evaluation of arsenic field test kits for drinking water analysis. J. Am. Water Works Assoc. 98 (12), 97–105.

Sracek, O., et al., 2010. Discrimination between diffuse and point sources of arsenic at Zimapan, Hidalgo state, Mexico. J. Environ. Monit. 12 (1), 329–337.

States, S., et al., 2003. Utility-based analytical methods to ensure public water supply security. J. Am. Water Works Assoc. 95 (4), 103.

Stocker, J., et al., 2003. Development of a set of simple bacterial biosensors for quantitative and rapid measurements of arsenite and arsenate in potable water. Environ. Sci. Technol. 37 (20), 4743–4750.

Stoytcheva, M., Sharkova, V., Panayotova, M., 1998a. Electrochemical approach in studying the inhibition of acetylcholinesterase by arsenate(III): analytical characterisation and application for arsenic determination. Anal. Chim. Acta 364 (1), 195–201.

Stoytcheva, M., Sharkova, V., Magnin, J.-P., 1998b. Electrochemical approach in studying the inactivation of immobilized acetylcholinesterase by arsenate(III). Electroanalysis 10 (14), 994–998.

Sutton, M.A., et al., 2013. Our Nutrient World: The Challenge to Produce More Food and Energy With Less Pollution. NERC/Centre for Ecology & Hydrology.

Tanabe, S., Ramu, K., 2012. Monitoring temporal and spatial trends of legacy and emerging contaminants in marine environment: results from the environmental specimen bank (es-BANK) of Ehime University, Japan. Mar. Pollut. Bull. 64 (7), 1459–1474.

Trang, P.T.K., et al., 2005. Bacterial bioassay for rapid and accurate analysis of arsenic in highly variable groundwater samples. Environ. Sci. Technol. 39 (19), 7625–7630.

Upadhyay, L.S.B., Kumar, N., Chauhan, S., 2018. Minireview: whole-cell, nucleotide, and enzyme inhibition-based biosensors for the determination of arsenic. Anal. Lett. 51 (9), 1265–1279.

van der Meer, J.R., 2016. Towards improved biomonitoring tools for an intensified sustainable multi-use environment. J. Microbial. Biotechnol. 9 (5), 658–665.

Van der Oost, R., Beyer, J., Vermeulen, N.P., 2003. Fish bioaccumulation and biomarkers in environmental risk assessment: a review. Environ. Toxicol. Pharmacol. 13 (2), 57–149.

van Geen, A., et al., 2005. Reliability of a commercial kit to test groundwater for arsenic in Bangladesh. Environ. Sci. Technol. 39 (1), 299–303.

Van Geen, A., et al., 2007. Monitoring 51 community wells in Araihazar, Bangladesh, for up to 5 years: implications for arsenic mitigation. J. Environ. Sci. Health A 42 (12), 1729–1740.

van Geen, A., et al., 2014. Comparison of two blanket surveys of arsenic in tubewells conducted 12 years apart in a 25 km² area of Bangladesh. Sci. Total Environ. 488-489, 484–492.

van Green, A., et al., 2003. Spatial variability of arsenic in 6000 tube wells in a 25 km² area of Bangladesh. Water Resour. Res. 39(5). https://doi.org/10.1029/2002WR001617.

Van, G.A., et al., 2002. Promotion of well-switching to mitigate the current arsenic crisis in Bangladesh. Bull. World Health Organ. 80 (9), 732–737.

Vandeweerd, V., et al., 2006. Meeting the Commitments on Oceans, Coasts, and Small Island Developing States Made at the 2002 World Summit on Sustainable Development: How Well Are We Doing? Global Forum on Oceans, Coasts, and Islands.

Verma, N., Singh, M., 2005. Biosensors for heavy metals. Biometals 18 (2), 121–129.

Vinebrooke, D.R., et al., 2004. Impacts of multiple stressors on biodiversity and ecosystem functioning: the role of species co-tolerance. Oikos 104 (3), 451–457.

Wang, J., et al., 1997. Microfabricated electrochemical sensor for the detection of radiation-induced DNA damage. Anal. Chem. 69 (7), 1457–1460.

Wang, T., et al., 2016. Laccase inhibition by arsenite/arsenate: determination of inhibition mechanism and preliminary application to a self-powered biosensor. Anal. Chem. 88 (6), 3243–3248.

Ward, M.H., et al., 1996. Drinking water nitrate and the risk of non-Hodgkin's lymphoma. Epidemiology, 465–471.

Waste, H., 2010. World Water Quality Facts and Statistics. Pacific Institute, Oakland, CA. https://www.pacinst.org/wp-content/uploads/2013/02/water_quality_facts_and_stats3.pdf. [{Accessed 25 February 2019}].

Welch, A.H., et al., 2000. Arsenic in ground water of the United States: occurrence and geochemistry. Ground Water 38 (4), 589–604.

Whelton, A.J., et al., 2015. Residential tap water contamination following the freedom industries chemical spill: perceptions, water quality, and health impacts. Environ. Sci. Technol. 49 (2), 813–823.

Whelton, A., et al., 2017. Case study: the crude MCHM chemical spill investigation and recovery in West Virginia USA. Environ. Sci.: Water Res. Technol. 3 (2), 312–332.

Willner, I., Zayats, M., 2007. Electronic aptamer-based sensors. Angew. Chem. Int. Ed. 46 (34), 6408–6418.

Wilson, I.B., Silman, I., 1977. Effects of quaternary ligands on the inhibition of acetylcholinesterase by arsenite. Biochemistry 16 (12), 2701–2708.

Wu, C.-M., Lin, L.-Y., 2004. Immobilization of metallothionein as a sensitive biosensor chip for the detection of metal ions by surface plasmon resonance. Biosens. Bioelectron. 20 (4), 864–871.

Wu, Y., et al., 2012a. Ultrasensitive aptamer biosensor for arsenic(III) detection in aqueous solution based on surfactant-induced aggregation of gold nanoparticles. Analyst 137 (18), 4171–4178.

Wu, Y., et al., 2012b. Nanoparticles assembled by aptamers and crystal violet for arsenic(iii) detection in aqueous solution based on a resonance Rayleigh scattering spectral assay. Nanoscale 4 (21), 6841–6849.

Wu, Y., et al., 2012c. Cationic polymers and aptamers mediated aggregation of gold nanoparticles for the colorimetric detection of arsenic(iii) in aqueous solution. Chem. Commun. 48 (37), 4459–4461.

Yoo, E.H., Lee, S.Y., 2010. Glucose biosensors: an overview of use in clinical practice. Sensors 10 (5), 4558–4576.

Zheng, Y., Zheng, Y., 2017. Lessons learned from arsenic mitigation among private well households. Curr. Environ. Health Rep. 4 (3), 373–382.

Ziegler, B.A., et al., 2017a. A mass balance approach to investigate arsenic cycling in a petroleum plume. Environ. Pollut. 231, 1351–1361.

Ziegler, B.A., Schreiber, M.E., Cozzarelli, I.M., 2017b. The role of alluvial aquifer sediments in attenuating a dissolved arsenic plume. J. Contam. Hydrol. 204, 90–101.

CHAPTER

12

Investigating the Missing Link: Effects of Noncompliance and Aging Private Infrastructure on Water-Quality Monitoring

Adam Cooper[a],*, Alexa Fortuna[a], Satinder Ahuja[b]

[a]Stetson University, DeLand, FL, United States [b]Ahuja Consulting for Water Quality, Calabash, NC, United States

*Corresponding author: E-mail: awcooper@ucsd.edu

1 INTRODUCTION

The Flint water crisis (FWC) changed the way our nation prioritizes and respects water quality. Before Flint, many communities were under the impression that their water was safe regardless of where they lived. They were not fully aware of the harms of having lead in their drinking water. Alas, Flint opened the door to revolutionary research and increased accountability when it came to our country's water supply. The FWC was due to the switch in water supply from Lake Huron to the Flint River, to save money. The water was not treated with an anticorrosion chemical to prevent lead particles and solubilized lead from being released from the interior of water pipes, particularly those from lead service lines or those with lead solder. Flint opened the door to deliberation of policy changes and kick-started additional attention toward the issue of lead release in the United States.

Lead poisoning is a silent killer because it is tasteless, odorless, and mostly colorless. However, the effects of lead poisoning are extremely harmful and long-lasting. Lead can cause immediate acute poisoning but the subacute, moderate, long-term exposure impact of concern in Flint is more common, and much more insidious. Levels that may be relatively harmless in a single dose can cause long-term damage if ingested regularly, which is what happens when a water source contains high levels of lead. Effects of long-term lead exposure include,

Separation Science and Technology, Volume 11
https://doi.org/10.1016/B978-0-12-815730-5.00012-0

mainly, neurological decay, which severely affects children and the elderly. This is a poor warning sign for lead exposure in a community, because any resulting behavioral disturbance or loss of intellectual function would probably not be immediately linked by their physicians or families to lead poisoning, and instead accepted as an unrelated symptom (Abadin et al., 2007). (Additionally, the adverse effects from this event may take years to surface, as most negative health effects from low-level lead exposure develop slowly (National Toxicology Program (NTP), n.d.).) Following the repercussions of the high levels of lead found in the water supply in Flint, Michigan, the need to have better control of public water supplies was highlighted. Chemists Without Borders, a nonprofit dedicated to solving humanitarian issues through the mobilization of the international chemistry community, began a project entitled US Water Quality Study, under the direction of Dr. Sut Ahuja. This project plans to monitor water quality in various counties in the United States in an effort to raise the awareness of water quality in the country. This chapter explores the first community-based project that investigated the local water supply of DeLand, FL, and discusses possible remediation strategies to overcome issues found within their sample-collection methodology.

This project sought a partnership with the community organizations within Spring Hill, a neighborhood in DeLand. Spring Hill was chosen because of a high potential that this area possesses the right criteria for having unsafe drinking water. Spring Hill and Flint have a very similar demographic makeup, as well as Spring Hill's lack of a large water management department. Additionally, Spring Hill is geographically close to Stetson University, the location of Chemists Without Borders volunteers, Adam Cooper, and Alexa Fortuna. Because of these factors, it was decided that Spring Hill would be an excellent system to investigate and to explore possible solutions to the problem of lead release on a community level.

By analyzing and understanding previous literature, it is important to note that most of the literature/studies on lead contamination in the United States are either from the late 1980s and 1990s or post 2014. One can infer that this is because research on the topic of environmental racism and injustice follows almost exclusively after a major disaster strikes. There is an apparent lack of continuous research and testing being conducted by outside organizations as well as the federal government. The recent resurgence in research into lead contamination comes from the FWC, which began in 2014. Because of this reactionary research practice, there are significant gaps and lapses in continuous research to study long-term effects of blood lead levels, how policies have affected the progress in changing lead pipes to PVC pipes in poorer communities, and tracking which homes have tested positive for lead. However, these factors do not necessarily discount previous research. Specifically, a study found a significant correlation between the number of commercial hazardous waste facilities in a zip code and the percentage of people of color in the zip code's population (United Church of Christ Commission for Racial Justice, 1987). The Environmental Justice Movement began making a solidified foundation in 1994 with an executive order by President Clinton establishing the Environmental Justice Interagency Working Group (US EPA, 2013). There is both a notable internal and external perspective from this movement. The internal perspective looks at the group from the "ground up"—from the experience of communities that struggle daily with environmental degradation and with their disenfranchisement from the institutions and structures that control their living environments. The external perspectives cast a critical eye on the political economy of environmental degradation, including the structure of environmental decision-making in disaffected communities (Cole, 1992). We believe both perspectives are crucial to understanding the scope of the problem and shaping the solutions.

2 CASE STUDY: COMMUNITY-BASED LEAD TESTING

The Safe Drinking Water Act holds local and state governments accountable for providing acceptably clean water to the nation's citizens. Currently, action levels are set at 15 ppb for lead with mandatory action required if more than 10% of households test above this level. Regarding pipes and other vessels that contain or transport water, new infrastructure has a requirement that it must be made of "lead-free" materials, i.e., less than 0.25% lead for pipes, fittings, and fixtures and 0.2% lead for solder and flux (Environmental Protection Agency, 2007).

Analyzing this data from a local perspective, we decided to look at previous literature and primary data from DeLand in Volusia County, FL. The graph below (Fig. 1) depicts lead in Volusia County drinking water. It is apparent that DeLand, in comparison to other cities in the county, has an elevated level of lead compared to its counterparts. In past testing cycles, DeLand continuously tests at the threshold or right below the 15 ppb standard set by the EPA.

In 2014, 90% of households tested under the action level for lead. This result is based on the testing of a sample comprised of 30 households out of DeLand's 13,000. There were 10% of the homes that tested over the EPA standard action level at 19, 61, and 110 ppb. For these three homes, letters were sent out informing these residents of their elevated levels of lead, recommending that they flush water in the mornings, and to encourage the use of cold water for cooking. None of these households began a discussion with DeLand utilities about replacing their infrastructure (Konoval, 2014). After the FWC occurred, there were numerous studies conducted by both public and private companies and institutions. Among those was Chemists Without Borders members who concluded that there are several warning signs we can foresee in order to prevent the "next Flint." These signs are to be cautious with low-income areas, medium-sized cities, and old infrastructure (Kirkwood, n.d.). Dated infrastructure is increasingly more likely to contain pipes that test positive for alarming levels of lead.

Due to its proximity to Stetson University and factors aligning with that of an environmental

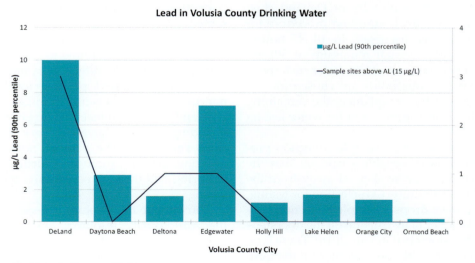

FIG. 1   Lead in Volusia County drinking water.

justice community, the Spring Hill community was used for this study. Spring Hill is a functionally segregated neighborhood found geographically in DeLand but has not been formally incorporated by the city. The community originally developed as a location for black citrus workers when DeLand was a primarily citrus-growing community. After Florida's Great Freeze in 1895, the workers were without jobs and did not have the mobility to pursue other work. During the Jim Crow era in the South, Spring Hill was formally segregated and the ramifications of this remain clear today. This is similar to the lingering effects of segregation found prominently across the United States (Gillmor and Doig, 1992). The 2016 Spring Hill Needs Assessment conducted by local Stetson University found that 89% of those that identify as Spring Hill residents are African American (Carey et al., 2016). This is in stark contrast to DeLand's 75% Caucasian population (Census Bureau, n.d.). Several factors (such as an unincorporated area next to a city, lacking sidewalks and public utilities, and designation as a food desert, etc.) led to the selection of Spring Hill as a test community for the community-based water testing (Carey et al., 2016).

In an interview with City of DeLand chief chemist Alex Konoval, sampling procedure for water testing was discussed. In all, 45 households (150% of the legal minimum of 30) are selected from a database of city water customers. Out of that, 30 households are chosen for initial testing. On noncompliance, the remaining 15 households are approached for water testing (Konoval, 2017).

Because of DeLand's sample-collection procedure, there is reasonable suspicion that the water collected may be skewed away from the Spring Hill community. On selection for testing, a city employee drops a sample-collection container and a letter of instruction on the doorsteps of selected customers (Grevatt, 2016). There is no personal contact with the customer (in the form of a phone call or in-person conversation),

besides the letter of instruction. This can lead to noncompliance in key populations of the city, including those functionally illiterate and untrusting of the government. Political trust is a key factor in this equation. Scholars have defined political trust as a basic evaluative orientation toward the government founded on how well the government is operating according to people's normative expectations. Declining political trust contributes to dissatisfaction with incumbents and institutions, creating an environment in which it is difficult for those in government to succeed. To bridge this trust gap, we focused on approaching the population of Spring Hill through our partnership with the Spring Hill Community Center, a trusted resource and functional city hall for the unincorporated neighborhood. Other community centers, such as local churches and schools, were also considered to involve community leaders with this process.

Another population that may have higher rates of noncompliance are those anxious about lowered property value. If elevated lead levels are detected, the cost of replacing private infrastructure almost always falls on the homeowner. These projects can cost anywhere from $2000 to $20,000. There is little to no support from the government to help alleviate this newly presented financial burden. Additionally, if that homeowner wants to sell or rent his or her home, the owner is legally required to get consent from tenants or a buyer that they were informed that the house has tested positive for lead. This noncompliance has been seen in DeLand in the past, and attempts at incentivizing participation, including payment, have seen very little success (Konoval, 2017). The rate of noncompliance and potential bias motivated further screening of Spring Hill through community-based water testing.

Throughout this project, it became apparent that various city resources are failing certain demographics. To address this, nearby Stetson University's Center for Community Engagement is beginning to become more and more involved in providing support and awareness in the Spring Hill community regarding environmental health. From helping create a neighborhood garden to working with centers in the community with education and donations of supplies, the university can partner with local organizations to break the "college bubble" and utilize the university's resources to help those who are lacking resources. By mobilizing local university faculty, students, and citizen scientists, we can utilize low-cost screening devices in the community. We worked closely with the Spring Hill Resource Center in collaboration with Director Shirletha Dixon. Through conversations with numerous community members, it became apparent that one of the best ways to get participation from the community for lead testing is to involve community members in the process. This encourages ownership over projects rather than the feeling of being the subjects of a scientific survey (Spring Hill Resource Center, n.d.). In collaboration with Shilretha Dixon, a flyer (Fig. 2) and a sign-up sheet were created. A total of 20 community members signed up for testing and were followed through with multiple phone calls. Eight followed through with testing. Continued testing will be conducted in future years by volunteers with Stetson's Center for Community Engagement.

3 METHODS

The eXact LEADQuick with Bluetooth Photometer developed by Industrial Test Systems was chosen for this study because of its low cost, high analytical metrics, and simple methodology. It adapts a colorimetric determination from benchtop to field analysis. The assay is conducted with sequential addition of four reagents (Industrial Test Systems, 2015).

First, a 50-mL sample is collected from a water tap. Then, 5 drops of 18% nitric acid are added. This serves to solubilize any present lead particulate matter. Of this solution, 4 mL is transferred to the spectrophotometer cell, which reduces the number of further reagents needed. Five drops of AMP/tris buffer are added to the cell to ensure an alkali environment. Then, a reagent strip is swished in solution, delivering *meso*-tetra(*N*-methyl-4-pyridyl)prophine tetratosylate (TMPYP). This porphyrin indicator forms a colorimetric complex with soluble Pb^{2+} in solution. After 1 min, the device calibrates to the solution's intensity. Then, a reagent strip containing EDTA is swished in solution. The EDTA breaks apart the colorimetric TMPYP-Pb^{2+} complex, and the difference in intensity is used to calculate the concentration of lead, using Beer's law. Unfortunately, the device manufacturers do not provide information on the exact quantity of solid reagents or information about the spectrographic cell (Industrial Test Systems, 2015).

As with any colorimetric assay, inhibiting and competing metals are the main concern when trying to avoid interference. Water exhibiting hardness above 400 ppm shows interference through inhibiting the formation of the colorimetric TMPYP-Pb^{2+} complex (Industrial Test Systems, 2015). Fortunately, Volusia County exhibits water hardness at less than 120 ppm (Shampine, 1975). Two metal contaminants of note, mercury and cadmium, show interference through forming similar colorimetric complexes with TMPYP. The current methodology has a tolerance of concentrations of 70 ppb for cadmium and 10 ppb for mercury. However, these values are rarely seen within safe water systems, and interference would lead to high results during screening tests, thus highlighting them for further analysis (Industrial Test Systems, 2015).

Industrial Test Systems promotes very good analytical metrics. The LEADQuick device is advertised to have a detection limit of 3 ppb with an upper threshold of 500 ppb. Accuracy is marketed to be 3 ppb for lower values and 6% for larger values. The device gives readouts in 1 ppb steps (Industrial Test Systems, 2015). To assure the quality of data collected, the device was tested against prepared standards at Stetson.

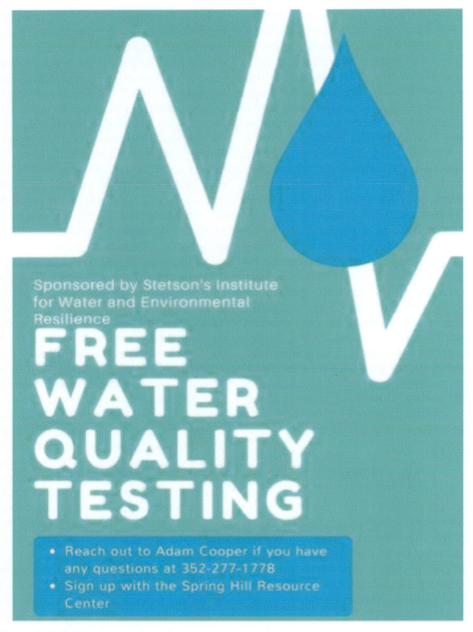

FIG. 2 Water screening flyer.

Solutions of lead were prepared through the dissolution of lead nitrate in deionized water, followed by serial dilution to arrive at desired concentrations. The methodology discussed earlier for LEADQuick was followed, with one modification: The step including addition of nitric acid was eliminated. This is due to the effect of nitric acid's causing lead nitrate to precipitate out of solution (Ostanova et al., 2002). The previously discussed procedure was followed, beginning with the addition of buffer to the system.

4 RESULTS AND DISCUSSION

First, commonly seen concentrations of lead above its action level were tested from 12.5 to 100 ppb (Fig. 3). The device performed admirably, with a relationship of 99.29% of expected concentration and a coefficient of determination of 0.9992. Then, levels of lead seen within EPA-compliant water samples were tested, ranging from 5 to 20 ppb (Fig. 4). The apparatus lost some accuracy, with a relationship of 100.67% of expected concentration and a coefficient of determination of 0.9313. The analytical metrics were deemed suitable enough for a field-based screening campaign, and so the next steps of community testing were followed.

All eight samples tested below the LEADQuick device's threshold of 3 ppb and thus gave an "LO" reading. This is not unexpected, as 26/30 samples collected in the 2014 DeLand survey were similarly below this threshold (Konoval, 2014). A map of sample sites can be seen in Fig. 5. Four residential locations were tested, as well as two after-school programs (the Boys & Girls Club, as well as the Chisholm Center) and two community centers (the Spring Hill Community Garden and the Spring Hill Resource Center). Samples were taken from either kitchen sinks at residential locations or at water fountains, as well as sinks at community centers.

Overall, our suspicions of elevated levels of lead in the Spring Hill community are luckily not supported by the data gathered. Further information will be collected in future years by other volunteers and interested Stetson chemistry students. The goal of this project was to establish a continued partnership with the Spring Hill Resource Center that will hopefully outlast this senior research project.

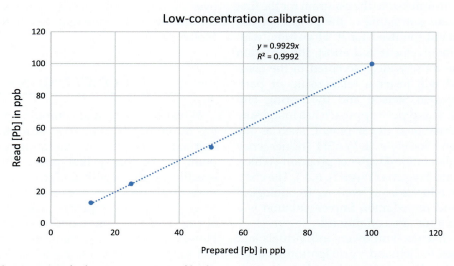

FIG. 3 Calibration curve for low concentrations of lead.

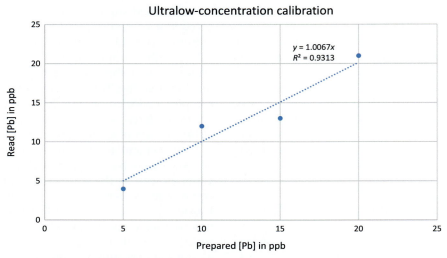

FIG. 4 Calibration curve for ultralow concentrations of lead.

Lead release is a community problem as much as it is a chemistry and policy problem. Working within a community is important to designing mitigation strategies. However, it is important to note that there was a relatively low involvement rate for the free water testing that was offered. This may very well be because citizens were being introduced to the program for the first time and were untrusting of the process or did not understand the need for it. Hopefully, with continued involvement and effort, the program will become more widely accepted in the community and will foster a relationship or partnership between Stetson and the Spring Hill community. This type of approach has found success in international development, such as the Himmotthan Society's E-WASH (Emergency Water and Sanitation-Hygiene) program (Water, Sanitation and Hygiene (WASH), n.d.). The combination of education as well as technological sampling and remediation implementation is a powerful tool for any community.

Part of what also empowers individuals and communities to demand participation in decisions that fundamentally affect their lives is the realization that power relationships within a decision-making structure are fluid and open to debate. Segmentation of housing markets, spatial mismatch of labor markets, and decentralization of metropolitan governance contribute to unequal access to economic opportunities, services, and the fragmentation of local control over land use and zoning in ways that affect community environmental health (Keister and Moller, 2000).

5 CONCLUSIONS

When analyzing and compiling possible policy recommendations, it is important to note that almost all previous literature and research point to the theory that poor communities—specifically those with a high concentration of minorities—are most at risk for occurrence of environmental injustice. The case study of Spring Hill was conducted to test this theory. However, our testing came up with negative test results. What does this tell us? The theory does not match the practice? Previous research

5 CONCLUSIONS

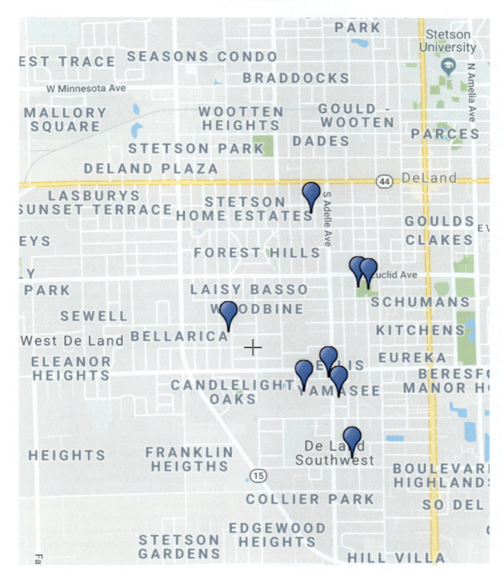

FIG. 5 Map of testing locations.

also pays most attention to major system failures, but not regulatory failures that occur on a local scale. Another consideration is that perhaps the threshold for required reporting lead levels (90% of consumer's water testing above 15 ppb) is too high. This threshold offers that a noticeably high level of lead can be present in drinking water before action must be taken. Shouldn't action be taken if there is any lead present? These unanswered questions lead us to offer the following policy recommendations:

(1) It would be advantageous if testing was conducted by a contracted private agency to the government. This would relieve the government of additional pressures while providing a third-party company the benefit of being unbiased.

(2) Lower the threshold for mandatory reporting on lead-level results. The current action level is a noticeably high amount of lead, allowing one-tenth of a population to be exposed to dangerous amounts. For the health and safety of citizens, this should be reconsidered.

(3) Continue with regulated testing more regularly. As prices for analytical methods fall and with the advent of affordable screening technologies, more thorough analysis in a community can be performed.

Throughout the course of our research and field-testing for this project, it became apparent that there are numerous stakeholders invested in this topic from numerous perspectives. Stakeholders are the backbone of a cause, and without it, action would be much more difficult, if not impossible, to achieve. Fortunately, now more than ever, we are seeing organizations and people come together to support the idea of providing safe, clean water to all people in our communities. Some front-runners for the cause are nonprofit organizations that make it their purpose and responsibility to connect communities with these resources. The Spring Hill Community Center, whom we partnered with on this project, provides the main representation for Spring Hill. Across the county, communities make requests to the Volusia County Council, which coordinates water distribution. Stetson University's Institute for Water and Environmental Resiliency is invested in the water quality of the community. In 2017, they hosted a panel regarding water quality in Volusia County. They hold an annual Water Summit, attended by elected officials from across the county. These meetings serve to increase dialogue on higher levels, which should result in further action being taken in underserved communities.

As citizens of our beloved planet, we should promote the idea: Everyone has a right to clean and safe drinking water. In 2010, the United Nations recognized "the right to safe and clean drinking water and sanitation as a human right that is essential for the full enjoyment of life and all human rights" (General Assembly Resolution 64/292, n.d.). Comprehensive and sound water monitoring is essential to assure the quality of water delivered in our nation and across the world.

Acknowledgments

It was a pleasure to conduct this research to provide an example for other similar projects that can be conducted by volunteers worldwide for Chemists Without Borders. We would like to thank Dr. Ramee Indralingam and Kevin Winchell for support and direction with the community testing project, collaborating organizations (Spring Hill Resource Center, Center for Community Engagement and Bonner Program), and the Stetson Institute for Water and Environmental Resilience for financial support.

References

Abadin, H., Ashizawa, A., Stevens, Y.-W., Llados, F., Diamond, G., Sage, G., Citra, M., Quinones, A., Bosch, S.J., Swarts, S.G., 2007. Toxicological Profile for Lead, Agency for Toxic Substances and Disease Registry (ATSDR) Toxicological Profiles. Agency for Toxic Substances and Disease Registry (US), Atlanta (GA).

Carey, E., Fernandez, G., Pollack, J., Schaefer, E., 2016. Public Health and Community Needs Assessment Report of Spring Hill Community 2015–2016.

Cole, L.W., 1992. Empowerment as the key to environmental protection: the need for environmental poverty law. Ecol. Law Q. 19, 619–683.

Environmental Protection Agency, 2007. Lead and Copper Rule.

General Assembly Resolution 64/292, The Human Right to Water and Sanitation, A/RES/64/292 (28 July 2010)

Gillmor, D., Doig, S.K., 1992. Segregation forever? Am. Demogr. 14, 48.

Grevatt, P., 2016. Clarification of Recommended Tap Sampling Procedures for Purposes of the Lead and Copper Rule (Memorandum). US EPA.

Industrial Test Systems, 2015. eXact LEADQuick Manual.

REFERENCES

Keister, L.A., Moller, S., 2000. Wealth inequality in the United States. Annu. Rev. Sociol. 26, 63–81. https://doi.org/10.1146/annurev.soc.26.1.63.

Kirkwood, Julie. "Preventing the Next Flint" (AACC. Nov. 1, 2016).

Konoval, A., 2014. Certification of Notification of Lead and Copper Tap Sample Results. Florida DEP. Print.

Konoval, A., 2017. Interview With DeLand Chief Chemist.

National Toxicology Program (NTP), NPT Monograph on Health Effects of Low-Level Lead. Available online, http://ntp.niehs.nih.gov/ntp/ohat/lead/final/monographhealtheffectslowlevellead_newissn_508.pdf.

Ostanova, S.V., Chubarov, A.V., Drozdov, S.V., Patrushev, V.V., Tatarenko, A.A., 2002. Solubility of Lead Nitrate in Aqueous Solutions of Nitric Acid and Zinc Nitrate 2.

Shampine, W.J., 1975. Hardness of Water From the Upper Part of the Floridan Aquifer in Florida [Cartographic Material]/by William J. Shampine; Prepared by United States Geological Survey in Cooperation With the Bureau of Geology. Florida Department of Natural Resources.

Spring Hill Resource Center [WWW Document], n.d. https://www.deland.org/departments/spring-hill-resource-center (accessed 4.15.18).

U.S. Census Bureau Quick Facts: DeLand City, Florida [WWW Document], n.d. https://www.census.gov/quickfacts/fact/table/delandcityflorida/PST045216 (accessed 2.7.18).

United Church of Christ Commission for Racial Justice, 1987. Toxic Wastes and Race in the United States.

US EPA, 2013. Summary of Executive Order 12898—Federal Actions to Address Environmental Justice in Minority Populations and Low-Income Populations [WWW Document]. US EPA. https://www.epa.gov/laws-regulations/summary-executive-order-12898-federal-actions-address-environmental-justice. [(Accessed February 2019)].

Water, Sanitation and Hygiene (WASH) [WWW Document], n.d.. Himmotthan. http://www.himmotthan.org/wash/whome.aspx (accessed 8.18.18).

CHAPTER 13

GenX Contamination of the Cape Fear River, North Carolina: Analytical Environmental Chemistry Uncovers Multiple System Failures

Lawrence B. Cahoon*

Department of Biology and Marine Biology, University of North Carolina Wilmington, Wilmington, NC, United States

*Corresponding author: E-mail: cahoon@uncw.edu

1 INTRODUCTION

Emerging contaminants (i.e., newly produced compounds, contaminants newly recognized in environmental matrices, and compounds newly understood to pose hazards) present major scientific and regulatory challenges to modern societies. More than 80,000 commercial compounds are now registered in the United States (https://ntp.niehs.nih.gov/about/index.html), but this figure does not include by-products that also derived from the synthesis and degradation of are commercial products (the Chemical Abstract Services [CAS] registry lists over 142 million organic and inorganic substances; http://support.cas.org/content/chemical-substances).

A proposal to produce a chemical for commercial purposes in the United States triggers the provisions of the Toxic Substances Control Act (TSCA) of 1976 and its follow-on legislation, the Frank R. Lautenberg Chemical Safety for the 21st-Century Act (2016). The Registration, Evaluation, Authorisation and Restriction of Chemicals (REACH) agreements apply in the European Union (REACH, 2006). These regulatory frameworks require some degree of chemical safety evaluation for the products themselves, but the by-products typically do not trigger such evaluations, as they are pathway-dependent and often kept secret as confidential business information. Toxicological research has so far generated a very short list of compounds considered sufficiently dangerous to warrant outright bans (e.g., dioxins) or routine monitoring and strict discharge limits under the US Clean Water Act. Consequently, the technological ability to create new chemicals far outstrips the ability to evaluate their potential for harm.

2 PFASs AS EMERGING CONTAMINANTS

Perfluoro- and polyfluoro-alkylated substances (PFASs) are one major group of emerging contaminants, although produced commercially for decades, partly because more uses for a broader array of these compounds have been developed relatively recently, and partly because newer compounds have been produced as others were phased out owing to toxicological concerns. The very strong C–F bonds in PFASs make these compounds highly inert and resistant to physical, chemical, and biological degradation. As a direct consequence PFASs have found wide application for many commercial uses, including nonstick coatings, nonconducting materials for electronics fabrication, fire-fighting foams, etc.

The particular chemical properties of PFASs make them an interesting group of emerging contaminants. The relatively unreactive chemistry, diversity of uses, and broad distribution of low concentrations of PFASs make them challenging to separate from environmental matrices, detect, identify, and quantify (van Leeuwen et al., 2006). PFASs degrade slowly in the environment, although this aspect of their behavior is poorly studied, so low levels of them appear ubiquitously (Lindstrom et al., 2011; Prevedouros et al., 2006). Considerable analysis remains to be done, but toxicological studies have found evidence that some PFASs are biologically problematic in various ways at environmentally relevant concentrations and for exposed populations (ATSDR, 2018; Lindstrom et al., 2011). Consequently, interest in analyses of these compounds in the environment and in human subjects is rising.

"GenX" is a trade name for a commercial PFAS (also termed "C3 dimer acid," perfluoro-2-propoxypropionic acid, and other names; CAS# 13252-13-6) produced as an emulsifying agent for production of nonstick coatings, e.g., Teflon, and is generated by hydration of "C3 dimer acid fluoride" (HFPO dimer, acid fluoride, CAS# 2062-98-8; DuPont, 2010). GenX is also created as a by-product of other PFAS syntheses, notably for polyvinyl fluorides (Chemours, 2017).

Commercial production of GenX was intended to replace a PFAS compound previously used in Teflon synthesis, perfluorooctanoic acid (PFOA, also commonly known as "C8," CAS# 335-67-1). Epidemiological studies revealed that PFOA levels in the bodies of people employed in its manufacture, living in communities adjacent to manufacturing sites, and who consumed drinking water contaminated by C8 were high enough to cause a variety of adverse health effects, leading to regulatory action at a global scale (Blum et al., 2015; Lindstrom et al., 2011; Scheringer et al., 2014; Wang et al., 2013). Significant litigation has also resulted: https://www.bloomberg.com/news/articles/2018-02-20/3m-is-said-to-settle-minnesota-lawsuit-for-up-to-1-billion; http://www.theintell.com/8dff6946-f1fc-11e6-91df-db15dd726e10.html. Consequently, C8 production has been phased out, although detectable levels of this "legacy" PFAS remain in many environmental matrices (ATSDR, 2018; Lau et al., 2007; Mudumbi et al., 2017). A related compound, perfluorooctane sulfonic acid (PFOS, CAS# 1763-23-1), has been widely used in fire-fighting foams, and has also been implicated in adverse health effects. Both compounds have now been placed under EPA drinking water health advisories (https://www.epa.gov/sites/production/files/2016-06/documents/drinkingwaterhealthadvisories_pfoa_pfos_updated_5.31.16.pdf) and have been phased out of production and import in the United States (ATSDR, 2018).

GenX is currently produced commercially by Chemours (The Chemours Company FC, LLC) at the Fayetteville Works plant downstream of Fayetteville, North Carolina, USA, along the banks of the Cape Fear River (Fig. 1), and used in production processes at a Chemours facility

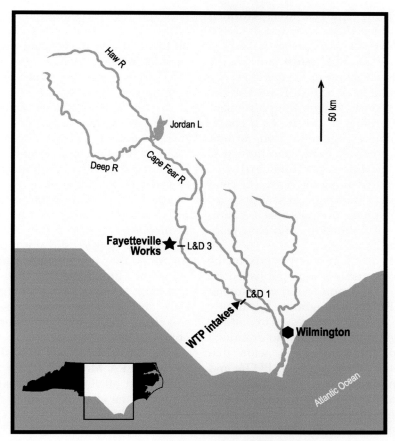

FIG. 1 Map of central North Carolina illustrating major features of the Cape Fear River watershed. "L&D 1" and "L&D 3" are Lock & Dam 1, just above which are intakes for water treatment plants serving the lower Cape Fear region, and Lock & Dam 3, just above which is located the Chemours discharge (outfall # 002) from Fayetteville Works, respectively.

in Dordrecht, The Netherlands. The Fayetteville Works facility was originally constructed and operated by I.E. DuPont de Nemours & Co. in the 1970s; in 2015, Chemours became a separate company headquartered in Wilmington, DE, and took over management of the Fayetteville Works, which includes facilities still operated by DuPont (now DowDuPont) and by Kuraray America, Inc. The plant discharges wastewater into the Cape Fear River under NPDES Permit #NC0003573 through an outfall (#002) that combines noncontact cooling water plus various process and domestic wastewater discharges from outfall #001.

3 PFAS ANALYSES IN THE CAPE FEAR RIVER

The Cape Fear River watershed is the largest river basin wholly contained within the state of North Carolina and has an area of 23,734 km^2

(9164 mi^2) (Fig. 1). The Cape Fear River is formed by the juncture of the Haw River and the Deep River, just below B. Everett Jordan Lake, and has a series of locks and dams built in the 1920s to facilitate shipping to Fayetteville. The river exits through the Cape Fear River Estuary directly to the Atlantic Ocean, the only major river in North Carolina to do so. As a consequence of the river's high flow volume and direct exit to the ocean, numerous industrial dischargers, in addition to municipal dischargers, were located along the river. The river is also the primary drinking water supply for numerous communities along its reach, culminating in the city of Wilmington and New Hanover County near its mouth.

The first study of PFAS compounds in the Cape Fear River watershed was published by Nakayama et al. (2007), and found widespread distribution of PFASs throughout, apparently deriving from diverse sources. This study focused on PFOA and PFOS, but also examined several other perfluorocarboxylic acids that were identifiable in their protocol. PFOA and PFOS had by this time become of substantial concern owing to their toxicity to humans and wildlife and their widespread use and presence in the environment (Scott et al., 2006; Yamashita et al., 2005). For example, DuPont's PFOA production at its Parkersburg, WV plant had caused widespread contamination of drinking water supplies, ultimately leading to regulatory action by US EPA (US EPA, 2006, 2010/2015). Notably, GenX and similar compounds (perfluoroalkyl ether carboxylic acids, PFECAs) were not reported by Nakayama et al. (2007).

Nakayama et al. (2007) presented a well-defined laboratory procedure for extraction, concentration, identification, and quantification of PFASs, which had been a significant challenge for environmental studies of these compounds (van Leeuwen et al., 2006). In addition to rigorous QA/QC protocols, e.g., isotopically labeled internal standards, field blanks, chromatography procedures, Nakayama et al. (2007) used positive pressure pumping through an HLB SPE cartridge and a Fluofix-II analytical HPLC column to concentrate organic compounds from relatively large volumes of prefiltered water and to separate other organic compounds from target PFASs, respectively. They were able to achieve limits of quantitation (LOQs) of 1 ng/L = 1 part per trillion (ppt) and thereby detected PFOA and PFOS in a large majority of samples using a combination of HPLC separation and triple quadrupole mass spectrometry in the electro-spray ionization mode (ESI-LC/MS) for targeted analyses. One important limitation in this study was the availability of isotopically labeled standard PFASs; the authors did not report compounds for which they had no such standards. Another significant limitation was that TQMS (tandem quadrupole mass spectrometry) cannot provide monoisotopic mass resolution and is therefore inadequate to identify nontarget compounds.

The US EPA (2009) published a standard method (EPA Method 537) for analysis of perfluorinated alkyl acids (PFAAs) that was very similar to the methods used by Nakayama et al. (2007) in its use of solid phase extraction, liquid chromatography and TQMS. The limitations of the method are similar, and it focused on relatively widely distributed, well known compounds that had been produced for many years, such as PFOA and PFOS, now collectively termed "legacy" or "traditional" PFAS compounds.

Subsequent studies of PFASs elsewhere and in the Cape Fear region employed additional advances in the methodologies used to detect, identify, and quantify these myriad compounds, yielding additional insights into PFAS contamination. Strynar et al. (2015) followed the sample collection, preparation, extraction, and chromatography protocols of Nakayama et al. (2007), and used TQMS analyses of traditional PFAS compounds to identify samples with relatively high overall PFAS concentrations. They then used high-resolution time-of-flight mass spectrometry (TOFMS) to obtain exact masses, eliminate known compounds or compounds of no investigative interest, and select compounds showing negative mass defects (monoisotopic mass < nominal mass, a common feature of poly- and perfluorinated compounds), whose mass spectra were then further investigated. Unknown compounds were identified based on exact masses, molecular formula generation, and natural isotope abundances and distributions.

4 DISCOVERY OF GenX IN THE CAPE FEAR RIVER

Strynar et al. (2015) reported on the discovery of a dozen novel PFECAs and perfluoroalky ether sulfonic acids (PFESAs; Table 1). GenX was among these, along with a range of 3- to 8-carbon mono- and poly-ether PFASs. The geographic sampling scheme used in this study focused on the DuPont/Chemours fluorocarbon

TABLE 1 PFASs Reported in the Cape Fear River Watershed

Class	Formula	CAS#	Name	Water	Sediment
Monoether PFECAs					
	$C_3HF_5O_3$	674-13-5	Perfluoro-2-methoxyacetic acid	X[1,2]	X[3]
	$C_4HF_7O_3$	377-73-1	Perfluoro-3-methoxypropanoic acid	X[1,2]	X[3]
	$C_5HF_9O_3$	863090-89-5	Perfluoro-4-methoxybutanoic acid	X[1,2]	
"GenX"	$C_6HF_{11}O_3$	13252-13-6	Perfluoro-2-propoxypropanoic acid	X[1,2,3]	
	$C_7HF_{13}O_3$	801212-59-9	Perfluoro-4-isopropoxybutanoic acid	X[1,2]	
	$C_8HF_{15}O_3$	919005-55-3	Perfluoro-7-oxanonanoic acid	X[1,2]	
Polyether PFECAs					
	$C_7HF_{13}O_7$	39492-91-6	Perfluoro-3,5,7,9,11-pentaoxadodecanoic acid	X[1,2]	
	$C_6HF_{11}O_6$	39492-90-5	Perfluoro-3,5,7,9-butaoxadecanoic acid	X[1,2]	X[3]
	$C_5HF_9O_5$	39492-89-2	Perfluoro-3,5,7-propaoxaoctanoic acid	X[1,2]	
	$C_4HF_7O_4$	39492-88-1	Perfluoro-3,5-dioxahexanoic acid	X[1,2]	X[3]
PFESAs					
Nafion by-product #1	$C_7HF_{13}O_5S$	29311-67-9	Perfluoro-3,6-dioxa-4-methyl-7-octene-1-sulfonic acid	X[1]	
Nafion by-product #2	$C_7H_2F_{14}O_5S$	749836-20-2	2-[1-[difluoro(1,2,2,2-tetrafluoroethoxy)methyl]-1,2,2,2-tetrafluoroethoxy]-1,1,2,2-tetrafluoro-ethanesulfonic acid	X[1]	X[3]
"Legacy" PFASs					
PFOA, "C8"	$C_8HF_{15}O_2$	335-67-1	Perfluorooctanoic acid	X[1,2]	
PFOS	$C_8HF_{17}SO_3$	1763-23-1	Perfluorooctane sulfonic acid	X[1,2]	
PFHxA	$C_6HF_{11}O_2$	307-24-4	Perfluorohexanoic acid	X[1,2]	

Sources: [1]Strynar et al. (2015); [2]Sun et al. (2016); [3]UNCW (2018).

EVALUATING WATER QUALITY TO PREVENT FUTURE DISASTERS

production facilities at Fayetteville Works, and demonstrated that this facility was the most likely source of these novel compounds, as opposed to the broader distributions of legacy compounds and likely sources reported by Nakayama et al. (2007). Strynar et al. (2015) questioned the lability of these novel compounds in the environment and the ability of conventional drinking water treatment systems to remove them, suggesting that they could be highly persistent and refractory to conventional treatment methods.

Sun et al. (2016) further examined the presence of legacy PFASs, PFECAs, and PFESAs in Cape Fear River water, and quantified the concentrations and loadings of PFOA, PFOS, and GenX (for which pure standards were available), among other PFASs (Table 1). They also demonstrated that these compounds were not removed by drinking water treatment processes used by the Cape Fear Public Utility Authority's (CFPUA) Sweeney Water Treatment Plant (WTP) in Wilmington, NC. The Sweeney WTP had been upgraded as of 2012 and used a combination of prefiltration, ozonation, ultraviolet irradiation, and chlorination to meet all federal drinking water treatment standards in effect at the time. Sun et al. (2016) reported that GenX concentrations in finished drinking water exceeded 400 ng/L (parts per trillion, ppt). Other PFECA compounds for which pure standards were not then available were also present in finished drinking water at concentrations likely much higher than those reported for GenX. Average daily mass flux of GenX in the Cape Fear River was estimated at 5.9 kg (range: 0.6–24 kg/day) at the water supply intake for the Sweeney WTP, approximately 40 km upstream of Wilmington (Fig. 1) and 70 km downstream of the Fayetteville Works outfall (Sun et al., 2016).

Concern for public health driven by discovery of diverse PFASs in finished drinking water prompted efforts to alert local and state authorities. The CFPUA was a participant in the study by Sun et al. (2016), but its senior administration and board apparently did not fully grasp the importance of the findings. Direct contact with the North Carolina Department of Environmental Quality's (NC DEQ) senior administration in November 2016 yielded no apparent response. No public or regulatory response to the study by Sun et al. (2016) occurred until their findings were headlined in the *Wilmington Star-News*, "Toxin Taints CFPUA Drinking Water," on June 8, 2017. A series of print and broadcast media coverage of the "GenX story" began after that, raising substantial public visibility and concern.

It is important to qualify this narrative throughout as an interpretation of the information available and not as a legal opinion or finding; such considerations are the subject of ongoing investigation and litigation. Investigations by the NC DEQ, US EPA, the NC State Bureau of Investigation (NC SBI), and the US Attorney for the Eastern District of North Carolina commenced very quickly after the story broke publicly. Several civil actions have also been initiated against Chemours/DowDuPont.

Chemours representatives met with local officials in Wilmington on June 15, 2017, and admitted that GenX was also produced as a by-product of the polyvinyl fluoride synthesis process and that it was discharged into the river (Chemours, 2017). Chemours corporate headquarters in Wilmington, DE, promised on June 22, 2017, to stop all GenX discharges immediately by trucking process wastewater to hazardous waste treatment facilities, as they had apparently done with AFPO process wastewater as of 2003, but ensuing investigations yielded evidence of substantial concentrations in the river, indicating ongoing discharges, including GenX.

Public disclosure of the presence of GenX and other PFASs in the drinking water supplies of communities downstream of Fayetteville Works led immediately to efforts to measure concentrations of these compounds in close to real time. Sampling efforts were mounted by NC DEQ, the

CFPUA, Dr. Detlef Knappe's laboratory at North Carolina State University, and the University of North Carolina Wilmington (UNCW), the latter under contract with the CFPUA and later with state funding under NC House Bill 56. Analyses of GenX and other PFASs were performed by Dr. Mark Strynar's laboratory at the EPA facility in Research Triangle Park, NC, as well as by EPA Region IV's laboratory in Athens, GA, and by researchers in the Department of Chemistry and Biochemistry at UNCW, using in-house high-resolution mass spectrometry, as well as by one or more private, certified laboratories using EPA Method 537 and pure GenX standards.

Analyses of GenX and nontarget PFASs in the Chemours wastewater discharge and in drinking water during the summer of 2017 employed high-resolution mass spectrometry, coupling liquid chromatographic separation with quadrupole time of flight mass spectrometry (LC/QTOFMS). GenX concentrations in Chemours' discharge (outfall 002, where treated process wastewater from outfall #001 had been diluted $\sim20\times$ with noncontact cooling water) and at the Sweeney WTP (finished drinking water) in Wilmington, NC, were 21,760 and 726 ng/L, respectively, in the first samples collected for this surveillance effort (US EPA, 2017). For reference, the NC Department of Health and Human Services (NC DHHS) established a "health goal" for drinking water of 140 ng/L (ppt) in July 2017 (https://ncdenr.s3.amazonaws.com/s3fs-public/GenX/NC%20DHHS%20Risk%20Assessment%20FAQ%20Final%20Clean%2007417%20PM.pdf). By the 6th week of the summer sampling effort, GenX concentrations had dropped to 102 and 70 ng/L at the discharge point and WTP, respectively. Estimates of the concentrations of most other PFASs similarly showed very large declines, with the exceptions of two PFESA compounds identified as "Nafion by-products 1 and 2" (Table 1). These two compounds apparently entered the discharge stream from Chemours' Nafion production facility. NC DEQ subsequently requested that Chemours cease discharge of these two compounds as well (https://edocs.deq.nc.gov/WaterResources/0/doc/611860/Page1.aspx).

Scientists in the Marine and Atmospheric Chemistry Research Laboratory (MACRL; Drs. Avery, Kieber, Mead, and Skrabal) at UNCW also discovered GenX in rainwater and determined that it was likely present in that matrix owing to atmospheric releases of HFPO dimer acid fluoride, which would have hydrated rapidly in the atmosphere by reacting with water vapor to form the carboxylic acid product, GenX (UNCW, 2018). Subsequent investigations by the NC DEQ confirmed this finding and determined that air releases were likely a significant source of GenX in the air shed around Fayetteville Works (~1000 kg year^{-1}). Chemours as of May 2018 has committed to significant expenditures and installation of advanced technology to control these emissions (Otterbourg, 2018).

MACRL investigators developed techniques for extracting, identifying, and analyzing GenX and similar PFASs from sediments in the Cape Fear River using high-resolution LC/QTOFMS (UNCW, 2018). In addition to quantifying GenX concentrations in sediments, they identified several of the same compounds reported by Strynar et al. (2015) and Sun et al. (2016) (Table 1). Some compounds reported in high quantities in river water by Sun et al. (2016), however, were not identified in the sediment samples, notably perfluoro-3,5,7-trioxa-octanoic acid and perfluorohexanoic acid (Table 1). These findings indicated that some number of PFASs could accumulate in sediments, forming a reservoir from which desorption processes could maintain long-term releases into the water column, but that not all PFASs may behave similarly in the environment.

MACRL investigators also reported discovery of two other compounds previously unreported in the literature that appeared to be perfluorinated and to contain chlorine (UNCW, 2018).

5 PFAS DISCHARGES FROM FAYETTEVILLE WORKS

Further work to identify those compounds was ongoing at this writing.

It is instructive to consider the publicly visible timeline of PFAS production and discharges at Fayetteville Works to understand how critical the development of new analytical techniques for PFASs in the environment was for the discovery of GenX in the Cape Fear River and in drinking water. The timeline that follows is based on examination of the NPDES Wastewater permit file for the Fayetteville Works facility, available at: https://edocs.deq.nc.gov/WaterResources/Browse.aspx?startid=589912&cr=1. Additional information helping inform this timeline comes from public statements by Chemours officials and from publicly available NC DEQ documents (all available on request).

Fluorocarbon production at Fayetteville Works dates back approximately 45 years, as the plant was constructed and began operating in the early 1970s. DuPont began producing poly-vinyl fluoride in 1980, which generated GenX as a by-product that was discharged in the plant's waste stream to the Cape Fear River (Chemours, 2017). DuPont began producing AFPO (ammonium perfluorooctanoate, CAS# 3825-26-1, the ammonium salt of PFOA) in November 2002. DuPont apparently trucked its AFPO wastewater to hazardous waste facilities. A subsequent renewal of DuPont's NPDES permit in January 2004, by the NC Department of Environment and Natural Resources (NC DENR, now NC DEQ) would have allowed discharge of AFPO wastewater without any monitoring requirement, but DuPont requested and was granted a revision in February 2004, which stipulated no AFPO wastewater discharge to the river. DuPont reported finding AFPO in groundwater at the plant in 2003 to NC DENR, but no regulatory response occurred. At this time there was no EPA-approved certifiable, standard method for quantification of AFPO or other PFASs in surface waters; EPA Method 537 was adopted in 2009 (US EPA, 2009). Owing to increasing health concerns about AFPO/PFOA, the US EPA reached an agreement in 2006 with chemical companies to phase out PFOA production (US EPA, 2010/15). DuPont's 2007 permit renewal specified sampling of both influent and effluent water for PFOA, but did not specify discharge limits (https://edocs.deq.nc.gov/WaterResources/0/doc/591737/Page1.aspx). DuPont ceased AFPO production at Fayetteville Works in April 2013 (https://edocs.deq.nc.gov/WaterResources/0/doc/591822/Page1.aspx).

DuPont sought and was granted consent in 2009 under the TSCA to produce GenX as a replacement product for AFPO at Fayetteville Works, with the stipulation that no discharge of GenX to surface water was allowed, and began production shortly thereafter (http://1lbxcx1bcuig1rfxaq3rd6w9.wpengine.netdna-cdn.com/wp-content/uploads/2017/07/EPA_DuPont_ConsentOrder.pdf). DuPont's NPDES permits in 2004, 2007, and 2012 (in force at this writing), did not list GenX nor "C3 dimer acid fluoride" as constituents of the plant's permitted discharge to the river. C3 dimer acid fluoride hydrolyzes rapidly in contact with water to form the carboxylic acid form, GenX, so it would have appeared at the plant's outfall as GenX. The sampling program by Strynar et al. (2015) that first identified GenX in the Cape Fear River collected samples in 2012. Sampling by Sun et al. (2016) in 2013 and 2014 detected GenX and other PFASs in the river and in finished drinking water downstream of DuPont's discharge.

6 HOW DID THE GenX PROBLEM HAPPEN?

The "discovery" of GenX and related PFASs in a public drinking water supply begs the obvious question about how this happened despite laws and regulations intended to protect the public. The water-consuming public was, of course, the last party to know about the presence of potentially toxic PFAS compounds in their drinking water. The public might never have known if Sun et al. (2016) and Strynar et al. (2015) had not directly disclosed their findings to the public through the media. Public drinking water utilities in the lower Cape Fear River watershed knew something about these findings, having collaborated initially in the study by Sun et al. (2016), but apparently did not grasp their full importance, were unsure of any easily accessible and affordable treatment options, and had no regulatory authority. Ultimate responsibility for the safety of public drinking water supplies rests with the regulatory agency with authority to enforce the Clean Water Act and the Safe Drinking Water Act, in this case NC DENR/DEQ, and with the discharger of any potentially toxic pollutants, DuPont/Chemours.

NC DENR/DEQ had/has full-delegated authority to implement the provisions of the Clean Water Act, including issuance of permits, regulation, and monitoring of discharges, and enforcement actions to secure compliance. The US EPA has oversight authority, but generally acts in consultation with NC DEQ. The difficulties posed by emerging contaminants, however, expose critical weaknesses in this regulatory scheme.

First, discharge permit applications require the good faith disclosure of the constituents expected to be present in the proposed discharges. The composition of industrial discharges can include numerous compounds at very low concentrations in which case a full listing of such constituents would be burdensome and could disclose confidential business information (synthesis pathways, etc.), considerations that must be balanced against the need to protect the public from toxic discharges. Accordingly, the CWA requires disclosure of the presence of any toxic substance believed to be present frequently or routinely in a discharge at concentrations above 100 μg/L, or 100 ppb, and of toxic substances infrequently present

within discharge water at concentrations above 500 μg/L. There is a clear presumption that a discharger has a duty to know what is likely to be present in its proposed discharge. The burden of such knowledge and disclosure lies with the discharger, as regulatory agencies logically cannot analyze actual discharges before permitting them.

State regulatory agencies with delegated CWA authority generally lack the resources to determine the toxicity of emerging compounds, instead relying on toxicity testing by the producers under TSCA and evaluations reported to the US EPA, as well as published toxicological research studies. North Carolina was in exactly this position with regard to GenX. TSCA toxicity testing requirements apply only to commercial products, however, not to the by-products. GenX was produced and discharged into the Cape Fear River beginning in 1980 as a byproduct of DuPont's polyvinyl fluoride process (Chemours, 2017), but TSCA consent for its commercial production, and the nondischarge requirement included in that consent, did not occur until 2009. Nor was toxicity testing of GenX conducted until its commercial production was actively considered (Caverly Rae et al., 2015). The numerous PFAS by-products tentatively identified in Cape Fear River water by Strynar et al. (2015) and Sun et al. (2016) have had no toxicity testing. Limited toxicity testing also provides opportunities for disagreements about proper toxicity determinations, particularly for mixtures of compounds. Toxicity evaluations for regulatory purposes usually must await the US EPA's action, which can take a long time. The 2018 ASTDR report on toxicity of PFOA and PFOS, for example, illustrates the long time lag between production of a compound and, in the case of these PFASs, even a decision to halt production, vs a fuller appreciation of a compound's toxic potential.

Discovery and quantification of emerging contaminants in the environment is a challenge above and beyond the resources of most if not all state regulatory agencies. Emerging contaminants present in the environment, as in the case of GenX, are almost always constituents of a cocktail of compounds, many of them chemically similar enough to require sophisticated analytical procedures to separate, identify, and quantify. The studies by Strynar et al. (2015) and Sun et al. (2016) highlight the analytical challenges of the discovery process, and reinforce the magnitude of the technical challenges any regulatory agency would have to face, even assuming that availability of funding, technical expertise and political will were not limiting.

An additional complication is that regulatory agencies rely on the EPA-approved standard methods for their compliance monitoring and enforcement duties. Such standard methods must meet very strict QA/QC requirements and must be performed by laboratories certified to conduct those analyses. Adoption of standard methods and certification of analytical laboratories is a slow process. Adoption of EPA's Method 537 for analysis of PFOA and related compounds in 2009, many years after commercial production of these compounds commenced and 3 years after EPA requested that their production be stopped (US EPA, 2006), illustrates the time lag problem inherent in detecting and quantifying emerging contaminants by methods that can withstand legal challenges.

Another critical point of failure in the regulatory system with regard to emerging contaminants is the technical capability of a discharger to identify and quantify the constituents, both products and by-products, in a proposed waste stream when standard method development is a challenge. A document in the Chemours permit file sheds some light on what are apparently significant differences between full analysis and disclosure vs DuPont's actual statements regarding fluorocarbon discharges at Fayetteville Works. The following is the unedited text of a certified letter from DuPont to the NC Division of Water Quality (NC DWQ) on April 23, 2002 (https://edocs.deq.nc.gov/WaterResources/0/doc/611947/Page1.aspx):

This letter requests that your office clarify a requirement found in Part III of the subject North Carolina issued NPDES permit.

The DuPont Company-Fayetteville Works facility manufactures many fluorocarbon compounds. Each of these processes creates a wastewater that is ultimately treated in and discharged from the on-site wastewater treatment plant (WWTP).

As with all chemical processes, side reactions to the desired product reaction create dozens or hundreds of byproducts in very low concentrations. The fluorochemistry involved in this processes is exceptionally complicated, and most of the byproducts are unknown compounds. There is no standard method to identify these compounds, so a research methodology utilizing nuclear magnetic resonance (NMR) spectroscopy must be employed by an on-site DuPont chemist to qualify and quantify an unknown fluorocarbon compound.

DuPont is considering a research effort to identify and quantify some of the unknown fluorocarbon byproducts in the various processes at the Fayetteville Works facility. Samples would be taken from the wastewater discharge nearest to the process so as to maximize the possibility of a detectable concentration.

In Part III(C) of the subject NPDES permit, there is a requirement for the permittee to notify the Division of Water Quality "as soon as it knows or has reason to believe… that an activity has occurred or will occur which would result in the discharge, on a routine or frequent basis, of any toxic pollutant which is not limited in the permit, if that discharge will exceed… one hundred micrograms per liter (100 μg/L)".

The question to the Division is whether or not the subject permit requires, pursuant to Part III(C), reporting of compounds that are detectable only in the discharge of the manufacturing process, and that would not be detectable exiting the site's WWTP?

For example, assume a wastewater sample is taken from the discharge of a manufacturing process and using NMR spectroscopy, Compound A is detected at a concentration of 20 mg/L. The NMR detection limit for Compound A is determined to be 1 mg/L, meaning any concentration less than 1 mg/L cannot be detected nor quantified. Assume that the process wastewater stream is added to the many other wastewater streams sent to the WWTP and that it represents 1% of the total WWTP influent. This stream would be diluted 100 times with the other wastewaters, so that the concentration of Compound A entering the WWTP is now 0.2 mg/L (200 μg/L) and cannot be detected using the NMR spectroscopy method.

In the above example, Compound A is entering the WWTP at a calculated concentration of 200 µg/L. There is no literature available to indicate if Compound A is degraded in an activated sludge biological treatment system. If one assumes that little of the material is biodegraded, then it follows that there is as much as 200 µg/L of Compound A exiting the WWTP through the permitted Outfall 001. Per the requirement of Part III(C), if the discharge exceeds the 100 µg/L "notification level", then the Division of Water Quality would have to be notified. However, analysis of Outfall 001 shows no detectable concentration of Compound A because the calculated concentration of 0.2 mg/L is less than the detection limit (1 mg/L) of the only known analytical method for detecting Compound A.

In the above example, would a permitee be deemed to know or have reason to believe that a toxic substance is being discharged above the "notification level" and therefore be required to notify the Division of Water Quality of the discharge of Compound A pursuant to Part III(C) of its NPDES permit?

Interpretation of this correspondence must be cautious, as there are issues of context and intent in addition to an obvious lack of specifics. Nevertheless, it is clear that DuPont knew:

(1) some fluorocarbon compounds DuPont proposed to discharge were toxic;
(2) those compounds were not removed or broken down by normal waste treatment processes at Fayetteville Works;
(3) one or more toxic fluorocarbons would be present in the final discharge (#001) at concentrations above 100 µg/L;
(4) DuPont was required to disclose the presence of such compounds at those concentrations, even if they were technically by-products.

It is curious that DuPont claimed to be limited to use of nuclear magnetic resonance (NMR) spectroscopy as an analytical tool for quantifying their discharge constituents, which included many fluorocarbon compounds, a challenging NMR task. They admitted that NMR lacked the sensitivity to quantify many of the PFASs likely present in their discharge, but it is odd that they would employ NMR at all for that purpose, when the advantages of mass spectrometry should have been obvious to DuPont's chemists. As stated above, dischargers have a duty to know and to disclose what constituents are present in their discharges for permitting purposes. Triple quadrupole mass spectrometry had been developed in 1978 (Griffiths, 2008) and must have been available as an analytical tool for targeted analyses in 2002 within the company. Even though TQMS does not support nontarget analysis, it would have offered µg/L sensitivity for quantitation of PFASs known to DuPont. As a major chemical manufacturer, DuPont would also have had access to substantial in-house analytical capabilities as part of their research and development programs. Chemours officials also made the same claim, however, in June 2017, about having to use NMR to quantify fluorocarbons in their discharge as in DuPont's 2002 letter, with the same limitations on detection ability (Chemours, 2017). One can plausibly argue that a major chemical manufacturer should have had access to analytical capabilities sufficient to serve its duty to know and disclose the constituents of its discharges.

It is correct that there was no standard, certifiable laboratory procedure for analysis of PFASs in 2002, so NC DENR would not have been able to direct DuPont to monitor AFPO/PFOA or other PFASs in their discharge for regulatory compliance purposes at that time. The option to dispose of PFAS process wastewater by shipment to a hazardous waste treatment facility was certainly one that NC DENR could have required in the absence of a standard method for analyses of such discharge constituents. That was, in fact, what DuPont chose to do with its AFPO process wastewater (see p. 34 in: https://edocs.deq.nc.gov/WaterResources/0/doc/611879/Page1.aspx).

The discoveries by Strynar et al. (2015) and Sun et al. (2016) of discharges of GenX and related PFASs at Fayetteville Works and the absence of any mention of such PFAS discharges in Chemours' NPDES permit documents highlight many of the same issues raised in DuPont's 2002 letter to NC DWQ. DuPont knew from the 2009 TSCA Consent Order that GenX was not to be released into the river. DuPont/Chemours knew from in-house studies that GenX had toxic properties (Caverly Rae et al., 2015), although human health limits for exposure to GenX are still being debated (Gomis et al., 2018). DuPont/Chemours knew that Strynar et al. (2015) reported GenX in the Cape Fear River (https://edocs.deq.nc.gov/WaterResources/0/doc/591707/Page1.aspx), so they knew it was not removed or broken down by waste treatment processes at Fayetteville Works and that it did not break down rapidly in the environment. Calculations from the GenX loading data provided by Sun et al. (2016) and wastewater budgets from the DuPont/Chemours permit files show that average GenX concentrations in the treated wastewater discharge at Outfall 001 would have been >1 mg/L, exceeding the 100 and 500 µg/L reporting limits for frequent and infrequent discharges of toxic substances, respectively. No such disclosure was made, however, by DuPont/Chemours in its April 27, 2016 NPDES permit renewal application, which states that wastewater from the DuPont polyvinyl fluoride manufacturing area would be discharged to Chemours' WWTP and thence to Outfall #001 (https://edocs.deq.nc.gov/WaterResources/DocView.aspx?dbid=0&id=482844&page=1&searchid=c8df0ae8-6399-4928-b46a-5bbd2e85db95&cr=1). The current permit, last modified as of July 1, 2015, >5 years after commercial production of GenX had commenced, similarly makes no mention of any PFAS compound except PFOA, which was to be monitored in the plant's water intake, but not in its discharge (https://files.nc.gov/ncdeq/GenX/NC0003573%20Ownership%20Change2015.pdf).

7 SYSTEM FAILURES

The legal and regulatory system constructed by the Clean Water Act, the Safe Drinking Water Act, the Clean Air Act, and other major environment legislation was intended to protect human and environmental health as its top priority, while permitting discharges to the environment that would not cause unreasonable harm. That system appears to have failed in the case of GenX and its related PFASs, as these compounds were discharged into the drinking water supply for hundreds of thousands of citizens without their knowledge or informed consent. The permitting process relies on good faith disclosure of probable discharge composition by a permit applicant as well as good faith disclosure of all relevant toxicological information. Regulatory agencies must establish discharge limits when appropriate, but these limits reflect the state of knowledge about proposed discharge constituents. Noncommercial compounds, i.e., by-products, are usually not well studied (dioxins, PAHs, and PCBs are notable exceptions), so their regulation is more problematic. Discharge monitoring is financially beyond the scope of regulatory agencies' means and should be a regular cost of doing business, so discharge monitoring is legally the responsibility of dischargers themselves. Monitoring work is held to high standards (state-certified laboratory analyses of samples obtained through rigorous QA/QC practices), but requires that certified methods and laboratories be available. Each of these aspects of the environmental regulatory system in the United States has obvious limitations and weaknesses, as the narrative here shows. Thus, the regulatory system responds slowly to emerging contaminants at best. New compounds are discovered faster than they can be evaluated for safety, new substances are produced before the totality of their human and environmental health effects can be understood, and substances can find their way into the environment before the analytical methods to find and identify them have been developed and vetted.

The story of GenX in the Cape Fear River exemplifies all these weaknesses and more in the regulatory system designed to protect public drinking water. The difficulties attendant to emerging contaminants in general imply that this may not be at all a unique situation. It is not the intent of this article to ascribe motives or assign legal responsibilities, but to highlight how more timely, specific analysis and communication might have minimized the problems posed by GenX and other PFASs in the Cape Fear River. In particular, the intent is to highlight the significant contributions made by advances in analytical environmental chemistry as a means of discovering the facts and by the science community in communicating their findings to the public. Without cutting-edge separation and analytical capabilities (both human and technical), these pollutants could not be detected, identified, quantified, or regulated.

Acknowledgments

Melissa D. Smith assisted with preparation of Fig. 1. The author thanks Drs. R. Mead and M. Mallin for critical comments.

References

ATSDR, 2018. Toxicological Profile for Perfluoroalkyls; Draft for Public Comment. Agency for Toxic Substances and Disease Registry. US Department of Health and Human Services. 852 pp.

Blum, A., Balan, S.A., Scheringer, M., Goldenman, G., Trier, X., Cousins, I., Diamond, M., Fletcher, T., Higgins, C., Lindeman, A.A., Peaslee, G., de Voogt, P., Wang, Z., Weber, R., 2015. Madrid Statement. Environ. Health Perspect. 123, A107.

Caverly Rae, J.M., Craig, L., Slone, T.W., Frame, S.R., Buxton, L.W., Kennedy, G.L., 2015. Evaluation of chronic toxicity and carcinogenicity of ammonium 2,3,3,3-tetrafluoro-2-(heptafluoropropoxy)-propanoate in Sprague–Dawley rats. Toxicol. Rep. 2, 939–949.

Chemours, 2017. Notes From Chemours Meeting With Local, State Officials. June 15, 2017, New Hanover County Government Center, Wilmington, NC.

DuPont, 2010. DuPont GenX Processing Aid for Making Fluoropolymer Resins. http://www2.dupont.com/Industrial_Bakery_Solutions/en_GB/assets/downloads/DuPont_GenX_Brochure_Final_07July2010.pdf. 2010.

Gomis, M.I., Vestergren, R., Borg, D., Cousins, I.T., 2018. Comparing the toxic potency in vivo of long-chain perfluoroalkyl acids and fluorinated alternatives. Environ. Int. 113, 1–9.

Griffiths, J., 2008. A brief history of mass spectrometry. Anal. Chem. 80, 5678–5683.

Lau, C., Anitole, K., Hodes, C., Lai, D., Pfahles-Hutchens, A., Seed, J., 2007. Perfluoroalkyl acids: a review of monitoring and toxicological findings. Toxicol. Sci. 99 (2), 366–394.

Lindstrom, A.B., Strynar, M.J., Libelo, E.L., 2011. Polyfluorinated compounds: past, present, and future. Environ. Sci. Technol. 45 (19), 7954–7961.

Mudumbi, J.B.N., Ntwampe, S.K.O., Matsha, T., Mekuto, L., Itoba-Tombo, E.F., 2017. Recent developments in polyfluoroalkyl compounds research: a focus on human/environmental health impact, suggested substitutes and removal strategies. Environ. Monit. Assess. 189, 402–431.

Nakayama, S., Strynar, M.J., Helfant, L., Egeghy, P., Ye, X., Lindstrom, A.B., 2007. Perfluorinated compounds in the Cape Fear drainage basin in North Carolina. Environ. Sci. Technol. 41, 5271–5276.

Otterbourg, K., 2018. Teflon's River of Fear: Chemical Giant Chemours Is Facing Off Against Residents of North Carolina in a Battle Over a Potentially Harmful Compound Used to Make Nonstick Pans. Fortune. May 28, 2018. http://fortune.com/longform/teflon-pollution-north-carolina/.

Prevedouros, K., Cousins, I.T., Buck, R.C., Korzeniowski, S.H., 2006. Sources, fate and transport of perfluorocarboxylates. Environ. Sci. Technol. 40, 32–44.

REACH (Registration, Evaluation, Authorisation and Restriction of Chemicals), 2006. Regulation (EC) No. 1907/2006 of the European Parliament and of the Council of 18 December 2006 Concerning the Registration, Evaluation, Authorisation and Restriction of Chemicals (REACH), Establishing a European Chemicals Agency, Amending Directive 1999/45/EC and Repealing Council Regulation (EEC) No 793/93 and Commission Regulation (EC) No 1488/94 as Well as Council Directive 76/769/EEC and Commission Directives 91/155/EEC, 93/67/EEC, 93/105/EC and 2000/21/EC (EC 1907/2006). http://ec.europa.eu/environment/chemicals/reach/reach_en.htm.

Scheringer, M., Trier, X., Cousins, I.T., de Voogt, P., Fletcher, T., Wang, Z., Webster, T.F., 2014. Helsingør Statement on poly- and perfluorinated alkyl substances (PFASs). Chemosphere 114, 337–339.

Scott, B.F., Spencer, C., Mabury, S.A., Muir, D.C.G., 2006. Poly and perfluorinated carboxylates in North American precipitation. Environ. Sci. Technol. 40, 7167–7174.

Strynar, M., Dagnino, S., McMahen, R., Liang, S., Lindstrom, A., Andersen, E., McMillan, L., Thurman, M., Ferrer, I., Ball, C., 2015. Identification of novel perfluoroalkyl ether carboxylic acids (PFECAs) and sulfonic acids (PFESAs) in natural waters using accurate mass time-of-flight mass spectrometry (TOFMS). Environ. Sci. Technol. 49, 11622–11630.

Sun, M., Arevalo, E., Strynar, M., Lindstrom, A., Richardson, M., Kearns, B., Pickett, A., Smith, C., Knappe, D., 2016. Legacy and emerging perfluoroalkyl substances are important drinking water contaminants in the Cape Fear River watershed of North Carolina. Environ. Sci. Technol. Lett. 3, 415–419.

UNCW, 2018. Report to the Environmental Review Commission From the University of North Carolina at Wilmington Regarding the Implementation of Section 20.(a)(2) of House Bill 56 (S.L. 2017-209). NC Legislature—House Select Committee on North Carolina Water Quality.https://www.ncleg.net/documentsites/committees/house2017-185/Meetings/7%20-%20April%2026%202018/2018-April%20HB%2056%20UNCW%20Rpt.pdf.

US EPA, 2006. EPA-DuPont 2006 Order on Consent. http://www.epa.gov/region03/enforcement/dupont_order.pdf.

US EPA, 2009. Method 537. Determination of Selected Perfluorinated Alkyl Acids in Drinking Water by Solid Phase Extraction and Liquid Chromatography/Tandem Mass Spectrometry (LC/MS/MS). EPA Document #: EPA/600/R-08/092. 45268,National Exposure Research Laboratory, Office of Research and Development, US Environmental Protection Agency, Cincinnati, OH 50 pp.

US EPA, 2010/15. PFOA Stewardship Program. http://www.epa.gov/oppt/pfoa/pubs/stewardship/index.html.

US EPA, 2017. Memorandum: Laboratory PFAS Results for NC DEQ Cape Fear Watershed Sampling: Preliminary Non-Targeted Analysis. August 31, 2017.

van Leeuwen, S.P., Kärrman, A., van Bavel, B., de Boer, J., Lindström, G., 2006. Struggle for quality in determination of perfluorinated contaminants in environmental and human samples. Environ. Sci. Technol. 2006 (40), 7854–7860.

Wang, Z., Cousins, I.T., Scheringer, M., Hungerbühler, K., 2013. Fluorinated alternatives to long-chain perfluoroalkyl carboxylic acids (PFCAs), perfluoroalkane sulfonic acids (PFSAs) and their potential precursors. Environ. Int. 60, 242–248.

Yamashita, N., Kannan, K., Taniyasu, S., Horii, Y., Petrick, G., Gamo, T.A., 2005. Global survey of perfluorinated acids in oceans. Mar. Pollut. Bull. 51, 658–668.

Further Reading

NC HB 56.n.d. Session Bill 2017–209; House Bill 56. An Act to Amend Various Environmental Laws. Section 20. (a) GenX Response Measures.

CHAPTER 14

Analysis of GenX and Other Per- and Polyfluoroalkyl Substances in Environmental Water Samples

Qin Tian[a], Mei Sun[b],*

[a]National Research Center for Geoanalysis, Beijing, China [b]Department of Civil and Environmental Engineering, University of North Carolina at Charlotte, Charlotte, NC, United States

*Corresponding author: E-mail: msun8@uncc.edu

1 INTRODUCTION

Per- and polyfluoroalkyl substances (PFAS), historically referred as perfluorinated chemicals (PFC), are a class of persistent, toxic, and bioaccumulative compounds (ITRC, 2017). Until the year 2000, long-chain PFAS, especially perfluoroalkyl carboxylic and sulfonic acids (PFCA and PFSA) were predominantly used in the production of fluoroplastics, firefighting foams, water/stain repellents, and commercial products treated with water/stain repellent coatings (Buck et al., 2011). Due to their toxicity and carcinogenicity, in 2016 the US Environmental Protection Agency (EPA) issued a 70 ng/L health advisory level for the sum concentration of perfluorooctane sulfonic acid (PFOS) (US EPA, 2016b) and perfluorooctanoic acid (PFOA) (US EPA, 2016c), the two most widely used and studied legacy PFAS species. Ecotoxicological and human health concerns as well as increasingly stringent regulations on

long-chain PFASs have led to a transition to short-chain PFAS and fluorinated alternatives (Barzen-Hanson and Field, 2015; Scheringer et al., 2014; Wang et al., 2013a, 2015). The presence of both legacy and emerging PFAS in environmental matrices is receiving increased attention. Some emerging PFAS may be more toxic than the legacy species they are replacing (Gomis, 2017), making it critical to assess the occurrence of emerging PFAS in the environment. Among the emerging PFAS, per- and polyfluoroalkyl ether acids (PFEA) are a family receiving growing attention. Some PFEA species are used as replacement of PFOS and PFOA, and have been detected in various water bodies. For example, GenX ($CF_3CF_2CF_2OCF(CF_3)COO^-NH_4^+$, CAS No. 62037-80-3) is the DuPont trade name for the ammonium salt of hexafluoropropylene oxide dimer acid (HFPO-DA), and serves as a replacement for PFOA (Wang et al., 2013a). HFPO-DA has been detected in surface water and drinking

---

Separation Science and Technology, Volume 11
https://doi.org/10.1016/B978-0-12-815730-5.00014-4

Copyright © 2019 Elsevier Inc. All rights reserved.

water in the United States, Europe, and Asia in multiple studies (Gebbink et al., 2017; Heydebreck et al., 2015; Mulabagal et al., 2018; Pan et al., 2018; Strynar et al., 2015; Sun et al., 2016), together with an amalgamation of other structurally similar PFEA such as HFPO trimer acid (Pan et al., 2017, 2018; Song et al., 2018), HFPO tetramer acid (Song et al., 2018), and other mono- and polyether carboxylic and sulfonic acids (Strynar et al., 2015; Sun et al., 2016). Similarly, ADONA ($CF_3OCF_2CF_2CF_2OCHFCF_2$ $COO^-NH_4{}^+$, CAS No. 958445-44-8) is the 3M/Dyneon replacement of PFOA (Wang et al., 2013a), and has been detected in the US (Mulabagal et al., 2018) and German (Pan et al., 2018) surface water samples. Two other types of PFEA salts serving as PFOS replacements, F-53 (salts of $C_6F_{13}OCF_2CF_2SO_3H$, CAS No. 754925-54-7) and F-53B (Cl-C_6F_{12} $OCF_2CF_2SO_3K$, CAS No. 73606-19-6), have only been detected in China (Pan et al., 2018; Wang et al., 2013b). More types of emerging PFAS identified in recent years have been summarized in a recent review (Ruan and Jiang, 2017).

The identification of emerging PFAS typically relies on the application of nontargeted analysis by high-resolution mass spectrometry (HRMS). Once the chemical structures have been confirmed and analytical standards are available, the quantification can be done by targeted analysis using regular liquid chromatography–mass spectrometry (LC/MS) in the same manner as legacy PFAS. Up to now, US EPA Method 537 (US EPA, 2009a) is the only standard method specifically developed for PFAS analysis in United States, and it only focuses on the analysis of 14 PFCA and PFSA species in drinking water. Meanwhile, a couple of EPA standard operation procedures (SOP) are developed for a broader range of PFAS analysis in surface water and groundwater, including SOP EMAB-113.0 for sample collection (US EPA, 2009b), D-EMMD-PHCB-062-SOP-01 (US EPA, 2017) for analysis of 5 PFEA species, and D-EMMD-PHCB-043-SOP-03 (US EPA, 2016a) for analysis of 24 other

PFAS species. Laboratories all over the world have developed a large variety of other PFAS analytical methods. This chapter will discuss generic information that applies to both legacy and emerging PFAS analysis, mass spectrometry methods for targeted and nontargeted analysis separately, the evolving analytical techniques for total PFAS measurement, and in the end a case study of GenX and other PFEA in North Carolina with its implications.

2 SAMPLE COLLECTION, STORAGE AND CONSERVATION

All types of fluoropolymer plastics, including polytetrafluoroethylene (PTFE) and fluoroelastomer materials should be avoided during sample collection, storage, and preparation. Method 537 recommends polypropylene (PP) bottles while SOP EMAB-113.0 recommends high-density polyethylene (HDPE) bottles. However, one study reported traces of PFOA from PP bottles (Yamashita et al., 2004). Thus, precleaning is important when targeting low concentrations in water samples. It is noticed that prior to use, Method 537 does not require a cleaning step on the sample bottles, while SOP EMAB-113.0 requires a methanol rinse of sample bottles and lids. While it is typically believed glassware should be avoided as PFAS sorb to glass surfaces (Holm et al., 2004; Martin et al., 2004), one study suggested PFAS concentrations in blood samples were unchanged when stored in glass containers for at least 4 months at $-20°C$ and at least 9 days at $20°C$ (Kärrman et al., 2006). Because the majority of previous PFAS studies focus on anionic species with low vapor pressure, evaporative loss is typically not a concern and headspace in sample bottles is allowed; however, if the target analytes include volatile neutral species such as fluorotelomer alcohols, headspace should be avoided.

While samples storage in refrigerators ($<6°C$) is the requirement of Method 537 and the practice of many researchers (Gebbink et al., 2017; Mulabagal et al., 2018; US EPA, 2009a), others have kept samples at room temperature (Pan et al., 2018; Strynar et al., 2015; Sun et al., 2016; US EPA, 2009b) due to the persistent nature of many legacy PFAS and PFEA species, while storage in freezers is also reported (Barzen-Hanson and Field, 2015; Heydebreck et al., 2015; Pan et al., 2017, 2018). Samples sometimes are acidified by formic acid or nitric acid to inhibit microbial activities (Kallenborn et al., 2004; McCord et al., 2018). The practice of acidification can be important for PFAS species that undergo biotransformation (e.g., fluorotelomer-derived precursors (Harding-Marjanovic et al., 2015) or electrochemical fluorination-derived precursors (Mejia-Avendaño et al., 2016)), or for biologically active samples such as sewage sludge. Different types of antibiotics, such as formalin (Schultz et al., 2004), silver nitrate, penicillin, and 2-bromoethanesulfonate (Allred et al., 2015), have been found to suppress MS responses during analysis. In addition, Method 537 requires addition of Trizma, a premixed blend of Tris [Tris(hydroxymethyl)amino-methane] and Tris–HCl [Tris(hydroxymethyl) aminomethane hydrochloride], to serve as a buffer and removes free chlorine in chlorinated finished waters.

3 SAMPLE PREPARATION

Water samples may be filtered to remove suspended solids if required by subsequent analysis; for example, glass microfiber filters are recommended in SOP D-EMMD-PHCB-062-SOP-01 and D-EMMD-PHCB-043-SOP-03. However, some researchers found certain filter materials, including glass fiber, nylon, cellulose acetate, and polyethersulfone, either adsorbed or released PFAS (Schultz et al., 2006; Yamashita et al., 2005). Thus, centrifuge could be used as an alternative to remove suspended solids.

Solid phase extraction (SPE) is used for concentrating PFAS in water samples in Method 537, SOP D-EMMD-PHCB-062-SOP-01 and D-EMMD-PHCB-043-SOP-03 as well as the methods chosen by most PFAS studies referenced in this chapter and many more. Various SPE cartridges have been explored, with the weak anion-exchange (Oasis WAX or Strata XAW) (Gebbink et al., 2017; Heydebreck et al., 2015; McCord et al., 2018; Pan et al., 2017, 2018; Song et al., 2018; Strynar et al., 2015; Sun et al., 2016) and hydrophile–lipophile-balance (HLB) (Mulabagal et al., 2018; US EPA, 2009a; Wang et al., 2013b) cartridges used most often. While HLB cartridges perform well in capturing long-chain PFAS, weak anion-exchange cartridges recover short-chain anions, especially PFEA more effectively (Hopkins et al., 2018). If cationic and nonionic PFAS are analytes of interest, cation-exchange (e.g., Oasis WCX) and nonpolar (e.g., C18) cartridges, respectively, should be explored (D'Agostino and Mabury, 2014). When a collection of PFAS species with different properties are of interest, sequential elution with different cartridges stacked in series may be desirable; however, such efforts have not been published on PFAS analysis. Also, it is crucial to make sure the cartridges do not introduce PFAS to the background (Yamashita et al., 2004). Online SPE is another option that saves time and labor but requires special instrumentation not available in most laboratories.

Liquid–liquid extraction (LLE) is another option and does not require sample filtration, thus can be used to analyze the sum of PFAS dissolved and sorbed to suspended particulate matter. However, conventional LLE typically requires large sample volume (up to 1000 mL) as well as large solvent volume (which needs further concentration). Thus, microextraction techniques have been explored via liquid-phase microextraction (Allred et al., 2014; Martín et al., 2015; Papadopoulou et al., 2011) and solid-phase microextraction (Huang et al., 2019) or micro-SPE (Lashgari et al., 2015).

As mass spectrometers have become more sensitive in recent years, large volume direct injection becomes possible (Sun et al., 2016; Xiao et al., 2017). By eliminating the extraction step, sample preparation becomes faster and cheaper and analyte loss/contamination can be minimized. However, application of large volume injection is limited to relatively clean matrices (e.g., drinking water, groundwater, surface water), and application in samples with high concentrations of salts and other organic components that interfere with the MS analysis (e.g., wastewater, landfill leachate) is rare.

4 MISCELLANEOUS WAYS TO IMPROVE PRECISION AND ACCURACY

Due to the ubiquity of PFAS, as well as the need to analyze PFAS in low ng/L levels, several measures have been considered to improve analytical precision and accuracy.

Although PP and HDPE are generally considered suitable materials to use during PFAS analysis, Method 537, SOP D-EMMD-PHCB-062-SOP-01 and D-EMMD-PHCB-043-SOP-03 still require a methanol-rinse of the sample bottle to recover PFAS from container walls after storing the samples. In Method 537, the methanol used to rinse the sample bottles also serves as the SPE eluent. In contrast, the two SOP employ the methanol rinse prior to sample filtration and SPE, where the methanol rinse is combined with the aqueous sample.

In many PFAS analytical methods using SPE, including Method 537, the SPE eluate is evaporated to dryness. However, this may lead to analyte losses. As an alternative, SOP D-EMMD-PHCB-062-SOP-01 and D-EMMD-PHCB-043-SOP-03 require evaporating SPE eluates to 0.5–1.0 mL.

PTFE and fluoroethylene polymer (FEP) parts are extensively used in regular LC–MS systems, which can lead to background PFAS in blank samples. This problem can be solved by replacing PTFE or FEP filters, solvent plumbing, and pump seals with parts make with stainless steel, PE, or polyetheretherketone. Also, an additional column can be installed between the pump-mixing chamber and the analytical column to serve as a delay column; that way PFAS leaching from the LC system and eluents will be retained on the delay column and be eluted at different peak times with these of the analytes from the samples. Glass vials with PTFE lined septa which are commonly used in the analysis of many other organic compounds should be avoided and replaced with PP vials and caps.

The presence of co-extracted matrix or interfering compounds can cause unexpected signal enhancement or suppression of the analytes, referred to as the matrix effect. Matrix effect can significantly affect the precision and the accuracy of the method, and thus it is important to evaluate if matrix effect is significant in samples to be analyzed. Matrix effect can be evaluated by comparing the slope ratios ($K_{matrix}/K_{solvent}$) of calibration curves developed in the matrix vs those in pure solvent, or the signal ratio ($S_{matrix}/S_{solvent}$) of standards in the matrix to standards in pure solvent. Some matrix components can be removed by cleanup procedures during sample preparation (e.g., using ENVI-Carb products to clean the SPE eluate). To account for the matrix effect that cannot be eliminated, the isotope dilution approach (using mass-labeled analogs of target analytes as internal standards) is recommended, because matrix effect should affect the efficiency of ionization of the analyte and its isotope-labeled internal standards in a similar manner.

5 TARGETED ANALYSIS

LC–MS and LC–MS/MS have been so far the most commonly used instruments for the analysis of nonvolatile PFAS, especially PFCA and PFSA, and are the techniques of choice by Method 537, SOP D-EMMD-PHCB-062-SOP-01 and D-EMMD-PHCB-043-SOP-03. Ion trap, orbitrap, triple quadrupole, and time-of-flight MS have all been used in a large number of studies summarized in this chapter and many more. C18 and C8 reverse phase columns are typically used for LC separation, and they work well for most of the anionic PFAS species. However, their performance is not satisfactory for molecules with less than four carbon atoms; as fluorochemical manufacturers shift their production to short-chain products, analytical methods for these small molecules are needed.

The major challenge for targeted analysis (e.g., LC–MS) is that it relies on calibration using available authentic standards of the analytes. To date, there are more than 4700 PFAS with registered CAS numbers (OECD, 2018), while only a small fraction (<100) of this large compound family have analytical standards available. When standards are not available for certain analytes, semiquantitative estimates can be made using surrogate standards with similar LC elution peak times (and preferably similar chemical structures). Such estimates, although can possibly be off by as much as a factor of 10, can inform public health decisions, and have been used in the analysis of PFEA (Hopkins et al., 2018).

6 NONTARGETED ANALYSIS

In addition to the large number of PFAS registered in CAS, there are also PFAS by-products formed during industrial synthesis processes, as well as PFAS transformation products generated through known and unknown abiotic/biotic reactions. Thus, monitoring only a limited number of PFAS species via targeted analysis may not capture the species most important in the environment. The application of HRMS using suspect screening and nontargeted approaches can provide more comprehensive characterization of PFAS in environmental matrices. Using HRMS, the mass-to-charge ratio (m/z) of ionizable compounds can be determined with up to 10^{-6} accuracy, e.g., m/z can be reported as

accurate as four decimal points for an m/z in the order of 100, and three decimal points for an m/z in the order of 1000. With the accurate mass of the parent molecule, its fragments in sequential MS analysis, and its isotopic pattern, environmental chemists can identify the molecular formula and structure of unknown PFAS.

For PFAS analysis, the Organisation for Economic Co-operation and Development (OECD) PFAS list ($\sim$4700 compounds) and the Master List of PFAS Substances in the EPA Chemistry Dashboard ($\sim$5000 compounds) are useful databases for screening suspect compounds. If no matched ions can be found in these databases (i.e., an unidentified compound is detected), data-dependent acquisition for other precursor ions from survey scans can be conducted in the order of decreasing spectral abundance. Examples of how suspect screening and nontargeted analysis are used to identify emerging PFAS species are demonstrated in a few recent studies (Barzen-Hanson et al., 2017; Gebbink et al., 2017; Newton et al., 2017; Strynar et al., 2015; Wang et al., 2018).

7 TOTAL PFAS ANALYSIS

Even with nontargeted analysis, there is uncertainty of whether all PFAS present will be captured and at which concentrations they are present. Therefore, methods for the determination of total organofluorine, or ideally total PFAS concentrations, are also helpful to fully capture the extent of PFAS contamination in the environment. The methods currently applied to analyze total PFAS in environmental samples have been illustrated in Fig. 1.

The total oxidizable precursor (TOP) assay is developed assuming that many unidentified PFAS species are oxidizable precursors of PFCA and PFSA. This analysis combines heat-activated persulfate oxidation and targeted product analysis by LC–MS/MS, which captures a substantially larger fraction of the total PFAS concentration than targeted analysis alone

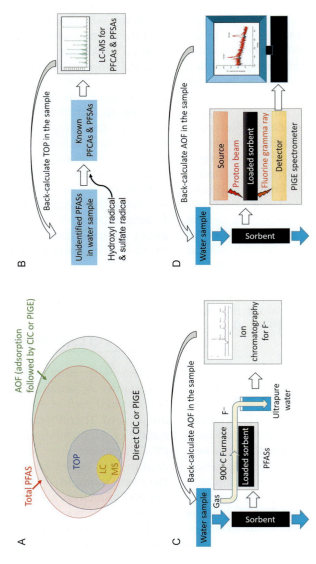

FIG. 1 Schematic representation of different analytical methods for total PFAS determination in aqueous environmental samples: (A) comparison of current options, (B) TOP, (C) CIC, and (D) PIGE.

(Houtz and Sedlak, 2012). Although this method does not reveal identifications of the precursors, it provides important information on how much PFAS is present for treatment purpose such as contaminated subsurface remediation. However, in many cases, a sizeable percentage of the total PFAS concentration remains unaccounted for (Field et al., 2017). Such phenomena may be explained by (1) the inability of heat-activated persulfate to oxidize some PFAS such as GenX (Zhang et al., 2018), (2) the formation of oxidation products that are not traditionally targeted by LC–MS/MS methods (Zhang et al., 2018), and/or (3) competition for persulfate from co-contaminants (Banzhaf et al., 2017; Robel et al., 2017).

An alternative option is combustion ion chromatography (CIC). The direct CIC method, as described in ASTM D7359, measures total fluorine as follows: samples are directly combusted at 900°C to convert organic fluorine to fluoride, which is captured by a water trap and quantified by ion chromatography. Two limitations of the ASTM method are: (1) the volume of sample that can be directly combusted is limited (10–100 μL), leading to a high method reporting limit ($\sim$0.1 mg/L as F) and (2) because organic fluorine is not separated from fluoride in the original samples, sensitivity and accuracy of the analysis are heavily affected by the inorganic fluoride content of the samples. To overcome these limitations, an adsorption step has been introduced ahead of the CIC analysis (Wagner et al., 2013), giving the results as the adsorbable organic fluorine (AOF) concentrations. In such analysis, PFAS in water samples are captured by a sorbent, followed by rinse off inorganic fluoride and combustion of the sorbent. Adsorbed organofluorine is converted to fluoride and quantified by ion chromatography. A few studies reported that the AOF analysis captured a much larger concentration of total organofluorine in environmental samples than targeted LC–MS/MS analysis of routine PFAS compounds (D'Agostino and Mabury, 2017a,b;

Miyake et al., 2007; Willach et al., 2016), e.g., <5% of AOF measured in German surface water samples could be explained by speciated PFAS analysis (Willach et al., 2016). In a study combining total organofluorine measurements and nontargeted HRMS, 12 novel, and 10 infrequently reported classes of PFASs were reported (D'Agostino and Mabury, 2014).

Another method for total fluorine analysis is particle-induced gamma-ray emission (PIGE) spectroscopy, which has been applied to analyze textiles, paper, and food packaging (Lang et al., 2016; Ritter et al., 2017; Robel et al., 2017; Schaider et al., 2017). PIGE uses accelerated protons to excite the ^{19}F nucleus, which then emits characteristic gamma-rays proportional to the amount of ^{19}F atoms and can be measured to give quantitative and unambiguous identification of total fluorine contents in a sample. This spectroscopic technique is rapid (a few minutes per sample) and nondestructive. However, PIGE is a surface analysis technique that only measure fluorine content <0.22-mm penetration depth, and is not capable of differentiating organic and inorganic fluorine without proper pretreatment. To apply PIGE analysis on water samples, it is necessary to capture the PFAS molecules by SPE, and spread the SPE media on a thin film. This pretreatment in combination with PIGE analysis, essentially is another way of analyzing AOF. Given that not all organofluorine belongs to the PFAS family, and that the sensitivity of AOF analysis (by either CIC or PIGE) is lower than that of LC–MS, AOF analysis is a supplemental tool rather than a replacement for LC–MS analysis.

8 CASE STUDY: GenX AND EMERGING PFEA IN NORTH CAROLINA, USA

In recent years, high concentrations of emerging PFEA, including GenX, were identified in the Cape Fear River (CFR) water in North Carolina via nontargeted analysis (Strynar et al., 2015), and later detected in drinking water derived from the river water by targeted analysis (Sun et al., 2016). Fig. 2 shows concentrations of GenX comparing to legacy PFAS in surface water, while Fig. 3 shows concentrations of PFAS in the drinking water treatment plant in the impacted area. The chemical identities of these PFEAs are shown in Table 1. These compounds had been released to the environment by a fluorochemical manufacturer located near the river since 1980 (Hopkins et al., 2018). Average GenX concentration in the river downstream of the manufacturer was 631 ng/L in a 3-month sampling campaign, with the highest GenX concentration observed at 4560 ng/L. Meanwhile, three other PFEA species (without analytical standards at the time) showed chromatographic peak areas up to 15 times that of GenX. Similar to other recalcitrant legacy PFAS, none of the PFEA were removed by conventional or advanced water treatment processes (coagulation/flocculation/sedimentation, raw and settled water ozonation, biological activated carbon filtration, and disinfection by medium-pressure UV lamps and free chlorine) in the impacted water utilities. Impacted residents include over 250,000 people relying on drinking water served by three major public water utilities downstream of the manufacturer (served population counts from EPA Safe Drinking Water Information System), a few thousands relying on drinking water from other small public water utilities, and unknown number of people relying on private wells in vicinity and downstream of the manufacturer.

These findings were published in two papers in 2015 (Strynar et al., 2015) and 2016 (Sun et al., 2016) and were disseminated to the public through news media in 2017. Great concerns were raised from residents, local officials, and state regulators. Since very limited, if any, information existed for these emerging compounds, there were no federal or state regulations on any of these species. In 2017, North Carolina Department of Environmental Quality developed a provisional health goal for GenX at 140 ng/L in drinking water as a "non-regulatory, non-enforceable level of contamination below which no adverse health effects would be expected over a lifetime of exposure." Multiple measures have been deployed to address the PFEA problems in the area, with source control the most effective one. As soon as the manufacturer stopped discharging wastewater containing

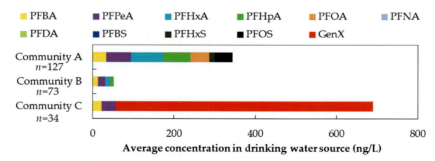

FIG. 2 PFAS occurrence at drinking water intakes in the CFR watershed. Concentrations represent averages of samples collected between June and December 2013. The first seven legends are legacy PFCA with 4–10 carbon atoms in the molecules; the next three legends are legacy PFSA with 4, 6, and 8 carbon atoms in the molecules. *Reprinted with permission from Sun, M., Arevalo, E., Strynar, M., Lindstrom, A., Richardson, M., Kearns, B., Pickett, A., Smith, C., Knappe, D.R.U., 2016. Legacy and emerging perfluoroalkyl substances are important drinking water contaminants in the Cape Fear River watershed of North Carolina. Environ. Sci. Technol. Lett. 3 (12), 415–419. Copyright (2016) American Chemical Society.*

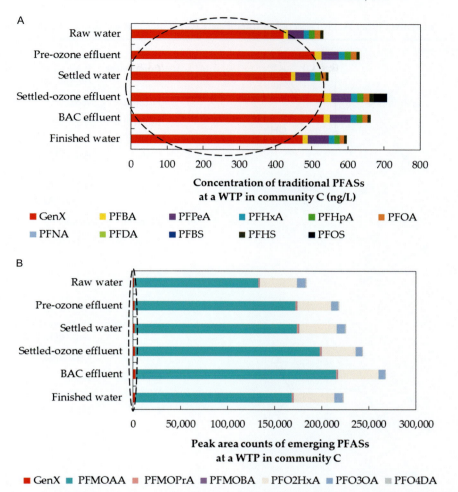

FIG. 3 Fate of (A) legacy PFAS and GenX and (B) PFEA through a full-scale water-treatment plant. Because authentic standards were not available for emerging PFECAs, chromatographic peak area counts are shown in (B). GenX data are shown in both panels and highlighted in *dashed ovals* for reference. Compound information for (B) can be found in Table 1. *Reprinted with permission from Sun, M., Arevalo, E., Strynar, M., Lindstrom, A., Richardson, M., Kearns, B., Pickett, A., Smith, C., Knappe, D.R.U., 2016. Legacy and emerging perfluoroalkyl substances are important drinking water contaminants in the Cape Fear River watershed of North Carolina. Environ. Sci. Technol. Lett. 3 (12), 415–419. Copyright (2016) American Chemical Society.*

GenX and other PFEA, their concentrations in the downstream river water (Fig. 4) and drinking water derived from the river water (Fig. 5) both dropped dramatically, and the GenX concentrations have been below 140 ng/L most of the time (Fig. 6). However, some residents near the manufacturer with drinking water derived from groundwater are still exposed to GenX levels >140 ng/L. As of September 2018, total concentrations of 35 PFAS species in long-term monitoring wells adjacent to the manufacturer along the CFR were still at ~370,000 ng/L, with the major contributions from a few PFEA species, including GenX (Geosyntec, 2018).

14. ANALYSIS OF GenX AND OTHER PER- AND POLYFLUOROALKYL SUBSTANCES

TABLE 1 Perfluoroalkyl Ether Acids Detected in Cape Fear River Water and in Finished Drinking Water

Mono-ether carboxylic acids

Compound	Perfluoro-2-methoxyacetic acid (PFMOAA)
Formula	180.0
Molecular weight	$C_3HF_5O_3$
CAS #	674-13-5

Compound	Perfluoro-2-methoxypropanoic acid (PFMOPrA)[a]
Formula	230.0
Molecular weight	$C_4HF_7O_3$
CAS #	13140-29-9

Compound	Perfluoro-2-ethoxypropanoic acid (PFEOPrA)[b]
Formula	$C_5HF_9O_3$
Molecular weight	280.0
CAS #	N/A

Compound	HFPO-DA (GenX), Perfluoro-2-propoxypropanoic acid (PFPrOPrA)
Formula	$C_6HF_{11}O_3$
Molecular weight	330.1
CAS #	13252-13-6

EVALUATING WATER QUALITY TO PREVENT FUTURE DISASTERS

8 CASE STUDY: GenX AND EMERGING PFEA IN NORTH CAROLINA, USA

TABLE 1 Perfluoroalkyl Ether Acids Detected in Cape Fear River Water and in Finished Drinking Water—cont'd

Multi-ether carboxylic acids

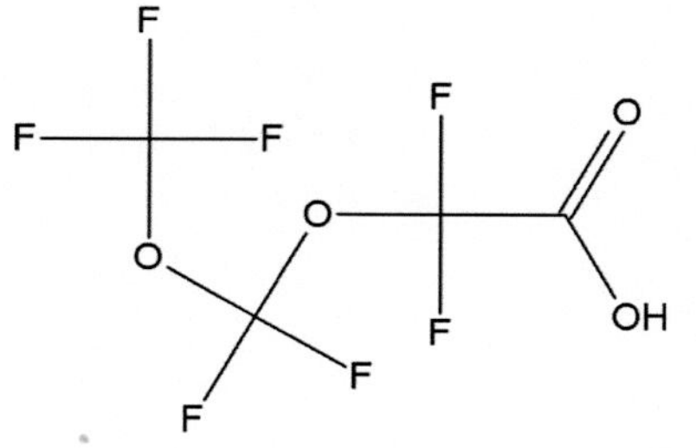

Compound	Perfluoro(3,5-dioxahexanoic) acid (PFO2HxA)
Formula	$C_4HF_7O_4$
Molecular weight	246.0
CAS #	39492-88-1

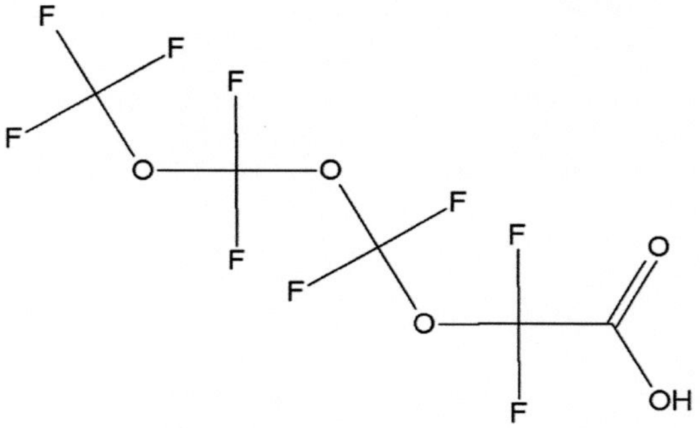

Compound	Perfluoro(3,5,7-trioxaoctanoic) acid (PFO3OA)
Formula	$C_5HF_9O_5$
Molecular weight	312.0
CAS #	39492-89-2

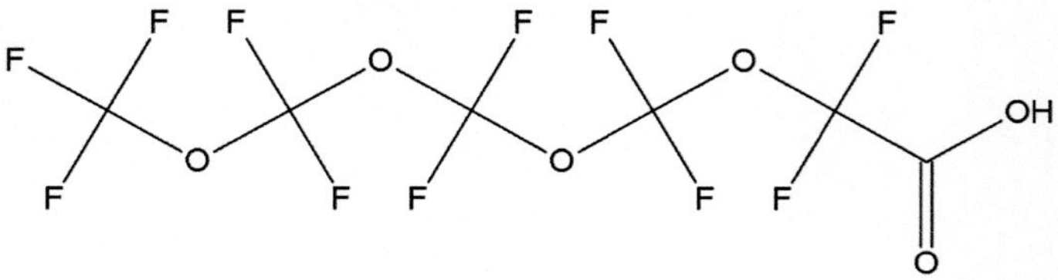

Compound	Perfluoro(3,5,7,9-tetraoxadecanoic) acid (PFO4DA)
Formula	$C_6HF_{11}O_6$
Molecular weight	378.1
CAS #	39492-90-5

Multi-ether sulfonic acids

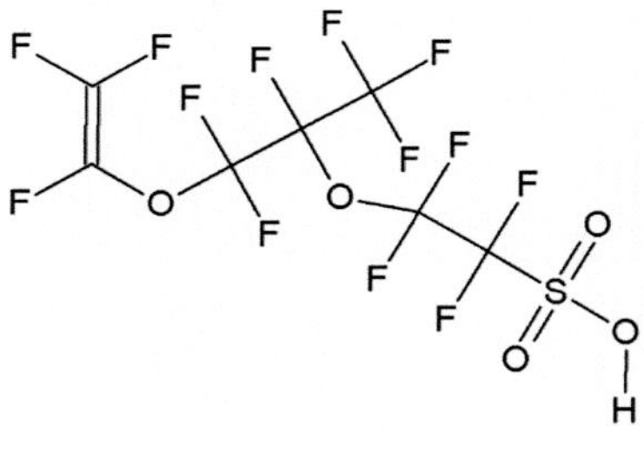

Compound	Ethanesulfonic acid, 2-[1-[difluoro[(1,2,2-trifluoroethenyl)oxy]methyl]-1,2,2,2-tetrafluoroethoxy]-1,1,2,2-tetrafluoro- (Nafion by-product 1)
Formula	$C_7HF_{13}O_5S$
Molecular weight	443.9
CAS #	29311-67-9

[a] In the Sun et al. (2016) paper, this compound was presented as the linear perfluoro-3-methoxypropanoic acid isomer. However, it is more likely that environmental samples contain the branched isomer shown here and in the Strynar et al. (2015) paper.

[b] In the Sun et al. (2016) paper, this compound was presented as the linear perfluoro-4-methoxybutanoic acid (PFMOBA) isomer. However, it is more likely that environmental samples contain the branched isomer shown here.

Reprinted with permission from Hopkins, Z.R., Sun, M., DeWitt, J.C., Knappe, D.R.U., 2018. Recently detected drinking water contaminants: GenX and other per- and polyfluoroalkyl ether acids. J. Am. Water Works Assoc. 110 (7): 13–28. Copyright (2018) John Wiley and Sons.

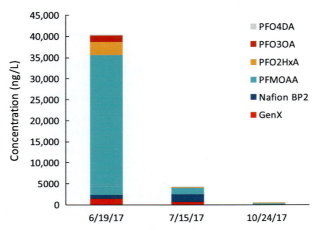

FIG. 4 Evolution of PFEA concentrations in Cape Fear River water at a drinking water intake located ~90 miles downstream from the fluorochemical manufacturer. *Reprinted with permission from Hopkins, Z.R., Sun, M., DeWitt, J.C., Knappe, D.R.U., 2018. Recently detected drinking water contaminants: GenX and other per- and polyfluoroalkyl ether acids. J. Am. Water Works Assoc. 110 (7): 13–28. Copyright (2018) John Wiley and Sons.*

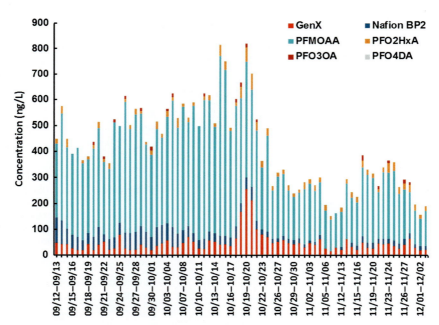

FIG. 5 Time series of PFEA concentrations in finished drinking water following conventional surface water treatment. Each bar represents results for a 24-h composite. *Reprinted with permission from Hopkins, Z.R., Sun, M., DeWitt, J.C., Knappe, D.R.U., 2018. Recently detected drinking water contaminants: GenX and other per- and polyfluoroalkyl ether acids. J. Am. Water Works Assoc. 110 (7): 13–28. Copyright (2018) John Wiley and Sons.*

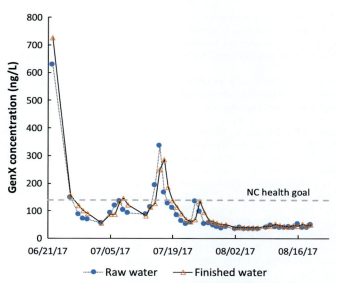

FIG. 6 Concentrations of GenX in raw and finished water of a treatment plant employing raw water ozonation, coagulation/flocculation/sedimentation, settled water ozonation, GAC biofiltration, medium-pressure UV disinfection, and free chlorine disinfection. *Reprinted with permission from Hopkins, Z.R., Sun, M., DeWitt, J.C., Knappe, D.R.U., 2018. Recently detected drinking water contaminants: GenX and other per- and polyfluoroalkyl ether acids. J. Am. Water Works Assoc. 110 (7): 13–28. Copyright (2018) John Wiley and Sons.*

9 CONCLUSIONS

Analyzing PFAS in environmental water samples is challenging because of the ubiquitous existence of PFAS, the low toxicity thresholds, and the large number of unidentified species. As instruments become more sensitive, best practice is established and applied, more chemical structures are revealed and more analytical standards become available, our understanding to PFAS occurrence is improving. As reflected in the North Carolina case, the improvement in analytical techniques is important to identify emerging contaminants of environmental and public health concerns, reveal contamination problems, inform stakeholders, and promote effective regulations. While such development raises awareness of current water-quality problems and contributes to avoiding further environmental disasters, more work is still needed to advance all types of analytical tools for a more accurate and comprehensive understanding of the scope of environmental contaminants, including but not limited to PFAS.

Acknowledgment

The authors thank China Scholarship Council for financial support for Qin Tian (201708110001).

References

Allred, B.M., Lang, J.R., Barlaz, M.A., Field, J.A., 2014. Orthogonal zirconium diol/C18 liquid chromatography–tandem mass spectrometry analysis of poly and perfluoroalkyl substances in landfill leachate. J. Chromatogr. A 1359, 202–211.

Allred, B.M., Lang, J.R., Barlaz, M.A., Field, J.A., 2015. Physical and biological release of poly- and perfluoroalkyl substances (PFASs) from municipal solid waste in anaerobic model landfill reactors. Environ. Sci. Technol. 49 (13), 7648–7656.

Banzhaf, S., Filipovic, M., Lewis, J., Sparrenbom, C.J., Barthel, R., 2017. A review of contamination of surface-, ground-, and drinking water in Sweden by perfluoroalkyl and polyfluoroalkyl substances (PFASs). Ambio 46 (3), 335–346.

Barzen-Hanson, K.A., Field, J.A., 2015. Discovery and implications of C2 and C3 perfluoroalkyl sulfonates in aqueous film-forming foams and groundwater. Environ. Sci. Technol. Lett. 2 (4), 95–99.

Barzen-Hanson, K.A., Roberts, S.C., Choyke, S., Oetjen, K., McAlees, A., Riddell, N., McCrindle, R., Ferguson, P.L., Higgins, C.P., Field, J.A., 2017. Discovery of 40 classes of per- and polyfluoroalkyl substances in historical aqueous film-forming foams (AFFFs) and AFFF-impacted groundwater. Environ. Sci. Technol. 51 (4), 2047–2057.

Buck, R.C., Franklin, J., Berger, U., Conder, J.M., Cousins, I.T., de Voogt, P., Jensen, A.A., Kannan, K., Mabury, S.A., van Leeuwen, S.P.J., 2011. Perfluoroalkyl and polyfluoroalkyl substances in the environment: terminology, classification, and origins. Integr. Environ. Assess. Manag. 7 (4), 513–541.

D'Agostino, L.A., Mabury, S.A., 2014. Identification of novel fluorinated surfactants in aqueous film forming foams and commercial surfactant concentrates. Environ. Sci. Technol. 48 (1), 121–129.

D'Agostino, L.A., Mabury, S.A., 2017a. Certain perfluoroalkyl and polyfluoroalkyl substances associated with aqueous film forming foam are widespread in Canadian surface waters. Environ. Sci. Technol. 51 (23), 13603–13613.

D'Agostino, L.A., Mabury, S.A., 2017b. Aerobic biodegradation of 2 fluorotelomer sulfonamide–based aqueous film–forming foam components produces perfluoroalkyl carboxylates. Environ. Toxicol. Chem. 36 (8), 2012–2021.

Field, J., Sedlak, D., Alvarez-Cohen, L., 2017. Characterization of the Fate and Biotransformation of Fluorochemicals in AFFF-Contaminated Groundwater at Fire/Crash Testing Military Sites. (SERDP Project ER-2128).

Gebbink, W.A., van Asseldonk, L., van Leeuwen, S.P.J., 2017. Presence of emerging per- and polyfluoroalkyl substances (PFASs) in river and drinking water near a fluorochemical production plant in the Netherlands. Environ. Sci. Technol. 51 (19), 11057–11065.

Geosyntec, 2018. Assessment of the Chemical and Spatial Distribution of PFAS in the Cape Fear River. Geosyntec Consultants of North Carolina, Inc.

Gomis, M.I., 2017. From Emission Sources to Human Tissues: Modelling the Exposure to Per- and Polyfluoroalkyl Substances. Stockholm University.

Harding-Marjanovic, K.C., Houtz, E.F., Yi, S., Field, J.A., Sedlak, D.L., Alvarez-Cohen, L., 2015. Aerobic biotransformation of fluorotelomer thioether amido sulfonate

(lodyne) in AFFF-amended microcosms. Environ. Sci. Technol. 49 (13), 7666–7674.

Heydebreck, F., Tang, J., Xie, Z., Ebinghaus, R., 2015. Alternative and legacy perfluoroalkyl substances: differences between European and Chinese river/estuary systems. Environ. Sci. Technol. 49 (14), 8386–8395.

Holm, A., Wilson, S.R., Molander, P., Lundanes, E., Greibrokk, T., 2004. Determination of perfluorooctane sulfonate and perfluorooctanoic acid in human plasma by large volume injection capillary column switching liquid chromatography coupled to electrospray ionization mass spectrometry. J. Sep. Sci. 27 (13), 1071–1079.

Hopkins, Z.R., Sun, M., DeWitt, J.C., Knappe, D.R.U., 2018. Recently detected drinking water contaminants: GenX and other per- and polyfluoroalkyl ether acids. J. Am. Water Works Assoc. 110 (7), 13–28.

Houtz, E.F., Sedlak, D.L., 2012. Oxidative conversion as a means of detecting precursors to perfluoroalkyl acids in urban runoff. Environ. Sci. Technol. 46 (17), 9342–9349.

Huang, Y., Lu, M., Li, H., Bai, M., Huang, X., 2019. Sensitive determination of perfluoroalkane sulfonamides in water and urine samples by multiple monolithic fiber solid-phase microextraction and liquid chromatography tandem mass spectrometry. Talanta 192, 24–31.

ITRC, 2017. PFAS Fact Sheets. Interstate Technology and Regulatory Council.

Kallenborn, R., Berger, U., Järnberg, U., 2004. Perfluorinated Alkylated Substances (PFAS) in the Nordic Environment. Nordic Council of Ministers.

Kärrman, A., van Bavel, B., Järnberg, U., Hardell, L., Lindström, G., 2006. Perfluorinated chemicals in relation to other persistent organic pollutants in human blood. Chemosphere 64 (9), 1582–1591.

Lang, J.R., Allred, B.M., Peaslee, G.F., Field, J.A., Barlaz, M.A., 2016. Release of per- and polyfluoroalkyl substances (PFASs) from carpet and clothing in model anaerobic landfill reactors. Environ. Sci. Technol. 50 (10), 5024–5032.

Lashgari, M., Basheer, C., Kee Lee, H., 2015. Application of surfactant-templated ordered mesoporous material as sorbent in micro-solid phase extraction followed by liquid chromatography–triple quadrupole mass spectrometry for determination of perfluorinated carboxylic acids in aqueous media. Talanta 141, 200–206.

Martin, J.W., Kannan, K., Berger, U., Voogt, P.D., Field, J., Franklin, J., Giesy, J.P., Harner, T., Muir, D.C.G., Scott, B., Kaiser, M., Järnberg, U., Jones, K.C., Mabury, S.A., Schroeder, H., Simcik, M., Sottani, C., Bavel, B.V., Kärrman, A., Lindström, G., Leeuwen, S.V., 2004. Peer reviewed: analytical challenges hamper perfluoroalkyl research. Environ. Sci. Technol. 38 (13), 248A–255A.

REFERENCES

Martín, J., Santos, J.L., Aparicio, I., Alonso, E., 2015. Determination of hormones, a plasticizer, preservatives, perfluoroalkylated compounds, and a flame retardant in water samples by ultrasound-assisted dispersive liquid–liquid microextraction based on the solidification of a floating organic drop. Talanta 143, 335–343.

McCord, J., Newton, S., Strynar, M., 2018. Validation of quantitative measurements and semi-quantitative estimates of emerging perfluoroethercarboxylic acids (PFECAs) and hexfluoroprolyene oxide acids (HFPOAs). J. Chromatogr. A 1551, 52–58.

Mejia-Avendaño, S., Vo Duy, S., Sauvé, S., Liu, J., 2016. Generation of perfluoroalkyl acids from aerobic biotransformation of quaternary ammonium polyfluoroalkyl surfactants. Environ. Sci. Technol. 50 (18), 9923–9932.

Miyake, Y., Yamashita, N., Rostkowski, P., So, M.K., Taniyasu, S., Lam, P.K.S., Kannan, K., 2007. Determination of trace levels of total fluorine in water using combustion ion chromatography for fluorine: a mass balance approach to determine individual perfluorinated chemicals in water. J. Chromatogr. A 1143 (1), 98–104.

Mulabagal, V., Liu, L., Qi, J., Wilson, C., Hayworth, J.S., 2018. A rapid UHPLC-MS/MS method for simultaneous quantitation of 23 perfluoroalkyl substances (PFAS) in estuarine water. Talanta 190, 95–102.

Newton, S., McMahen, R., Stoeckel, J.A., Chislock, M., Lindstrom, A., Strynar, M., 2017. Novel polyfluorinated compounds identified using high resolution mass spectrometry downstream of manufacturing facilities near Decatur, Alabama. Environ. Sci. Technol. 51 (3), 1544–1552.

OECD, 2018. Toward a New Comprehensive Global Database of Per- and Polyfluoroalkyl Substances (PFASs): Summary Report on Updating the OECD 2007 List of Per- and Polyfluoroalkyl Substances (PFASs). Organisation for Economic Co-operation and Development, Paris, France.

Pan, Y., Zhang, H., Cui, Q., Sheng, N., Yeung, L.W.Y., Guo, Y., Sun, Y., Dai, J., 2017. First report on the occurrence and bioaccumulation of hexafluoropropylene oxide trimer acid: an emerging concern. Environ. Sci. Technol. 51 (17), 9553–9560.

Pan, Y., Zhang, H., Cui, Q., Sheng, N., Yeung, L.W.Y., Sun, Y., Guo, Y., Dai, J., 2018. Worldwide distribution of novel perfluoroether carboxylic and sulfonic acids in surface water. Environ. Sci. Technol. 52 (14), 7621–7629.

Papadopoulou, A., Román, I.P., Canals, A., Tyrovola, K., Psillakis, E., 2011. Fast screening of perfluorooctane sulfonate in water using vortex-assisted liquid–liquid microextraction coupled to liquid chromatography–mass spectrometry. Anal. Chim. Acta 691 (1), 56–61.

Ritter, E.E., Dickinson, M.E., Harron, J.P., Lunderberg, D.M., DeYoung, P.A., Robel, A.E., Field, J.A., Peaslee, G.F., 2017. PIGE as a screening tool for per- and polyfluorinated substances in papers and textiles. Nucl. Instrum. Methods Phys. Res. B 407, 47–54.

Robel, A.E., Marshall, K., Dickinson, M., Lunderberg, D., Butt, C., Peaslee, G., Stapleton, H.M., Field, J.A., 2017. Closing the mass balance on fluorine on papers and textiles. Environ. Sci. Technol. 51 (16), 9022–9032.

Ruan, T., Jiang, G., 2017. Analytical methodology for identification of novel per- and polyfluoroalkyl substances in the environment. TrAC Trends Anal. Chem. 95, 122–131.

Schaider, L.A., Balan, S.A., Blum, A., Andrews, D.Q., Strynar, M.J., Dickinson, M.E., Lunderberg, D.M., Lang, J.R., Peaslee, G.F., 2017. Fluorinated compounds in U.S. fast food packaging. Environ. Sci. Technol. Lett. 4 (3), 105–111.

Scheringer, M., Trier, X., Cousins, I.T., de Voogt, P., Fletcher, T., Wang, Z., Webster, T.F., 2014. Helsingør statement on poly- and perfluorinated alkyl substances (PFASs). Chemosphere 114, 337–339.

Schultz, M.M., Barofsky, D.F., Field, J.A., 2004. Quantitative determination of fluorotelomer sulfonates in groundwater by LC MS/MS. Environ. Sci. Technol. 38 (6), 1828–1835.

Schultz, M.M., Barofsky, D.F., Field, J.A., 2006. Quantitative determination of fluorinated alkyl substances by large-volume-injection liquid chromatography tandem mass spectrometry characterization of municipal wastewaters. Environ. Sci. Technol. 40 (1), 289–295.

Song, X., Vestergren, R., Shi, Y., Huang, J., Cai, Y., 2018. Emissions, transport, and fate of emerging per- and polyfluoroalkyl substances from one of the major fluoropolymer manufacturing facilities in China. Environ. Sci. Technol. 52 (17), 9694–9703.

Strynar, M., Dagnino, S., McMahen, R., Liang, S., Lindstrom, A., Andersen, E., McMillan, L., Thurman, M., Ferrer, I., Ball, C., 2015. Identification of novel perfluoroalkyl ether carboxylic acids (PFECAs) and sulfonic acids (PFESAs) in natural waters using accurate mass time-of-flight mass spectrometry (TOFMS). Environ. Sci. Technol. 49 (19), 11622–11630.

Sun, M., Arevalo, E., Strynar, M., Lindstrom, A., Richardson, M., Kearns, B., Pickett, A., Smith, C., Knappe, D.R.U., 2016. Legacy and emerging perfluoroalkyl substances are important drinking water contaminants in the Cape Fear River watershed of North Carolina. Environ. Sci. Technol. Lett. 3 (12), 415–419.

US EPA, 2009a. Method 537: Determination of Selected Perfluorinated Alkyl Acids in Drinking Water by Solid Phase Extraction and Liquid Chromatography/Tandem Mass Spectrometry (LC/MS/MS). Washington, DC.

US EPA, 2009b. EMAB-113.0: The Sample Collection Protocol for PFCs in Surface and Well Water. Research Triangle Park, NC.

US EPA, 2016a. D-EMMD-PHCB-043-SOP-03: Standard Operating Procedure for the Improved Method for Extraction and Analysis of Perfluorinated Compounds (PFCs) From Surface Waters and Well Water by Ultra-High Performance Liquid Chromatography (UPLC)-Tandem Mass Spectrometry (MS/MS). Research Triangle Park, NC.

US EPA, 2016b. Drinking Water Health Advisory for Perfluorooctane Sulfonate (PFOS). 822-R-16-004, US Environmental Protection Agency.

US EPA, 2016c. Drinking Water Health Advisory for Perfluorooctanoic Acid (PFOA). 822-R-16-005, US Environmental Protection Agency.

US EPA, 2017. D-EMMD-PHCB-62-SOP-01: Method for Extraction and Analysis of Perfluoroethercarboxylic Acids (PFECAs) From Surface Water, Well Water and Waste Water by Ultra-Performance Liquid Chromatography (UPLC)-Tandem Mass Spectrometry (MS/MS). Research Triangle Park, NC.

Wagner, A., Raue, B., Brauch, H.-J., Worch, E., Lange, F.T., 2013. Determination of adsorbable organic fluorine from aqueous environmental samples by adsorption to polystyrene-divinylbenzene based activated carbon and combustion ion chromatography. J. Chromatogr. A 1295, 82–89.

Wang, Z., Cousins, I.T., Scheringer, M., Hungerbühler, K., 2013a. Fluorinated alternatives to long-chain perfluoroalkyl carboxylic acids (PFCAs), perfluoroalkane sulfonic acids (PFSAs) and their potential precursors. Environ. Int. 60, 242–248.

Wang, S., Huang, J., Yang, Y., Hui, Y., Ge, Y., Larssen, T., Yu, G., Deng, S., Wang, B., Harman, C., 2013b. First report of a Chinese PFOS alternative overlooked for 30 years: its toxicity, persistence, and presence in the environment. Environ. Sci. Technol. 47 (18), 10163–10170.

Wang, Z., Cousins, I.T., Scheringer, M., Hungerbuehler, K., 2015. Hazard assessment of fluorinated alternatives to long-chain perfluoroalkyl acids (PFAAs) and their precursors: status quo, ongoing challenges and possible solutions. Environ. Int. 75, 172–179.

Wang, Y., Yu, N., Zhu, X., Guo, H., Jiang, J., Wang, X., Shi, W., Wu, J., Yu, H., Wei, S., 2018. Suspect and nontarget screening of per- and polyfluoroalkyl substances in wastewater from a fluorochemical manufacturing park. Environ. Sci. Technol. 52, 11007–11016.

Willach, S., Brauch, H.-J., Lange, F.T., 2016. Contribution of selected perfluoroalkyl and polyfluoroalkyl substances to the adsorbable organically bound fluorine in German rivers and in a highly contaminated groundwater. Chemosphere 145, 342–350.

Xiao, X., Ulrich, B.A., Chen, B., Higgins, C.P., 2017. Sorption of poly- and perfluoroalkyl substances (PFASs) relevant to aqueous film-forming foam (AFFF)-impacted groundwater by biochars and activated carbon. Environ. Sci. Technol. 51 (11), 6342–6351.

Yamashita, N., Kannan, K., Taniyasu, S., Horii, Y., Okazawa, T., Petrick, G., Gamo, T., 2004. Analysis of perfluorinated acids at parts-per-quadrillion levels in seawater using liquid chromatography-tandem mass spectrometry. Environ. Sci. Technol. 38 (21), 5522–5528.

Yamashita, N., Kannan, K., Taniyasu, S., Horii, Y., Petrick, G., Gamo, T., 2005. A global survey of perfluorinated acids in oceans. Mar. Pollut. Bull. 51 (8), 658–668.

Zhang, C., Hopkins, Z., Strynar, M., Lindstrom, A., McCord, J., Knappe, D., 2018. Enhancing the total oxidizable precursor assay for environmental samples containing per- and polyfluoroalkyl ether acids. In: The 256th American Chemical Society National Conference. Boston, MA.

CHAPTER 15

Sustainable Magnetically Retrievable Nanoadsorbents for Selective Removal of Heavy Metal Ions From Different Charged Wastewaters

Sriparna Dutta, R.K. Sharma*

Department of Chemistry, Green Chemistry Network Centre, University of Delhi, New Delhi, India
*Corresponding author: E-mail: rksharmagreenchem@hotmail.com

1 INTRODUCTION

1.1 Global Water Contamination Challenges

Water, which is one of our most precious resources right after air has been at the core of sustainable and socioeconomic development. Without water, as we all know possibly no form of life could have ever existed and consequently survived on earth. But unfortunately, today, clean drinking water has become a scarce resource in many parts of the world. The degradation in the quality of water has directly translated into environmental, social, and economic problems. It is believed that uncontrolled human activities have resulted in such deleterious consequences. According to geographical distribution of water, only 0.06% of freshwater (that comprises only 3% of the total water available

to humans shown in Fig. 1) is accessible which itself is exposed to variety of contaminants, many arising from the unsafe production, utilization, and disposal of inorganic and organic compounds (Ahuja, 2015).

The implication is clear: one cannot drink this water unless it is subjected to a proper clean-up process. And in case, this water is consumed as such without purification it can lead to serious health hazards. In fact, as per the reports of World Health Organization (WHO), the gravity of the situation is such that every year 2 million people are dying from diarrheal diseases, attributed to unsafe water, sanitation, and hygiene (Kumar, 2006). Also, as per estimates of United Nation, by 2025, 2.7 billion people will not have access to safe drinking water (Courtland, 2008). This is indeed a very serious problem which calls for urgent attention. The need to reverse

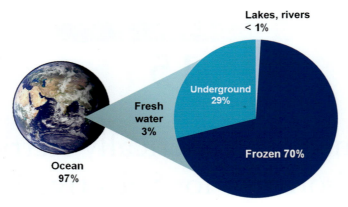

FIG. 1 Distribution of water on earth.

this damage is reflected in the 2030 Agenda for Sustainable Development Goals (SDGs) that brings water-quality issues to the forefront of international action (Le Blanc, 2015). Goal 6 specifically aims at *Ensuring availability and sustainable management of water and sanitation for all* to respond to the pressing challenges posed by water-quality issues.

Did you know?

Water-Quality Global Facts (World Health Organization, WHO/UNICEF Joint Water Supply, & Sanitation Monitoring Programme, 2015):

➢ *One in nine people worldwide consume drinking water from unsafe sources.*
➢ *Contaminated water can transmit diseases such diarrhea, cholera, dysentery, typhoid, and polio and is estimated to cause 502,000 diarrheal deaths each year.*
➢ *Water-related diseases remain the number one cause of death for children under five worldwide.*
➢ *According to estimates of UNEP, UN-Habitat 2010, up to 90% of wastewater globally gets dumped into water bodies untreated. In other words, 90% of sewage in developing countries is discharged untreated directly into water bodies.*
➢ *Every day 2 million tons of sewage and other effluents drain into the world's water.*
➢ *Industry discharges an estimated 300–400 megatons of waste into water bodies every year.*
➢ *Pollution levels in rivers and lakes now put more than 320 million people in Asia, Africa, and Latin America at risk for cholera and typhoid.*
➢ *In the United States, 12–18 million cases of waterborne diseases occur annually; 16.5 million Americans have detectable levels of PFAS in their water; 16 million have perchlorate in their water and lead contamination is on the rise.*
➢ *Despite recent preliminary assessments of the current worldwide water-quality situation, the magnitude of the challenge is still unknown.*
➢ *300 million rural people in developing countries may be at particular health risk from contact with polluted waters.*

Some aspects of the global water-quality challenge were articulated in a recent report, Snapshot of the World's Water Quality, published by UN Environment (UNEP) with the support of UN-Water (UNEP, 2016) which presents a rapid, preliminary assessment of the current water-quality situation, in particular in rivers and in developing countries (UNEP, 2016). It reveals that although water contamination problem has become a global menace the magnitude of which is increasing with every passing day in face of growing population, accelerated urbanization and rapid industrialization,

1 INTRODUCTION

but, it is important to note that this problem is particularly more serious in developing countries like India where facilities of drinking water treatment before supply are not always available or possible. Here, the water resources are unevenly distributed both temporally as well as spatially and increasingly overexploited by the human population. According to the assessment of *WaterAid*, an international organization working for water sanitation and hygiene, an alarming 80% of India's surface water is polluted (Ahuja, 2016) and as per report in *Economic Times* (*an English-language, Indian daily newspaper published by the Bennett, Coleman & Co. Ltd*), dated December 21, 2017, 10 crore (1 crore = 10 million) people are drinking contaminated water in India (https://economictimes.indiatimes.com/news/politics-and-nation/10-crore-people-drinking-contaminated-water-in-india/articleshow/62191841.cms). If we consider the situation of Yamuna and Ganges river water (which are the sacred rivers of India), these water bodies have got so severely contaminated that despite efforts by the Delhi government and various agencies to reduce the contamination levels in the river, the pollution levels have refused to come down (Malik et al., 2014). This unmet crisis of safe drinking water may be attributed to three major factors (Goel, 2006):

1) untreated municipal and domestic sewage;
2) untreated industrial effluents; and
3) agricultural run-off.

As quite evident, it is we human beings who are constantly dumping wastes into water and polluting it. The untreated effluents primarily contain pathogens, heavy metals, organic pollutants, and micro-pollutants which come from the economic activities like agriculture, industry, mining, and other sectors such as pharmacy (Fig. 2) (Akpor et al., 2014).

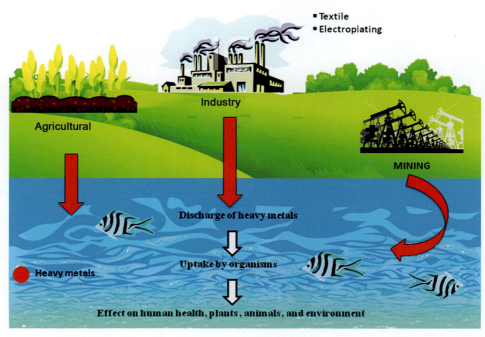

FIG. 2 Schematic illustration showing how water gets contaminated.

Indeed untreated wastewater is one of the biggest sources of water pollution.

1.2 Heavy Metals: The Undeniable Cause Behind Water Contamination

The quality of water has been endangered by noxious contaminants, of which heavy metals "a class of inorganic pollutants" represents an area of increasing concern because they have the ability of dissolving in wastewaters and when discharged into surface waters, they tend to get concentrated and travel up the food chain or seep into groundwater, hence contaminating drinking water (Tchounwou et al., 2012). They often tend to bioaccumulate in the food chain. Over the past several decades, heavy metals such as cadmium, chromium, copper, iron, lead, cobalt, manganese, mercury, nickel, zinc have been inevitably associated with every area of modern consumerism (Fig. 3). For instance, lead has been used for at least 5000 years as building materials, pigments for glazing ceramics, and pipes for transporting water; arsenic in wood preservatives, cadmium as stabilizers in PVC products, color pigment, several alloys and, now most commonly, in rechargeable nickel-cadmium batteries (Nagajyoti et al., 2010; Singh et al., 2011). Although heavy metals have been used by society for a long time yet this has caused deterioration of the quality of potable water, resulting in adverse health effects to man, other organisms and the environment itself. Studies have revealed that the major threats to human health from heavy metals are associated with exposure in form of either ingestion or inhalation to lead, cadmium, mercury, arsenic, and chromium. In order to investigate the toxic effects of these metals, their effects on human health have been regularly reviewed by international bodies such as the WHO. Recently, a review from the Environmental Working Group (EWG) stated that *"75% of America's drinking water contains the compound hexavalent chromium. 200 million Americans—that's 62% of the total population—drink from this 'pool' of contaminated water daily"*

FIG. 3 Uses of heavy metals.

1 INTRODUCTION

(https://www.davidwolfe.com/75-drinking-water-contaminated-heavy-metal/).

So before proceeding further, let us understand what heavy metals are actually?

The term "heavy metal" cannot altogether be clearly defined and a lot of debate exists over it, but chemists categorize heavy metal as any metal that exhibits toxicity, irrespective of their atomic mass or density. In other words *"Any metal (or metalloid) species may be considered a 'contaminant' if it occurs where it is unwanted, or in a form or concentration that causes a detrimental human or environmental effect."* Metals/metalloids include lead (Pb), cadmium (Cd), mercury (Hg), arsenic (As), chromium (Cr), copper (Cu), selenium (Se), nickel (Ni), silver (Ag), and zinc (Zn). Other less common metallic contaminants include aluminum (Al), cesium (Cs), cobalt (Co), manganese (Mn), molybdenum (Mo), strontium (Sr), and uranium (U). The heavy metals have often been referred as persistent environmental toxins because unlike other contaminants like pharmaceuticals and dyes, these cannot be metabolized to carbon dioxide and water or easily rendered harmless by chemical or biological remediation processes (Khan et al., 2011). According to the U.S. Environmental Protection Agency (EPA) and the International Agency for Research on Cancer (IARC), heavy metals have been classified as human carcinogens (Jaishankar et al., 2014).

1.2.1 Sources of Heavy Metals

Even though heavy metals are found throughout the earth's crust, yet most environmental contamination results from weathering and volcanic eruptions (leaching from bedrocks and atmospheric deposition), industrial sources (metal processing in refineries, coal burning in power plants, petroleum combustion, nuclear power stations, plastics, textiles, microelectronics, wood preservation and paper processing plants), agricultural runoffs, etc. (Fig. 4) (Verma and Dwivedi, 2013).

Specific sources of some of the heavy metals have been outlined in the following:

- Chromium (Cr)—Mining, industrial coolants, chromium salts manufacturing, leather tanning.
- Lead (Pb)—Lead-acid batteries, paints, *E*-waste, smelting operations, coal-based thermal power plants, ceramics, bangle industry.
- Mercury (Hg)—Chlor-alkali plants, thermal power plants, fluorescent lamps, hospital waste (damaged thermometers, barometers, sphygmomanometers), electrical appliances, etc.
- Arsenic (As)—Geogenic/natural processes, smelting operations, thermal power plants, fuel.
- Copper (Cu)—Mining, electroplating, smelting operations.
- Vanadium (V)—Spent catalyst, sulfuric acid plant.
- Nickel (Ni)—Smelting operations, thermal power plants, battery industry.
- Cadmium (Cd)—Zinc smelting, waste batteries, e-waste, paint sludge, incinerations and fuel combustion.
- Molybdenum (Mo)—Spent catalyst.
- Zinc (Zn)—Smelting, electroplating.

But the million dollar question is how do heavy metals exhibit toxicity?

1.3 What Makes Heavy Metal Toxic?

1.3.1 Health Impacts of Heavy Metals

Heavy metal toxicity has proven to be a major threat as there are a number of health hazards associated with it. Once absorbed by living organisms, it can result in damaged or reduced mental and central nervous function, lower energy levels, and damage to blood composition, lungs, kidneys, liver, and other vital organs (Inoue, 2013). Besides, long-term exposure to these metals may also result in slow progressing physical, muscular, neurological degenerative processes that mimic Alzheimer's

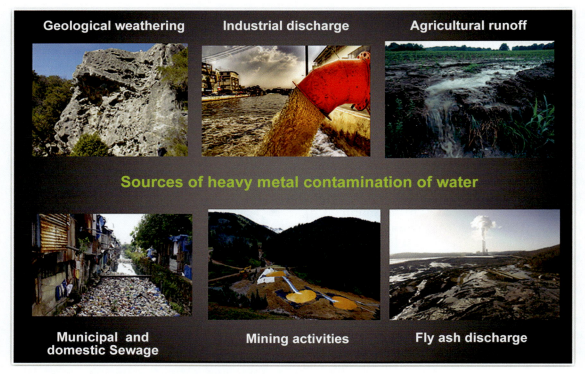

FIG. 4 Schematic illustration showing the sources of heavy metals that contaminate water directly or indirectly.

disease, Parkinson's disease, muscular dystrophy, multiple sclerosis, gangrene, diabetes mellitus, hypertension, and ischemic heart disease. Several renowned organizations working in the area of water quality have specified maximum contaminant limits (MCLs) in drinking water (Table 1) (Yamamura et al., 2003).

Off late, India has seen a spurt in health cases pertaining to physical, muscular, and neurological degenerative diseases that resemble Alzheimer's, Parkinson's disease, muscular dystrophy, and multiple sclerosis. Fig. 5 shows the average variation in heavy metal content in the waste water effluents released from different industries of Mumbai—"one of the industrially developed states of India" (during the assessment periods 1999) (Lokhande et al., 2011). This experimental data clearly reveal the high level of pollution along Taloja Industrial estate of Mumbai, India.

Similarly, many other regions of the country are affected with metal contamination issues. Although several adverse health effects of heavy metals have been known for a long time, exposure to heavy metals continues, and is even increasing in some parts of the world, in particular in less developed countries, though emissions have declined in most developed countries over the last 100 years. Therefore, utmost removal of heavy metals from wastewater prior to their discharge in the environment is an issue of high significance. Also, depletion of metal resources has presented an alarming situation. The following periodic table highlights all the elements that are at risk of depletion (Fig. 6) (https://inhabitat.com/this-periodic-table-shows-which-of-earths-elements-are-at-risk-for-depletion/). The elements in red are at greatest risk while those in yellow are at lowest risk.

TABLE 1 Maximum Contaminant Limit (MCL) Contaminants in Drinking Water

Type of Contaminant	Toxicities	HK WSD	MCL (µg/L) China MOH	US EPA	WHO
Arsenic	Cancer in liver, lungs, kidneys, and skin lesions	10	10	10	10
Mercury	Neurotoxins, kidney damage	1	1	2	1
Lead	Kidney problems, high blood pressure	10	10	15	10
Chromium	Allergic dermatitis	50	50	100	50

HK WSD = Hong Kong's Water Supply Department; China MOH = Ministry of Health of the People's Republic of China; US EPA = U.S. Environmental Protection Agency; WHO = World Health Organization.

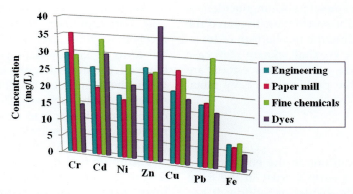

FIG. 5 Average variation in heavy metal content in the waste water effluents released from different industries of Taloja Industrial Estate of Mumbai for the assessment year 1999.

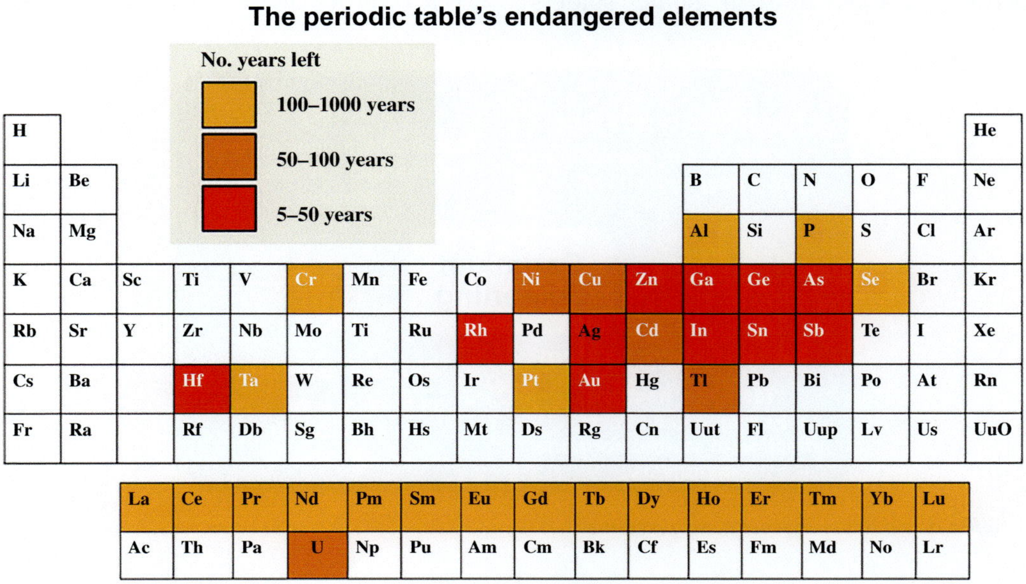

FIG. 6 Periodic table highlighting Earth's elements are at risk for depletion.

So clearly, there is a need for technological innovation to transform the way we treat, distribute, use, and reuse water toward a distributed, differential water treatment, and reuse paradigm. So, the question that arises is how can this challenge be met sustainably in the future?

2 HEAVY METAL POISONING CASE STUDIES

A string of serious metal poisoning incidences in different parts of the world have triggered the need for understanding the mechanism of toxicity exhibited by them. We shall take up a few case studies to emphasize why serious efforts are needed to combat the problem of heavy metal poisoning.

2.1 Cadmium

Cadmium is ubiquitous in the environment and exists mostly in earth's crust in the form of cadmium chloride, oxide, or sulfide. It has found widespread applicability in several items, including electroplating devices, storage batteries, pigments, vapor lamps, neutron absorbent in nuclear reactors, and in some solders (Bernhoft, 2013). In the United States, most of the cadmium used is obtained as a by-product of the smelting of zinc, lead, or copper ores. However, during the past century, cadmium emissions have increased dramatically, the major reason being that Cd containing products are not only dumped as such along with the household goods, specifically waste materials but also very rarely recycled. Also, phosphate fertilizers used by farmers in agriculture constitute a major source of Cd pollution. Recent investigations have shown that adverse health effects of cadmium may occur at lower exposure levels than previously anticipated, primarily in the form of kidney damage but possibly also bone effects and fractures. Long-term exposure to this metal may also result in cancer and organ system toxicity such as skeletal, urinary, reproductive, cardiovascular, central and peripheral nervous, and respiratory systems (Fig. 7) (Nordberg et al., 2014).

Cadmium toxicity came into limelight for the first time when in 1946, the inhabitants of the Jinzu River basin of North-West Japan got severely afflicted with the itai-itai or ouch-ouch disease-, characterized by severe pain and bone fractures (Aoshima, 2016; Rahimzadeh et al., 2017). The root cause of this disease was found

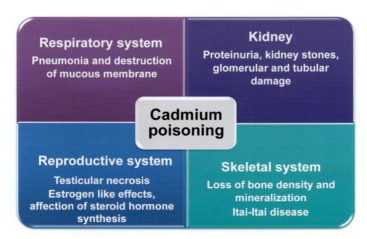

FIG. 7 Effect of cadmium on several organ systems.

to be the consumption of water and rice contaminated with Cd from discharges of a local zinc-mining operation in Toyama Prefecture (the leading industrial prefecture on the Japan sea coast). Actually, Kamioka Mines in Toyama had become the most productive in the world after a new mining technology arrived from Europe postworld war I (WWI) and way back in 1910, through the mining operations, significant quantities of Cd was released into the Jinzu river which continued for more than 40 years. This river water was not only being used for irrigation of rice fields, but also for drinking, washing, fishing, and other uses by the downstream populations. Surprisingly, the farmers and peasants who used this river water for agriculture and fishermen who fished those same waters noticed an unusual decrease in crop yield as well as catch of fish. Soon the local residents started complaining of a very "strange disease" they considered could never be cured once affected. This disease was termed "itai-itai" disease which became endemic among the inhabitants of the Jinzu River basin (Fig. 8) (https://people.ucsc.edu/~flegal/migrated/etox80e/SpecTopics/itaiitaipics.html). It leads to osteomalacia (softening of the bones) and osteoporosis (loss of bone mass and weakness). The first case of this disease was documented in 1912. In the 1940s, over 200 elderly women living in the Jinzu Valley being mothers of multiple children were disabled by the disease. Later on, the affected residents successfully sued the polluters, the Mitsui company, to make their claims public and receive compensation for damages.

Since then, many other cases of Cd poisoning have been reported. Very recently, there was a report in the Telegraph dated June 17, 2018 stating that "Twenty tons of the cancer-causing metal cadmium have been discharged into a river in southern China in one of the worst chemical spills of its kind that could affect up to 4 million people (https://www.telegraph.co.uk/news/earth/environment/9053671/20-tons-of-cadmium-poisoning-vital-Chinese-river.html)." Thus, at present as a fast-developing country, China has been facing an increasingly serious problem of Cd contamination in the environment.

Reason for toxicity of cadmium (Sharma, 2005):

➢ Cd(II) strongly binds to sulfhydryl groups of cysteine residues of enzymes, for example, Carbonic anhydrase, dipeptidase, carboxy peptidase, etc., affecting the active confirmation of biomolecules due to the strong binding.

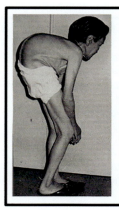

Did you know?

Itai-Itai disease was documented in case of mass Cadmium poisoning in Toyoma Prefecture, Japan starting around 1912. Cadmium was released into rivers by mining companies. The Cd poisoning caused softening of bones and kidney failure [27]. The mining companies were successfully sued for the damage.

FIG. 8 Person suffering from Itai-Itai disease.

➢ Cd is similar to zinc. Therefore, Cd(II) can displace Zn(II) in many zinc containing enzymes. Besides, Cd competes with calcium for binding sites.

2.2 Mercury

Mercury is a naturally occurring element that exists mainly in either elemental (Hg^0—a silver-gray liquid at room temperature that vaporizes readily when heated), organic (methylmercury, the soluble form of mercury), or inorganic (Hg^{1+} and Hg^{2+} that form numerous salts). For more than 3000 years, mercury has been used in different industrial as well as pharmacological sectors such as in insects formulas, agriculture products, lamps, batteries, paper, dyes, electrical/electronic devices, jewelry, and in dentistry (Bolger and Schwetz, 2002). But unfortunately, this increasing use has led to irreparable consequences. Although all the different forms of mercury have been reported to be hazardous to human health, it is interesting to note that their toxic manifestations have been found to depend on their routes of absorption and clinical findings. Depending on the route of exposure, this metal has affected neurologic, gastrointestinal, and renal systems. Methylmercury, the organic soluble form of mercury has particularly been found to be neurotoxic (Castoldi et al., 2003). Elemental mercury vapors have accounted for most of the occupational exposures. As investigated by the EPA, elevated mercury concentrations have been detected in approximately 25% of the groundwater and surface-water samples from 2783 hazardous waste sites (Rall and Pope, 1995).

Severe outbreak of mercury poisoning occurred in several places in Japan during the 1950s in people who had ingested fish and shellfish contaminated by MeHg discharged in waste water from a chemical plant named Chisso Co. Ltd. located at the Minamata City, south-west region of Japan's Kyushu Island. Approximately, 111 cases of mercury poisoning were reported; among them 45 people died and at least 30 suffered from infantile cerebral palsy (Harada, 1995). Investigations were carried out and it was found that the sea fish in the Minamata bay were found to be containing 27–102 ppm of Hg in the form of methyl mercury. Actually, the source of contamination was the effluent discharged into the bay from a factory of Chisso Co. Ltd. that was using a mercury catalyst to make vinyl chloride. People affected with this disease known as the Minamata disease suffered from symptoms like sensory disturbances (glove and stocking type), ataxia, dysarthria, constriction of the visual field, auditory disturbances, and tremor (Fig. 9). This incidence was followed by another tragic report of mercury

Did you know?

The Minamata disease was identified 50 years ago in Minamata city, Kumamoto, Japan. It was caused by release of methyl mercury in industrial wastewater from Chisso Corporation's chemical factory which continued from 1932-1968. It resulted in poisoning from fish contaminated by mercury discharges to the surrounding sea. But even today, unfortunately, thousands of victims are still fighting for compensation from the company.

FIG. 9 Person suffering from chronic Hg poisoning (Minamata disease).

poisoning from Iraq in 1972, where 450 villagers died after eating wheat that was dusted with mercury containing pesticides. Both of these tragic incidences boosted our awareness of mercury as a pollutant; ultimately resulting in it being studied more extensively than any other trace elements. As a result of the Minamata mishappening, Japan follows strictest environmental laws in the industrialized world.

Reason for toxicity of mercury (Sharma, 2005):

➢ Like Cd(II), Hg(II) also binds strongly with the thiol(–SH) group of various metalloproteins and enzymes as it has a very strong affinity for the deprotonated forms of thiol ligand. This binding changes the confirmation of protein about the active site.
➢ Mercury may also interfere with cellular metabolism by binding to amine and phosphoryl groups.

2.3 Lead

Lead is one such naturally occurring valuable metal that has been utilized by human beings since ancient times either as a pigment and binder in paints, or a sweetener and preservative for wine, or as a component of several alloys such as bronze and pewter (Ziegfeld, 1964). Today, it has found the most important use is in the production of lead-acid storage batteries. Although found freely in nature, it may also be easily mined and refined from ore. Examples include galena (PbS), anglesite ($PbSO_4$), cerussite ($PbCO_3$), and minium (Pb_3O_4). Albeit its extensive use, lead has proven to be a dangerous metal worldwide as it interferes with a number of body functions primarily affecting the central nervous, hematopoietic, hepatic, and renal system producing serious disorders (Winecker et al., 2002). The potential toxicity of this metal requires governmental regulations, high-risk and occupational screening, and clinical intervention when necessary. The common route of lead exposure of the general population includes air and water in roughly equal proportions. The last century has witnessed a tremendous hike in lead emissions that have resulted in considerable pollution, mainly due to lead emissions from petrol.

In early August 2009, a serious case of lead poisoning in Fengxiang County in north western Shaanxi Province was reported in which 174 children from three nearby villages were diagnosed with lead poisoning (Jia, 2009). Operation of a smelting plant was blamed for this which eventually saw violence among the outraged parents, who reportedly smashed trucks and tore down fences to protest at the government's slow response to the incident. Thankfully, an unscheduled test was carried out that showed that 20 out of 30 children had high levels of lead in their blood (Hanna-Attisha et al., 2016). The smelting plant has been suspended and is under investigation by the environmental experts. The topic of lead poisoning again grabbed serious national attention in 2016, when a pediatrician in Flint, Michigan, USA noticed an abrupt and unusual doubling in the number of children having elevated lead levels. On detailed investigation, it was discovered that these kids had one thing in common which was the "source of their drinking water." Actually what has happened was that the City of Flint had switched their source of potable water from Lake Huron to the Flint River which was of poorer quality. This Flint water had not been treated well with anticorrosive agents like orthophosphate that had resulted in widespread pipe corrosion and consequently lead contamination (Fig. 10) (https://edition.cnn.com/2016/01/11/health/toxic-tap-water-flint-michigan/). In fact, finding from this observation lead to a cascade of water testing by other municipalities and organizations that identified lead concentrations above the currently accepted drinking water standard of 15 parts per billion (ppb).

Presently, governments worldwide are devoting substantial efforts to reduce the amount of lead in the environment to minimize toxic exposure.

EVALUATING WATER QUALITY TO PREVENT FUTURE DISASTERS

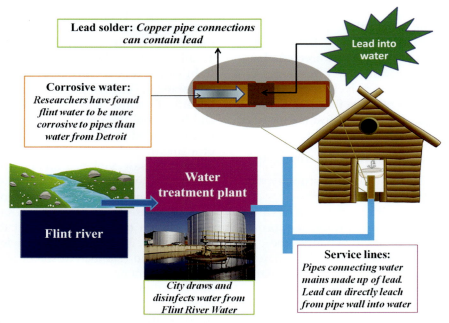

FIG. 10   Flint water toxicity: How lead gets in?

Reason for toxicity of lead (Sharma, 2005)

➢ Similar to Hg(II) and Cd(II), lead also inhibits SH-enzyme but less strongly.
➢ Primarily, lead exhibits toxicity by interfering with heme synthesis as it inhibits several of key enzymes involved in the overall process of heme synthesis.

2.4 Arsenic

Arsenic is an element known since antiquity. It is widely distributed throughout the environment in both organic (arsenic atoms bonded with carbon) and inorganic (no carbon) forms and has a number of allotropes: the most common one being metallic gray, followed by yellow and then black. According to the Minerals Education Coalition, As is typically found in minerals, such as arsenopyrite, realgar, and orpiment. Since the time of the Roman Empire all the way to the Victorian era, this element has been considered as the "king of poisons" as well as the "poison of kings." Apart from this, arsenic compounds have been frequently employed in wood preservation, insecticides, laser diodes, LEDs, in lead alloys for ammunition (Flora, 2014). But despite its widespread utility, As [As(III) and As(V)] has been considered the most toxic metal from both environmental as well as human health standpoints. In fact, As is so toxic that it ranks no. 1 on the ATSDR/EPA priority list of hazardous substances (Hughes et al., 2011). Now, it is recognized as a type of carcinogen that is especially dangerous since it does not have a taste or odor, so humans can be exposed to it without knowing it. Broadly speaking, the exposure to Arsenic occurs via intake of food and drinking water and studies have revealed that long-term exposure to this metal in drinking water may lead to increased risks of not only skin cancer, but also some other cancers, as well as other skin lesions such as hyperkeratosis and pigmentation changes (Fig. 11) (Haley, 2016).

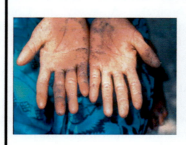

> **Did you know?**
> Arsenic is one of WHO's 10 chemicals of major public health concern. Although the current recommended limit of arsenic in drinking-water is 10 μg/L, millions of people around the world are exposed to arsenic at concentrations much higher than the guideline value (100 μg/L or greater).
> In 1984, the occurrence of a large number of cases of arsenic-induced skin lesions was reported from Kolkata, West Bengal.

FIG. 11 Arsenic keratoses on the palms of a patient who ingested arsenic from a contaminated well over a prolonged time period.

Arsenic pollution in groundwater has been envisaged as a problem of significant global concern. Reports suggest that many countries like Taiwan, China, Argentina, Chile, Mexico, Cambodia, Thailand, Myanmar, Nepal, and United States are suffering from this problem, but the situation is even more grave in countries such as India and Bangladesh. The prime cause of arsenic toxicity throughout the world has been found to be the well water contaminated by natural sources such as arsenic-containing bedrock. Several years ago around 1983, significant arsenic contamination in groundwater was detected in West Bengal, India when some of the villagers were diagnosed to be suffering from arsenicosis because of consumption of arsenic contaminated water (Bhowmick et al., 2018; Pal et al., 2002).

In 1988, a fatal case of arsenic poisoning was reported at a Swedish Copper Smeltry, where a worker got buried under arsenic trioxide in an accident (Gerhardsson et al., 1988). Actually, in this smeltry, where this incident had occurred, huge quantities of crude arsenic (containing more than 90% of arsenic trioxide) were stored in silos and accidentally in a 25 m high As silo, arsenic trioxide got unloaded from below by the use of explosives. During the operation, a fragile vault was formed and a worker was just standing outside the open door when the vault suddenly collapsed and buried him. Seeing this, a nearby worker rushed into the spot and immediately tried to free his head within 1–2 min, but unfortunately his filter mask got clogged. In order to get air, the half-suffocated worker removed the mask in panic which leads to massive inhalation of arsenic dust. Although he was released within 5–10 min and immediately taken to the occupational health service center where he was given intensive treatment, yet he could not be saved. These cases clearly show why effective steps need to be taken to prevent As pollution as it leads to serious health hazards, even death in certain cases.

Reason for toxicity of As (Ralph, 2008)

➢ Arsenic, especially As(III) binds to single, but with higher affinity to vicinal sulfhydryl groups, thus inhibits the activity of several –SH containing enzymes. For instance, it inhibits not only the formation of acetyl-CoA but also the enzyme succinic dehydrogenase.
➢ Arsenate can replace phosphate in many reactions.

2.5 Chromium

Chromium is an essential trace element that exists predominantly in the environment in two valence states, trivalent chromium: Cr(III) and hexavalent chromium Cr(VI). Undeniably,

Cr(III) is the most stable and common form that occurs naturally in chromite ore. Cr(VI) is the next stable form which rarely occurs in nature and is essentially produced from industrial activities. In acidic environment that has a high organic content, reduction of Cr(VI) to nontoxic Cr(III) may occur. Owing to their exceptional physicochemical properties, chromium compounds have been used in diverse areas (i) manufacture of catalysts, dyes, pigments (ii) chrome plating, (ii) in ceramic, glass, leather, and tanning industries, (iv) wood preservation, (v) treatment of cooling tower water, and (vi) corrosion control. However, there is a flip side of the coin too that is, the large-scale industrial applicability of this metal has resulted in environmental pollution. Among the two predominant forms of chromium, the hexavalent chromium has been reported to be more toxic (Shekhawat et al., 2015). In fact, the IARC has declared Cr(VI) as a human carcinogen through the exposure via inhalation route while the National Toxicology Program (NTP) has released a 2-year study demonstrating that ingested Cr(VI) was carcinogenic in rats and mice (Mulware, 2013). Among the different environmental compartments, the aquatic life has been most severely affected. In particular, drinking water has been found to be contaminated with chromate. A study in China reported an increase of stomach cancer mortality in regions of elevated Cr(VI) concentrations in drinking water (Beaumont et al., 2008).

Very recently, Cr(VI) toxicity engrossed the attention of public, when there was a report of U.S. Steel plant in Portage, Indiana spilling wastewater containing this potentially cancer-causing chemical into Burns Waterway, a tributary about 100 yards from Lake Michigan (Paustenbach, 2015). Apparently, what had happened was that the wastewater discharge occurred due to a pipe failure that contained hexavalent chromium (chromium-6), which was being used for industrial processes. The toxic chemical was made famous by the environmental activist and 2000 movie of the same name, "Erin Brockovich (Fig. 12) (Harris et al., 1994; https://www.ewg.org/research/chromium-six-found-in-us-tap-water#.W0hBde-FPIU)." This leak not only impelled the shutdown of the water intake of this company, but also prompted the closure of four beaches and a riverwalk at the Indiana Dunes National Lakeshore, and Indiana American Water in Ogden Dunes. Thus, this Chromium spill near Lake Michigan brought new attention to cancer-causing pollutant.

Did you know?

"It's INEXCUSABLE that the government has done so little to protect us from this chemical that has been shown to cause cancer at even insanely low levels." - Erin Brockovich

Indeed, Cr poisoning got public attention in the 2000 film "Erin Brockovich [52]." In this film, the environmental crusader confronts the lawyer of a power company that polluted the tap water of Hinkley, Calif., with a carcinogenic chemical called chromium-6.

When the lawyer picks up a glass of water, Brockovich says: "We had that water brought in 'specially for you folks. Came from a well in Hinkley." The lawyer sets down the glass and says, "I think this meeting's over."

FIG. 12   Erin Brockovich-Chromium poisoning.

Several health groups say hexavalent chromium is so dangerous that a national standard is long overdue. "Even a single gallon of hexavalent chromium could contaminate billions of gallons of drinking water," said David Andrews, senior scientist at the EWG (http://www.chicagotribune.com/news/ct-chromium-pollution-lake-michigan-met-2-20170412-story.html). "The lack of a drinking water standard … is just one more example of our failed drinking water regulations."

2.5.1 Mechanism of Toxicity

After Cr(VI) enters the cells, it eventually gets reduced to Cr(III), that in turn mediates its toxicity through induction of oxidative stress during the reduction, while the Cr intermediates react with protein and DNA. Further, Cr(III) shows the propensity to form adducts with DNA that may lead to mutations (Cohen et al., 1993).

3 STRATEGIES DEVISED TO COMBAT METAL CONTAMINATION ISSUES

The toxicity, persistence, recalcitrance, and bioaccumulative nature of the heavy metals have raised serious public concerns across the globe as exemplified through the heavy metal poisoning case studies described in the previous section. Thus, the past decades have witnessed an increasing interest toward the development of specialized techniques that can either reduce the toxicity of these metals prior to their discharge in order to meet technology-based treatment standards or remove these metals from the waste discharges. Some of the conventional technologies that rely on physicochemical removal processes (Pugazhenthiran et al., 2016) and have been utilized for accomplishing the removal of heavy metals from industrial wastewater are summarized as follows.

3.1 Chemical Precipitation

Chemical precipitation is one of the most widely adopted methodologies that have been employed for heavy metal removal from inorganic effluents in industries (Abid et al., 2011; Fu and Wang, 2011). It principally involves the transformation of a soluble compound into an insoluble form (insoluble precipitates of heavy metals as hydroxide, sulfide, carbonate, and phosphate) via the addition of chemicals (precipitants) (Barakat, 2011). Once the heavy metals precipitate to form solids, they can be removed easily.

Removal efficiency of this methodology can be significantly improved by either adjusting the pH to basic conditions (around 9–11) or changing other major parameters such as temperature, initial concentration, charge of the ions, etc. Owing to outstanding advantages such as relative simplicity, low cost of precipitant, and ease of automatic pH control, the hydroxide treatment stands out to be the most commonly utilized precipitation technique. However, the major drawback associated with the chemical precipitation technique is requirement of enormous amounts of chemicals to reduce metals to an acceptable level for discharge. Besides, factors such as excessive sludge production that further requires post treatment, slow metal precipitation, aggregation tendency of metal precipitates, and long-term environmental impacts of sludge disposal limit the large-scale applicability of this methodology.

3.2 Ion-Exchange

Another method that has been successfully utilized for removal of heavy metals from different effluents in the water treatment industries is the "ion-exchange method." This technique principally relies on a chemical reaction in which heavy metal ions from wastewater are exchanged for a similarly charged ion attached to a solid support (cations or anions containing special ion exchanger); thus the soluble ions from the liquid phase get attracted to the solid phase (Gunatilake, 2015a). Synthetic organic ion-exchange resins have been generally employed as ion exchangers. These are water-insoluble solid substances that can absorb positively or negatively charged ions from an electrolyte solution and release other ions with the same charges into the solution in an equivalent amount. It is important to mention here that although ion-exchange is a cost-effective and convenient methodology, yet it has proven to be effective only in the treatment of water with low concentration of heavy metals. In case of highly concentrated metal solution, the matrix gets fouled by organics and other solids present in wastewater. Above and beyond, it is nonselective and also highly sensitive to the pH of the solution which renders it unsuitable for commercial applications.

3.3 Membrane Separation Process and Ultrafiltration

In recent years, membrane filtration technology has drawn the intensive interest of the research community as it shows enormous capability of not only removing suspended solids and organic substances but also efficiently removes inorganic contaminants such as heavy metals (Maximous et al., 2010). Different types of membrane filtration techniques such as ultrafiltration (UF), nanofiltration (NF), and reverse osmosis (RO) have been employed for the removal of heavy metals from contaminated sources depending on the size of the particle that can be retained (Le and Nunes, 2016). The UF technique uses a permeable membrane for the separation of heavy metals, macromolecules and suspended solids from inorganic solution on the basis of:

➢ pore size (5–20 nm) and
➢ molecular weight of the separating compounds (1000–100,000 Da).

Juang and Shiau reported a chitosan-enhanced membrane filtration technique for the removal of Cu(II) and Zn(II) ions from synthetic wastewater and found that in comparison to the membrane alone, chitosan could significantly enhance the metal removal at least by 6–10 times (Juang and Shiau, 2000). The reason behind this enhancement of metal removal could be attributed to the presence of amine groups on the surface of chitosan that served as coordination site for metal binding. Although the UF membranes shows high removal efficiency (in fact, depending on the membrane characteristics, UF can achieve more than 90% of removal efficiency with a metal concentration ranging from 10 to 112 mg/L), the only problem it suffers from is the relatively high operational cost because of possibility of membrane fouling. Besides UF, RO technique that uses pressure to force a solution through a membrane and retains the solute on one side, allowing the pure solvent to pass to the other side has also gained attention in present times (Bakalár et al., 2009). Using this technique, 98% and 99% removal efficiency for copper and cadmium could be accomplished by Qdais and Moussa (2004). Among all the membrane separation techniques, NF has emerged as the most promising and innovative technologies that has been widely applied in drinking water and industrial effluents treatment. NF membranes require lower operational pressure and energy consumption as they have a less dense structure compared to RO membranes (Mikulášek and Cuhorka, 2016). Recently, a commercial NF membrane (NF270) was employed for removal of heavy metals and surprisingly it showed 99% efficiency in removing cadmium (Al-Rashdi et al., 2013).

3.4 Electrochemical Techniques

Electrochemical techniques such as electrodeposition, electrocoagulation, electro-floatation, and electrooxidation have also been employed quite often for the treatment of wastewater containing heavy metals (Gunatilake, 2015b). Electricity was utilized for treating water for the very first time in United Kingdom a century ago, and since then it has been considered a highly reliable method for wastewater treatment. In this method, electricity is used to pass current through an aqueous metal-bearing solution containing a cathode plate and an insoluble anode which leads to the precipitation of the heavy metals in weak acidic or neutralized catholyte as hydroxides. It is usually performed at temperatures lower than the non-electrochemical approaches, which results in increased energy efficiency. Besides, varying the applied potential, designing appropriate electrochemical cells and choosing the right kind of electrodes can help in minimizing power losses; voltage drops prevent unwanted side reactions. Among the several electrochemical techniques, electrocoagulation that forms coagulants in situ via electrolytic oxidation of an appropriate anode material has emerged as the most popular heavy metal precipitation method. It has successfully been applied for the treatment of treatment of potable water, urban wastewater, heavy metal laden wastewater, restaurant wastewater, and colored water containing heavy metals such as As, Cr, Cd, Zn, Ni, and Hg have been removed (Bazrafshan et al., 2015).

4 ADSORPTION: ULTIMATE CHOICE FOR HEAVY METAL REMOVAL FROM WASTEWATER

All the conventional techniques are witnessing a downfall because they are either not economical or eco-friendly and suffer from various drawbacks like lack of sensitivity and selectivity, use of large amount of toxic organic solvents which have deleterious effect on human health and environment. Adsorption has aroused the interest of the researchers in recent times as it is a potential alternative of the other conventional technologies used for wastewater laden with heavy metals (Bobade and Eshtiagi, 2015;

Lakherwal, 2014). The key benefits associated with the adsorption methodology include cost effectiveness due to low operation cost, unproblematic design, and less requirement of control systems. It basically involves a mass transfer process wherein the dissolved solids from liquid get deposited onto the surface of a solid material because of some sort of physical or chemical interaction. To put it in simple words, adsorption occurs whenever a gaseous or liquid solute molecule accumulates on the surface of a solid or a liquid (adsorbent), forming a molecular or atomic film (the adsorbate). The adsorbent could be of mineral, organic, or biological origin. Although activated carbon is one of the extensively employed adsorbents that have found use in diverse applications, but over the years, a number of low-cost adsorbent materials derived from agricultural waste, industrial by-product, natural material, or modified biopolymers have been designed, synthesized, and subsequently applied for the removal of heavy metals from metal contaminated wastewater (Tripathi and Ranjan, 2015). In general, three major steps are involved in the pollution sorption onto solid sorbent, as illustrated in Fig. 13.

The adsorption process could be either batch, semibatch, or continuous. At molecular level, adsorption occurs because of attractive forces between a surface and the group being absorbed. Depending on the types of intermolecular attractive forces adsorption could be of following types:

a) *Physical adsorption*: In this process, the binding of an adsorbate on the surface of an adsorbent occurs via weak van der Waals forces of attraction.

b) *Chemical adsorption*: Also known as chemisorption or activated adsorption, this method involves a chemical reaction between the adsorbent and the adsorbate. Thus, strong interaction (covalent or ionic bonds are formed) occurs between the adsorbate and the substrate.

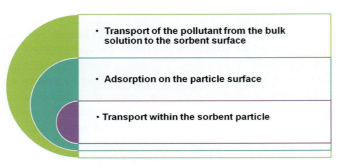

FIG. 13 Steps involved in the pollution sorption onto solid sorbent.

FIG. 14 Advantages of adsorption technology.

In today's scenario, solid-phase extraction (SPE)/adsorption has become the ultimate choice in heavy metal removal due to reasons best listed in Fig. 14 (Li et al., 2017). Thus, it provides an economic, persuasive, greener, and sustainable alternative to other traditional methodologies.

Notably, selection of a suitable adsorbent to treat inorganic effluents depends on the technical applicability as well as its cost effectiveness. A brief comparison of different frequently employed technologies for heavy metal removal highlighting the advantages as well as disadvantages was reported by Farooq et al., which has been presented in Table 2 (Farooq et al., 2010).

5 SYNERGISTIC INTEGRATION OF ADSORPTION TECHNOLOGY WITH NANOTECHNOLOGY

Although these existing technologies contribute inertia against a paradigm shift, the current water challenges call for a change toward

5 SYNERGISTIC INTEGRATION OF ADSORPTION TECHNOLOGY WITH NANOTECHNOLOGY

TABLE 2 Comparison of Technologies for Heavy Metal Removal From Wastewater

Method	Advantages	Disadvantages
Chemical precipitation	• Simple • Inexpensive • Most of the metals can be removed	• Large amount of sludge produced • Disposal problems
Chemical coagulation	• Sludge settling • Dewatering	• High cost • Large consumption of chemicals
Ion-exchange	• High regeneration of materials • Metal selective	• High cost • Less number of metal ions removed
Electrochemical methods	• Metal selective • No consumption of chemicals • Pure metals can be achieved	• High capital cost • High running cost • Initial solution pH and current density
Membrane process and ultrafiltration	• Less solid waste produced • Less chemical consumption • High efficiency (>95% for single metal)	• High initial and running cost • Low flow rates • Removal percentage decreases with the presence of other metals
Adsorption using activated carbon	• Most of metals can be removed • High efficiency	• Cost of activated carbon • No regeneration • Performance depends upon the adsorbent
Adsorption using natural zeolite	• Most of metals can be removed • Relatively less costly materials	• Low efficiency

integrated management of water which in turn demarcates the need for the development of new technologies that can provide: (i) high efficiency, (ii) multiple functionality, and (iii) high flexibility in system size. Nanotechnology, which is related to the manipulation of materials at the nanometer scale integrates all these features and, therefore may offer leapfrogging opportunities in this transformation (Qu et al., 2012). In other words, the essence of nanotechnology lies in its ability to fabricate and engineer the materials and systems with the desired structures and functionalities using the nanosized building blocks. Consequently, quality research has been conducted on individual nanotechnology enabled treatment processes, many of which have shown enhanced performance over conventional technologies. In fact, interest in nanoscale materials has increased in the last decade due to their unique characteristics resulting from their finite size, high specific surface area, tunable pore size, and surface reactivity, a large number of active sites on the surface and strong adsorption ability (Li et al., 2015). It is important to mention that it is the high specific surface area that is mainly responsible for their high adsorption capacity. Also, the possibility of tuning the surface of the nanomaterials using suitable linkers to target specific contaminants helps in achieving high selectivity. Moreover, nanoadsorbents can be readily integrated into existing treatment processes such as slurry reactors, filters, or adsorbers (e.g., by coating filter media or loading into porous granules). These wondrous features of nanomaterials have allowed new products and

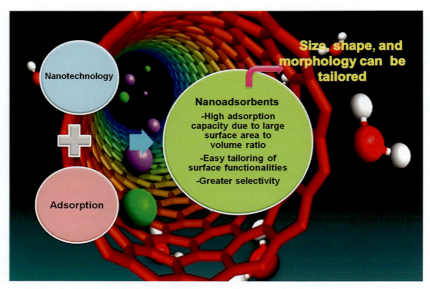

FIG. 15 The benefits of synergistic integration of Adsorption Technology with Nanotechnology.

solutions to be implemented for the social problems related to natural resources, drinking water, energy generation, and conservation.

Considering the environmental threat of heavy metals, lately the synergistic integration of SPE with nanotechnology (SPE using different nanostructures) has been realized as a clean adsorptive technology to simultaneously defend the human community from the awful ecotoxicological impacts of heavy metals and to broaden the horizon of their recovery and recyclability (Fig. 15) (Hung et al., 2016). Nanosized adsorbents not only help the social demands for environmental clean-up, but also play a crucial part in sustainability by reducing the metal resource shortages.

6 MAGNETIC NANOADSORBENTS

Although a number of efficient nanoadsorbents have been developed till date for the removal of heavy metals, yet their profitable utilization has been hindered due to drawbacks like cumbersome synthesis and separation strategy, secondary waste generation, and lack of stability and reusability. With the rapid development of nanotechnology, magnetic nanoadsorbents have come to the forefront as they offer unprecedented opportunities in wastewater treatment for the sustainable development of growing society (Aghaei et al., 2017). In addition to exhibiting excellent adsorption capacities toward different types of organic as well as inorganic contaminants because of their nanosized dimensions, they avoid the generation of secondary waste, produce no contaminants, are capable of treating large amount of wastewater within a very short time, show good selectivity and exceptional reusability (they can be recycled and used easily in an industrials scale). These features impart them an enormous advantage over the traditional methodologies.

6.1 Magnetic Nanoparticles: What and Why?

Nowadays, substantial interest of the scientific fraternity has been directed toward the synthesis of magnetic nanoparticles (MNPs). Fig. 20 illustrates "What makes MNPs interesting

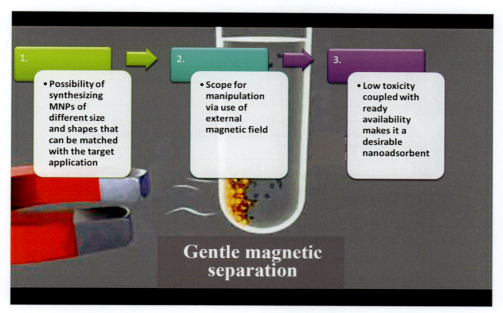

FIG. 16 What makes MNPs so interesting?

(Sharma et al., 2016)?" They have found extensive applications in diverse fields due to their multifunctional properties such as small size, superparamagnetism, and low toxicity. The growing interest can be reflected from the increase in number of scientific publications focusing on magnetic materials. So, it is important to firstly understand what MNPs actually are. MNPs are generally composed of magnetic elements, such as iron, cobalt, nickel, or their oxides like magnetite (Fe_3O_4), maghemite (c-Fe_2O_3), nickel ferrite ($NiFe_2O_4$), cobalt ferrite ($CoFe_2O_4$), etc., that have diameters ranging from 1 to 100nm (Pradeep, 2003). They exhibit superparamagnetism, a fascinating property that makes them behave as a giant paramagnetic atom with a fast response to the external magnetic field and with negligible remanence and coercivity. As a result of the superparamagnetic properties, these NPs can be easily attracted to an applied magnetic field and separated easily. Fe_3O_4 NPs have particularly emerged as efficient sorbents in SPE due to their good dispersion in solution, their very high specific surface area, and their ability to be controlled and separated with an external magnetic field (Fig. 16) (Akbarzadeh et al., 2012).

Advantages of magnetic solid-phase extraction have been elucidated in Fig. 17 (Wan Ibrahim et al., 2015).

6.2 Synthetic Approaches for Preparation of MNPs

In view of the outstanding advantages, it is not surprising that over the past few decades, a variety of suitable methods have been designed for synthesis of MNPs of different shapes, sizes, and compositions, yet their successful applications are highly dependent on the stability of these particles under varied reaction conditions. Some of the most popular routes that have been commercially utilized to synthesize high-quality MNPs include coprecipitation, thermal decomposition, microemulsion, and hydrothermal processes (Laurent et al., 2008; Lu et al., 2007; Majidi et al., 2016). Each of these techniques is discussed

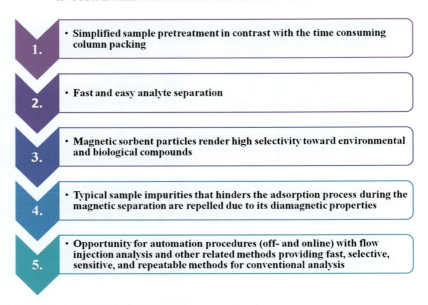

FIG. 17   Advantages of magnetic solid-phase extraction.

FIG. 18   Synthesis of magnetite nanoparticles via the coprecipitation method.

in brief to provide a basic overview to the readers about the synthetic approaches for preparing MNPs:

6.2.1 Coprecipitation

The coprecipitation method is one of the most widely adopted methodologies that is extremely facile and involves mild synthetic conditions (Petcharoen and Sirivat, 2012). It is principally based on the precipitation of iron oxide nanoparticles from an aqueous Fe^{2+}/Fe^{3+} solution via the addition of a suitable base (Fig. 18). The requirement of an inert (nonoxidizing) environment seems to be a pre-requisite for the coprecipitation method because of the tendency of the magnetite to get oxidized to maghemite or iron hydroxides under oxidative conditions. However, the most fascinating aspect of this method that causes its frequent utility is the control that can be exerted over the morphology and the composition of the MNPs via the variation in the type of salts used (e.g., nitrates, sulfates, and chlorides), Fe^{2+}/Fe^{3+} ratio, reaction temperature, pH, and ionic strength of the medium.

6.2.2 Thermal Decomposition

In recent years, the thermal decomposition technique has gained tremendous popularity since it gives rise to MNPs of high crystallinity possessing a narrow size distribution (Angermann and Töpfer, 2008). The syntheses of nanocrystals via this approach involve heating of organometallic precursors in a high boiling point organic solvent in the presence of suitable capping agents (surfactants) such as oleic acid, oleylamine, and hexadecylamine (Park et al., 2004; Sun et al., 2004). Organometallic species because of their meta-stable nature can readily undergo decomposition by mild agitation such as heat, light, or even sound. It has been observed that this process can be carried out by either a progressive heating of the reactants in an appropriate media to a desired temperature or alternatively through a more rapid process termed as hot injection. Thermal decomposition of organometallic precursors in which the metal is in zero oxidation state [such as $Fe(CO)_5$] initially leads to the formation of metal nanoparticles, but on being subjected to atmospheric conditions, get eventually oxidized to form high-quality metal oxide nanoparticles. On the other hand, decomposition of precursors with cationic metal centers [such as $Fe(acac)_3$] leads to the synthesis of metal oxide NPs directly. Parameters that primarily govern the size, shape, and morphology of the particles are the ratio of the starting material used that is, organometallic precursors, surfactants, and solvents, reaction time, and temperature (Abdulwahab et al., 2013).

6.2.3 Microemulsion

The microemulsion technique involves dispersing a fine microdroplet (typically 1–50 nm in size) of an aqueous phase in a continuous oil phase (Liu et al., 2004). A soluble metal salt is then added in this microemulsion which would reside in the aqueous microdroplets; this is followed by the addition of a base that would cause the precipitation of the MNPs. Surfactant-stabilized nanocavities provide a confinement effect that limits the particle nucleation, growth, and agglomeration. AOT (aerosol OT or sodium dioctylsulfosuccinate) was the first and most characterized surfactant system used in the synthesis of MNPs (Blanco-Andujar and Thanh, 2010). Other systems, such as CTAB (cetyltrimethylammonium bromide), SDS (sodium dodecylsulfate), and polyethoxylates have also been used (Lu et al., 2013). The beauty of this method lies in the ability to control or vary the average size of the nanoparticles by suitably adjusting the reaction temperature, metal concentration, base concentration, and the water/surfactant ratio; thereby leading to the formation particles of desired morphologies. However, some of the disadvantages of this method that limits its wide scale use are low yield, requirement of large amounts of solvents and complicated purifying procedures required for the removal of surfactants.

6.2.4 Hydrothermal

The hydrothermal synthetic method is based on the principle that almost all inorganic substances dissolve in water at elevated temperatures and pressures. Subsequently, they crystallize from the dissolved material of the fluid. Water at elevated temperature plays a role of transforming the precursor materials due to its excessive vapor pressure. In fact, this technique can generate a large number of nanostructured materials due to the potential benefits that includes narrow particle size distribution, formation of monodisperse particles, high yield, etc., via careful tuning of synthetic parameters like water pressure, reaction time, etc. (Daou et al., 2006). Very recently, Yang and coworkers have reported the one-step hydrothermal synthesis of highly water-soluble secondary structural Fe_3O_4 nanoparticles using ferric acetylacetonate as the sole iron source and poly(acrylic acid) (PAA) as the stabilizer (Fig. 19) (Yang et al., 2012).

6.2.5 Recent Synthetic Approaches

Besides, these conventional techniques, MNPs have also been synthesized using continuous flow (CF) and microwave (MW) reactors in recent years (Jiao et al., 2015; Kalyani et al., 2015). The utilization of these reactors has changed the way synthetic chemistry is performed at the research and industrial levels. While CF chemistry enables improved thermal

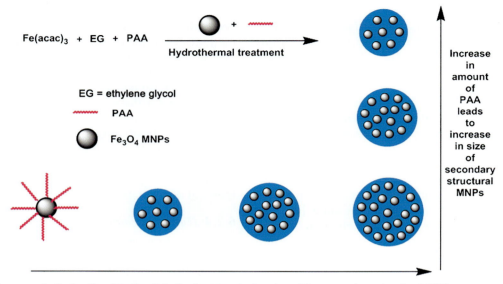

FIG. 19 Hydrothermal synthesis for the preparation of MNPs.

management, enhanced mass transfer and mixing control at the nanoliter scale which in turn helps in tuning the nanoparticle sizes and properties, MW chemistry allows uniform heating, fast reaction under a controlled environment, products with high purity and yield. Apart from this, with the rising environmental concerns and the current stress on adopting greener pathways, some research groups have also come up with green procedures for the synthesis of MNPs that seem to be highly promising in their approach. For instance, Venkateswarulu accomplished the synthesis of iron oxide nanoparticles from plantain leaf extract. This straightforward procedure simply involved mixing of iron(III) chloride hexahydrate and sodium acetate in freshly prepared plantain peel extract solution containing carbohydrates at 70°C, followed by washing of the resultant solution with ethanol (Venkateswarlu et al., 2013). Likewise, many other greener sustainable protocols have been developed for the eco-friendly and economical synthesis of MNPs.

6.3 Surface Modification of MNPs

Even though MNPs are utilized in multidisciplinary fields, but one obvious problem that diminishes their overall efficacy is their susceptibility to undergo agglomeration or oxidation that may alter their magnetic characteristics. Thus, the synthesis of MNPs while maintaining the long-term stability is indeed a key challenge for the researchers. Nevertheless, suitable protecting agents such as surfactants and polymers, inorganic components like silica, carbon, precious metals such as Ag or Au, metal oxides, and hydroxides that form a shell around the magnetic core have been explored for stabilizing them (Bystrzejewski et al., 2009; Gawande et al., 2015; Ge et al., 2012; He et al., 2008; Wang et al., 2008). The coating by such type of materials helps in improving chemical stability, resistance to oxidation and increases the selectivity toward target metal ions. Also, to further render the coated MNPs suitable for water treatment, their surface is modified/tailored with the aid of either organic molecules, inorganic ions or some functional groups (Ojemaye et al., 2017). Literature reports have revealed that functional groups –COOH, –NH$_2$, and –SH groups have been used for providing active sites for exchange of metal ions (Shi et al., 2015; Xin et al., 2012; Yantasee et al., 2007). Physical interactions (such as electrostatic and van der Waal's interactions) and chemical interactions (such as chemical binding, complex formation and modified ligand combination) are responsible for the adsorption of metal ions on the adsorbent surface. For instance mercapto-functionalized nano-Fe$_3$O$_4$ magnetic polymers (SH-Fe$_3$O$_4$-NMPs) were prepared for the removal of Hg(II) from wastewater (Pan et al., 2012).

Surface modification of MNPs imparts:

➢ long-term stability by preventing them from oxidation and
➢ selectivity and specificity for uptake of heavy metal ions by providing specific reaction sites and functional groups.

6.4 Characterization Techniques Employed for Gaining an Insight Into the Morphological and Structural Characteristics of the Magnetic Nanoadsorbents

One of the greatest challenges of nanotechnology research lies in utilizing the right characterization techniques that have the optimum capabilities for studying the minute details pertaining to the structural, electrical, magnetic, and optical properties of developed nanomaterials. This is because once the dimensions of materials are reduced to nanoscale, there is a drastic enhancement in their different physicochemical properties. Thus, a wide range of advanced spectroscopic and microscopic techniques, such as Fourier transform infrared spectroscopy (FTIR), powder X-ray diffraction (PXRD), transmission electron microscopy (TEM), field emission scanning electron microscopy (FESEM), energy dispersive X-ray spectroscopy (EDS),

vibrating sample magnetometry (VSM), energy dispersive X-ray spectroscopy (EDXRF), X-ray photoelectron spectroscopy (XPS), flame atomic absorption spectroscopy (FAAS), and inductively coupled plasma spectroscopy mass spectrometry (ICP-MS) have been employed for gaining an insight into the internal structure, composition, and morphologies of the developed nanoadsorbents (Liao and Chen, 2002). FTIR is an important technique that is frequently employed to confirm the surface modification of the MNPs. PXRD assists the researchers in acquiring knowledge about the crystallographic structure, chemical composition, and physical properties of the nanocomposites. Electron microscopy tools such as TEM and SEM help in analyzing the fine morphological details of the synthesized nanosorbents. EDS and EDXRF are generally utilized for elemental analysis. The XPS technique provides information about the oxidation state of the adsorbed metal. VSM is used to examine the magnetic properties of the bare as well as surface engineered MNPs.

6.5 Illustrative Examples of Superior Magnetic Nanoadsorbents: How They Are Designed for Effective Removal of Heavy Metals?

Broadly speaking, three different approaches have been utilized for the fabrication of MNPs-based adsorbents (Fig. 20):

(i) The surface of the bare MNPs is coated with a shell material that imparts the desired functionality while the magnetic core allows easy particle separation. The coated MNPs are thereafter employed directly for the heavy metal removal. But, the drawback associated with this approach is that it does not permit the selective recovery of the target metal species.
(ii) Either the bare or coated MNPs are functionalized using suitable linkers (functionalizing agents) that can bind with metals, thereby enabling their removal.

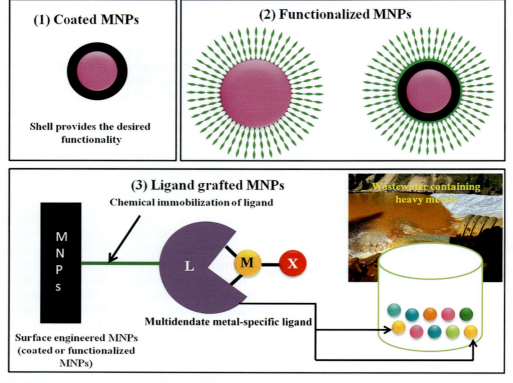

FIG. 20 Approaches for heavy metal removal using magnetic nanoadsorbents.

(iii) A suitable ligand that has affinity for a particular metal ion is grafted/chemically immobilized onto the surface of the engineered (coated or functionalized MNPs). This approach allows the selective removal of the target metal from the wastewater stream. Indeed, it can be considered to be "double green" approach, as it reduces the hazardous waste with simultaneous reduction in the usage of virgin resources.

6.5.1 Magnetic Nanoadsorbents for Removal of Pd(II)

The ever increasing demand for precious metals such as palladium that have found extensive use in catalytic and other related industries has led to growing interest in the recovery of these strategic elements. However, the separation of Pd(II) particularly is highly problematic because of its complex chemistry and overlapping properties. Uheida and coworkers utilized bare Fe_3O_4 NPs for the adsorption of Pd(II) from diluted hydrochloric acid solutions. Through adsorption studies, they found that although the equilibrium time was attained within 20 min, however selective adsorption could not be accomplished as these NPs showed stronger affinity for other metal ions such as Rh(III) rather than for Pd(II) (Uheida et al., 2006). Zhou et al. developed ethylenediamine modified magnetic cross-linking chitosan NPs (EMCN) for studying the removal of Pd(II) and Pt(IV) from aqueous solutions

(Zhou et al., 2010). A w/o microemulsion methodology was utilized for preparing the chitosan-coated MNPs which was followed by the grafting of ethylenediamine using epichlorohydrin as the cross-linking agent. Chitosan was used for the coating purpose because it had previously shown good efficiency in the removal of metal ions from aqueous solution due to the presence of amino and hydroxyl functional groups that served as chelation sites. The EMCN NPs showed good kinetic characteristics, that is, equilibrium time of less than 1 h and excellent adsorption loading capacities for Pt(IV) and Pd(II) at pH 2 (i.e., 171 and 138 mg g^{-1}, respectively). Regeneration studies were carried out for checking the reusability of the developed sorbent and it was found that the adsorption capacities of the EMCN for both metals could be retained at over 90% level up to fifth cycle.

Subsequently, Lin and Lien synthesized a novel magnetic nanoadsorbent comprising of magnetite nanoparticles coated with thiourea (MNP-Tu) for studying the adsorption of Pd(II), Au(III), and Pt(IV) from acidic aqueous solution (Lin and Lien, 2013). This nanoadsorbent showed high selectivity toward the precious metals especially Au(III) with the coexistence of other impurities. Also, a multicycle adsorption desorption experiment was conducted which showed that MNP-Tu NPs could be reused several times without much appreciable loss in its activity.

Recently Sharma and his research group have fabricated a highly efficient and magnetically retrievable core-shell structured nanoadsorbent (DAPTS@NH$_2$@SiO$_2$@Fe$_3$O$_4$) for the rapid and selective extraction of Pd^{2+} ions from various water samples (catalytic converter samples, water samples from commercial drinking water bottles, etc.) (Sharma et al., 2015). The synthesis of the nanoadsorbent was accomplished using a covalent immobilization approach wherein a selective ligand "2,6-diacetylpyridine monothiosemicarbazide (DAPTS)" was grafted onto the surface of amine functionalized silica-coated Fe$_3$O$_4$ NPs (Fig. 21).

TEM image of the bare Fe$_3$O$_4$ NPs showed the high dispersibility of NPs, while on the other hand, the TEM image of SiO$_2$@Fe$_3$O$_4$ NPs suggested the homogenous coatings of amorphous SiO$_2$ over the Fe$_3$O$_4$ NPs with uniform thickness of around 5 nm (Fig. 22). Studies revealed that as compared to the previous reports, the DAPTS@NH$_2$@SiO$_2$@Fe$_3$O$_4$ nanoadsorbent not only displayed outstanding adsorption characteristics by rapidly, efficiently, and selectively adsorbing Pd^{2+} ions from the bulk solution containing the complex matrix of multiple interfering ions but also exhibited higher reusability. Besides, it obeyed Langmuir adsorption isotherm model of adsorption.

6.5.2 Magnetic Nanoadsorbents for Removal of Pb(II)

Sarkar et al. have synthesized polyethylene glycol (PEG-4000)-coated Fe$_3$O$_4$ magnetic NPs for the selective removal of toxic Pb(II) ion from wastewater. The removal of Pb^{2+} from wastewater could be successfully accomplished via its adsorption onto Fe$_3$O$_4$ NPs within 10 min of equilibration time at pH 6 (Khayat Sarkar and Khayat Sarkar, 2013). Apart from Fe$_3$O$_4$, mesoporous amine functionalized magnesium ferrite nanoparticles (MgFe$_2$O$_4$–NH$_2$ NPs) (Fig. 23) have also been reported for the removal of Pb^{2+} from simulated wastewater (Nonkumwong et al., 2016). The selectivity experiments were carried out which showed that the MgFe$_2$O$_4$–NH$_2$ nanoadsorbent provides higher selectivity coefficient values for Pb^{2+} in comparison to other metal ions such as Ca^{2+}, Cd^{2+}, Zn^{2+}, Cu^{2+}, and Ni^{2+}.

Magnetic chitosan nanocomposites have also been applied for the extraction of Pb^{2+} ions from water (Liu et al., 2008). It was found that the interaction between chitosan and heavy metal ions is reversible, implying that those ions can be removed from chitosan in weak acidic deionized water with the assistance of ultrasound radiation. Covalent immobilization approach has also been employed for the fabrication of magnetic adsorbents that have proven to be useful for removal of Pb(II) ions. A magnetic

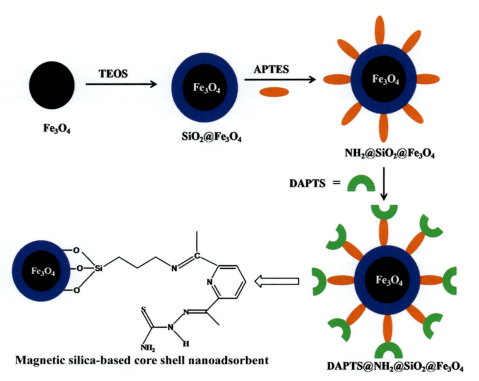

FIG. 21 Fabrication of DAPTS@NH$_2$@SiO$_2$@Fe$_3$O$_4$ nanoadsorbent. *Reprinted with permission from Sharma, R.K., Kumar, H., Kumar, A., 2015. A highly efficient and magnetically retrievable functionalized nanoadsorbent for ultrasonication assisted rapid and selective extraction of Pd^{2+} ions from water samples. RSC Adv. 5 (54), 43371–43380. Copyright 2014 Royal Society of Chemistry.*

nanoadsorbent was synthesized by Zargoosh et al., by the covalent immobilization of poly (acrylic acid) (PAA) and a ligating agent 4-phenyl-3-thiosemicarbazide (TSC) on the surface of Fe$_3$O$_4$ nanoparticles (Zargoosh et al., 2014). The developed nanomaterial showed good efficacy in removal of Pb(II) ions (maximum adsorption capacity was found to be 181.1 emu/g).

Behbahani and coworkers chemically modified the surface of magnetite nanoparticles by coating covalently with 3-aminopropyltriethoxysilane (APTES) and PEG as shown in Fig. 24 (Behbahani et al., 2014).

The synthesized MNPs-APTES-PEG NPs exhibited high uptake capability for Pb(II) ions.

Sharma et al. have developed an efficient, selective, and reusable acetoacetanilide (AAA) functionalized Fe$_3$O$_4$ nanoadsorbent for selective and cyclic removal of Pb^{2+} ions from different charged wastewaters (Sharma et al., 2014). The sorbent was fabricated using a similar approach as discussed previously for the removal of Pd^{2+} (Fig. 25) and its applicability was tested in rain water, river water, and mycorrhizal treated ash samples. Results revealed a very good quantitative recovery of Pb^{2+}. The AAA-NH$_2$-Si@MNPs adsorbent displayed advantages such as cost effectiveness of embedded MNPs, good material stability, high adsorption capacity, ease of fabrication, and superior reusability over multiple adsorption/desorption cycles.

Very recently, magnetic mesoporous carbon composites derived from in situ MgO template were employed for the fast removal of Pb^{2+}

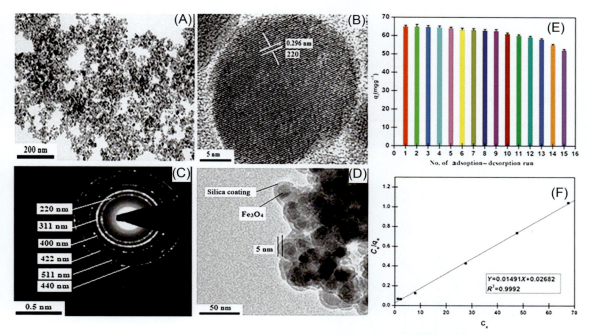

FIG. 22 (A) TEM image of synthesized Fe₃O₄ NPs, (B) HRTEM image of the single nanocrystal entity with lattice fringe, (C) SAED pattern of Fe₃O₄ and (D) TEM image of SiO2@Fe₃O₄, (E) reusability test of the adsorbent and (f) Linearized plot of Langmuir isotherm for adsorption of Pd^{2+} on nanoadsorbent. The schematic illustration of MgFe₂O₄-NH₂ synthesis and adsorption of Pb^{2+} ions. *Reprinted with permission from Sharma, R.K., Kumar, H., Kumar, A., 2015. A highly efficient and magnetically retrievable functionalized nanoadsorbent for ultrasonication assisted rapid and selective extraction of Pd^{2+} ions from water samples. RSC Adv. 5 (54), 43371–43380. Copyright 2014 Royal Society of Chemistry.*

(Zhang et al., 2018). For this, first calcination of the mixture of magnesium citrate and silica-coated Fe₃O₄ NPs were carried out in an inert atmosphere (Fig. 26).

Thereafter, a high content of Fe₃O₄@SiO₂ and MgO was in situ embedded in a carbon matrix and after this the MgO template was removed using diluted acid and finally the resulting material (Fe₃O₄@SiO₂@mC) was subjected to further oxidation treatment. This material presented fast adsorption dynamics (<1 min) and high adsorption capacity of 156 mg g^{-1} for Pb(II). The oxygen containing functional groups within the sorbent provided sites for the adsorption of the analyte material.

In light of the fact that the critical need of the hour is to design more efficient, environmentally friendly adsorbents, Luo and research group fabricated millimeter-scale magnetic cellulose-based beads possessing micro and nanopore structure using an optimal extrusion dropping technology from NaOH/urea aqueous solution (Fig. 27) (Luo et al., 2016).

The composite beads consisting of carboxyl (–COOH) decorated magnetite nanoparticles and nitric acid modified activated carbon showed effective removal performance for Pb(II), Cu(II), and Zn(II). The adsorption of the metal ions occurred primarily via an electrostatic interaction between the adsorbents surface and heavy metals. Recognizing the superlative qualities of Fe₃O₄ NPs and MnO₂, Chang et al. synthesized hierarchical MnO₂-coated magnetic nanocomposites (Fe₃O₄/MnO₂) using a mild

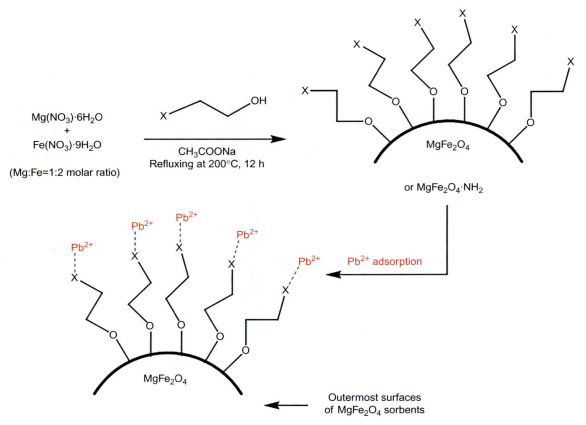

FIG. 23 The schematic illustration of MgFe$_2$O$_4$-NH$_2$ synthesis and adsorption of Pb^{2+} ions.

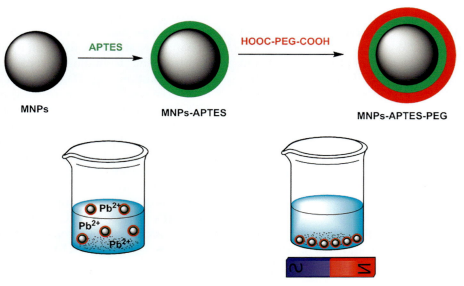

FIG. 24 Synthesis of MNPs and their surface modification by APTES and HOOC-PEG-COOH.

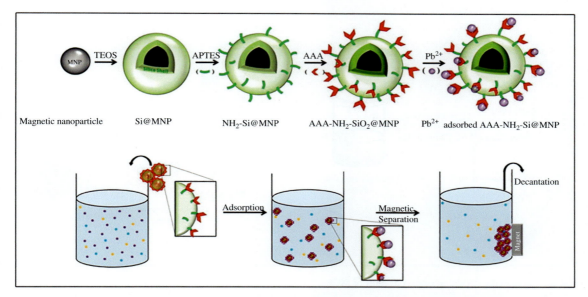

FIG. 25 Schematic illustration of the synthetic route of AAA-NH$_2$-Si@MNPs followed by Pb^{2+} removal sequence. *Reprinted with permission from Sharma, R.K., Puri, A., Monga, Y., Adholeya, A., 2014. Acetoacetanilide-functionalized Fe$_3$O$_4$ nanoparticles for selective and cyclic removal of Pb^{2+} ions from different charged wastewaters. J. Mater. Chem. A 2 (32), 12888–12898. Copyright 2014 Royal Society of Chemistry.*

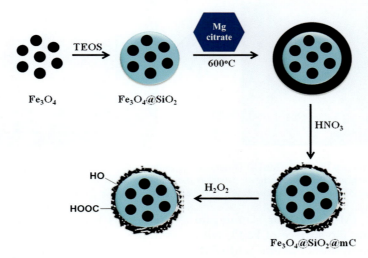

FIG. 26 Synthesis of the nanoadsorbent.

hydrothermal process (Fig. 28) which showed enhanced removal capacity toward four separate heavy metals [Cd(II), Cu(II), Pb(II), and Zn(II)] (Kim et al., 2013). The composites could be reused for five consecutive cycles without any substantial decrease in its adsorption capacity.

Du and coworkers have synthesized magnetic hydroxyapatite (HAP) entrapped agarose

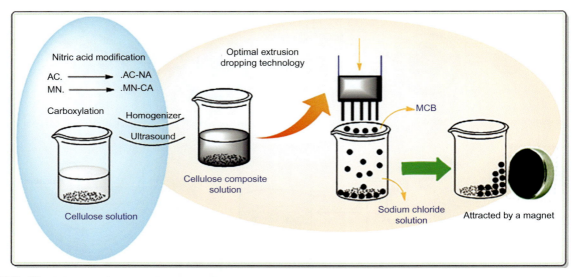

FIG. 27 Schematic depiction of preparation of MCB.

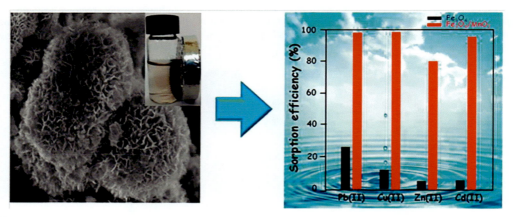

FIG. 28 Sorption efficiency of (Fe$_3$O$_4$/MnO$_2$) illustrated. *Reprinted with permission from Kim, E.J., Lee, C.S., Chang, Y.Y., Chang, Y.S., 2013. Hierarchically structured manganese oxide-coated magnetic nanocomposites for the efficient removal of heavy metal ions from aqueous systems. ACS Appl. Mater. Interfaces 5 (19), 9628–9634. Copyright 2013 American Chemical Society.*

composite beads (M-HAP/Agar composite beads) by modifying the surface of Fe$_3$O$_4$ with N (phosphonomethyl)-iminodiacetic acid (PM-IDA) and followed by coating with HAP (Fig. 29) (Zhang et al., 2017).

The resulting MHAP/Agar composite beads exhibited outstanding performance in adsorption of Pb^{2+} showing maximum binding capacities as high as 842.6 mg g^{-1}, respectively.

The results suggested that the magnetic NPs entrapped in beads had a positive effect on drastically improving the adsorption capacity.

6.5.3 Magnetic Nanoadsorbents for Removal of Cd(II)

The diverse industrial applications, unremitting reservoir exhaustion, environmental fate, and toxicity associated with Cd(II) necessitates

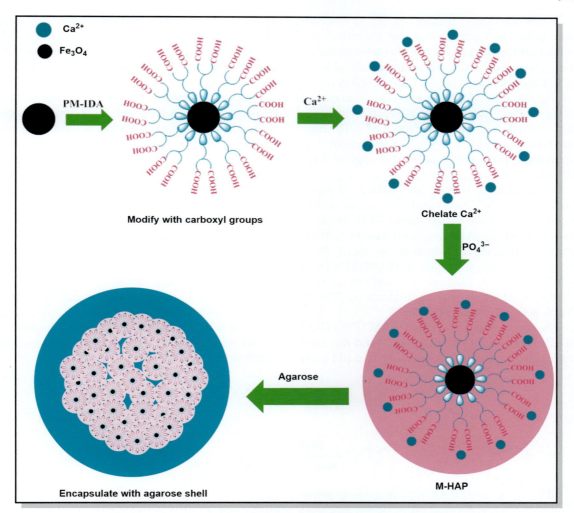

FIG. 29 Schematic illustration of M-HAP/Agar composite beads fabricated by modification with PM-IDA.

the establishment of sustainable analytical protocols devoted for rapid identification, separation, and recovery of this element. Magnetic separation by far has proven to be the most facile and profitable technology for the purpose of removal of Cd(II). Within this perspective, a number of low-cost polymer-coated MNP adsorbents have been reported for water treatment and environmental remediation. Gong and coworkers developed shellac-coated iron oxide NPs for Cd(II) ion removal from water (Gong et al., 2012). Shellac is a natural, economic, biodegradable, and renewable resin comprising of abundant hydroxyl and carboxylic groups. It was suggested that carboxyl groups on the surface of the developed nanosorbent participated in the reaction with cadmium ions and both electrostatic attraction as well as chemical adsorption were involved in the removal mechanism.

Ehrampoush and research group synthesized iron oxide nanoparticles using tangerine peel extract and used it as an adsorbent for cadmium ions removal from contaminated solution (Ehrampoush et al., 2015). A coprecipitation methodology was used for the preparation of iron oxide nanoparticles. The tangerine peel extract was used to prevent the aggregation of these NPs. The highest removal efficacy (90%) of cadmium ions occurred at pH 4 and adsorbent dosage of 0.4 g/ 100 mL.

Subsequently using an *Ananas comosus* peel pulp extract, Yoon and coworkers synthesized water-dispersible diethyl-4-(4-amino-5-mercapto-4*H*-1,2,4-triazol-3-yl)phenyl phosphonate (DEAMTPP)-capped biogenic Fe_3O_4 magnetic nanocomposites (DEAMTPP@Fe_3O_4 MNPs) that was utilized as a sorbent material for the rapid removal of cadmium ions (Fig. 30) (Venkateswarlu and Yoon, 2015).

The adsorption capacity of DEAMTPP@ Fe_3O_4 MNPs was found to be 49.2 mg g^{-1} and the maximum adsorption efficiency occurred at pH 6 and at 60 mg/L of Cd(II) concentration. The proposed mechanism has been depicted in Fig. 31. The ferromagnetic nature of the composite material allowed its facile recovery as well as recyclability.

Graphene oxide (GO) possesses the potential to remove heavy metal ions and ionic dyes in wastewater, but it suffers from separation inconvenience. To render the separation process easier, GO has been made magnetically recoverable by dispersing Fe_3O NPs. Xian et al. developed polyacrylic acid (PAA) functionalized water-soluble magnetic graphene nanocomposites that were utilized as nanoadsorbents for recyclable removal of Cd^{2+} from aqueous solution (Zhang et al., 2013). A copper catalyzed azide-alkyne cycloaddition reaction was employed for the first time for the synthesis of graphene oxide/Fe_3O_4 nanocomposites. The developed nanomaterial not only showed extraordinary removal efficiency over 85%, but was also easy to separate and could also be recycled due to the superparamagnetism exhibited by Fe_3O_4.

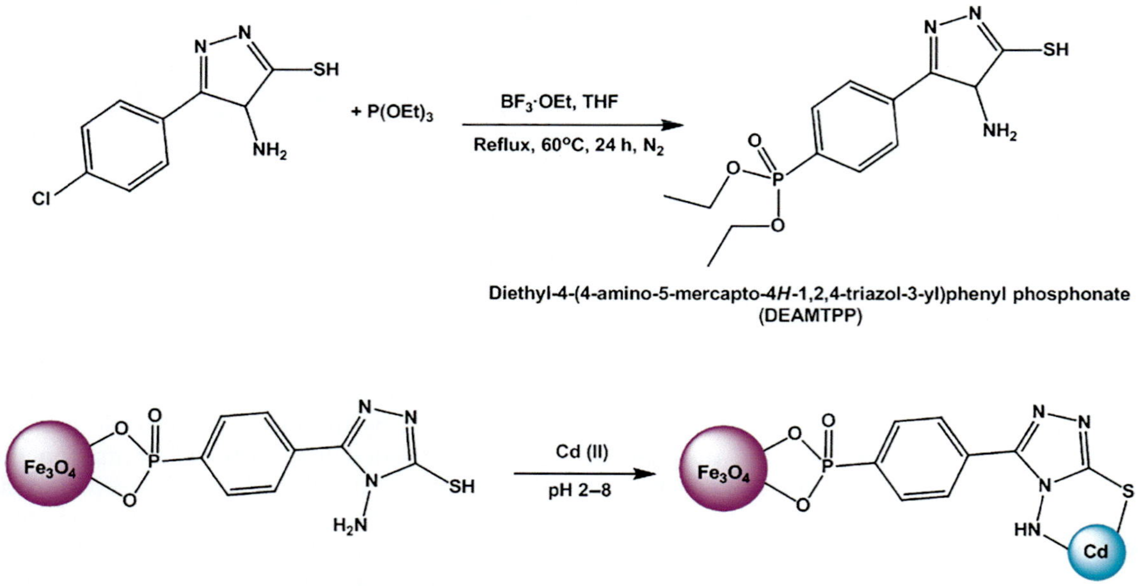

Diethyl-4-(4-amino-5-mercapto-4*H*-1,2,4-triazol-3-yl)phenyl phosphonate (DEAMTPP)

FIG. 30 The procedure for synthesis of DEAMTPP and Fe_3O_4 MNPs functionalized with DEAMTPP.

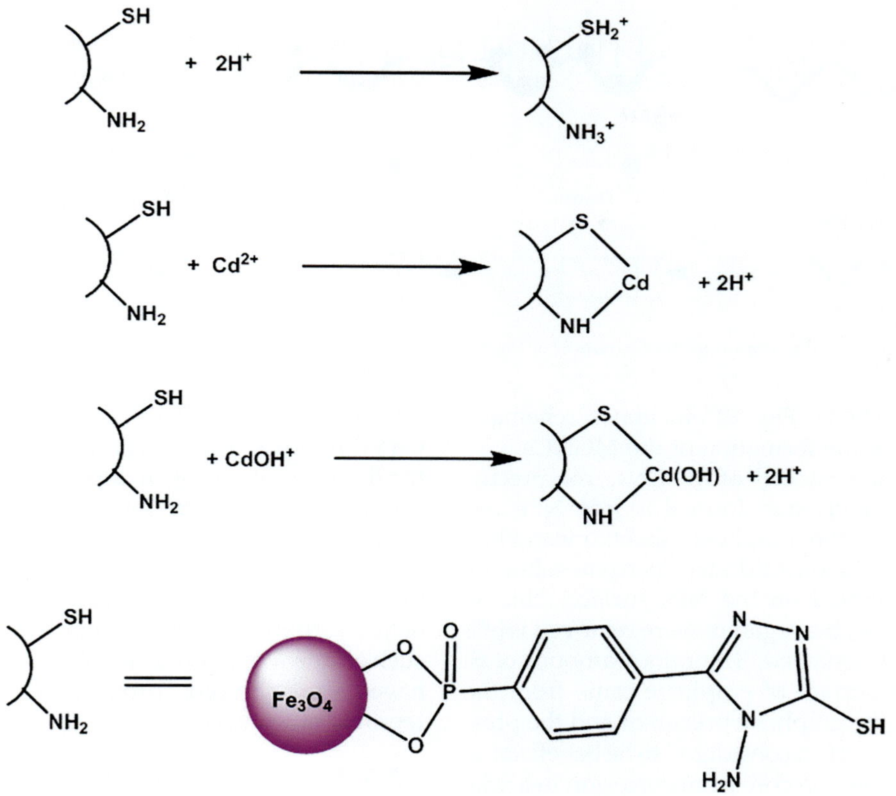

FIG. 31 Proposed mechanism for removal of Cd^{2+}.

Recently, maghemite decorated multiwalled carbon nanotubes (MWCNTs) were prepared using a versatile and cost-effective chemical route and subsequently investigated in the removal of Cd by Gatabi et al. (2016). The optimum adsorption capacity of cadmium ions was obtained at a contact time of 30 min and pH 8.0. The Freundlich adsorption isotherm was found to be the best fit that suggested the applicability of multilayer coverage of cadmium on magnetic MWCNTs surface.

6.5.4 Magnetic Nanoadsorbents for Removal of Cr(III) and Cr(VI)

As described in the previous sections, Cr(III) and Cr(VI) are the two most common contaminants found in industrial wastewater effluents. To tackle challenges of chromium contamination, magnetic nanocomposites have been developed and utilized as efficient sorbents for the clean-up of these metal contaminants especially Cr(VI). For instance Sinha et al. fabricated superparamagnetic starch functionalized maghemite nanoparticles (SMhNPs) using an in situ functionalized-based coprecipitation method which showed good efficacy in the removal of Cr(VI) from aqueous waste (Singh et al., 2015).

Recently, Zhu and his other research members demonstrated the synthesis of magnetic graphene nanocomposites decorated with core@double-shell NPs using a facile thermodecomposition process that enabled fast Cr(VI) removal from the wastewater with a very high removal efficiency and within 5 min

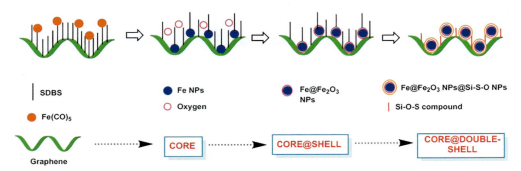

FIG. 32 Schematic illustration of the formation of the MGNCs.

(Zhu et al., 2011). Fig. 32 elucidates schematic illustration of the formation of the MGNCs.

During the synthesis of MGCNs, iron precursor got eventually transformed to iron NPs and adhered onto the graphene sheet; meanwhile the dissipated sodium dodecylbenzenesulfonate (SDBS) assembled on the NPs surface. Finally, after annealing, the organics were removed while the S and O remained. The minor amount of Si detected on surface of graphene came from the impurity of the graphene precursor and the presence of Si was in fact considered to be beneficial as it protected the iron core from corrosion in acidic medium. The energy-filtered TEM (EFTEM) was utilized to clarify the specific component of the core@double-shell structure, especially the outer shell. The presence of Fe, O, S, Si, and C confirm an iron core covered with a crystalline iron oxide inner shell and an amorphous outer shell made of sulfur and silicon elements (Fig. 33).

Lately, nitrogen-doped magnetic carbons have proven their efficiency as adsorbents for the heavy metal removal. Doping with heteroatoms, such as nitrogen, fluorine, and sulfur, can increase the negative charge density on the surface of the materials, which enhance the probability of metal adsorption. Huang et al. developed fluorine- and nitrogen-codoped magnetic carbons (FN-MC) via a simple one-step thermal pyrolysis method using polyvinylidene fluoride (PVDF) and melamine as F and N precursor without postprocessing (Huang et al., 2018). The material prove to be highly beneficial for Cr(VI) removal due to synergistic effect between the fluorine and nitrogen dopant on the magnetic carbons (as high as 188.7 mg g^{-1} and 2.28 mg m^{-2} in neutral solution, while 740.7 mg g^{-1} and 8.96 mg m^{-2} in acid solution). Huang et al. have prepared fibrillar and particulate structure magnetic carbons from inexpensive precursors such as polyacrylonitrile and Fe(NO$_3$)$_3$·9H$_2$O that have also showed promising efficiency in removal of Cr(VI) (Huang et al., 2017).

6.5.5 Magnetic Nanoadsorbents for Removal of As(III)

Arsenic contamination has affected several countries worldwide as already discussed. According to extensive literature survey carried out, the order of toxicities of arsenic species have been found to be: arsenite > arsenate > monomethyl arsenic acid (MMA) > dimethyl arsenic acid (DMA). Consequently efforts are being directed toward combating this problem. Zhu and his team have synthesized zerovalent iron supported onto active carbon for arsenic removal from drinking water (Zhu et al., 2009). Langmuir adsorption isotherms revealed that the adsorption capacities of arsenite and arsenate were 18.2 and 12.0 mg g^{-1}, respectively. Through studies, authors found that the presence of phosphate and silicate decreased the removal of both arsenite and arsenate, while on the other hand, the effect of other anions and humic acid remained

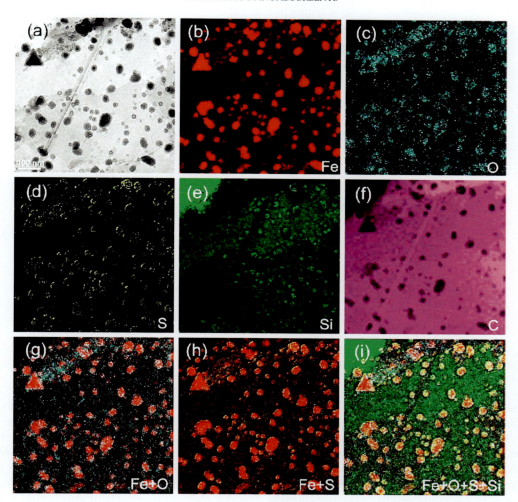

FIG. 33 EFTEM of the MGNCs (A) zero-loss image, (B) Fe map, (C) O map, (D) S map, (E) Si map, (F) C map, (G) Fe + O map, (H) Fe + S map, and (I) Fe + O + S + Si map. *Reprinted with permission from Zhu, J., Wei, S., Gu, H., Rapole, S.B., Wang, Q., Luo, Z., … Guo, Z., 2011. One-pot synthesis of magnetic graphene nanocomposites decorated with core@ double-shell nanoparticles for fast chromium removal. Environ. Sci. Technol. 46 (2), 977–985. Copyright 2012 American Chemical Society.*

insignificant. Subsequently, Mostafa et al. carried out adsorption study of As(V) onto nanosized iron oxide-coated quartz (IOCQ) (Mostafa et al., 2010). This protocol enabled 100% arsenic removal within 5.0 min; thus the reported method was found to be suitable for the removal of arsenate from drinking water. Magnetite maghemite nanoparticles have also been employed for arsenic removal from aqueous solution by Chowdhury and Yanful (2010). They found that 96%–99% arsenic uptake could be accomplished using these NPs under controlled pH conditions. While the maximum arsenic sorption was found to be 3.69 mg g^{-1} for As(III), while for As(V), it was found to be 3.71 mg g^{-1}. But the only limitation of this method was reduction in arsenic uptake in presence of competing anion phosphate.

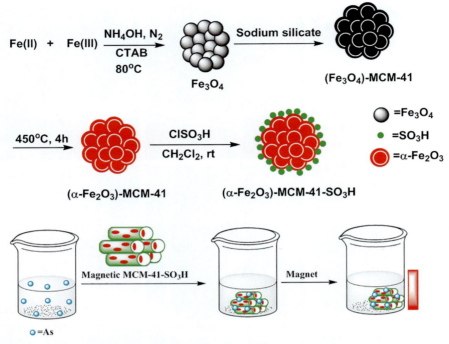

FIG. 34 Synthesis process of MCM-41-SO$_3$H and schematic illustration of As (V) adsorption by MMCM-41-SO$_3$H in optimal conditions.

Apart from this, chitosan-coated Fe$_3$O$_4$ nanoparticles were also used for the purpose of removal of arsenic from aqueous solution (Abdollahi et al., 2015). For the development of the nanosorbent, chitosan was first carboxymethylated and then covalently grafted on the surface of Fe$_3$O$_4$ nanoparticles via carbodiimide activation. This nanomaterial contributed significantly to the enhancement of the adsorption capacity because of the strong ability of multiple functional groups like hydroxyl and amino on chitosan to bind arsenic. This methodology could reduce arsenic concentration under the allowed limit as declared by WHO.

Recently, mesoporous (organo) silica decorated with MNPs [specifically, magnetic mobile crystalline material-41 (MCM-41) functionalized by chlorosulfonic acid (MMCM-41-SO$_3$H)] prove to be an efficient and reusable nanoadsorbent for arsenic removal from water samples (Fig. 34) (Hasanzadeh et al., 2015). Results revealed that the coefficient of determination (R^2) was more than 0.999 while the limit of detection (LOD) was around 0.061 ppb. In view of these results, MMCM-41-SO$_3$H was considered to be a potent material for the removal of As(V) contaminants and it showed prospects of application in large-scale wastewater treatment plants.

Saiz and coworkers have reported magnetic silica/magnetite nanoparticles functionalized with aminopropyl groups incorporating Fe^{3+} for adsorption of As^{5+} and As^{3+} from polluted groundwater (Saiz et al., 2014). Fig. 35 depicts the schematic representation of the adsorbent preparation methodology. Batch adsorption process was employed along with the magnetic recovery of the adsorbent to carry out the removal of arsenic. Maximum arsenic adsorption capacities of material S1-F3 was found to be 14.7 ± 0.3 mg As^{3+} g^{-1} and 121 ± 4.1 mg As^{5+} g^{-1}, respectively.

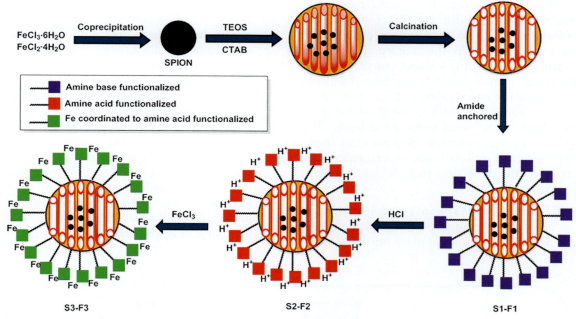

FIG. 35 Synthesis of solid adsorbent S1, S1-F1, S1-F2, and S1-F3.

Recently, magnetic nanocrystalline barium hexaferrite has also been utilized for the first time in arsenic removal from wastewater and it showed better (75%) arsenic removal than magnetite of the similar sizes (Patel et al., 2012).

Apart from these materials, magnetic metal organic framework (MOF) [CoFe$_2$O$_4$@MIL-100(Fe) hybrid MNPs] has been reported to exhibit fast and selective adsorption of inorganic arsenic (iAs) with high adsorption capacity (Yang and Yin, 2017). The hybrid adsorbent had a core-shell-mesoporous structure. Through adsorption experiments it was found that the maximum adsorption capacities were 114.8 mg g^{-1} for As(V) and 143.6 mg g^{-1} for As(III), which were much higher as compared to the most of the previously reported adsorbents. The MIL-100(Fe) shell played a very crucial role in improving the anti-interference capacity for As adsorption through the electrostatic repulsion and size exclusion effects. Adsorption mechanism was studied and it was found that Fe-O-As structure was formed on CoFe$_2$O$_4$@MIL-100(Fe) via hydroxyl substitution with the deprotonated iAs species. Batch magnetic separation mode and a portable filtration strategy also validated the high efficiency and excellent pH- and interference-tolerance capacities of the nanoadsorbent that allowed effective iAs removal from natural water samples.

7 NEED FOR LARGE-SCALE INDUSTRIAL APPLICABILITY: INTRODUCING REACTORS FOR ONLINE AND CYCLIC RECOVERY OF HEAVY METALS

The fruitful utility of any developed material can be realized only when it is capable of performing the desired task on a large scale. Since many of the synthesized adsorbents have integrated efficiency and sustainability in a constructive manner, it is anticipated that they can be fostered to control environmental pollution and to balance resource utilization on a large scale. In this regard, it is worth mentioning that Sharma and coworkers have designed a novel batch mode solid-phase reactor for the online, efficient, and fast extraction of chromium from tannery waste using recyclable silica-based metal adsorbents. This technology was patented (Sharma et al., n.d.). The designed online sorbent extraction and preconcentration system (Fig. 36) has following components:

(i) *Upper segment*: The invented system's injection part(upper segment) is equipped with four sequential glass inputs, which can be permuted and combined as per the need for fluidic delivery of digested media, buffer (to maintain pH), washing, and eluting media through inlet propylene tubes into the batch extractor (vessel). These propylene tubes are fitted with knobs, valves. and pump for controlled and regulated flow. One of the inputs is generally being left for air passage. It also has one centered pH electrode port, where electrode can be placed which is outwardly controlled by a pH controller. Any change during propagation of agitation can be sensed by pH electrode and it automatically leads to inflow pumping of acidic/basic solution and

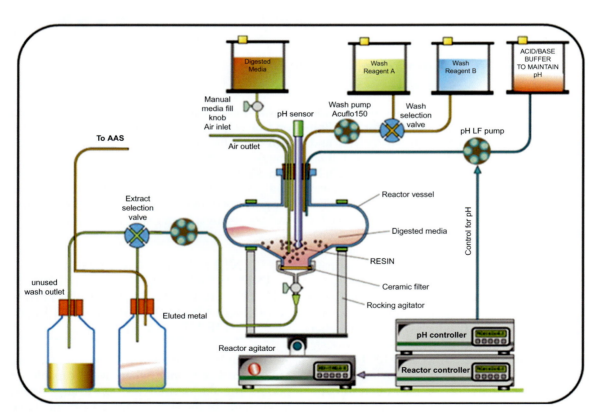

FIG. 36 Reactor for online recovery of heavy metals.

adjusted as per the environment in the vessel to the desired pH.

(ii) *Batch extractor*: The batch extractor, which is heart of the system, has a silica porous disc and is connected to the motor to provide it a special motion for contacting the aqueous medium with resin for a period of time sufficient to allow the resin to extract specified metal.

(iii) *Elution unit*: The lower segment of the extractor is further hooked up with elution unit containing suction pump attached to the extractor through polypropylene tube in order to pump out the solution from extractor at a controlled flow rate.

7.1 Working of the Reactor

Initially, the batch extractor is charged with an optimized amount of chelating resin and then the digested effluent, containing the metal ion to be sorbed, was delivered into it through one of the sequential glass input. The pH of the solution was adjusted to required value with the help of pH adjusting unit and thereafter the mixture in the extractor was subjected to agitation for an appropriate time. After the adsorption of metal ion on the resin, the solution was pumped out through a polypropylene tube at the bottom of the extractor. Then for desorption of metal ion from the resin, selected eluting agent entered through one of the valve of the extractor for the elution. The concentration of metal ion was determined directly by aspirating the desorbed metal ion solution into flame atomic absorption spectrophotometer.

7.2 Advantages

The advantage associated with this designed reactor is that large volume of samples can be injected at a time, and the silica-based organic–inorganic hybrid materials/nanomaterials can be used continuously for longer time without change in its adsorption properties. The group

is currently working on design of a magnetic reactor that can extract target metal ions from large volumes of analyte.

8 CONCLUSIONS AND FUTURE OUTLOOK

The onset of industrialization has undeniably brought development and prosperity; however the flip side of the coin is that it has given rise to serious environmental issues in the society by disturbing the ecosystem. One of the most visible impacts of its irreparable consequences is water pollution arising due to discharge of heavy metals. The critical need for combating the challenges of heavy metal pollution and sustainable development has been realized and this has in fact spawned wide-ranging environmental legislation in most developed countries. Today, substantial research efforts are thus underway for the development of promising water treatment technologies. In this context, it is important to mention that nanotechnology coupled with the adsorption technology has offered leapfrogging opportunities for the heavy metal clean-up process. Magnetic nanoadsorbents have in particular shown their efficacy in this regard, as they possess unique superparamagnetic properties that allow easy separation from water in a weak magnetic field in addition to their other intrinsic benefits they offer which include high adsorption capacity, faster removal rates, excellent selectivity, greater reusability, etc. These features render them a superior alternative to most of the conventional sorbents. Although, a growing interest in the use of surface engineered MNPs-based adsorbents has been witnessed off late, but the large-scale industrial application of these methods is still in the early stage and more research in the field is certainly necessary. Especially, research on the environmental fate and health impact of these nanomaterials is required. Nevertheless, the problems related to poor water quality in many countries of the world make their future quite promising.

References

Abdollahi, M., Zeinali, S., Nasirimoghaddam, S., Sabbaghi, S., 2015. Effective removal of As (III) from drinking water samples by chitosan-coated magnetic nanoparticles. Desalin. Water Treat. 56 (8), 2092–2104.

Abdulwahab, K., Malik, M.A., O'Brien, P., Govender, K., Muryn, C.A., Timco, G.A., ... Winpenny, R.E., 2013. Synthesis of monodispersed magnetite nanoparticles from iron pivalate clusters. Dalton Trans. 42 (1), 196–206.

Abid, B.A., Brbooti, M.M., Al-Shuwaiki, N.M., 2011. Removal of heavy metals using chemicals precipitation. Eng. Technol. J. 29 (3), 595–612.

Aghaei, E., Alorro, R.D., Encila, A.N., Yoo, K., 2017. Magnetic adsorbents for the recovery of precious metals from leach solutions and wastewater. Meta 7 (12), 529.

Ahuja, S., 2015. Overview of global water challenges and solutions. In: Water Challenges and Solutions on a Global Scale. American Chemical Society, pp. 1–25.

Ahuja, S., 2016. Chemistry and Water: The Science Behind Sustaining the World's Most Crucial Resource. Elsevier.

Akbarzadeh, A., Samiei, M., Davaran, S., 2012. Magnetic nanoparticles: preparation, physical properties, and applications in biomedicine. Nanoscale Res. Lett. 7 (1), 144.

Akpor, O.B., Otohinoyi, D.A., Olaolu, T.D., Aderiye, B.I., 2014. Pollutants in wastewater effluents: impacts and remediation processes. Int. J. Environ. Res. Earth Sci. 3, 050–059. No. 3. March, 2014.

Al-Rashdi, B.A.M., Johnson, D.J., Hilal, N., 2013. Removal of heavy metal ions by nanofiltration. Desalination 315, 2–17.

Angermann, A., Töpfer, J., 2008. Synthesis of magnetite nanoparticles by thermal decomposition of ferrous oxalate dihydrate. J. Mater. Sci. 43 (15), 5123–5130.

Aoshima, K., 2016. Itai-itai disease: renal tubular osteomalacia induced by environmental exposure to cadmium-historical review and perspectives. Soil Sci. Plant Nutr. 62 (4), 319–326.

Bakalár, T., Búgel, M., Gajdošová, L., 2009. Heavy metal removal using reverse osmosis. Acta Montan. Slovaca 14 (3), 250.

Barakat, M.A., 2011. New trends in removing heavy metals from industrial wastewater. Arab. J. Chem. 4 (4), 361–377.

Bazrafshan, E., Mohammadi, L., Ansari-Moghaddam, A., Mahvi, A.H., 2015. Heavy metals removal from aqueous environments by electrocoagulation process—a systematic review. J. Environ. Health Sci. Eng. 13 (1), 74.

Beaumont, J.J., Sedman, R.M., Reynolds, S.D., Sherman, C.D., Li, L.H., Howd, R.A., ... Alexeeff, G.V., 2008. Cancer mortality in a Chinese population exposed to hexavalent chromium in drinking water. Epidemiology 19, 12–23.

Behbahani, N.S., Rostamizadeh, K., Yaftian, M.R., Zamani, A., Ahmadi, H., 2014. Covalently modified magnetite nanoparticles with PEG: preparation and characterization as nano-adsorbent for removal of lead from wastewater. J. Environ. Health Sci. Eng. 12 (1), 103.

Bernhoft, R.A., 2013. Cadmium toxicity and treatment. Sci. World J. 2013, Article ID 394652, 7 pages. https://doi.org/10.1155/2013/394652.

Bhowmick, S., Pramanik, S., Singh, P., Mondal, P., Chatterjee, D., Nriagu, J., 2018. Arsenic in groundwater of West Bengal, India: a review of human health risks and assessment of possible intervention options. Sci. Total Environ. 612, 148–169.

Blanco-Andujar, C., Thanh, N.T., 2010. Synthesis of nanoparticles for biomedical applications. Annu. Rep. Prog. Chem., Sect. A: Inorg. Chem. 106, 553–568.

Bobade, V., Eshtiagi, N., 2015. Heavy metals removal from wastewater by adsorption process: A review. In: Asia Pacific Confederation of Chemical Engineering Congress 2015: APCChE 2015, incorporating CHEMECA 2015. Engineers Australia, p. 312.

Bolger, P.M., Schwetz, B.A., 2002. Mercury and health. N. Engl. J. Med. 347 (22), 1735–1736.

Bystrzejewski, M., Pyrzyńska, K., Huczko, A., Lange, H., 2009. Carbon-encapsulated magnetic nanoparticles as separable and mobile sorbents of heavy metal ions from aqueous solutions. Carbon 47 (4), 1201–1204.

Castoldi, A.F., Coccini, T., Manzo, L., 2003. Neurotoxic and molecular effects of methylmercury in humans. Rev. Environ. Health 18 (1), 19–32.

Chowdhury, S.R., Yanful, E.K., 2010. Arsenic and chromium removal by mixed magnetite–maghemite nanoparticles and the effect of phosphate on removal. J. Environ. Manag. 91 (11), 2238–2247.

Cohen, M.D., Kargacin, B., Klein, C.B., Costa, M., 1993. Mechanisms of chromium carcinogenicity and toxicity. Crit. Rev. Toxicol. 23 (3), 255–281.

Courtland, R., 2008. Enough water to go round. Nature. https://doi.org/10.1038/news.2008.678.

Daou, T.J., Pourroy, G., Bégin-Colin, S., Greneche, J.M., Ulhaq-Bouillet, C., Legaré, P., ... Rogez, G., 2006. Hydrothermal synthesis of monodisperse magnetite nanoparticles. Chem. Mater. 18 (18), 4399–4404.

Ehrampoush, M.H., Miria, M., Salmani, M.H., Mahvi, A.H., 2015. Cadmium removal from aqueous solution by green synthesis iron oxide nanoparticles with tangerine peel extract. J. Environ. Health Sci. Eng. 13 (1), 84.

Farooq, U., Kozinski, J.A., Khan, M.A., Athar, M., 2010. Biosorption of heavy metal ions using wheat based biosorbents—a review of the recent literature. Bioresour. Technol. 101 (14), 5043–5053.

Flora, S.J.S., 2014. Handbook of Arsenic Toxicology. Academic Press.

REFERENCES

Fu, F., Wang, Q., 2011. Removal of heavy metal ions from wastewaters: a review. J. Environ. Manag. 92 (3), 407–418.

Gatabi, M.P., Moghaddam, H.M., Ghorbani, M., 2016. Efficient removal of cadmium using magnetic multi-walled carbon nanotube nanoadsorbents: equilibrium, kinetic, and thermodynamic study. J. Nanopart. Res. 18 (7), 189.

Gawande, M.B., Monga, Y., Zboril, R., Sharma, R.K., 2015. Silica-decorated magnetic nanocomposites for catalytic applications. Coord. Chem. Rev. 288, 118–143.

Ge, F., Li, M.M., Ye, H., Zhao, B.X., 2012. Effective removal of heavy metal ions $Cd2+, Zn2+, Pb2+, Cu2+$ from aqueous solution by polymer-modified magnetic nanoparticles. J. Hazard. Mater. 211, 366–372.

Gerhardsson, L., Dahlgren, E., Eriksson, A., Lagerkvist, B., Lundstrom, J., Nordberg, G.F., 1988. Fatal arsenic poisoning—a case report. Scand. J. Work Environ. Health 14, 130–133.

Goel, P.K., 2006. Water Pollution: Causes, Effects and Control. New Age International.

Gong, J., Chen, L., Zeng, G., Long, F., Deng, J., Niu, Q., He, X., 2012. Shellac-coated iron oxide nanoparticles for removal of cadmium (II) ions from aqueous solution. J. Environ. Sci. 24 (7), 1165–1173.

Gunatilake, S.K., 2015a. Methods of removing heavy metals from industrial wastewater. J. Multidiscip. Eng. Sci. Stud. 1 (1), 14–15.

Gunatilake, S.K., 2015b. Methods of removing heavy metals from industrial wastewater. Methods 1 (1), 14.

Haley, R., 2016. Evaluation of a Colorimetric Assay for the Detection of Arsenic in Water. https://www.researchgate.net/profile/Ryan_Haley/publication/323525844_Evaluation_of_a_colorimetric_assay_for_the_detection_of_arsenic_in_water/links/5a99b51445851586a2a9f77c/Evaluation-of-a-colorimetric-assay-for-the-detection-of-arsenic-in-water.pdf.

Hanna-Attisha, M., LaChance, J., Sadler, R.C., Champney Schnepp, A., 2016. Elevated blood lead levels in children associated with the Flint drinking water crisis: a spatial analysis of risk and public health response. Am. J. Public Health 106 (2), 283–290.

Harada, M., 1995. Minamata disease: methylmercury poisoning in Japan caused by environmental pollution. Crit. Rev. Toxicol. 25 (1), 1–24.

Harris, H.J., Wenger, R.B., Harris, V.A., Devault, D.S., 1994. A method for assessing environmental risk: a case study of green Bay, Lake Michigan, USA. Environ. Manag. 18 (2), 295–306.

Hasanzadeh, M., Farajbakhsh, F., Shadjou, N., Jouyban, A., 2015. Mesoporous (organo) silica decorated with magnetic nanoparticles as a reusable nanoadsorbent for arsenic removal from water samples. Environ. Technol. 36 (1), 36–44.

He, Q., Zhang, Z., Xiong, J., Xiong, Y., Xiao, H., 2008. A novel biomaterial—Fe3O4: TiO2 core-shell nano particle with magnetic performance and high visible light photocatalytic activity. Opt. Mater. 31 (2), 380–384.

Huang, J., Cao, Y., Shao, Q., Peng, X., Guo, Z., 2017. Magnetic nanocarbon adsorbents with enhanced hexavalent chromium removal: morphology dependence of fibrillar vs particulate structures. Ind. Eng. Chem. Res. 56 (38), 10689–10701.

Huang, J., Li, Y., Cao, Y., Peng, F., Cao, Y., Shao, Q., Liu, H., Guo, Z., 2018. J. Mater. Chem. A. https://doi.org/10.1039/C8TA02861C.

Hughes, M.F., Beck, B.D., Chen, Y., Lewis, A.S., Thomas, D.J., 2011. Arsenic exposure and toxicology: a historical perspective. Toxicol. Sci. 123 (2), 305–332.

Hung, Y.T., Wang, L.K., Wang, M.H.S., Shammas, N.K., Chen, J.P., 2016. Remediation of Heavy Metals in the Environment. CRC Press.

Inoue, K.I., 2013. Heavy metal toxicity. J. Clinic Toxicol. S3, 2161-0495.

Jaishankar, M., Tseten, T., Anbalagan, N., Mathew, B.B., Beeregowda, K.N., 2014. Toxicity, mechanism and health effects of some heavy metals. Interdiscip. Toxicol. 7 (2), 60–72.

Jia, H., 2009. Heavy Metal Poisoning Sparks Protests in China. Chemistry World.

Jiao, M., Zeng, J., Jing, L., Liu, C., Gao, M., 2015. Flow synthesis of biocompatible Fe3O4 nanoparticles: insight into the effects of residence time, fluid velocity, and tube reactor dimension on particle size distribution. Chem. Mater. 27 (4), 1299–1305.

Juang, R.S., Shiau, R.C., 2000. Metal removal from aqueous solutions using chitosan-enhanced membrane filtration. J. Membr. Sci. 165 (2), 159–167.

Kalyani, S., Sangeetha, J., Philip, J., 2015. Microwave assisted synthesis of ferrite nanoparticles: effect of reaction temperature on particle size and magnetic properties. J. Nanosci. Nanotechnol. 15 (8), 5768–5774.

Khan, M.S., Zaidi, A., Goel, R., Musarrat, J., 2011. Biomanagement of Metal-Contaminated Soils. Vol. 20. Springer Science & Business Media.

Khayat Sarkar, Z., Khayat Sarkar, F., 2013. Selective removal of lead (II) ion from wastewater using superparamagnetic monodispersed iron oxide (Fe3O4) nanoparticles as a effective adsorbent. Int. J. Nanosci. Nanotechnol. 9 (2), 109–114.

Kim, E.J., Lee, C.S., Chang, Y.Y., Chang, Y.S., 2013. Hierarchically structured manganese oxide-coated magnetic nanocomposites for the efficient removal of heavy metal ions from aqueous systems. ACS Appl. Mater. Interfaces 5 (19), 9628–9634.

Kumar, A., 2006. Ecobiology of Polluted Waters. Daya Books.

Lakherwal, D., 2014. Adsorption of heavy metals: a review. Int. J. Environ. Res. Dev. 4 (1), 41–48.

EVALUATING WATER QUALITY TO PREVENT FUTURE DISASTERS

Laurent, S., Forge, D., Port, M., Roch, A., Robic, C., Vander Elst, L., Muller, R.N., 2008. Magnetic iron oxide nanoparticles: synthesis, stabilization, vectorization, physico-chemical characterizations, and biological applications. Chem. Rev. 108 (6), 2064–2110.

Le Blanc, D., 2015. Towards integration at last? The sustainable development goals as a network of targets. Sustain. Dev. 23 (3), 176–187.

Le, N.L., Nunes, S.P., 2016. Materials and membrane technologies for water and energy sustainability. Sustain. Mater. Technol. 7, 1–28.

Li, R., Zhang, L., Wang, P., 2015. Rational design of nanomaterials for water treatment. Nanoscale 7 (41), 17167–17194.

Li, Z., Yu, B., Cong, H., Yuan, H., Peng, Q., 2017. Recent development and application of solid phase extraction materials. Rev. Adv. Mater. Sci. 48, 87–111.

Liao, M.H., Chen, D.H., 2002. Preparation and characterization of a novel magnetic nano-adsorbent. J. Mater. Chem. 12 (12), 3654–3659.

Lin, T.-L., Lien, H.-L., 2013. Effective and selective recovery of precious metals by thiourea modified magnetic nanoparticles. Int. J. Mol. Sci. 14, 9834–9847.

Liu, Z.L., Wang, X., Yao, K.L., Du, G.H., Lu, Q.H., Ding, Z.H., … Xi, D., 2004. Synthesis of magnetite nanoparticles in W/O microemulsion. J. Mater. Sci. 39 (7), 2633–2636.

Liu, X., Hu, Q., Fang, Z., Zhang, X., Zhang, B., 2008. Magnetic chitosan nanocomposites: a useful recyclable tool for heavy metal ion removal. Langmuir 25 (1), 3–8.

Lokhande, R.S., Singare, P.U., Pimple, D.S., 2011. Toxicity study of heavy metals pollutants in waste water effluent samples collected from Taloja Industrial Estate of Mumbai, India. Resour. Environ. 1 (1), 13–19.

Lu, A.H., Salabas, E.E., Schüth, F., 2007. Magnetic nanoparticles: synthesis, protection, functionalization, and application. Angew. Chem. Int. Ed. 46 (8), 1222–1244.

Lu, T., Wang, J., Yin, J., Wang, A., Wang, X., Zhang, T., 2013. Surfactant effects on the microstructures of Fe3O4 nanoparticles synthesized by microemulsion method. Colloids Surf. A Physicochem. Eng. Asp. 436, 675–683.

Luo, X., Lei, X., Cai, N., Xie, X., Xue, Y., Yu, F., 2016. Removal of heavy metal ions from water by magnetic cellulose-based beads with embedded chemically modified magnetite nanoparticles and activated carbon. ACS Sustain. Chem. Eng. 4 (7), 3960–3969.

Majidi, S., Zeinali Sehrig, F., Farkhani, S.M., Soleymani Goloujeh, M., Akbarzadeh, A., 2016. Current methods for synthesis of magnetic nanoparticles. Artif Cells Nanomed Biotechnol. 44 (2), 722–734.

Malik, D., Singh, S., Thakur, J., Singh, R.K., Kaur, A., Nijhawan, S., 2014. Heavy metal pollution of the Yamuna river: an introspection. Int. J. Curr. Microbiol. App. Sci. 3 (10), 856–863.

Maximous, N.N., Nakhla, G.F., Wan, W.K., 2010. Removal of heavy metals from wastewater by adsorption and membrane processes: a comparative study. World Acad. Sci. Eng. Technol. 64, 594–599.

Mikulášek, P., Cuhorka, J., 2016. Removal of heavy metal ions from aqueous solutions by nanofiltration. Chem. Eng. 47, 379–384.

Mostafa, M.G., Chen, Y.H., Jean, J.S., Liu, C.C., Teng, H., 2010. Adsorption and desorption properties of arsenate onto nano-sized iron-oxide-coated quartz. Water Sci. Technol. 62 (2), 378–386.

Mulware, S.J., 2013. Trace elements and carcinogenicity: a subject in review. 3 Biotechnol. 3 (2), 85–96.

Nagajyoti, P.C., Lee, K.D., Sreekanth, T.V.M., 2010. Heavy metals, occurrence and toxicity for plants: a review. Environ. Chem. Lett. 8 (3), 199–216.

Nonkumwong, J., Ananta, S., Srisombat, L., 2016. Effective removal of lead (II) from wastewater by amine-functionalized magnesium ferrite nanoparticles. RSC Adv. 6 (53), 47382–47393.

Nordberg, G.F., Fowler, B.A., Nordberg, M. (Eds.), 2014. Handbook on the Toxicology of Metals. Academic Press.

Ojemaye, M.O., Okoh, O.O., Okoh, A.I., 2017. Surface modified magnetic nanoparticles as efficient adsorbents for heavy metal removal from wastewater: progress and prospects. Mater. Express 7 (6), 439–456.

Pal, T., Mukherjee, P.K., Sengupta, S., Bhattacharyya, A.K., Shome, S., 2002. Arsenic pollution in groundwater of West Bengal, India—an insight into the problem by subsurface sediment analysis. Gondwana Res. 5 (2), 501–512.

Pan, S., Shen, H., Xu, Q., Luo, J., Hu, M., 2012. Surface mercapto engineered magnetic Fe3O4 nanoadsorbent for the removal of mercury from aqueous solutions. J. Colloid Interface Sci. 365 (1), 204–212.

Park, J., An, K., Hwang, Y., Park, J.G., Noh, H.J., Kim, J.Y., … Hyeon, T., 2004. Ultra-large-scale syntheses of mono-disperse nanocrystals. Nat. Mater. 3 (12), 891.

Patel, H.A., Byun, J., Yavuz, C.T., 2012. Arsenic removal by magnetic nanocrystalline barium hexaferrite. In: Nanotechnology for Sustainable Development. Springer, Cham, pp. 163–169.

Paustenbach, D.J. (Ed.), 2015. Human and Ecological Risk Assessment: Theory and Practice (Wiley Classics Library). John Wiley & Sons.

Petcharoen, K., Sirivat, A., 2012. Synthesis and characterization of magnetite nanoparticles via the chemical co-precipitation method. Mater. Sci. Eng. B 177 (5), 421–427.

Pradeep, T., 2003. A Textbook of Nanoscience and Nanotechnology. Tata McGraw-Hill Education.

Pugazhenthiran, N., Anandan, S., Ashokkumar, M., 2016. Removal of heavy metal from wastewater. In: Handbook of Ultrasonics and Sonochemistry. Springer, Singapore, pp. 813–839.

Qdais, H.A., Moussa, H., 2004. Removal of heavy metals from wastewater by membrane processes: a comparative study. Desalination 164 (2), 105–110.

Qu, X., Brame, J., Li, Q., Alvarez, P.J., 2012. Nanotechnology for a safe and sustainable water supply: enabling integrated water treatment and reuse. Acc. Chem. Res. 46 (3), 834–843.

Rahimzadeh, M.R., Rahimzadeh, M.R., Kazemi, S., Moghadamnia, A.A., 2017. Cadmium toxicity and treatment: an update. Caspian J Intern Med 8 (3), 135.

Rall, D.P., Pope, A.M. (Eds.), 1995. Environmental Medicine: Integrating a Missing Element Into Medical Education. National Academies Press.

Ralph, S.J., 2008. Arsenic-based antineoplastic drugs and their mechanisms of action. Metal-Based Drugs 2008, Article ID 260146, 13 pages, https://doi.org/10.1155/2008/260146.

Saiz, J., Bringas, E., Ortiz, I., 2014. Functionalized magnetic nanoparticles as new adsorption materials for arsenic removal from polluted waters. J. Chem. Technol. Biotechnol. 89 (6), 909–918.

Sharma, R.K., 2005. Bioinorganic Chemistry E-Book Module. National Science Digital Library. http://niscair.res.in/ispui/handle/123456789/230.

Sharma, R.K., Puri, A., Monga, Y., Adholeya, A., 2014. Acetoacetanilide-functionalized Fe3O4 nanoparticles for selective and cyclic removal of Pb 2+ ions from different charged wastewaters. J. Mater. Chem. A 2 (32), 12888–12898.

Sharma, R.K., Kumar, H., Kumar, A., 2015. A highly efficient and magnetically retrievable functionalized nanoadsorbent for ultrasonication assisted rapid and selective extraction of Pd^{2+} ions from water samples. RSC Adv. 5 (54), 43371–43380.

Sharma, R.K., Dutta, S., Sharma, S., Zboril, R., Varma, R.S., Gawande, M.B., 2016. Fe_3O_4 (iron oxide)-supported nanocatalysts: synthesis, characterization and applications in coupling reactions. Green Chem. 18 (11), 3184–3209.

Sharma, R. K., Adholeya, A., Kumar, A. n.d. Registration No.-n3287/DEL/2013.

Shekhawat, K., Chatterjee, S., Joshi, B., 2015. Chromium toxicity and its health hazards. Int. J. Adv. Res. 3 (7), 167–172.

Shi, J., Li, H., Lu, H., Zhao, X., 2015. Use of carboxyl functional magnetite nanoparticles as potential sorbents for the removal of heavy metal ions from aqueous solution. J. Chem. Eng. Data 60 (7), 2035–2041.

Singh, R., Gautam, N., Mishra, A., Gupta, R., 2011. Heavy metals and living systems: an overview. Indian J. Pharm. 43 (3), 246.

Singh, P.N., Tiwary, D., Sinha, I., 2015. Chromium removal from aqueous media by superparamagnetic starch functionalized maghemite nanoparticles. J. Chem. Sci. 127 (11), 1967–1976.

Sun, S., Zeng, H., Robinson, D.B., Raoux, S., Rice, P.M., Wang, S.X., Li, G., 2004. Monodisperse MFe_2O_4 (M = Fe, Co, Mn) nanoparticles. J. Am. Chem. Soc. 126 (1), 273–279.

Tchounwou, P.B., Yedjou, C.G., Patlolla, A.K., Sutton, D.J., 2012. Heavy metal toxicity and the environment. In: Molecular, Clinical and Environmental Toxicology. Springer, Basel, pp. 133–164.

Tripathi, A., Ranjan, M.R., 2015. Heavy metal removal from wastewater using low cost adsorbents. J. Bioremed. Biodegr. 6 (1000315), 5.

Uheida, A., Iglesias, M., Fontàs, C., Hidalgo, M., Salvadó, V., Zhang, Y., Muhammed, M., 2006. Sorption of palladium (II), rhodium (III), and platinum (IV) on Fe3O4 nanoparticles. J. Colloid Interface Sci. 301 (2), 402–408.

UNEP (United Nations Environment Programme), 2016. A Snapshot of the World's Water Quality: Towards a Global Assessment. UNEP, Nairobi. https://uneplive.unep.org/media/docs/assessments/unep_wwqa_report_web.pdf.

Venkateswarlu, S., Yoon, M., 2015. Rapid removal of cadmium ions using green-synthesized Fe3O4 nanoparticles capped with diethyl-4-(4 amino-5-mercapto-4 H-1, 2,4-triazol-3-yl) phenyl phosphonate. RSC Adv. 5 (80), 65444–65453.

Venkateswarlu, S., Rao, Y.S., Balaji, T., Prathima, B., Jyothi, N.V.V., 2013. Biogenic synthesis of Fe3O4 magnetic nanoparticles using plantain peel extract. Mater. Lett. 100, 241–244.

Verma, R., Dwivedi, P., 2013. Heavy metal water pollution—a case study. Recent Res. Sci. Technol. 5 (5). https://updatepublishing.com/journal/index.php/rrst/article/view/1075.

Wan Ibrahim, W.A., Nodeh, H.R., Aboul-Enein, H.Y., Sanagi, M.M., 2015. Magnetic solid-phase extraction based on modified ferum oxides for enrichment, preconcentration, and isolation of pesticides and selected pollutants. Crit. Rev. Anal. Chem. 45 (3), 270–287.

Wang, L., Park, H.Y., Stephanie, I., Lim, I., Schadt, M.J., Mott, D., … Zhong, C.J., 2008. Core@ shell nanomaterials: gold-coated magnetic oxide nanoparticles. J. Mater. Chem. 18 (23), 2629–2635.

Winecker, R.E., Ropero-Miller, J.D., Broussard, L.A., Hammett-Stabler, C.A., 2002. The toxicology of heavy metals: getting the lead out. Lab. Med. 33 (12), 934–947.

World Health Organization, WHO/UNICEF Joint Water Supply, & Sanitation Monitoring Programme, 2015. Progress on Sanitation and Drinking Water: 2015 Update and MDG Assessment. World Health Organization.

Xin, X., Wei, Q., Yang, J., Yan, L., Feng, R., Chen, G., … Li, H., 2012. Highly efficient removal of heavy metal ions by amine-functionalized mesoporous Fe3O4 nanoparticles. Chem. Eng. J. 184, 132–140.

Yamamura, S., Bartram, J., Csanady, M., Gorchev, H.G., Redekopp, A., 2003. Drinking Water Guidelines and Standards. Arsenic, Water, and Health: The State of the Art. https://www.who.int/water_sanitation_health/dwq/arsenicun5.pdf.

Yang, J.C., Yin, X.B., 2017. $CoFe_2O_4$@MIL-100 (Fe) hybrid magnetic nanoparticles exhibit fast and selective adsorption of arsenic with high adsorption capacity. Sci. Rep. 7, 40955.

Yang, X., Jiang, W., Liu, L., Chen, B., Wu, S., Sun, D., Li, F., 2012. One-step hydrothermal synthesis of highly water-soluble secondary structural Fe3O4 nanoparticles. J. Magn. Magn. Mater. 324 (14), 2249–2257.

Yantasee, W., Warner, C.L., Sangvanich, T., Addleman, R.S., Carter, T.G., Wiacek, R.J., … Warner, M.G., 2007. Removal of heavy metals from aqueous systems with thiol functionalized superparamagnetic nanoparticles. Environ. Sci. Technol. 41 (14), 5114–5119.

Zargoosh, K., Zilouei, H., Mohammadi, M.R., Abedini, H., 2014. 4-Phenyl-3-thiosemicarbazide modified magnetic nanoparticles: synthesis, characterization and application for heavy metal removal. CLEAN–Soil, Air, Water 42 (9), 1208–1215.

Zhang, W., Shi, X., Zhang, Y., Gu, W., Li, B., Xian, Y., 2013. Synthesis of water-soluble magnetic graphene nanocomposites for recyclable removal of heavy metal ions. J. Mater. Chem. A 1 (5), 1745–1753.

Zhang, Q., Dan, S., Du, K., 2017. Fabrication and characterization of magnetic hydroxyapatite entrapped agarose composite beads with high adsorption capacity for heavy metal removal. Ind. Eng. Chem. Res. 56 (30), 8705–8712.

Zhang, Q., He, M., Chen, B., Hu, B., 2018. Magnetic mesoporous carbons derived from in situ MgO template formation for fast removal of heavy metal ions. ACS Omega 3 (4), 3752–3759.

Zhou, L., Xu, J., Liang, X., Liu, Z., 2010. Adsorption of platinum (IV) and palladium (II) from aqueous solution by magnetic cross-linking chitosan nanoparticles modified with ethylenediamine. J. Hazard. Mater. 182 (1–3), 518–524.

Zhu, H., Jia, Y., Wu, X., Wang, H., 2009. Removal of arsenic from water by supported nano zero-valent iron on activated carbon. J. Hazard. Mater. 172 (2–3), 1591–1596.

Zhu, J., Wei, S., Gu, H., Rapole, S.B., Wang, Q., Luo, Z., … Guo, Z., 2011. One-pot synthesis of magnetic graphene nanocomposites decorated with core@ double-shell nanoparticles for fast chromium removal. Environ. Sci. Technol. 46 (2), 977–985.

Ziegfeld, R.L., 1964. Importance and uses of lead. Arch. Environ. Health 8 (2), 202–212.

Further Reading

Jernelöv, A., Landner, L., Larsson, T., 1975. Swedish perspectives on mercury pollution. J. Water Pollut. Control Fed. 47, 810–822.

CHAPTER

16

Lessons Learned From Water Disasters of the World

Satinder Ahuja*

Ahuja Consulting for Water Quality, Calabash, NC, United States
*Corresponding author: E-mail: sutahuja@atmc.net

Water sustains life: without water, life on Earth would not be possible. Even though water is a valuable commodity, we frequently fail to prevent its contamination. Webster dictionary defines disaster as a sudden contamination event bringing great damage or loss. A disaster generally affects a large number of people. Over the years, numerous water disasters have occurred worldwide (a list of water disasters taken from the internet was provided in Chapter 1). Table 1 includes some other disasters that occurred earlier or more recently, but were omitted in that listing.

Table 1 shows a variety of contaminants that can enter our water supply and lead to significant crises that affect human health. Three disasters that span almost 4 decades were selected for further discussion in this chapter, based on the author's personal interests and his efforts to help remediate the situations:

1. Arsenic contamination of groundwater in Bangladesh.
2. Lead contamination of surface water in Flint, Michigan.
3. GenX contamination of drinking water in Wilmington, North Carolina.

These situations reflect a variety of ways that contamination can occur:

- Poor consideration of groundwater quality in Bangladesh.
- Inability to recognize the impact on water quality from a different source in Flint, Michigan.
- Failure of Chemours to assure the safety of people in three counties in North Carolina when it added GenX from its waste stream to the Cape Fear River.

Discussion in this chapter will focus on lessons we need to learn from these water disasters. Most importantly, it would become clear that these disasters could have been avoided with the help of separation-based techniques of analytical chemistry that would have warned of the possibilities of potential occurrence of these problems.

1 ARSENIC CONTAMINATION OF GROUNDWATER

Arsenic (As) contamination of groundwater is a horrendous problem that went undetected for many years and involved many countries

TABLE 1 Water Disasters Around the World

Location	Contaminant(s)	Date
Yamuna River, India	Sewage, garbage, chemicals	Since 1909
Hinkley, California	Chromium	1952–66
Camp Lejeune, Jacksonville, North Carolina	PCE and TCE	1953
Woburn, Massachusetts	PCE and TCE	1970s
Bangladesh	Arsenic	Found in the 1980s
Lanzhou, China	Benzene	1987
Prince William Sound, Alaska	Crude oil	1989
Walkerton, Ontario, Canada	*Escherichia coli*	2000
Ghana	Cyanide	2001
Gulf of Mexico, near the Mississippi River	Crude oil	2010
Mutare, Zimbabwe	Chromium and nickel	2012
Minneapolis, Minnesota	TCE	2013
Elk River, West Virginia	MCHM	2014
Flint, Michigan	Lead	2014
Wilmington, North Carolina	Gen X	2017

TCE = trichloroethylene; PCE = perchloroethylene; MCHM = 4-methylcyclohexanemethanol.

(Ahuja, 2008a). Almost 200 million people have been affected worldwide. Arsenic contamination has been discovered in regional water supplies of many developing and developed countries in Asia, Africa, Europe, North America, and South America. Groundwater contamination of arsenic can occur from various anthropogenic sources such as pesticides, wood preservatives, glass manufacture, and other miscellaneous sources. The natural content of arsenic in soil is mostly in a range below 10 mg/kg; however, it can cause major havoc when it is present in groundwater at levels over $10 \mu g/L$ or 10 parts per billion (ppb) (Ahuja, 2008a). Bangladesh has faced the brunt of this problem. The worst case of groundwater contamination was discovered there in the 1980s (Ahmed et al., 2006; Ahuja, 2005a, 2005b, 2006, 2008b; Ahuja and Malin, 2004, 2006): a large number of shallow tube wells (10–40 m) installed in the 1970s with the help of UNICEF were found to be contaminated with arsenic.

1.1 Why This Disaster Occurred

The arsenic contamination crisis in Bangladesh occurred because the main focus of the problem was on providing water free of microbial contamination, a problem that was commonly encountered in surface water (Ahuja, 2009). Unfortunately, potential contamination from naturally occurring arsenic was ignored; apparently, the project failed to include adequate testing to detect arsenic. This unfortunate calamity could have been avoided, as analytical methods that can test for arsenic down to the ppb levels have been available for many years (Ahuja, 1986).

EVALUATING WATER QUALITY TO PREVENT FUTURE DISASTERS

Arsenic contamination was reported as early as 1938; however, skin lesions and cancers attributable to arsenic were rare and ignored until new evidence emerged from Taiwan in 1977. The serious health effects of arsenic exposure that include lung, liver, and bladder cancers were confirmed shortly thereafter by studies of exposed populations in Argentina, Chile, and China. In 1984, Dr. K.C. Saha and colleagues at the School of Tropical Medicine in Kolkata, India, attributed lesions observed on the skin of villagers in the state of West Bengal to the elevated arsenic content of groundwater drawn from shallow tube wells. Of the various countries affected by the pollution, Bangladesh and India experienced the most serious groundwater arsenic problem, and the situation in Bangladesh has been described as "the worst mass poisoning in human history." The magnitude of the problem in India has been investigated for a number of years by Chakraborti and others (for more details, see chapter 5 in Ahuja, 2009) who have analyzed 225,000 tube well water samples from the Ganga-Meghna-Brahamaputra plain, covering an area of $569,749 \, km^2$ and a population of more than 500 million. The investigators found that Bangladesh and a number of states in India (Uttar Pradesh, Bihar, West Bengal, Jharkhand, and Assam) are affected by a concentration of arsenic at $50 \, \mu g/L$. On the average, about 50% of the water samples contained arsenic above $10 \, \mu g/L$ and 30% were above $50 \, \mu g/L$.

It should be noted that groundwater pollution by arsenic is found even in the developed countries such as Australia, the United Kingdom, and the United States. Over 31,000 samples analyzed over about a 30-year period revealed that a large number of states in the United States are affected by this contamination. In the United States, nearly 10% of groundwater resources exceed the maximum contamination level (MCL) of $10 \, \mu g/L$. Recognizing the fact that inorganic arsenic is a documented human carcinogen, the World Health Organization in 1993 set a standard at no more than $10 \, \mu g/L$ (or 10 ppb) of arsenic in drinking water as the MCL. This standard was finally adopted by the United States in 2006. However, the MCL remains at $50 \, \mu g/L$ (or 50 ppb) in Bangladesh and other developing countries. Furthermore, it should be mentioned that these guidelines do not consider different arsenic species, even though it is well established that the toxicity of arsenic may vary enormously with its speciation, as discussed in the following section.

1.2 The Impact of the Disaster

Arsenic is a well-known poison, with a lethal dose in humans at about 125 mg. It is a semimetal or metalloid that is stable in several oxidation states ($-III$, 0, $+III$, $+V$). Arsenic is a natural constituent of Earth's crust and ranks 20th in abundance in relation to the other elements. It should be noted that the $+III$ and $+V$ states of arsenic are most common in natural systems. Arsine($-III$), a compound with extremely high toxicity, can be formed under high reducing conditions, but its occurrence in gases emanating from anaerobic environments in nature is relatively rare. The relative toxicity of arsenic depends mainly on its chemical form and is dictated in part by the valence state. Trivalent arsenic has a high affinity for thiol groups, as it readily forms kinetically stable bonds to sulfur. Hence, reaction with As(III) induces enzyme inactivation, as thiol groups are important to the functions of many enzymes. Arsenic affects the respiratory system by binding to the vicinal thiols in pyruvate dehydrogenase and 2-oxoglutarate dehydrogenase, and it has also been found to affect the function of glucocorticoid receptors. Pentavalent arsenic has a poor affinity toward thiol groups, resulting in more rapid excretion from the body. However, it is a molecular analog of phosphate and can uncouple mitochondrial oxidative phosphorylation, resulting in failure of the energy metabolism system. The effects of the oxidation state on chronic toxicity are confounded by the redox conversion of As(III) and As(V) within human cells and tissues. Methylated arsenicals such as monomethylarsenic acid (MMAA) and dimethylarsenic acid (DMAA) are less harmful than inorganic arsenic compounds. Clinical symptoms of arsenicosis may take about 6–24 months or more to appear, depending on the quantity of arsenic ingested and also on the nutritional status and immunity level of the individual. Untreated arsenic poisoning results in several stages; for example, various effects on the skin with melanosis and keratosis; dark spots on the chest, back, limbs, and gums; and enlargement of the liver, kidneys, and spleen. Subsequently, patients may develop nephropathy, hepatopathy, gangrene, or cancers of the skin, lungs, or bladder.

It should be noted that a number of toxicologists consider a 10-ppb level of arsenic to be too high because even at 1 ppb, the risk of getting cancer is one in 3000. In the long term, one in every 10 people could die of arsenic poisoning if they continue using water with high concentrations of arsenic.

Arsenic toxicity has no known effective treatment, but drinking arsenic-free water can help the arsenic-affected people who are at the preliminary stage of their illness alleviate the symptoms of arsenic toxicity. Hence, provision of arsenic-free water is urgently needed for the mitigation of arsenic toxicity and the protection of the health and well-being of people living in areas where the arsenic problem is acute.

1.2.1 The Impact of Arsenic-Laced Irrigation Water on the Food Chain

The fact that arsenic poisoning in the world's population is not consistent with the level of water intake has raised questions on the possible pathways of arsenic transfer from groundwater to the human system. Even if an arsenic-safe drinking water supply could be ensured, the same groundwater may continue to be used for irrigation purposes, leaving a risk of soil accumulation of this toxic element and eventual exposure to the food chain through plant uptake and animal consumption. Studies on arsenic uptake by crops indicate that there is great potential for the transfer of groundwater arsenic to crops. The fate of arsenic in irrigation water and its potential impact on the food chain, especially as it occurs in Bangladesh and other similar environments, has been discussed at length (see chapter 2 in Ahuja, 2009).

Green leafy vegetables have been found to act as arsenic accumulators, with arum (kochu), gourd leaf, *Amaranthus*, *Ipomea* (kalmi) at the top of the list. Arum, a green vegetable commonly grown and used in almost every part of Bangladesh, seems to be unique in that the concentration of arsenic can be high in every part of the plant. Arsenic in rice seems to vary widely. Speciation of Bangladeshi rice shows the presence of As(III), DMAA, and As(V); more than 80% of the recovered arsenic is in the inorganic form. It has been reported that more than 85% of the arsenic in rice is bioavailable, compared to only about 28% of arsenic in leafy vegetables. It is thus pertinent to assess the dietary load of arsenic from various arsenic-contaminated foods. A person consuming 100 g of arum daily, with an average arsenic content of 2.2 mg/kg, 600 g of rice with an average arsenic content of 0.1 mg/kg, and 3 L of water with an average arsenic content of 0.1 mg/L would ingest 0.56 mg/day, which exceeds the calculated threshold value based on the US EPA model. Over long periods of time, even a small amount of arsenic can cause harm; the chances of damage are slightly lower in women than in men.

1.3 Lessons Learned

It is extremely important to monitor water quality before a change in the water supply is implemented. This was apparently not done when the decision was made to change to groundwater from surface water. A detailed testing of water quality would have alerted officials to the potential occurrence of this problem. It is not very difficult to determine arsenic at 10 µg/L or at 10 ppb in water. A number of methods can be used for determining arsenic in water at the ppb level.

- Flame atomic absorption spectrometry.
- Graphite furnace atomic absorption spectrometry.
- Inductively coupled plasma–mass spectrometry (ICP–MS).
- Atomic fluorescence spectrometry.
- Neutron activation analysis.
- Differential pulse polarography.
- High-performance liquid chromatography (HPLC)–ICP–MS.

The speciation of arsenic requires separations based on the solvent extraction, chromatography, and selective hydride generation (HG). With the use of HG, a method that has been known for many decades, arsenic can be determined by a relatively inexpensive atomic absorption spectrometer or atomic fluorescence spectrometer (AFS) at single-digit micrograms-per-liter concentrations (see chapter 7 in Ahuja, 2009). Its generation is prone to interference from other matrix components, and as a result, different matrices can present various analytical problems. In this technique, arsenic compounds are converted to volatile derivatives by reaction with a hydride transfer reagent, usually tetrahydroborate III. HG can be quite effective as an interface between HPLC separation and element-specific detection. In fact, it is possible to get the same performance from HG–AFS as from ICP–MS. Therefore, as the former detector represents significant savings in both capital and operation costs compared with the latter, there is considerable interest in this technique in developing countries.

Very low detection limits for arsenic, down to 0.0006 µg/L, can be obtained with ICP–MS. HPLC coupled with ICP–MS is currently the best technique available for the determination of inorganic and organic species of arsenic;

however, the cost of the instrumentation is prohibitive. For the developing countries that confront this problem, the improvement of low-cost, reliable instrumentation, and reliable field test kits would be very desirable.

2 LEAD CONTAMINATION OF DRINKING WATER IN FLINT, MICHIGAN

The ill effects of lead have been known for a very long time. Common sources of this contaminant were soil (dirt) and paints. Its impact on children is especially severe because it can adversely affect their developing brains. Although blood lead levels have long been declining nationwide, many trouble spots remain. Right now in Michigan, 8.8% of children in Detroit, 8.1% of children in Grand Rapids, and almost 14% of children in Highland Park surpass the Centers for Disease Control reference level of lead.

High levels of lead were found in the water supply in Flint, Michigan, in February 2015. To save money, Flint had begun drawing its water from the Flint River in April 2014, instead of buying Lake Huron water from the city of Detroit (Ahuja, 2017). Residents started complaining about burning skin, hand tremors, hair loss, and even seizures. Scores of children were diagnosed with anemia. Because it has a large impact on their rapidly growing brains, lead is particularly harmful to children. For almost 19 months, the problem was ignored while the Flint River water corroded the city's decades-old pipes and leached lead into the water supply. The crisis exposed as many as 8000 children under age 6 to unsafe levels of lead, and it may be the most serious contamination threat facing the country's water supplies. Five months passed before the city told pregnant women and children not to drink the water, and the city shut down taps and fountains in schools.

Flint is hardly the only city with a lead-contaminated water supply. We failed to learn from the lead problems in tap water observed earlier in city after city: Washington, DC, in 2001; Columbia, SC, in 2005; Durham and Greenville, NC, in 2006. The problems in Jackson, MS (in June 2015), and Sebring, OH (August 2015) surfaced after the discovery of lead-contaminated water in Flint. Over a period of time—January 2013 to September 2015—local water systems across the United States have delivered nearly 3.6 million people water that exceeded the lead standards set by the federal government.

2.1 Why This Disaster Occurred

Soluble lead can be released from lead-containing pipes, solder, and corrosion products. Orthophosphate is an economical and widespread corrosion inhibitor that provides passivation (see Chapter 12). The end product for multiple routes of reactivity between soluble Pb^{2+} and orthophosphate is chloropyromorphite [$Pb_5(PO_4)_3Cl$]. This forms a small layer on the inside of the water delivery pipe. The Flint (MI) water crisis in 2014 was a result of the decision of a state-appointed emergency manager to change to the Flint River as a corrosive drinking water source. And, the Michigan Department of Environmental Quality (MDEQ) failed to enforce federally mandated corrosion-control treatment of the water. This created a variety of drinking water issues that resulted in violations for disinfection by-products and bacterial contamination. It caused unprecedented corrosion of iron water mains as well as high levels of lead and *Legionella* bacteria, the source of Legionnaires' disease, in the water. The distressing fact is that for 18 months, about 100,000 Flint residents were exposed to unsafe drinking water, while city and state officials publicly insisted the water was safe to drink.

Research collaboration between Virginia Tech's Flint Water Study Team and Flint residents in August to September 2015 revealed citywide lead in the water, as well as high levels of *Legionella* bacteria in large buildings. Here is an historical account of the lead crisis from a recent publication of Pieper et al. (2018): their results contradicted official claims of "no problem": the 90th percentile was 26.8 µg/L—almost double the lead and copper rule (LCR) action level of 15 µg/L. Back calculations of an LCR sampling pool with 50% lead pipes indicated an estimated 90th percentile lead value of 31.7 µg/L ($\pm$4.3 µg/L). Four subsequent sampling efforts were conducted to track reductions of lead in water after the change back to Lake Huron water and enhanced corrosion control. The incidence of lead contamination of water varied with service line material. Between August 2015 and November 2016, the median lead level in water decreased from 3.0 to <1 µg/L for homes with copper service lines, 7.2–1.9 µg/L with galvanized service lines, and 9.9–2.3 µg/L with lead service lines. As of the summer of 2017, the 90th percentile of 7.9 µg/L of lead in the water no longer differed from official results, indicating that Flint's water lead levels were below the action level.

2.2 Impacts of the Disaster

Pediatricians from the Hurley Medical Center in Flint said that the water switch had doubled instances of childhood lead poisoning in the city: The percentage of Flint children with elevated blood-lead levels may have risen from about 2.5% in 2013 to as much as 5% in 2015. The change of the source of water is also a possible cause of an outbreak of Legionnaires' disease in Genesee County—the home of Flint—that killed 10 people and affected another 77.

On January 5, 2016, the city was declared to be in a state of emergency by the Governor of Michigan, Rick Snyder. President Barack Obama declared it to be a federal state of emergency, authorizing additional help from federal agencies soon thereafter. Four government officials—one from the city of Flint, two from the Michigan Department of Water Quality, and one from the EPA—resigned over the mishandling of the crisis, and one additional MDEQ staff member was fired. There have also been 15 criminal cases filed against local and state officials in regard to the crisis.

Governor Snyder issued an apology to the citizens and promised to fix the problem; he then sent $28 million to Flint for supplies, medical care, and infrastructure upgrades, and later budgeted an additional $30 million to Flint that will give water bill credits of 65% for residents and 20% for businesses. Another $165 million for lead pipe replacements and water bill reimbursements was approved by Snyder in 2016. A $170 million stopgap spending bill for repairing and upgrading the city of Flint's water system and helping with healthcare costs was approved by the US House of Representatives on December 8, 2016. The Senate approved it the next day. In all, $100 million of the bill was for infrastructure repairs, $50 million for healthcare costs, and $20 million to pay back loans related to the crisis. On January 6, 2017, Snyder signed a bill that accelerates the public notice requirement for lead in drinking water to three business days, from the previous time of 30 days.

On January 24, 2017, the MDEQ told Flint Mayor Karen Weaver that the lead content of Flint water had fallen below the federal limit: the 90th percentile of lead concentrations in Flint was 12 ppb from July 2016 through December 2016—below the "action level" of 15 ppb. It had been 20 ppb in the prior 6-month period. On January 25, Flint spokeswoman Kristin Moore said that anywhere from 18,000 to 28,000 homes in the city still needed service line replacements, and that the city was planning to complete these in 6000 homes per year through 2019.

2.3 Lessons Learned

President Obama pointed out that Flint is a reminder of why you can't shortchange basic services that are provided to people. Currently, the action levels for lead are set at 15 ppb, with mandatory actions necessary if more than 10% of households test above this level. Methods are available for testing lead at this level. According to Governor Snyder, the crisis was brought about by criminal neglect and non-addition of essential orthophosphate corrosion inhibitors to an acidic water source.

Some warning signs for potential water disasters (Cooper and Ahuja, 2017) include small, underfunded water management districts and aging infrastructure. New infrastructure should be made of "lead-free" materials, that is, less than 0.25% lead for pipes, fittings, and fixtures, and 0.2% lead for solder and flux. It would be desirable to have community-engaged water testing that monitors a variety of water contaminants. Here are some important considerations:

- lead release is a community problem as much as it is a scientific and policy problem,
- working within a community is important to design mitigation strategies,
- two hurdles exist to fixing lead contamination in drinking water,
- noncompliance of individuals in water testing, and.
- replacement of private infrastructure.

The solutions involve working with the community and using public funds to help alleviate the financial burden of replacing private infrastructure.

3 GenX CONTAMINATION OF DRINKING WATER IN WILMINGTON, NORTH CAROLINA, AND OTHER COUNTIES IN THE STATE

The discovery of GenX and numerous other perfluorinated alkylated substances (PFASs) in the Cape Fear River, North Carolina, was reported in June 2017 by the *Star News*, Wilmington, NC. The fact that these compounds are not removed by standard drinking water treatment processes is a great deal of concern for the public. GenX is a trade name for "C3 dimer acid," perfluoro-2-propoxypropionic acid. It is produced as an emulsifying agent for the production of nonstick coatings such as Teflon (see Chapters 13 and 14 for more details). GenX is currently produced commercially by Chemours, a 2015 spin-off of DuPont, at the Fayetteville Works plant near Fayetteville, North Carolina, along the banks of the Cape Fear River. The compound is also used in the production processes at a Chemours facility in Dordrecht, The Netherlands. GenX is a by-product of other perfluoro- and polyfluoro-alkylated substances (PFASs) syntheses, notably for polyvinyl fluorides. The commercial production of GenX replaced perfluorooctanoic acid (PFOA), previously used in Teflon synthesis because epidemiological studies revealed that PFOA levels in the bodies of people employed in its manufacture, living in communities adjacent to manufacturing sites, and consuming drinking water contaminated by C8 were high enough to cause a variety of adverse health effects, leading to regulatory action at a global scale (Blum et al., 2015; Lindstrom et al., 2011; Scheringer et al., 2014; Wang et al., 2013).

3.1 Why This Disaster Occurred

Strynar et al. (2015) first identified GenX in the Cape Fear River in 2015, based on the samples collected in 2012 (see Chapters 13 and 14 for more details). Sampling by Sun et al. (2016) in 2013 and 2014 detected GenX and other PFASs in the river and in finished drinking water downstream of Chemours plant discharge. It should be noted that the first study to monitor PFAS compounds in the Cape Fear River watershed was published by Nakayama et al. (2007). It found widespread distribution of PFASs throughout, apparently derived from diverse sources. This study focused on the PFOA and perfluorooctane sulfonate (PFOS), but also examined several other perfluorocarboxylic acids that were identifiable in their protocol. PFOA and PFOS were of substantial concern, owing to their apparent toxicity to humans and wildlife and

their widespread use and the presence in the environment. For example, DuPont's PFOA production at its Parkersburg, West Virginia, plant had caused widespread contamination of drinking water supplies, ultimately leading to regulatory action by the federal government in 2006 and 2010/2015.

These studies should have alerted DuPont to carefully monitor all potential contaminants in their waste stream. Finally, the study by Sun et al. (2016) highlighted the presence of [PFASs, perfluoroether carboxylic acids (PFECAs), and perfluoroether sulfonic acids (PFESAs)] in the Cape Fear River water, and quantified the concentrations and loadings of PFOA, PFOS, and GenX. They also demonstrated that these compounds were not removed by drinking water treatment processes used by the Cape Fear Public Utility Authority's (CFPUA) Sweeney Water Treatment Plant (WTP) in Wilmington. GenX concentrations in finished drinking water exceeded 400 ng/L (parts per trillion or ppt). Other PFECA compounds for which pure standards were not then available were also present in finished drinking water at concentrations likely much higher than those reported for GenX. The average daily mass flux of GenX in the Cape Fear River was estimated at 5.9 kg (range: 0.6–24 kg/day) at the water supply intake for the Sweeney WTP, approximately 40 km upstream of Wilmington and 70 km downstream of the Fayetteville Works outfall (Sun et al., 2016).

It is important to note that analyses of GenX and nontargeted PFASs in the Chemours wastewater discharge and in drinking water during the summer of 2017 employed liquid chromatographic separations coupled with high-resolution quadrupole time-of-flight mass spectrometry (LC/QTOFMS). GenX concentrations in Chemours' discharge outfall [#002], where treated process wastewater from outfall [#001] had been diluted ~20× with noncontact cooling water, and at the Sweeney WTP (finished drinking water) in Wilmington were 21,760 and 726 ng/L, respectively, in the first samples collected for this surveillance effort.

Two PFESA compounds identified as "Nafion by-products 1 and 2" have also been found in water. These two compounds have been identified as perfluoro-3,6-dioxa-4-methyl-7-octene-1-sulfonic acid and 2-[1-[difluoro(1,2,2,2-tetrafluoroethoxy)methyl]-1,2,2,2-tetrafluoroethoxy]-1,1,2,2-tetrafluoro-ethanesulfonic acid and apparently entered the discharge stream from Chemours' Nafion production facility.

Public or regulatory response to this study occurred after the findings were headlined in the *Star News*, "Toxin Taints CFPUA Drinking Water," on June 8, 2017. In February 12, 2018, issue of *Chemical and Engineering News*, Satinder Ahuja pointed out that Chemours has been acting irresponsibly for a long time. Its parent DuPont had been releasing GenX-related chemicals since 1980. This led to the pollution of drinking water of several counties, affecting hundreds of thousands of people.

The North Carolina Department of Environmental Quality (NC DEQ) subsequently requested that Chemours cease discharge of Nafion by-products 1 and 2 as well.

3.2 Impacts of the Disaster

The drinking water supply of several counties was affected, with no clear indication of toxicity of GenX and Nafion products found in drinking water supplies in these counties. Around the Fayetteville plant, air, soil, and groundwater are contaminated.

The University of North Carolina-Wilmington (UNCW) investigators also discovered GenX in rainwater and determined that it was likely present in that matrix because of atmospheric releases of hexafluoropropylene oxide (HFPO) dimer acid fluoride, which would have hydrated rapidly in the atmosphere by reacting with water vapor to form the carboxylic acid product, GenX (UNCW, 2018). Subsequent investigations by the NC DEQ confirmed this finding and determined that air releases were likely a significant source of GenX shed from the air around the Fayetteville Works

(~1000 kg year). Chemours has recently committed to significant expenditures and installation of advanced technology to control these emissions.

A *Star News* editorial of July 26, 2018, summarized the status as follows: "But for all the ado over the prevalence of chemicals emitted from the Chemours plant on the Cumberland-Bladen County line, there's been little serious attempt to document the actual health effects of GenX and other poly- and perfluorinated compounds that Chemours, and earlier Dupont, let slip in our environment." Dupont and Chemours settled a class-action lawsuit over C8 contamination around its Parkersburg, WV, plant in 2017, agreeing to a payout $670 million. A scientific panel, formed as part of a settlement of an earlier suit, concluded that there's a likely link between C8 and multiple cancers and pregnancy-induced hypertension. Testing of GenX on animals has shown some similar health effects, although there's been little research into GenX effects on humans.

3.3 Lessons Learned

Contamination by GenX and related PFASs in a public drinking water supply raises an important question: Why did this occur in spite of various regulations to protect the public? The public might never have known if Sun et al. (2016) and Strynar et al. (2015) had not directly disclosed their findings to the public through the media. The ultimate responsibility for the safety of public drinking water supplies lies with the regulatory agency with authority to enforce the Clean Water Act (CWA) and the Safe Drinking Water Act, and with the discharger of any potentially toxic pollutants, DuPont/Chemours. NC DEQ has full authority to implement the provisions of the CWA, including issuance of permits, regulation and monitoring of discharges, and enforcement actions to secure compliance. The US EPA has oversight authority, but generally acts in consultation with the NC DEQ.

The emerging pollutants expose significant weaknesses in the regulatory processes. Discharge permit applications require the good-faith disclosure of the constituents expected to be present in proposed discharges. The composition of industrial discharges can include numerous compounds at very low concentrations. The manufacturers are reluctant to provide a full listing of such constituents because it would disclose confidential information (synthesis pathways, etc.). These considerations must be balanced against the need to protect the public from toxic discharges. To accommodate these considerations, the CWA requires disclosure of the presence of any toxic substance believed to be present frequently or routinely in a discharge at concentrations above $100\,\mu g/L$, or 100 ppb, and of toxic substances infrequently present at concentrations above $500\,\mu g/L$. This requires that the discharger determines what is likely to be present in the proposed discharge. State regulatory agencies with delegated CWA authority generally lack the resources to determine the toxicity of emerging compounds. They rely on toxicity testing by the producers under Toxic Substances Control Act (TSCA) and evaluations reported to the US EPA, as well as published toxicological research studies.

Unique chemical properties of PFASs should raise significant concerns. Their poor reactivity, diversity of uses, and broad distribution at low concentrations make them challenging to separate, detect, identify, and quantify from environmental matrices. PFASs do not breakdown rapidly in the environment, so low levels of them appear ubiquitously. Toxicological studies have found evidence that some PFASs, although relatively inert, are biologically problematic at low concentrations. They should not be added to air, water, or soil until their toxicity is well established to assure that no harm will occur to living creatures. The methodology to monitor them was known as early as 2007 (Nakayama et al., 2007). In any event, it is inexcusable to add to a water supply any chemical if satisfactory analytical methods are not available for monitoring them. The right thing to do is to develop suitable monitoring methods, perform toxicity studies, and determine the safe level of disposal of these chemicals in the waste stream.

4 CONCLUSIONS

The three examples of water disasters discussed here occurred over a period of nearly 4 decades. They occurred in underdeveloped as well as developed countries. A number of lessons that can be learned from these disasters have been discussed in this chapter. The important lesson from arsenic contamination was to monitor water quality carefully before making a change to an alternate source; that is, from surface water to groundwater. The Flint disaster clearly suggests that economic considerations should not overrule good science. Again, monitoring water quality of the new water source prior to making the change would have alerted the parties involved to rule out the change or to take adequate steps in terms of prevention of corrosion of delivery pipes. The GenX disaster shows the importance of analytical separations in elucidating the problem. Separation science can offer solutions to all of these problems. It becomes clear that analysis based on separations can help identify potential contaminants that may be disastrous to humans. Furthermore, sorption-based methods can help remove arsenic, lead, or GenX from water.

References

Ahmed, M.F.S., et al., 2006. Science 314, 1687–1688.

Ahuja, S., 1986. Ultratrace Analysis of Pharmaceuticals and Other Compounds of Interest. Wiley, New York.

Ahuja, S., 2005a. International Workshop on Arsenic Contamination and Safe Water, 11–13 December 2005, Dhaka.

Ahuja, S., 2005b. Origins and remediation of groundwater contamination by arsenic: objectives and recommendations. In: International Workshop on Arsenic Contamination and Safe Water, 11–3 December, Dhaka.

Ahuja, S., 2006. American Chemical Society Meeting, 26–30 March 2006, Atlanta, GA.

REFERENCES

Ahuja, S., 2008a. Arsenic Contamination of Groundwater: Mechanism, Analysis, and Remediation. Wiley, NY.

Ahuja, S., 2008b. UNESCO Conference on Water Scarcity, Global Changes, and Groundwater Management Responses, 1–5 December, 2008, Irvine, CA. Arsenic contamination of groundwater.

Ahuja, S., 2009. Handbook of Water Purity and Quality. Chapter 2, Academic Press.

Ahuja, S., 2017. American Chem. Soc. Meeting, Washington, DC, August.

Ahuja, S., Malin, J., 2004. International Conference on Chemistry for Water, 21–23 June 2004, Paris.

Ahuja, S., Malin, J., 2006. Chem. Int. 28 (3), 14–17.

Blum, A., Balan, S.A., Scheringer, M., Goldenman, G., Trier, X., Cousins, I., Diamond, M., Fletcher, T., Higgins, C., Lindeman, A.A., Peaslee, G., de Voogt, P., Wang, Z., Weber, R., 2015. Madrid statement. Environ. Health Perspect. 123, A107.

Cooper, A., Ahuja, S., 2017. Presented at American Chem. Soc. Meeting, Washington, DC, August.

Lindstrom, A.B., Strynar, M.J., Libelo, E.L., 2011. Polyfluorinated compounds: past, present, and future. Environ. Sci. Technol. 45 (19), 7954–7961.

Nakayama, S., Strynar, M.J., Helfant, L., Egeghy, P., Ye, X., Lindstrom, A.B., 2007. Perfluorinated compounds in the cape fear drainage basin in North Carolina. Environ. Sci. Technol. 41, 5271–5276.

Pieper, K.J., Martin, R., Tang, M., Walters, L.A., Parks, J., Devine, S.R.C., Edwards, M.A., 2018. Evaluating water lead levels during the Flint water crisis. Environ. Sci. Technol. 52 (15), 8124–8132.

Scheringer, M., Trier, X., Cousins, I.T., de Voogt, P., Fletcher, T., Wang, Z., Webster, T.F., 2014. Helsingør statement on poly- and perfluorinated alkyl substances (PFASs). Chemosphere 114, 337–339.

Strynar, M., Dagnino, S., McMahen, R., Liang, S., Lindstrom, A., Andersen, E., McMillan, L., Thurman, M., Ferrer, I., Ball, C., 2015. Identification of novel perfluoroalkyl ether carboxylic acids (PFECAs) and sulfonic acids (PFESAs) in natural waters using accurate mass time-of-flight mass spectrometry (TOFMS). Environ. Sci. Technol. 49, 11622–11630.

Sun, M., Arevalo, E., Strynar, M., Lindstrom, A., Richardson, M., Kearns, B., Pickett, A., Smith, C., Knappe, D., 2016. Legacy and emerging perfluoroalkyl substances are important drinking water contaminants in the Cape Fear River watershed of North Carolina. Environ. Sci. Technol. Lett. 3, 415–419.

UNCW, 2018. Report to the Environmental Review Commission from the University of North Carolina at Wilmington Regarding the Implementation of Section 20. (a)(2) of House Bill 56 (S.L. 2017–209). NC Legislature—House Select Committee on North Carolina Water Quality. https://www.ncleg.net/documentsites/committees/house2017-185/Meetings/7%20-%20April%2026%202018/2018-April%20HB%2056%20UNCW%20Rpt.pdf.

Wang, Z., Cousins, I.T., Scheringer, M., Hungerbühler, K., 2013. Fluorinated alternatives to long-chain perfluoroalkyl carboxylic acids (PFCAs), perfluoroalkane sulfonic acids (PFSAs) and their potential precursors. Environ. Int. 60, 242–248.

Index

Note: Page numbers followed by "*f*" indicate figures, "*t*" indicate tables, and "*b*" indicate boxes.

A

Abiotic process, 5–6, 172
ABS. *See* ArsR binding site (ABS)
AC. *See* Alternating current (AC)
Accessory pigments, 240–242
Accidental spills, 288–290
Acetylcholinesterase inhibition assay, 137
Acidic aqueous solution, 397
Acidic environmental condition, 24
Acidification, 23–25
Acidithiobacillus ferrooxidans, 312
Acid mine drainage (AMD), 291, 303–304
Acid-producing bacteria (APB), 208–209
Acoustic Doppler velocity meters (ADVMs), 243
Acoustic sensors, 243
Acute exposure procedure, 178–182
ACZA. *See* Ammoniacal copper zinc arsenate (ACZA)
Adsorbable organic fluorine (AOF), 361
Adsorption, 388–390, 411
  chemical, 387
  for heavy metal removal from wastewater, 387–388
  intermolecular attractive forces, 387–388
  with nanotechnology, 388–390
  physical, 387
ADVMs. *See* Acoustic Doppler velocity meters (ADVMs)
Aeromonas, 289
Affinity-binding elements, 309–310
Agar composite bead, 401–402
Ag core-shell nanoparticles, 312
Aging private infrastructure, 9
Air quality contamination, 201–202

Air release, 347, 425
Air temperature, 231, 251, 271
Al. *See* Aluminum (Al)
Algal bloom, 2–3
Alkalinity, 204
Alkaloid, 129
Alkaloid neurotoxins, 129–130
Alliance for Coastal Technologies (ACT), 226
Alternating current (AC), 244
Aluminum (Al), 375
Aluminum-rich mineral, 118
AMD. *See* Acid mine drainage (AMD)
American Public Health Association, 222, 234–235, 239
American Water Works Association, 222, 235, 239
Amino acid, 127–129, 131
Amino acids L-leucine, 127–129
Ammoniacal copper zinc arsenate (ACZA), 291–292
Amperometric biosensors, 311
AMRIT, 106*b*
AMS. *See* Asset management system (AMS)
Amyloid plaque, 131
Analog strip chart recorder, 260
Ananas comosus, 404
Anatoxin-a, 130
Anatoxin-a(S), 130
Anglogold ashanti mine, 302–303
Anions, 155–156
Anodic stripping voltammetry (ASV), 307–308
Anthropogenically polluted area, 298–299
Anthropogenic industrial activity, 291
Antibiotic resistance bacteria (ARB), 93
Antigen affinity interaction, 310

Antimicrobial resistance (AMR), 92–93
AOF. *See* Adsorbable organic fluorine (AOF)
APB. *See* Acid-producing bacteria (APB)
Aptamers, 310
Aquatic ecosystems, 54–59
Aquatic endocrine disruption, 5–6
Aquatic environmental salinity, 19
Aquatic nuisance species, 159
Aquatic organism, 136–137
ARB. *See* Antibiotic resistance bacteria (ARB)
Arid coastal region, 101–104, 103*b*
Arid rural area, 101–104, 103*b*
Arsenic (As), 375, 417–421
  biosensors for monitoring
    affinity-binding elements, 309–310
    bioaffinity, 312–313
    biorecognition elements, 309–310
    components, 308–309, 309*f*
    enzymatic, 313–314, 314*t*
    enzymes, 310
    technology, 308–311
    transduction methods, 310–311
    whole cells, 310, 314–316
  contamination, 7, 113–116, 417–421
  detection, 308
  factors influence concentration of, 296–308
  field test tools and applications, 307–309
  filter, 104–108, 108–109*b*
  hot spots, 299
  human exposure, 296, 296*f*
  magnetic nanoadsorbents for removal of, 406–409
  management, 293–296

430 INDEX

Arsenic (As) *(Continued)*
  mobilization, 296–298
  nonpoint source pollution, 298–302
  point source pollution, 302–307
  poisoning, 383
  sequestration, 296–298
  toxicity, 383
Arsenic-laced irrigation water, 420
Arsenicosis, 113–114
Arsenic-rich tailing, 302–303
Arsenic sensitive promoter, 314–315
Arsenite
  efflux pump, 314–315
  reductase, 314–315
  sensitive promotor, 314–315
  sensitive whole-cell biosensors, 315
Arsenopyrite, 291
ArsR binding site (ABS), 314–315
Artemia salina, 136–137
Aryl hydrocarbon receptor nuclear
  translocator (ARNT), 16–17
As. *See* Arsenic (As)
Asset management system (AMS), 62
Atmospheric diazotrophic nitrogen,
  126
Atomic fluorescence spectrometer
  (AFS), 421
Au@ag core-shell nanoparticles, 312
Automated data-processing system
  (ADAPS), 264
Automated meter reading (AMR), 65
Autonomous robotic instrument,
  269–270
Autonomous underwater vehicles
  (AUVs), 256–259
  propeller-driven, 257–259
Azide-alkyne cycloaddition reaction,
  404

B

Bacteria, 94–95
Baseline physiological performance, 24
Batch extractor, 411
Bathing water quality, 249–250
Bathing water standard, 250
Battery-powered propeller system, 257
Beckman Model G pH meter, 232
Beneficiation, 302–303
Bermuda Test-Bed Mooring (BTM),
  254–255
Biochemical assays, 137
Biodegradable organic material, 55–56

Biodegradation, 173
Biological assays, 136–137
Biological oxygen demand (BOD),
  55–56, 205, 286–287
Biorecognition element, 311
  affinity-binding elements, 309–310
  enzymes, 310
  whole cells, 310
Bioreporter cell, 317
Bioreporter test method, 317
Biosensors
  for arsenic detection in water,
    311–316
    bioaffinity, 312–313
    enzymatic, 313–314, 314*t*
    whole-cell, 296–308
  biorecognition elements
    affinity-binding elements,
      309–310
    enzymes, 310
    whole cells, 310
  components, 308–309, 309*f*
  for field monitoring of arsenic,
    316–318
    biorecognition elements, 309–310
    transduction methods, 308–311
  technology, 308–311
  transduction methods
    electrical properties, 311
    electrochemical methods, 311
    mass, 311
    optical methods, 311
    temperature, 311
Bivalve embryo-toxicity assay, 175
Blackwater branch, 304–305
Blanket groundwater testing, 292
Blockchain for the water sector, 64
Blood lead level, 330
Blood sample, 308, 356
BOD. *See* Biological oxygen demand
  (BOD)
Brackish water, 71–72, 76–77
British Petroleum (BP), 150
Burns Waterway (BW), 152

C

Cadmium (Cd), 375
  magnetic nanoadsorbents for
    removal of, 402–405
  toxicity, 378–380
Calcium-based treatment technology, 89
Callinectes sapidus, 20

Cancer, 114
Capacity building, 65, 74–75
Cape Fear Public Utility Authority
  (CFPUA), 346, 424
Cape Fear River (CFR), 344–346, 349,
  361–362
  GenX in, 345–348, 345*t*
  in North Carolina, 361–362
  watershed, 343*f*
Carbon, 358–359
Carbonate ion, 23
Carbon dioxide concentration, 140
Carbonized bone meal (CBM), 89–91
Carbon nanotubes (CNTs), 120
CBM. *See* Carbonized bone meal (CBM)
CBNERR. *See* Chesapeake Bay National
  Estuarine Research Reserve
  (CBNERR)
CCA. *See* Copper-chromated arsenicals
  (CCA)
Cd. *See* Cadmium (Cd)
CDA. *See* Command and data
  acquisition (CDA)
C3 dimer acid, 342
CDOM. *See* Colored dissolved organic
  matter (CDOM)
CEERI. *See* Central Electronics
  Engineering Research Institute
  (CEERI)
Cellatex chemical plant, 290
CellSens image analysis software,
  180–182
Central Electronics Engineering
  Research Institute (CEERI),
  94–95, 95*b*
Central Glass Ceramics Research
  Institute (CGCRI), 96*b*
Centralized wastewater treatment
  systems, 68–69
Central Public Health Environment
  and Engineering Organisation
  (CPHEEO), 87–88
Ceramic membrane size-based
  exclusion, 96–97
Ceramic membrane systems, 96–97, 96*b*
Ceramic micro-filtration membrane,
  96–97
Cesium (Cs), 375
CF. *See* Continuous flow (CF)
CFC. *See* Chlorofluorocarbon (CFC)
CFR. *See* Cape Fear River (CFR)
Chaetoceros muelleri, 178

INDEX

431

Chemical Abstract Services (CAS), 341
Chemical additives, 8, 202–203
Chemical adsorption, 387–388, 402–403
Chemical analyzer systems, 237–239, 274–275
Chemical contaminant, 68
Chemical Dispersant Corexit 9500A®, 173
  composition, 173
  mode of action, 173
Chemically enhanced water accommodated fractions (CEWAFs), 177
Chemically treated carbonized bone meal (CTBM), 89–91
Chemical oxygen demand (COD), 55–56, 205
Chemical parameter, 109–110, 109–110b
Chemical pollutant, 24–25, 321
Chemical precipitation, 385
Chemical processing procedure, 154
Chemical Response to Oil Spills: Ecological Effects Research Forum (CROSERF), 173–174
Chemical sensor, 308–309
Chemical spill, 219, 223, 272–273, 285–286, 288–289, 379–380
Chemical warfare agent, 313
Chesapeake Bay National Estuarine Research Reserve (CBNERR), 270
Chitosan-enhanced membrane filtration technique, 386
Chlorofluorocarbon (CFC), 59
Chromatographic separation, 424
Chromium (Cr), 98–99, 375
  magnetic nanoadsorbents for removal of, 405–406
  mechanism of toxicity, 385
Chromium levels detection, 99, 99b
Chromium VI, 98, 99b
Clean Water Act (CWA), 3, 204–205, 224, 286–287, 425
Climate, 51–52, 61
Climate change factor, 29
Climate change impact, 34–38, 53
Climate change-related factor, 32
Cloud-based control system, 76
CNTs. See Carbon nanotubes (CNTs)
Co. See Cobalt (Co)
Coal-based thermal power plant, 375
Coastal aquatic system, 15–16
Coastal marine water, 174–175

Coastal oceanic environment, 23
Coastal salinity, 100–103
Cobalt (Co), 375
COD. See Chemical oxygen demand (COD)
CoFe2O4@MIL-100(Fe) hybrid MNPs, 409
Colony-forming units (CFU), 70–71
Colorado River basin, 255
Color complex formation, 99, 99b
Colored dissolved organic matter (CDOM), 235
Columbia Water Center, 200
Combustion ion chromatography (CIC), 361
Command and data acquisition (CDA), 262f
Communication methods, 261–262
Community-based lead testing, 331–333
Community-based water testing, 331–332
Comprehensive characterization, 8
Concentrated animal feeding operations (CAFOs), 288
Conflict resolution mechanism, 75
Conservation measure, 65
Contaminants of emerging concern (CEC), 2, 4, 9–10
  classification, 318
  detection, 320–321
  exposure pathways for contaminants of, 318, 319f
  fate, 318–319
  prioritization scheme for, 319, 320t
  sources, 318–319
  transport, 318–319
Contamination of water
  by chromium, 98–99
  by ECs, 92–93
  through arsenic, 104–108
  through bacteria, 94–95
  through fluoride, 89–92
  through iron, 95–97
  through nitrate, 97–98
  through salinity and hardness, 100–103
  through silica, 97
  through uranium, 99–100
Continuous flow (CF), 393–394
Continuous water-quality measurements, 238, 245–250, 260–261, 270, 274

Continuous water-quality monitoring, 220–225
Copper (Cu), 375
Copper-chromated arsenicals (CCA), 291–292
Coprecipitation, 119
Cr. See Chromium (Cr)
Crassostrea gigas, 175
Crassostrea virginica, 170–171, 171b, 174
Cross-sectional water-quality profile, 245
Crowd sourcing, 64
Crude oils, 171–172
Cs. See Cesium (Cs)
CTBM. See Chemically treated carbonized bone meal (CTBM)
CWA. See Clean Water Act (CWA)
Cyanobacteria, 7, 125–126
  affecting waterbodies, 132–134
  blooms, 131–132, 140–141
  contaminated water, 133
  detection of toxins, 134–139
  exposure, 140–141
  nitrogen-fixing, 126
  removal, 139–140
  toxins, 127–131
Cybersecurity, 76
Cylindrospermopsin (CYN), 2–3, 129
Cylindrospermopsis raciborskii, 129
CYN. See Cylindrospermopsin (CYN)
Cytotoxic alkaloids, 129

D

DAPS. See Data Collection and Automated Processing System (DAPS)
DART. See DRDO arsenic removal technology (DART)
Data acquisition, 259–262, 262f
Data block, 64
Data Collection and Automated Processing System (DAPS), 261
Data-collection platforms (DCPs), 252, 262f
Data delivery, 260–262, 265–266
Data generation, 63
Data management, 62–64
  software, 263
Data-processing, 64
  software, 262–266, 275
Data recorder, 252
Data repository, 263, 275

432 INDEX

Data service, 266, 275
Data structure, 266
Data telemeter transmitter, 250–251
Data transmission, 252, 257, 261, 262f, 275
DC. *See* Direct-current (DC)
De-ammonification, 76
Decentralized treatment systems, 69
Decision analytics approach, 62
Decision support system (DSS), 61–62
Deep-cycle marine battery, 257
Deepwater Horizon (DWH), 8, 170
  oil, 1, 8
  oil spill incident, 170–171
Deepwater Horizon crude oil, 1, 8, 170–173
  Chemical dispersant Corexit 9500A®, 173
    composition, 173
    mode of action, 173
  on early life stages of Eastern Oysters
    brood stock collection and conditioning, 177–178
    embryo assay, 179
    fertilization assay, 179
    gamete collection, 178
    larvae in ecotoxicology, 175–176
    larval assay, 179–180
    larval rearing, 178
    lethal effects, 182
    model organism, 174–175
    settlement assay, 180
    spat assay, 180–182
    spawning, 178
    sublethal effects, 182–191
  exposure solutions, 176–177
  PAHs, 171–172
    bioavailability and uptake, 172–173
    composition, 171–172
    mechanisms of toxicity, 172
  water accommodated fraction, 173–174
  weathering processes, 173
Degenerative disease, 376
Dehydrobutyric acid (DHB), 127–129
Delaware River Estuary, 250–251
Depressed metabolic performance, 24
Dermatotoxins, 131
Desalination, 71–72
Desalination plant, 56–57, 60, 71–72, 74
Desalination technology, 71–72

Desalinator, 118
Dielectric barrier discharge (DBD), 94–95, 95b
Differential pulse polarography, 421
Diffraction grating, 240
Digestive tubules (DGTs), 180–182
Dihydrotestosterone (DHT), 21
Dimethyl arsenic acid (DMAA), 406–407, 419–420
Dioctyl sodium sulfosuccinate (DOSS), 173
Direct-current (DC), 231
Direct potable reuse (DPR), 68–69
Disinfection by-product, 294, 422
Disinfection technology, 95b
Dispersant toxicity, on oyster larvae, 176
Dissolved inorganic compounds, 204–205
Dissolved organic carbon (DOC), 205
Dissolved organic matter (DOM), 242
Dissolved oxygen (DO), 15–16, 55–56, 155–156, 179, 219
DMAA. *See* Dimethylarsenic acid (DMAA)
DNAs strand breakage, 175–176
DO. *See* Dissolved oxygen (DO)
DOC. *See* Dissolved organic carbon (DOC)
Domestic amrit filter, 106f
Domestic communications satellite (DOMSAT), 261, 262f
Domestic measures, 117–120
Doppler velocimetry log, 258
Dose response study, 116
DPHE-Danida arsenic mitigation project, 119
DPR. *See* Direct potable reuse (DPR)
DRDO arsenic removal technology (DART), 120
Drinking water, 288, 290–293, 318–319, 321, 417, 419–426
  distribution system, 289
  guidelines, 292–293
  health goalfor, 347
  limit, 99
  monitoring, 316
  needs, 101–104, 101–102b
  quality analysis, 109–110, 110b
  sources, 94, 96–97, 96b, 288
  standards, 98, 298–299

  supply, 289–290, 298–299, 304, 318–319, 343–344, 346–349, 352–353
  treatment plant, 101–104, 101–102b, 233, 235, 242, 294, 321, 361–362
Drought, 53
DSS. *See* Decision support system (DSS)
DWH. *See* Deepwater Horizon (DWH)

E

Early-warning systems (EWSs), 65, 290
Earth system, 219, 263
Eastern oyster *crassostrea virginica*, 174–176
  Deepwater Horizon crude oil toxicity
    brood stock collection and conditioning, 177–178
    embryo assay, 179
    fertilization assay, 179
    gamete, 182–185
    gamete collection, 177–178
    larval assay, 179–180
    larval rearing, 177–178
    lethal effects, 182
    as model organism, 174–175
    settlement assay, 180
    spat assay, 180–182
    spawning, 177–178
    sublethal effects, 182–191
  larvae in ecotoxicology, 175–176
Eco-friendly indigenous water filter, 91
Ecotoxicology research community, 38
ECs. *See* Emerging contaminants (ECs); Endocrine compounds (ECs)
EDCs. *See* Endocrine disrupting chemicals (EDCs)
Eenzyme-linked immunosorbent assay (ELISA), 209
EIA. *See* Environmental impact assessment (EIA)
Electrical resistance techniques, 226–231
  specific electrical conductance of water, 231
  water temperature, 226–231
Electrochemical DO sensors, 234–235
Electrochemical techniques, 386–387
  heavy metals removal, 386–387
Electrodialysis, 120
Electromotive force (EMF), 231
Electron donor availability, 97
Electron microscopy tool, 395

INDEX

433

Electro-regenerated anion exchange, 120
ELISA. *See* Enzyme-linked immunosorbent assay (ELISA)
Embryo assay, 179
Embryos membrane fluidity, 187–188
Emerging contaminants (ECs), 2, 4, 9–10, 88, 151, 162–165
EMF. *See* Electromotive force (EMF)
Endocrine compounds (ECs), 68
Endocrine disrupting chemicals (EDCs), 5–6, 13, 92
Endocrine disruption effect, 23
Endocrine disruptive flame retardants, 24–25
Endocrine disruptive persistent organic pollutant, 13–14
Energy
  consumption and emissions of greenhouse gases, 72
  dispersive X-ray spectroscopy, 395
  metabolism system, 419–420
  recovery device, 71
  resource, 17
  symbiotic relationship, 85
Energy-filtered TEM (EFTEM), 406
Energy-intensive water production, 74
Energy Policy Acts, 202–203
Engendering skin cancer, 114
Enhanced reductive dechlorination (ERD), 305–306
Entrapped agarose composite bead, 401–402
Environmental effect, 57
Environmental impact and mitigation measures, 71
Environmental Impact Assessment (EIA), 72
Environmental Justice Movement, 330
Environmental Management Website, 152
Environmental Protection Agency (EPA), 204, 224, 292–293, 331, 375
Environmental Sample Processor (ESP), 269
Environment and Public Health Organization (ENPHO), 120
Environment friendly technique, 120
Environment Protection Agency (EPA), 95–96

Environment regulatory requirement, 70–71
Enzymatic biosensors, 313–314
Enzyme
  acid phosphatase, 314
  arsenite oxidase, 313
  biosensor, 310
  catalyzed hydrolytic reaction, 311
  inhibition-based biosensors, 313
  succinic dehydrogenase, 383
Enzyme-linked immunosorbent assay (ELISA), 270
EPA. *See* Environmental Protection Agency (EPA)
EPA-prescribed permissible limit, 120
EPI. *See* Estimation Programs Interface (EPI)
Epigenetic effects, 30–32
ERD. *See* Enhanced reductive dechlorination (ERD)
Escherichia coli, 152, 154–155, 249, 288, 312
ESP. *See* Environmental Sample Processor (ESP)
Estimation Programs Interface (EPI), 21
Ethyl acetate extraction, 207–208
Ethylenediamine modified magnetic cross-linking chitosan NPs (EMCN), 396–397
Eutrophication, 287–288
EWSs. *See* Early-warning systems (EWSs)
Exposure solution, 173–174
Extensible markup language (xml), 266
Exxon oil spill, 1

F
FAA. *See* Flame atomic absorption (FAA)
Fayetteville Works, 342–343, 348, 350–351
FChl. *See* Fluorescent chlorophyll (FChl)
FDOM. *See* Fluorescent dissolved organic matter (FDOM)
Federal Water Pollution Control Act, 3, 222–223
Fertilization assay, 179
Fiber optic cable, 226
Filtered water sample, 155–156
Flame atomic absorption (FAA), 211, 421

Flight mass spectrometry, 347
Flint River water, 421
Flint water crisis (FWC), 329
Floods, 8–9
Flowback water, 200
Flow-through detection system, 240–241
Flow-through nitrate analyzer, 267
Fluoranthene (FL), 18*f*
Fluorescent chlorophyll (FChl), 226
Fluorescent dissolved organic matter (FDOM), 240
Fluoride, 89–92
Fluoride removal capacity, 89–91
Fluorine- and nitrogen-codoped magnetic carbons (FN-MC), 406
Fluoroethylene polymer (FEP), 358
Fluoropolymer plastics, 356
FNU. *See* Formazin Nephelometric Unit (FNU)
Food chain, 420
  arsenic-laced irrigation water on, 420
Food web, 33–34
Forced mass displacement, 77
Formazin Nephelometric Unit (FNU), 237
Forward osmosis (FO), 71
Fossil fuel, 162, 291
Fractured production well, 199–200
Freshwater
  aquatic biota, 162–164
  environment, 19, 257
  lake, 159
  organism, 21, 30
  stream system, 166
FWC. *See* Flint water crisis (FWC)

G
Gadus morhua, 25
Gaging station, 250–251, 264
Gamete collection, 178
Gamma ray, 307–308
Ganges River water, 372–373
Gas chromatography (GC), 207
Gas development stimulation process, 8
Gasoline combustion product, 1–2
Gastrointestinal (GI), 131
Gas vesicle protein, 125–126
GBNERR. *See* Grand Bay National Estuarine Research Reserve (GBNERR)

434 INDEX

GC. *See* Gas chromatography (GC)
GCC. *See* Global climate change (GCC)
GCC-altered abiotic factor, 13–14
GCC-related environmental stressors, 14
Gene cluster, 138
Genetic potential, 134
Genotoxic alkaloids, 129
Genotypic sex determination, 27
GenX, 348–352
  in Cape Fear River, 345–348, 345*t*
  contamination, 10
  contamination of drinking water in Wilmington, 423–426
  in environmental water samples
    conservation, 356–357
    nontargeted analysis, 359
    precision and accuracy, 358
    sample collection, 356–357
    sample preparation, 357–358
    storage, 356–357
    targeted analysis, 358–359
    total PFAS analysis, 359–361, 360*f*
  in NorthCarolina, 361–366
  water disaster, 1–2, 10
Geographic information system (GIS), 62, 264
Geological material, 118
GIS. *See* Geographic information system (GIS)
GIS mapping application, 62
Glass fiber filter paper, 154
Glass microfiber filter, 357
Glass tube, 236
Gliders, 259
Global arsenic emission, 302–303
Global climate, 5–6
Global climate change (GCC), 5–6, 13–14
Global positioning system (GPS), 243
Global warming, 53, 76
Glucose biosensor, 308
GO. *See* Graphene oxide (GO)
GoM. *See* Gulf of Mexico (GoM)
GPER. *See* G protein-couple estrogen receptor (GPER)
G protein-couple estrogen receptor (GPER), 17
GPS. *See* Global positioning system (GPS)
Gram-negative photosynthetic bacterium, 125–126

Grand Bay National Estuarine Research Reserve (GBNERR), 270
Graphene, 75
Graphene oxide (GO), 404
Graphite furnace atomic absorption (GFAA) spectrometry, 211, 300–301, 420–421
Gray water system, 69
Greener sustainable protocol, 393–394
Green fluorescent protein (GFP), 310, 312
Greenhouse gas, 5–6, 72, 74
Greenhouse gas emission, 15–16, 25–26
Green leafy vegetable, 420
Groundwater
  management, 58
  monitoring, 59
  resource, 57–61, 100
  sample, 116
  studies, 58
  systems modeling, 59
Groundwater, arsenic in, 417–421
  biosensors for monitoring
    affinity-binding elements, 309–310
    bioaffinity, 312–313
    biorecognition elements, 309–310
    components, 308–309, 309*f*
    enzymatic, 313–314, 314*t*
    enzymes, 310
    technology, 308–311
    transduction methods, 310–311
    whole cells, 310, 314–316
  factors influence concentration of, 296–308
  field test tools and applications, 307–308
  human exposure, 296, 296*f*
  management, 293–296
  mobilization, 296–298
  nonpoint source pollution, 298–302
  point source pollution, 302–307
  sequestration, 296–298
Growth hormone receptor, 20
Guanidine alkaloid toxin, 129
Gulf Cooperation Council (GCC), 69–71
Gulf of Mexico (GoM), 170

H
HABs. *See* Harmful algal blooms (HABs)
Handheld meters, 243–244
Harmful algal blooms (HABs), 219, 242, 257, 266, 268–270, 274, 287–288

Health effect, 300
Health hazard, 371–372, 375–376, 383
Health problems, due to arsenic contamination, 113–115
Health risk, 290, 302, 305–306
Heat shock protein gene, 21–22
Heat-trapping greenhouse gas, 5–6, 13–14
Heavier molecular weight (HMW), 172
Heavy metals, 385, 387, 390, 411
  arsenic, 382–383
  cadmium, 378–380
  chromium, 383–385
  health impacts of, 375–378
  lead, 381–382
  mechanism of toxicity, 385
  mercury, 380–381
  poisoning, 378–385
  reactors for online and cyclic recovery of, 409–411
  removal
    adsorption, 387–388
    adsorption technology with nanotechnology, 388–390
    chemical precipitation, 385
    electrochemical techniques, 386–387
    ion-exchange, 385–386
    magnetic nanoadsorbents, 390–409
    membrane separation process, 386
    ultrafiltration, 386
  sources, 375
  toxicity, 375–376
  water contamination, 374–375
Heavy metal ions, 10–11
Hemigrapsus oregonensis, 20
Hepatotoxins, 127–129
Hexafluoropropylene oxide (HFPO), 425
Hexafluoropropylene oxide dimer acid (HFPO-DA), 1, 355–356
HFOs. *See* Hydrous ferric oxides (HFOs)
HFPO. *See* Hexafluoropropylene oxide (HFPO)
High binding capacity, 121
High-density polyethylene (HDPE), 356, 358
High-end arsenic remediation technology, 117

INDEX

High-energy water accommodated fractions (HEWAFs), 177
High juvenile mortality, 32–33
High-performance liquid chromatography (HPLC), 135, 421
High-pressure liquid chromatography, 4
High-temporal frequency water-quality monitors, 269
HMW. *See* Heavier molecular weight (HMW)
Holocene alluvial aquifer, 300–301
Hoosier Environmental Council (HEC), 152
Horizontal unconventional well, 200
HPLC. *See* High-performance liquid chromatography (HPLC)
Human activity, 371
Hurricanes, 9, 270–271
Hyalella azteca, 21–22
Hybrid technology, 120
Hydride generation (HG), 421
Hydrocarbons, 208
Hydrocarbon-eating bacteria, 173
Hydrogel reduced arsenic concentration, 117–118
Hydrophobic endocrine-active compound, 34–38
Hydrothermal synthetic method, 393
Hydrous ferric oxides (HFOs), 106–109, 108–109b
Hydroxyapatite (HAP), 401–402
Hypertension-related cardiovascular disease, 106
Hyphenated methods, 292
Hypoxia, 15–19

I

IARC. *See* International Agency for Research on Cancer (IARC)
Illex illecebrosus, 24
Immunoassays, 137–138
Improve water security (IWA), 69–70
Indiana spilling wastewater, 384
Induced seismic event, 201–202
Inductively coupled plasma–mass spectrometry (ICP-MS), 421
Industrial effluent, 372–373, 386
Infantile cerebral palsy, 380–381
Inflammatory response, 190–191
Infrared (IR), 235

Ingested oil droplets, 190
Inhibition-based biosensor, 313
Inorganic arsenic (IAs), 409
Integrated water resources management (IWRM), 61
Internal osmotic composition, 20
International Agency for Research on Cancer (IARC), 114, 375
International Hydrological Program, 63
International Monetary Fund, 51
International Organization for Standardization (ISO), 294
International Union of Pure and Applied Chemistry (IUPAC), 75
International Water Association, 60
Internet of things (IOT), 64
IOB. *See* Iron-oxidizing bacteria (IOB)
Ion channel regulation, 19
Ion channel suppression, 17
Ion chromatography, 161–162, 361
Ion-exchange, 120, 385–386
Ion-exchange resin, 385–386
Ion-selective electrodes (ISEs), 231
IOT. *See* Internet of things (IOT)
IR. *See* Infrared (IR)
Iron, 95–97
Iron-based chemical coprecipitation, 119
Iron-based nanoparticles, 120
Iron-impregnated granular activated carbon, 117
Iron oxide, 297–298
    cadmium ions removal, 404
    coated quartz, 406–407
    nanoparticles, 392–394, 404
Iron-oxidizing bacteria (IOB), 208–209
Iron removal plant, 96–97, 96f, 96b
Iron removal technology, 96b
Iron sulfate minerals, 303–304
Irrigated crop, 66
Irrigation, 101–102b
    efficiency, 65–66
    of farmland, 65–66
    purpose, 67–68
    system, 65
    water demand, 65
ISEs. *See* Ion-selective electrodes (ISEs)
ISO. *See* International Organization for Standardization (ISO)

J

Jackson TU units (JTU), 236
Jinzu River basin, 378–379
JTU. *See* Jackson TU units (JTU)
Juvenile atlantic cod, 25, 26f

K

Kanawha valley water treatment, 290
Kanchan arsenic filter (KAF), 120

L

Lake Huron water, 2, 421–422
Lake Michigan coastal community, 7–8
Land-based water-quality sensor, 251
Large-scale industrial applicability, 10–11
Larval assay, 179–180
Larval rearing, 178
Larval settlement inhibition, 188
Laterite-based arsenic filter, 107–108b
LCR. *See* Lead and copper rule (LCR)
Leachate, 302–303
Lead (Pb), 375, 381–382
    and copper rule, 422
    contamination, 417, 421–423
    of drinking water in Flin, 421–423
    poisoning, 381
    toxicity, 382
Leak location detector, 75
Light crude oil, 173
Lignum nephriticum, 240
Lipopolysaccharide (LPS), 131
Liquid chromatography (LC), 207
Liquid–liquid extraction (LLE), 357
Liquid scintillation counting (LSC), 212
Log-linear regression approach, 246
Long-range strategic decision-making, 61–62
Long-term adverse health effect, 130
Long-term consistent program, 7–8
Lower molecular weight (LMW), 172
Lowest observed effect concentration (LO), 22f
Lowest observed effective concentration (LOEC), 190
LPS. *See* Lipopolysaccharide (LPS)
LSC. *See* Liquid scintillation counting (LSC)
Luminescent whole-cell biosensor, 316–317
Lung cancer, 114

M

MACRL. *See* Marine and Atmospheric Chemistry Research Laboratory (MACRL)
Macroinvertebrates, 156
Magnesium ferrite nanoparticles ($MgFe_2O_4$–NH_2 NPs), 397
Magnetic chitosan nanocomposites, 397–398
Magnetic mesoporous carbon composite, 398–399
Magnetic nanoparticles (MNPs), 390–391
  surface modification, 394–395
  synthetic approaches for preparation
    coprecipitation, 392
    hydrothermal, 393
    microemulsion, 393
    thermal decomposition, 392–393
Magnetic solid-phase extraction, 391
Magnetite-reduced graphene oxide, 121
Manganese (Mn), 375
Marine and Atmospheric Chemistry Research Laboratory (MACRL), 347
Marine recreational water, 224
Mass spectrometry, 135–139
Mass transfer process, 387
Matrix-assisted laser desorption, 135, 210
Matrix effect, 358
Maximum contaminant limits (MCLs), 375–376, 419
MBARI. *See* Monterey Bay Aquarium Research Institute (MBARI)
MBR. *See* Membranes biological reactors (MBR)
MCLs. *See* Maximum contaminant limits (MCLs)
MDEQ. *See* Michigan Department of Environmental Quality (MDEQ)
Mediterranean Sea monitoring, 57
Mekong River Delta region, 316–317
Melamine, 406
Membranes biological reactors (MBR), 68
Mercenaria mercenaria, 176
Mercury (Hg), 375, 380–381
  poisoning, 380–381
  toxicity, 381

MERHAB. *See* Monitoring and Event Response for Harmful Algal Blooms (MERHAB)
Metal organic framework (MOF), 409
Michigan Department of Environmental Quality (MDEQ), 422
Microcystis aeruginosa, 133
Microcystis bloom material, 136–137
Microextraction technique, 357
Microfluidic colorimetric chemical analyzer system, 238–239
Microwave (MW), 393–394
Middle East water resources, 6
Millivolts (mV), 233
Minamata disease, 380–381
Mineral salt, 70
Mississippi River basin, 266–268
Mitigation measure, 6, 72
Mitochondrial oxidative phosphorylation, 419–420
MMAA. *See* Monomethylarsenic acid (MMAA)
Mn. *See* Manganese (Mn)
MNPs. *See* Magnetic nanoparticles (MNPs)
Mo. *See* Molybdenum (Mo)
Mobile platforms, 256–259
  autonomous underwater vehicles, 257–259
  pontoon boats, 257
MOF. *See* Metal organic framework (MOF)
Molybdenum (Mo), 375
Monitoring and Event Response for Harmful Algal Blooms (MERHAB), 257
Mono-aromatic volatile organic compounds, 171–172
Monomethyl arsenic acid (MMAA), 406–407, 419–420
Monterey Bay Aquarium Research Institute (MBARI), 269–270
Moored observatories, 250, 252
Most probable number (MPN), 154–155
MSF. *See* Multistage flash (MSF)
Multicriteria analysis model, 61–62
Multicycle adsorption desorption experiment, 397
Multi-effect distillation (MED), 71, 101–104, 103*f*, 103*b*
Multiparameter sonde, 255, 257
Multiple stressors, 17, 30, 32

Multistage flash (MSF), 71
Multiwalled carbon nanotubes (MWCNTs), 405
Muscle tissue concentration, 26*f*
Mussel Watch Program, 174–175
mV. *See* Millivolts (mV)
MW. *See* Microwave (MW)
MWCNTs. *See* Multiwalled carbon nanotubes (MWCNTs)
Mytilus edulis, 24
Mytilus galloprovincialis, 24–25

N

Nanoadsorbents, 10–11
Nanofiltration (NF), 386
Nanomaterials, for chromium VI detection, 99*b*
Nanoparticles, use of, 120–121
Nanosized building block, 388–390
Nanotechnology, 388–390
National Ecological Observatory Network (NEON), 255–256
National Field Manual (NFM), 246
National Oceanic and Atmospheric Administration (NOAA), 252–253
National Research Council, 173, 209
National Sanitation Steering Committee, 116
National Toxicology Program (NTP), 329–330, 383–384
National Water-Information System web page (NWISweb), 262*f*
Natural geological material, 118
Naturally occurring radioactive material (NORM), 202, 211–212
Natural Resource Damage Assessment (NRDA), 170–171
Nature-Based Solutions (NBS), 61
NC Division of Water Quality (NC DWQ), 350–351
NCWR. *See* Nonconventional water resources (NCWR)
Near-future pollutant exposure scenario, 14, 15*f*
Nephelometric Turbidity Units (NTU), 70–71, 237
Neurological degenerative disease, 376
Neuropathy, 114–115
Neurotoxic amino acids, 131
Neutron activation analysis, 420–421
NF. *See* Nanofiltration (NF)
NFM. *See* National Field Manual (NFM)

INDEX

437

Nickel (Ni), 375
Nitrate removal efficiency, 98
Nitrates (NO₃), 97–98, 290
NMR. *See* Nuclear magnetic resonance (NMR)
NOAA. *See* National Oceanic and Atmospheric Administration (NOAA)
Nodularia spumigena, 127–129
Noncontact cooling water, 342–343, 347, 424
Nonconventional water resources (NCWR), 67
Nonpoint source pollution, 258, 298–302
Nonpolar organic compound, 318
Nonpurgeable organic carbon, 205
Nontargeted analysis, 356, 359, 361–362
NORM. *See* Naturally occurring radioactive material (NORM)
NRDA. *See* Natural Resource Damage Assessment (NRDA)
NTP. *See* National Toxicology Program (NTP)
NTU. *See* Nephelometric Turbidity Units (NTU)
Nuclear magnetic resonance (NMR), 4, 136, 350–351
Nucleic acid, 309–310
Nutrient-enriched flood water, 274
Nutrient-poor estuarine water, 267
Nutrient-poor warm water, 132
Nutrient-rich cold water, 132
NWISweb. *See* National Water-Information System web page (NWISweb)

O

OA. *See* Ocean acidification (OA)
Observatories Initiative (OOI), 255
Ocean acidification (OA), 23
Ocean-based low-temperature thermal desalination, 101–104
Ocean current, 254–255, 259
Ocean observatories initiative (OOI), 255
Off-shore seawater intake, 72
Ohio River, 223–224, 233, 290
Ohio River Valley Water Sanitation Commission (ORSANCO), 223–224
Oil pipe line, 58
Oil slick, 170, 173

Oil spill, 170–171, 171*b*, 174–175, 182, 188, 192
One-step hydrothermal synthesis, 393
On-site system wastewater infiltration, 93
On-site treatment plant, 93
Optical spectrophotometric method, 239
Organic micropollutants, 70
Organisation for Economic Co-operation and Development (OECD), 359
ORP. *See* Oxidation–reduction potential (ORP)
ORSANCO. *See* Ohio River Valley Water Sanitation Commission (ORSANCO)
Orthophosphate, 422
Oxidation, 118–119
Oxidation-based arsenic removal methodology, 118–119
Oxidation–reduction potential (ORP), 205–207, 223–224
Oxidize natural organic material, 154
Oxygen-limited metabolic stress, 18–19
Oxygen minimum zone (OMZ), 15–16
Oxygen partial pressure, 24
Oxygen permeable membrane, 234
Oyster larvae, 175–176, 188
  bivalve embryo-toxicity assay, 175
  dispersant toxicity, 176
  Oil/PAHs toxicity, 175–176
Oyster natural population, 8

P

PAA. *See* Polyacrylic acid (PAA)
Pachygrapsus crassipes, 20
Pacific killer whale, 33–34
Paper processing plant, 375
Paper strip chart, 259–260
PAR. *See* Photosynthetically active radiation (PAR)
Particle-induced gamma ray emission (PIGE), 361
Part-per-trillion (ppt), 344
Parts-per-billion (ppb), 4, 182, 381, 417–418
Parts-per-million (ppm), 182
Passive sedimentation, 119
Pb(II), magnetic nanoadsorbents for removal of, 397–402
PCR. *See* Polymerase chain reaction (PCR)

PDA. *See* Photodiode array detection (PDA)
Pd(II), magnetic nanoadsorbents for removal, 396–397
PEG. *See* Polyethylene glycols (PEG)
Pelecanus crispus, 133
Per- and polyfluoroalkyl substances (PFASs), 10
  in Cape Fear River, 343–345, 345*t*
  contaminants, 342–343
  discharges from Fayetteville Works, 348
  in environmental water samples conservation, 356–357
    nontargeted analysis, 359
    precision and accuracy, 358
    sample collection, 356–357
    sample preparation, 357–358
    storage, 356–357
    targeted analysis, 358–359
    total PFAS analysis, 359–361, 360*f*
Perca fluviatilis, 23–24
Perfluoroether carboxylic acids (PFECAs), 424
Perfluoroether sulfonic acids (PFESAs), 345–346, 424
Perfluorooctane sulfonic acid (PFOS), 342, 355–356, 424
Perfluorooctanoic acid (PFOA), 423–424
Permian Basin, 200
Persistent organic pollutants (POPs), 13–14, 25
Personal and Pharmaceutical Care Products (PPCPs), 88
Petroleum distillate, 173
Petroleum hydrocarbon pollution, 175
PFECAs. *See* Perfluoroether carboxylic acids (PFECAs)
PFESAs. *See* Perfluoroether sulfonic acids (PFESAs)
PFOA. *See* Perfluorooctanoic acid (PFOA)
PFOS. *See* Perfluorooctane sulfonic acid (PFOS)
Pharmaceuticals and personal care products (PPCPs), 68
Phenanthrene (PHE), 175–176, 187–188
Phenolic compound, 109–110, 110*b*
Phoeniconaias minor, 133
Phoenicopterus chilensis, 133
Phoenicopterus ruber, 133
Photodiode array detection (PDA), 135

438 INDEX

Photosynthetically active radiation (PAR), 251
Phycocyanin pigments, 246
Physical adsorption, 387
Physicochemical methods, 93, 135–136
Phytoplankton, 268
PIGE. *See* Particle-induced gamma ray emission (PIGE)
Planktothrix, 125–126
Plastic material, 151, 154, 162–165
Plastic ocean, 3
Platinized nickel electrode, 231
Plentiful raw laterite, 106–109, 107–108*b*
Point source pollution, 302–307
Poisonous blue alga, 57
Policy maker, 94
Polluted volcanic sediment, 295–296
Pollution control program, 3
Pollution point source management, 306
Pollution tolerance category, 156
Polyacrylic acid (PAA), 404
Polycyclic aromatic hydrocarbons (PAHs), 17, 171–177, 185–190, 192
Polyethylene glycols (PEG), 208
Polyfluoroalkyl ether acid, 10
Polymerase chain reaction (PCR), 134, 209–210
Polypropylene (PP), 356
Polypropylene glycols (PPG), 208
Polytetrafluoroethylene (PTFE), 356
Polyvinyl chloride (PVC), 251–252
Polyvinyl fluoride synthesis, 346, 349
Polyvinylidene fluoride (PVDF), 406
Pontoon boats, 256–257, 275
Populated coastal area, 152
Population level effects, 32–33
Power loss, 386–387
Power supply, 244, 252
PP. *See* Polypropylene (PP)
PPCPs. *See* Pharmaceuticals and personal care products (PPCPs)
PPG. *See* Polypropylene glycols (PPG)
Produced water, 200–213
  analysis, 204–212
    biological, 208–210
    bulk measurements, 204–207
    inorganic, 210–211
    naturally occurring radioactive material, 211–212

organics, 207–208
quality parameters, 204–207
Class II injections wells, 200–201
from oil-bearing shale, 202
Protein
  phosphatase, 127–129
  phosphatase inhibition assay, 137
  sequence, 16–17
  translation machinery, 139
PTFE. *See* Polytetrafluoroethylene (PTFE)
Purgeable organic carbon, 205
PVC. *See* Polyvinyl chloride (PVC)

Q

Quality Assurance Project Plan: *Deepwater Horizon* Laboratory Toxicity Testing, 173–174
Quartz crystal microbalance, 311

R

Radio transmitter, 252
Rainwater harvesting, 66, 85
Rapid Small Scale Column Test (RSSCT), 117
Reactive oxygen species (ROS), 185–187
Reactors, for online and cyclic recovery of heavy metals
  advantages, 411
  batch extractor, 411
  elution unit, 411
  upper segment, 410
  working, 411
Real-time (RT) data, 260
Real-time reliable monitoring, 110–111
Redox active product, 310
Redox reaction, 24–25
Relative fluorescence units (RFU), 241–242
Remediation technology, 116–117
Remote sensing (RS), 64
REMUS 600 AUV, 258–259
Research-based technology solution, 6–7
Resistive temperature devices (RTDs), 226
Resonance Rayleigh scattering, 312
Resources recovery system, 75
Respiratory disorders, 115
Responsive transcriptional repressor, 315

Reverse osmosis (RO), 59, 71, 89, 118, 386
  desalinator-type power survivor, 118
  membrane technology, 101–104, 101–102*b*
RFU. *See* Relative fluorescence units (RFU)
Rhodamine WT dyes, 240–241, 258
RO. *See* Reverse osmosis (RO)
RSSCT. *See* Rapid Small Scale Column Test (RSSCT)
RT. *See* Real-time (RT) data
RTDs. *See* Resistive temperature devices (RTDs)
Ruditapes philippinarum, 24–25
Rural water supply, 120
Rural Water Supply and Sanitation Support Programme (RWSSSP), 120

S

Safe Drinking Water Act (SDWA), 3
Saline aquifer area, 100
Salinity, 19–23, 100–103
  affected areas result, 101–104
  gradient, 5–6
  and hardness, 100–103
Saltwater disposal wells (SWD), 200–201
Sand filters, 119
SART. *See* Subterranean arsenic removal technology (SART)
Satellite systems, 252–253, 261
Saxitoxin, 137
SC. *See* Specific electrical conductance (SC)
SDGs. *See* Sustainable Development Goals (SDGs)
SDWA. *See* Safe Drinking Water Act (SDWA)
Se. *See* Selenium (Se)
Sea
  turtle species, 33
  urchin, 32
Seasonal floods, 266–270
  in Mississippi River, 266–268
  and Western Lake Erie algae blooms, 268–270
Seawater
  desalination, 71–72
    industry, 72
    market, 71

INDEX

439

plant, 71
program, 60
intrusion, 56
level, 56
temperature, 56
Selenium (Se), 375
Semivolatile organic compounds (SVOC), 207–208, 305
Sensitive biosensor, 9
Sensor
platforms, 250–259
recognition element, 310
system, 250–251
technology, 220, 234–235, 244, 274, 294, 320
Settlement assay, 180, 188
Sewage treatment plant, 3
Sewer system, 286–287
Sex steroid, 17, 19, 27
Shallow coastal water, 271
Shellac-coated iron oxide NPs, 402–403
Side-scan sonar imagery, 257
Silica (Si), 97
Single-atom thick carbon nano-sheets, 75
Siphonaria autralis, 32
Site-specific regression model, 248
Skin
cancer, 114–116
lesion, 113–116, 299–300, 382, 419
rash, 133
Slocum battery, 259
SM. *See* Standard method (SM)
Social outings (SC), 165
Sodium channel, 130, 137
Sodium dodecylbenzenesulfonate (SDBS), 406
Soil
erosion control, 54
moisture retention, 61
particle, 119–120
Solar-biomass-based multi-effect desalination, 101–104, 103*b*
Solar oxidation, 119
Solar power desalination, 103*b*
Solar thermal energy, 101–104, 103*b*
Solid-phase extraction (SPE), 138–139, 357, 388
Soluble lead, 422
Soluble organic matter, 239–240
Sono filter, 119

Sophisticated remediation technology, 117
Spat assay, 178–182
SPE. *See* Solid-phase extraction (SPE)
Specific electrical conductance (SC), 222
Spill mitigation response, 170
Spillway inflow, 267–268
SPR. *See* Surface plasmon resonance (SPR)
Spring Hill Community Center, 332, 338
Spring Hill Resource Center, 332–333, 335
Standard method (SM), 204–205
Starch functionalized maghemite nanoparticles (SMhNPs), 405
Stationary platforms, 250–256, 275
Steroid hormone, 26*f*
Steroid sex hormone, 320
Stomach cancer, 383–384
Storm runoff, 66–67
Stormwater runoff control measure, 66
Stream water, 156
Strontium (Sr), 375
Sublethal effects, 182–191
Submersible multiparameter sonde, 243–244
Subterranean arsenic removal technology (SART), 119–120
system supply, 119–120
water treatment plant, 119–120
Succinic dehydrogenase, 383
Sulfate-reducing bacteria (SRB), 208–209
Superior magnetic nanoadsorbents, 395–409
Surface
analysis technique, 361
cyanobacterial accumulation, 126
drainage system, 288
enhanced laser desorption ionization time, 135
freshwater source, 106
modification, 394–395
Surface plasmon resonance (SPR), 99, 309
Surface water, 219–221, 226, 235–236, 256, 291–292, 299–300, 303–307
environment, 162–164
monitoring, 150–151
quality analysis, 165

sample, 162–164
source, 4
Surfactants, 173
Surrogate measurements, 248–249
Surrogate statistical model, 248–249
Surrogate water-quality models, 246–250
Suspended-sediment concentration (SSC), 243, 248–249
acoustic Doppler technology for measurement of, 243
standardized methods for computing, 248–249
Suspended solids, 154, 156, 357, 386
Sustainable Development Goals (SDGs), 61, 371–372
Sustainable economic development, 94–95
Sustainable high-quality water, 76
SWD. *See* Saltwater disposal wells (SWD)
Sweeney water treatment plant, 346, 424
Systematic evolution of ligands by exponential enrichment (SELEX), 310

T

Tailings, 302–304
Targeted analysis, 356, 358–362
Tar sand, 150
Teflon synthesis, 342
Telemetry system, 251, 259, 261–262
Temperature, 25–29
Temperature compensation algorithm, 240
Temperature-dependent sex determination (TSD), 27
Temperature sensors, 226
Texas Commission on Environmental Quality, 289–290
Thermally stratified waterbodies, 125–126
Thiol group, 419–420
Thyroid hormone, 13, 30
Time-of-flight mass spectrometry (TOFMS), 344–345
Tisochrysis lutea, 177–178
Titanium-based nanoparticles, 120
Tivela stultorum, 23–24
TN. *See* Total nitrogen (TN)

440 INDEX

TOFMS. *See* Time-of-flight mass spectrometry (TOFMS)
Topographically flat plain, 117
Total dissolved solids (TDS), 204–207, 210–211
Total nitrogen (TN), 204, 247–248
Total organic carbon (TOC), 204–205, 207
Total oxidizable precursor (TOP), 359–361
Total petroleum hydrocarbon (TPH), 208
Total suspended solids (TSS), 70–71, 154, 204–205
Toxic Substances Control Act (TSCA), 341, 426
TPH. *See* Total petroleum hydrocarbon (TPH)
Trace element, 247–248, 254–255, 380–381, 383–384
Transboundary water, 63, 65, 72–75
Transboundary water management policy, 73
Transduction methods, biosensors, 308–311
    electrical properties, 311
    electrochemical methods, 311
    mass, 311
    optical methods, 311
    temperature, 311
Trichloroethylene (TCE), 290
Tropical cyclones, 270–271
Trucking process, 346
TSCA. *See* Toxic Substances Control Act (TSCA)
TSS. *See* Total suspended solids (TSS)
TU. *See* Turbidity (TU)
Tubule luminal ratio, 180–182
Turbidity (TU), 221
Turbulent coastal water, 271–272

U

UCI. *See* Urban Coastal Institute (UCI)
Ultrafiltration (UF), 386
Ultrahigh-performance liquid chromatography (UHPLC), 135
Ultratrace contaminants, 4–5
Ultraviolet (UV), 135, 235
UNCW. *See* University of North Carolina Wilmington (UNCW)

Underground storage space (USEPA), 66–67
UNDESA. *See* United Nations Department of Economics and Social Affairs (UNDESA)
UNICEF Joint Water Supply, 371–372, 372*b*
United Nations Department of Economics and Social Affairs (UNDESA), 285–286
United Nations Environmental Program (UNEP), 286–288
Universal serial bus (USB), 243
University of North Carolina Wilmington (UNCW), 346–347, 425
Unregulated contaminant monitoring rule, 2–3
Uranium, 99–100
Urban Coastal Institute (UCI), 261–262
Urban storm runoff management, 66
Urea aqueous solution, 399
USB. *See* Universal serial bus (USB)
US Geological Survey (USGS), 2, 63
Utilized precipitation technique, 385
UV. *See* Ultraviolet (UV)
UV light exposure, 30
Uv nitrate sensor, 246

V

Vanadium (Va), 375
van der Waals force, 387–388
Vapor compression (VC), 71
Vegetated infiltration basin, 66
Vegetated littoral habitat, 57
Vibrio cholerae, 131
Village green belt development, 101–104, 101–102*b*
Visible (VIS), 235
Vivo sensor measurement, 241
Volatile organic compounds (VOCs), 171–172, 207–208
Voltage-gated sodium channel, 130
Volusia County Council, 338

W

Wastewater
    discharge, 222, 347–348, 350–352
    discharge standard, 93
    treatment, 287–288, 318
        facility, 67–68, 75

plant, 73
    process, 68
    reuse systems, 67–70
Wastewater treatment plants (WWTPs), 93, 150, 152, 157–159, 162, 164–165, 286–287, 318, 350–351
Water
    body, 3, 10, 54–59, 66, 68, 288, 355–356, 371–373, 372*b*
    charges, 65
    clarity data, 156–157
    contaminants, 87–88, 92–93, 95, 100–101, 371–375
        global challenges, 371–374
        heavy metals, 374–375
    data quality, 64
    disasters, 1, 5, 11, 417
    environment federation, 76
    filters, 91, 107–108*b*
    intensive crop, 101–104, 103–104*b*
    need, 60, 67
    network situation, 62
    quantity and quality of, 85–87
    sample, 109–110, 109–110*b*
    scarcity, 86–87, 103–104*b*
    source, 93–94, 97, 101–104, 103–104*b*, 109–110, 109–110*b*
    supply need, 93
    taste astringent, 95–96
    technology, 94–95, 95*b*
    treatment system, 87, 106–109
    use efficiency, 85, 110–111
Water accommodated fraction (WAF), 173–176
WaterAid, 372–373
Water-borne disease, 94, 371–372, 372*b*
Water current profiling, 258
Water Data Storage and Retrieval System (WATSTORE), 263
Water demand management (WDM), 6, 64–65
Water disasters, 417, 426
    arsenic contamination of groundwater, 417–421
    arsenic-laced irrigation water on food chain, 420
    impact, 419–420
Water filters, 91, 107–108*b*
Water-information systems, 262–266, 275
Water lead level, 422
Water management, 6

INDEX

441

challenge, 61
principles, 60–61
reorganization, 64–66
tools, 61–62
Water partition coefficient, 172
Water pollution, 285–293, 317–318, 321
accidental spills, 288–290
natural pollution, 290–293
nutrient pollution, 287–288
problem, 7–8
regulations, 3–4
sources, 287–293
Water poverty line, 53–54
Water property, 223–224, 226, 233, 255–256, 258–259, 262, 274–275
Water quality, 350–351, 371–372, 372b
analytical laboratory, 109–110, 109–110b
challenges, 6–7
change, 159
evaluating, 4–5
global facts, 372
issues, 87, 109–110, 110b, 371–372
laboratory assistance, 109–110, 110b
models, 59
monitoring, 94
parameter, 154
to prevent disasters, 109–110
problems, 5, 86–108
sensor data, 8–9
sensors, 8–9
sondes, 243–244
testing, 109–110, 110b
Water-quality monitoring, 149–152
aging private infrastructure on, 9
anions, 155–156
awareness, 165–166
conductivity, 155–156
dissolved oxygen, 155–156
E. coli, 154–155
education, 165–166
field and lab water-quality data, 156–162
macroinvertebrates, 156
microplastics, 154
pH, 155–156
plastics, 162–165
stewardship, 165–166

study sites, 152–154
total suspended solids, 154
turbidity, 154–155
water temperature, 155–156
Water-quality sensors
acoustic sensors, 226–231
advantages, 272–274
continuous water-quality measurements
manuals and standardized methods, 231–235
surrogate models, 231–232
data acquisition and transfer, 259–260
data delivery, 260–262
data-processing software, 263–264
electrical resistance techniques
Sc of water, 231
water temperature, 226–231
electrometric techniques
Do sensors, 234–235
hydrogen ion concentration and ISEs, 232–233
ORP sensors, 233–234
specific electrical conductance, 232
water temperature, 231–232
handheld meters, 245–246
innovations in, 225–244
limitations, 272–274
mobile platforms
autonomous underwater vehicles, 257–259
pontoon boats, 257
monitoring
hurricanes, 270–271
seasonal floods, 266–270
storm-hardened platforms, 271–272, 272f
tropical cyclones, 270–271
optical techniques
chemical analyzer systems, 237–239
colorimetry, 237–239
direct optical spectrophotometry, 239–240
fluorescence spectroscopy, 240–243
spectrophotometry, 237–239

stationary platforms
on land, 250–252
Moored observatories in open water, 252–255
water-information systems, 263–264
water-quality sondes, 243–244
web applications, 265–266
WaterQualityWatch, 226, 239, 242, 252, 265, 265f
Water recharge volume, 55
Water resilience framework, 60
Water resource, 53–54, 59–67, 69, 72–76, 85, 88, 110–111, 140, 220, 222, 248, 263–264, 285–286, 292–293, 372–373
management decision, 320
Water reuse systems, 69–70
Water scarcity, 59–60
Watershed monitoring, 152, 162–167
Water solubility, 21
Water source, 125, 134
Water supply, 1–2, 4, 11, 53, 58, 60–62, 64–66, 69–71, 76–77
Water support meteorological sensor, 252
Water system, 58, 60, 69, 76, 285–286, 288–289, 294, 299, 422–423
Water temperature (WT), 155–159, 222
Water treatment
plant, 346, 424
technique, 140–141
Water users, 57, 65, 219
Water utility company, 2
WATSTORE. See Water Data Storage and Retrieval System (WATSTORE)
WDM. See Water demand management (WDM)
Weathering processes, 173
Web applications, 262–266, 275
Web service, 64
Wetland systems, 69
WHO. See World Health Organization (WHO)
WHOI. See Woods Hole Oceanographic Institute (WHOI)
Whole cells, 310

442 INDEX

W/O microemulsion methodology, 396–397
Woods Hole Oceanographic Institute (WHOI), 252–253
World Health Organization (WHO), 92–93, 233, 292–293, 371–372
World Wide Web, 265
World Wildlife Fund, 287–288

WT. *See* Water temperature (WT)
WTP. *See* Water treatment plant (WTP)
WWTP. *See* Wastewater treatment plants (WWTPs)

X

X-ray photoelectron spectroscopy (XPS), 395

Z

Zero oxidation state, 392–393
Zero-valent iron (ZVI), 106–109, 108–109*b*
Zinc (Zn), 375
Zirconium oxide nanoparticles, 120–121
Zn. *See* Zinc (Zn)

Printed in the United States
By Bookmasters